TK
5102.5
M63
1985

WITHDRAWN

3 0250 01062 1345

[redacted] DATE DUE

AUG 1

D1015545

MODERN SIGNAL PROCESSING

PROCEEDINGS OF THE ARAB SCHOOL ON SCIENCE AND TECHNOLOGY

A. H. El-Abiad (ed.)
Power Systems Analysis and Planning

T. Kailath (ed.)
Modern Signal Processing

G. Warfield (ed.)
Solar Electric Systems

FORTHCOMING

R. Descout (ed.)
Applied Arabic Linguistics and Information and Signal Processing

R. Risebrough (ed.)
Pollution and the Protection of Water Quality

D. M. Robinson (ed.)
Mini- and Microcomputers

MODERN SIGNAL PROCESSING

Edited by

THOMAS KAILATH
Stanford University
Stanford, California USA

HEMISPHERE PUBLISHING CORPORATION
Washington New York London

DISTRIBUTION OUTSIDE NORTH AMERICA

SPRINGER-VERLAG
Berlin Heidelberg New York Tokyo

MODERN SIGNAL PROCESSING

Copyright © 1985 by Hemisphere Publishing Corporation. All rights reserved. Printed in the United States of America. Except as permitted under the United States Copyright Act of 1976, no part of this publication may be reproduced or distributed in any form or by any means, or stored in a data base or retrieval system, without the prior written permission of the publisher.

1 2 3 4 5 6 7 8 9 0 B C B C 8 9 8 7 6 5 4

Library of Congress Cataloging in Publication Data

Modern signal processing.

(Proceedings of the Arab School on Science and Technology)

Proceedings from a school held in Zabadani, near Damascus, from Aug. 20–31, 1983 and sponsored by the Arab School on Science and Technology.

Includes bibliographies and index.

1. Signal processing—Digital techniques—Congresses. I. Kailath, Thomas. II. Arab School on Science and Technology. III. Series.

TK5102.5.M63 1985 621.38'0432 84-19289

ISBN 0-89116-386-7 Hemisphere Publishing Corporation

DISTRIBUTION OUTSIDE NORTH AMERICA:

ISBN 3-540-15074-9 Springer-Verlag Berlin

Contents

Contributors

P. Dewilde
Afdeling der Elektrotechniek
Delft University of Technology
Mekelweg 4, Postbus 5031
2628 CD Delft, The Netherlands

Anil K. Jain
Signal and Image Processing Laboratory
Department of Electrical Engineering and Computer
University of California
Davis, CA 95616

Thomas Kailath
Department of Electrical Engineering
Durand 117
Stanford University
Stanford, CA 94305

S. Y. Kung
Department of Electrical Engineering
Powell Hall, 426
University of Southern California
Los Angeles, CA 90089

Jae S. Lim
Department of Electrical Engineering and Computer Science
Massachusetts Institute of Technology
Cambridge, MA 02139

John Makhoul
Bolt, Beranek and Newman, Inc.
10 Moulton Street
Cambridge, MA 02238

S. Lawrence Marple, Jr.
Schlumberger Well Services Engineering
P.O. Box 4594
(500 Gulf Freeway)
Houston, TX 77210-4594

M. Mrayati
The Scientific Studies and Research Center
P.O. Box 4470, Damascus, Syria

John G. Proakis
Department of Electrical and Computer Engineering
Northeastern University
360 Huntington Ave.
Boston, MA 02115

H. J. Whitehouse
Naval Undersea Center
P.O. Box 19403
San Diego, CA 92119

Preface

Signal processing has always been a rather eclectic discipline, but modern signal processing can perhaps best be characterized by the rapidly growing extent to which it combines ideas and tools from signal analysis, system and operator theory, statistical methods, numerical analysis, computer science, and modern integrated circuit technology.

We shall not attempt a definition of signal processing—it is known by its fruits—nor shall we try to trace its development. But some perspective on its growth may be gained by reflecting that most of its development was in this century. Zobel and Cambell introduced selective wave filters in the early 1900s. Heaviside and Carson provided the analytical tools of impulse response, transfer functions, and transform methods in the first two decades, while the twenties saw the first network synthesis methods of Foster and Cauer. Brune, Cauer, and Darlington laid the proper mathematical foundations in the thirties, and Guillemin introduced these developments into the education of engineers. World War II saw the introduction of microwave circuits and of statistical concepts—and their embodiment in correlators, matched filters, and high fidelity recorders and receivers. Information theory held out the hope for dramatic gains in bandwidth compression. Then the field seemed to languish for a while until in the early sixties it became realized that the digital computer could be used to process signals more precisely and more completely than they had ever been before—one could come as close as one wished to implementing almost arbitrary transfer functions, which could only be imperfectly realized with physical inductors, capacitors, resistors, transformers, etc. The full awareness of this new freedom came in a rush with the rediscovery and effective application in the mid-sixties of Fast Fourier Transform (FFT) algorithms. This was soon followed by the development of digital filtering theory by Kaiser, Gold, Rader, Stockham, Rabiner, Oppenheim, and others, as has been ably summarized in the well known books of Gold and Rader (1969) and Oppenheim and Schafer (1975). Since then the field has seen an almost exponential growth, as witnessed by the rapid expansion of the ICASSP (International Conference on Acoustics, Speech, and Signal Processing), first held in Philadelphia in 1976. A wide range of mathematical techniques and implementation technologies, and an even wider range of applications, can be seen in the proceedings of these conferences, in the Transactions of the IEEE Society of the same name, and in related journals.

The reviews in this volume attempt to capture some of the most important and the most exciting currents in modern signal processing and are written by some of the major contributors in the field. There are chapters on the most recent developments in advanced filter design by Dewilde, high resolution spectral analysis and applications by Marple, speech processing by Makhoul and Mrayati, communications by Proakis, radar and sonar signal processing by Whitehouse, image processing by Jain. Two chapters by Whitehouse and by Kung cover the technology from sophisticated analog devices to the currently very active fields of specialized VLSI (systolic, data flow, and wavefront) parallel array processors. The basic mathematical tools required for the understanding of these different areas are presented in two introductory chapters, by Lim on deterministic techniques of signal analysis and by Kailath on the concepts and results from stochastic process theory.

Readers will find many links between the material in the different chapters, sometimes by explicit reference but more generally by common entries in their bibliographies. We shall give just one example.

In classical PCM systems, a voice signal is first processed by a low-pass " anti-aliasing" filter with a 3 kHz passband. These filters can be built with LC circuits, although with large inductors and capacitors because of the low frequency range of operation, or with active filters with opamps and, again, bulky capacitors. Such filters were not suitable for integrated circuit (IC) implementation. This led to an alternative design in which the voice signals are first processed by a crude analog filter with a 30 kHz cutoff (so that a less bulky RC circuit implementation can be achieved). The output is then digitized by a high rate (68 kHz) analog-to-digital (A/D) converter, followed by an accurate digital, lowpass filter. The high rate A/D converter and the digital, lowpass filter can be built with ICs.

As Dewilde points out in his chapter on advanced filter design, for similar reasons, digital filters have replaced analog ones in many applications. Furthermore, Dewilde goes on to describe a common choice for implementation, via what are called Wave Digital Filters, which have an almost minimal number of multipliers and therefore have low sensitivity and also use up only a small amount of chip area.

But the most common digital filter implementations are of the so-called FIR (finite impulse response) type, because among other reasons they are easy to design and can give linear phase, in contrast to the so-called recursive or IIR (infinite impulse response) filters. However, it is well known that IIR filters are generally much less bulky than equivalent FIR filters, a consideration that becomes increasingly important as chip area becomes scarcer in sophisitcated signal processing devices. Therefore, the traditional arguments for FIR vs. IIR filters can bear reexamination in each application.

In the same vein, high throughput is important in many signal processing applications and therefore an important consideration is 'pipelineability,' which is the property of having a throughput rate independent of the 'size' (appropriately measured) of the circuit; the name is based on the analogy with water flowing through a pipe. The rate of flow is independent of the length of the pipe. Now wave digital filters, despite their many nice properties, cannot always be made pipelined. Therefore, there is increasing interest in what are called 'systolic' array processors, introduced by Kung and Leiserson (1978). These are highly regular, locally interconnected, pipelined arrays of essentially identical processing modules, through which data flows in a regular rhythmic way: at each time instant, data flows into and out of every processing element (except perhaps those on the boundary). Such arrays can be designed to do a variety of signal processing tasks (especially those consisting of linear algebraic operations). They have attracted a great deal of attention—see the discussion and references in the articles by Whitehouse and by S. Y. Kung in this volume—in both the engineering and in the computer science communities. While this interest has been very beneficial in drawing (some parts of) these two communities closer together, it is reassuring, to engineers at least, to discover that linear systolic arrays are essentially rearrangements of the well known 'direct forms' of digital filtering (and linear system) theory. This is carefully explained by S. Y. Kung in Chapter 11. However, not only is this identification reassuring, but it immediately makes available (for the 'new' class of linear systolic arrays) the whole host of analyses that have been made of digital filters over the last two decades. One of the dearly-learned lessons of these analyses is that direct-form realizations, while easy to recognize from the transfer function, have very poor sensitivity properties and also display "overflow" and "limit cycle" oscillations as a result of finite precision and quantization effects in implementing the digital filter. It may have taken a while longer for these problems to be diagnosed and understood in the computer science context. In any case, to reduce coefficient sensitivity, digital filters are usually implemented as "cascades" of first- and second-order transfer functions, obtained by appropriately factoring the given transfer function. However, overflow and limit cycle oscillations can still arise in such implementations. Such undesirable oscillations can be suppressed by having cascades of "orthogonal" sections. In Chapter 4, Dewilde discusses the general theory of such orthogonal cascades, a field that he pioneered, and describes some of the nice results in this direction obtained at Delft and Stanford. Moreover, it might be mentioned that the mathematical theory underlying these orthogonal filters is closely related to the prediction theory of stationary stochastic processes, as presented in Chapter 2. Here, special attention is paid to the so-called lattice filter and transmission line implementations derived via the fast Levinson and Schur algorithms. These fliters and algorithms have also been very useful in modern high resolution spectral analysis, in speech synthesis, and in several communications applications, especially channel equalization, as briefly noted by Marple, Makhoul, and Proakis in their respective chapters.

Several other such examples of cross coupling can be traced through other chapters in this book. We shall leave to prospective readers the pleasure of tracking them themselves.

In conclusion, however, it may be appropriate to express the view that signal processing at this time stands at various thresholds—in design (IIR vs. FIR, pipelined orthogonal filters vs. wave digital filters, etc.), in technology (analog vs. digital, hybrids such as switched capacitors, etc.), and especially in applications (in telephony, mobile radio, videoconferencing, robotics). It is the hope of the authors, and of the organizers of the Arab summer school for which it was developed, that this book will provide some glimpses of these thresholds and provide some of the tools and background useful for properly seizing the opportunities thus created.

An account of the origins of this book may be of interest. For the last several years, the Arab School on Science and Technology has presented at least two short courses a year, on topics of current engineering and scientific significance, to an audience composed of teachers and researchers of Arab origin from around the world. At the invitation of Professor J. Cruz of Urbana, Illinois, I participated in a school of Modern

Control Theory, held in the mountain resort of Bloudan, near Damascus, Syria, in August 1981. It was a most worthwhile experience for lecturers and auditors alike: professionally, socially and culturally. The goodwill generated by it was of great help in assembling the present outstanding group of lecturers for a school on Advanced Signal Processing, held in Zabadani, near Damascus, from August 20-31, 1983. The school program committee, comprised of Drs. N. Harfouch, M. Mrayati and F. Haj Hassan, took considerable pains to try to secure a good balance between theory and application, academia and industry, formal lectures, panel discussions, and supplementary sessions. Moreover, Dr. Harfouch and his extremely well trained and efficient school staff spared no effort to provide, with courtesy and charm, the necessary infrastructure and assistance in all spheres, personal and professional. It has been a pleasure to have been associated with this project.

T. Kailath

Arab School on Science and Technology

SPONSORS

Scientific Studies and Research Center, Syria
Arab League Educational, Cultural, and Scientific Organization (ALECSO)

SUPERVISORY COMMITTEE

Abdalla W. Chahid, Director General
Scientific Studies and Research Center, Syria

Adnan Shehab-Eldin, Director General
Kuwait Institute for Scientific Research, Kuwait

Wafai Hakki, Senior Member
The Supreme Council of Sciences, Syria

SCHOOL COORDINATOR

Nabil Harfouch
Department of Electronics, Scientific Studies and Research Center, Syria

SCIENCE COMMITTEE

Mohammad Mrayati,
Scientific Studies and Research Center, Syria

Nabil Harfouch,
Scientific Studies and Research Center, Syria

Ferial Haj-Hassan
Scientific Studies and Research Center, Syria

ORGANIZING COMMITTEE

Huda Nahas Armanazi
Arab School on Science and Technology

Alia Kouatly
Arab School on Science and Technology

Bachar Ayoubi
Arab School on Science and Technology

OBJECTIVES OF THE ARAB SCHOOL ON SCIENCE AND TECHNOLOGY

The School through its Sessions aims to fulfill the following objectives:

1. Familiarizing Arab Scientists, Engineers, and University Professors with the latest advances in science and technology through intensive advanced post-graduate and highly specialized post-doctoral courses given by leading scientists.

2. Facilitating direct contact *between* Arab Scientists to establish a propitious atmosphere for joint cooperation in the field of science and technology.

3. Encouraging Arab Scientists working abroad to return to their countries by providing them with opportunities to contribute to the School's activities and closely reviewing the scientific resources of their countries.

4. Facilitating scientific cooperation among Arab and Muslim countries by direct contacts provided by the school.

5. Providing an overview of scientific activities in Arab countries by publishing Session Proceedings which cover scientific and technological developments in the Arab/Muslim World.

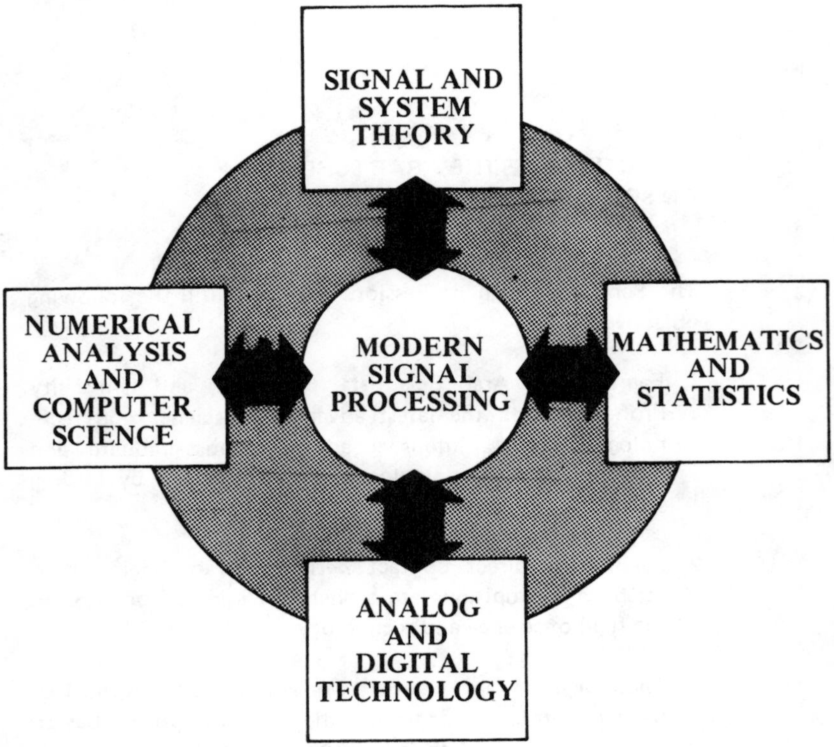
SIGNAL AND
SYSTEM
THEORY
NUMERICAL
ANALYSIS
AND
COMPUTER
SCIENCE
MODERN
SIGNAL
PROCESSING
MATHEMATICS
AND
STATISTICS
ANALOG
AND
DIGITAL
TECHNOLOGY

1

Fundamentals of Digital Signal Processing

JAE S. LIM

In this chapter, we discuss the fundamentals of digital signal processing. The topics covered are signals and systems, transform domain representation of signals and systems, fast Fourier transform algorithms, and digital filter design and implementation. Since many topics are covered and this chapter is intended to be a summary, we will concentrate on fundamental concepts in digital signal processing, and will not attempt to be complete in our treatment. A more complete and detailed treatment of the topics can be found in [1,2]. This chapter is aimed at the reader who has some familiarity with analog signals and systems and their frequency domain representations.

TABLE OF CONTENTS

1. SIGNALS AND SYSTEMS

1.1 Signals

Most signals can in practice be classified into three broad groups. One group of signals, "analog" or "continuous time" signals, are continuous in both time* and amplitude. A majority of signals fall into this group in practice. Examples of analog signals include speech, image, seismic, and radar signals. The second group of signals, "discrete time" signals, are discrete in time and continuous in amplitude. A common way to generate discrete time signals is by sampling analog signals. There exist, however, signals which are intrinsically discrete in time. An example is the monthly payment of a home mortgage. The third group of signals, "digital" or "discrete" signals, are discrete in both time and amplitude. One way digital signals can occur is by the amplitude quantization of discrete time signals. Discrete time signals and digital signals are also referred to as "sequences".

The only signals used by digital systems or computers are digital signals, which are discrete in both time and amplitude. The dev-

---

*Even though we refer to time, an analog signal can instead be variable in space, as in the case of image processing.

elopment of digital signal processing concepts based on digital signals, however, requires a detailed treatment of amplitude quantization, which is extremely difficult and tedious. In addition, many useful insights are lost in such a treatment because of the mathematical complexity. For this reason, digital signal processing concepts have been developed based on discrete time signals. Experience shows that the theories developed based on discrete time signals are often applicable to digital signals.

A discrete time signal (sequence) will be denoted by functions whose arguments are integers. For example, $x(n)$ represents a sequence which is defined for all integer values of n. Note that $x(n)$ for non-integer n is not zero, but is undefined. The notation $x(n)$ refers to the discrete time function x or to the value of the function x at a specific n. The distinction between these two will be obvious from the context.

There are some sequences and classes of sequences which play a particularly important role in digital signal processing. These sequences are discussed below.

Unit Sample Sequence. The unit sample sequence, denoted by $\delta(n)$, is defined as follows:

$$\delta(n) = \begin{cases} 1, & n=0 \\ 0, & \text{otherwise} \end{cases} \tag{1.1}$$

The sequence $\delta(n)$ plays a role similar to an impulse function in analog system analysis. In addition, any sequence $x(n)$ can be represented as a linear combination of delayed unit sample sequences as follows:

$$\begin{aligned} x(n) &= \cdots + x(-2)\cdot\delta(n+2) + x(-1)\cdot\delta(n+1) + x(0)\cdot\delta(n) \\ &\quad + x(1)\cdot\delta(n-1) + x(2)\cdot\delta(n-2) + \cdots \\ &= \sum_{k=-\infty}^{\infty} x(k)\cdot\delta(n-k) \end{aligned} \tag{1.2}$$

The representation of $x(n)$ by (1.2) is very useful in system analysis.

Unit Step Sequence. The unit step sequence, denoted by $u(n)$ is defined as follows:

$$u(n) = \begin{cases} 1, & n\geq 0 \\ 0, & \text{otherwise} \end{cases} \tag{1.3}$$

The sequence $u(n)$ is related to $\delta(n)$ as follows:

$$u(n) = \sum_{k=-\infty}^{n} \delta(k)$$

or (1.4)

$$\delta(n) = u(n) - u(n-1)$$

Exponential Sequence. Exponential sequnces play a role in digital signal processing similar to exponential functions in analog signal processing. An example of an exponential sequence is sketched in Figure 1-1.

$x(n) = (1/2)^n \cdot u(n)$

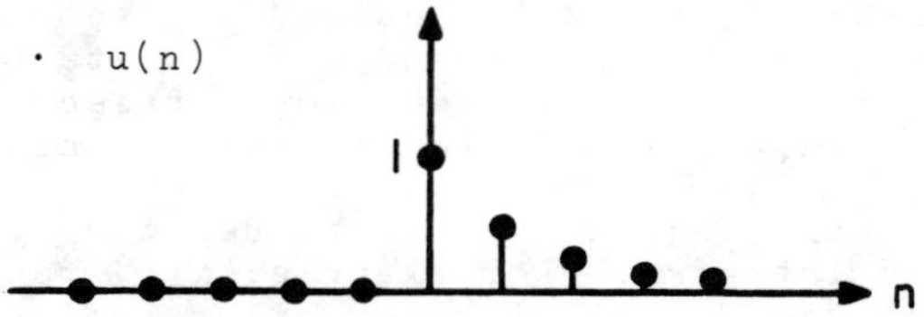

Figure 1-1. An exponential sequence.

Periodic Sequences. A sequence x(n) is called periodic with period of N if x(n) satisfies the following condition:

$$x(n) = x(n+N) \text{ for all } n \tag{1.5}$$

where N is an integer. For example, $\cos(\pi n)$ is a periodic sequence with period of 2, since $\cos \pi n = \cos \pi(n+2)$ for all n. The sequence cos n is not periodic, however, since cos n cannot be expressed as cos (n+N) for all n for any integer N. A periodic sequence is often denoted by adding "∿" (tilde) on the sequence, e.g. $\tilde{x}(n)$ to distinguish it from an aperiodic sequence.

1.2 Systems

If there is a unique output for any given input, the input-output relationship is called a system. A system T that relates an input x(n) to an output y(n) is represented by

$$y(n) = T[x(n)] \tag{1.6}$$

The above definition of a system is very broad. Without some restrictions, the characterization of a system requires a complete input-output relationship--knowing the output of a system to a certain set of inputs does not allow us to determine the output of the system to other sets of inputs. Two types of restrictions which greatly simplify the characterization and analysis of a system are linearity and shift-invariance. Fortunately, many systems can often be approximated in practice by a linear and shift-invariant system.

The linearity of a system T is defined as follows:

$$\text{linearity} \leftrightarrow T[a \cdot x_1(n) + b \cdot x_2(n)] = a \cdot y_1(n) + b \cdot y_2(n) \tag{1.7}$$

where $T[x_1(n)] = y_1(n)$, $T[x_2(n)] = y_2(n)$, a and b are any scalar constants, and $A \leftrightarrow B$ means that A implies B and B implies A. The condition in (1.7) is called the "principle of superposition."

The shift invariance (SI) of a system is defined as follows:

$$\text{shift invariance} \leftrightarrow T[x(n-n_0)] = y(n-n_0) \tag{1.8}$$

where $y(n) = T[x(n)]$ and n_0 is any integer.

Consider a linear system T. Using (1.2) and (1.7), the output y(n) for an input x(n) can be expressed as follows:

$$y(n) = T[x(n)] = T\left[\sum_{k=-\infty}^{\infty} x(k)\cdot\delta(n-k)\right]$$

$$= T[\cdots + x(-2)\cdot\delta(n+2) + x(-1)\cdot\delta(n+1) + x(0)\cdot\delta(n) + x(1)\cdot\delta(n-1) + x(2)\cdot\delta(n-2) + \cdots]$$

$$= \cdots + x(-2)\cdot T[\delta(n+2)] + x(-1)\cdot T[\delta(n+1)] + x(0)\cdot T[\delta(n)] + x(1)\cdot T[\delta(n-1)] + x(2)\cdot T[\delta(n-2)] + \cdots$$

$$= \sum_{k=-\infty}^{\infty} x(k)\cdot T[\delta(n-k)] \text{ for a linear system T} \qquad (1.9)$$

From (1.9), a linear system can be completely characterized by the response of the system to a unit sample sequence $\delta(n)$ and its delays $\delta(n-k)$. Specifically, if we know $T[\delta(n-k)]$ for all integer values of k, the output of the linear system to any input x(n) can be obtained from (1.9). For a non-linear system, knowledge of $T[\delta(n-k)]$ for all integer values of k does not tell us the output of the system when the input x(n) is $2\cdot\delta(n)$, $\delta(n)+\delta(n-1)$, and many others.

The system characterization is further simplified if we impose an additional restriction of shift invariance. Suppose we denote the response of a system T to an input $\delta(n)$ by h(n);

$$T[\delta(n)] \triangleq h(n) \qquad (1.10)$$

From (1.8) and (1.10),

$$h(n-n_0) = T[\delta(n-n_0)] \qquad (1.11)$$

for a shift-invariant system T.
For a linear and shift-invariant (LSI) system T, then, from (1.9) and (1.11), the input-output relation is given by

$$y(n) = T[x(n)] = \sum_{k=-\infty}^{\infty} x(k)\cdot h(n-k) \qquad (1.12)$$

Equation (1.12) states that an LSI system is completely characterized by the unit sample response of the system, h(n). Specifically, for an LSI system, knowledge of h(n) alone allows us to determine the output of the system to any input from (1.12). Equation (1.12) is referred to as "convolution", and is denoted by the convolution operator "*" as follows:

$$\text{For an LSI system, } y(n) = x(n)*h(n)$$

$$= \sum_{k=-\infty}^{\infty} x(k)\cdot h(n-k) \qquad (1.13)$$

Note that the unit sample response h(n), which plays an important role for an LSI system, loses its significance for a non-linear or shift-variant system. We also note that an LSI system can be completely characterized by the response of the system to one of many other sequences. The choice of $\delta(n)$ as an input in characterizing an LSI system is the simplest both conceptually and in practice.

1.3 Convolution

The convolution operator in (1.13) has a number of properties listed below.

Commutativity

$$x(n)*y(n) = y(n)*x(n) \tag{1.14}$$

Associativity

$$(x(n)*y(n))*z(n) = x(n)*(y(n)*z(n)) \tag{1.15}$$

Distributivity

$$x(n)*(y(n)+z(n)) = (x(n)*y(n))+(x(n)*z(n)) \tag{1.16}$$

Convolution With Dealyed Unit Sample Sequence

$$x(n)*\delta(n-n_0) = x(n-n_0) \tag{1.17}$$

The above properties can be obtained by a simple manipulation of (1.13). For example, the commutativity property can be shown as follows:

Letting m = n-k in (1.13),

$$x(n)*y(n) = \sum_{k=-\infty}^{\infty} x(k)\cdot y(n-k)$$

$$= \sum_{m=-\infty}^{\infty} y(m)\cdot x(n-m) \tag{1.18}$$

$$= y(n)*x(n)$$

The commutativity property states that the output of an LSI system is not affected when the roles of an input and the unit sample response are interchanged. The associativity property states that a cascade of two LSI systems with unit sample responses $h_1(n)$ and $h_2(n)$ has the same input-output relationship as one LSI system with its unit sample response given by $h_1(n)*h_2(n)$. The distributivity property states that a parallel combination of two systems with unit sample responses $h_1(n)$ and $h_2(n)$ has the same input-output relationship as one LSI system with unit sample response given by $h_1(n)+h_2(n)$. A special case of (1.17) when $n_0 = 0$ states that the unit sample response of an identity system is $\delta(n)$.

The convolution of two sequences x(n) and h(n) can be obtained by explicitly evaluating (1.13). It is often simpler and more instructive, however, to evaluate (1.13) graphically. Specifically, the convolution sum in (1.13) can be interpreted as multiplying two sequences x(k) and h(n-k), which are functions of the variable k, and summing the product over all integer values of k. The result, which is a function of n, is the result of convolving x(n) and h(n). One simple way to sketch h(n-k) as a function of k directly from h(n) is to first change the variable n to k, flip the sequence with respect to the origin, and then shift the result to the right by n points.

1.4 Stability and Casuality

For practical considerations, it is often appropriate to impose additional constraints on the class of systems we consider. These additional constraints are stability and causality.

A system is considered stable in the bounded-input-bounded-output (BIBO) sense if and only if a bounded input always leads to a bounded output. Stability is often a desirable constraint to impose, since an unstable system can generate an unbounded output, which can cause system overload or other difficulties. From this definition and (1.13), it can be shown that a necessary and sufficent condition for an LSI system to be stable is that its unit sample response h(n) is absolutely summable:

$$\text{For an LSI system,} \quad \text{stability} \leftrightarrow \sum_{n=-\infty}^{\infty} |h(n)| < \infty \qquad (1.19)$$

Because of (1.19), an absolutely summable sequence is defined to be a stable sequence. With this definition, then, a necessary and sufficient condition for an LSI system to be stable is that its unit sample response is a stable sequence.

A system is called causal if and only if the current output y(n) does not depend on any future values of input, e.g. x(n+1), x(n+2), x(n+3),... . Causality is often a desirable constraint to impose, since a non-causal system requires a delay, which is undesirable in many real time applications. From this definition and (1.13), it can be shown that a necessary and sufficient condition for an LSI system to be causal is that its unit sample response h(n) is zero for $n<0$:

$$\text{For an LSI system,} \quad \text{causality} \leftrightarrow h(n) = 0 \quad \text{for } n<0 \qquad (1.20)$$

Because of (1.20), any sequence which is zero for $n<0$ is defined to be a causal sequence. With this definition, then, a necessary and sufficient condition for an LSI system to be causal is that its unit sample response is a causal sequence.

2. FOURIER TRANSFORM

2.1 Fourier Transform Pair

It is a remarkable fact that any stable sequence x(n) can be obtained by appropriately combining complex exponentials of the form $X(\omega)\cdot e^{j\omega n}$. The function $X(\omega)$ which represents the amplitude associated with the complex exponential $e^{j\omega n}$ can be obtained from $x(n)$. The relationships between $x(n)$ and $X(\omega)$ are given by:

<u>Discrete Time Fourier Transform Pair</u>

$$X(\omega) = \sum_{n=-\infty}^{\infty} x(n)\cdot e^{-j\omega n} \tag{2.1}$$

$$x(n) = \frac{1}{2\pi} \int_{\omega=-\pi}^{\pi} X(\omega)\cdot e^{j\omega n}\cdot d\omega \tag{2.2}$$

Equation (2.1) shows how the amplitude $X(\omega)$ associated with the exponential $e^{j\omega n}$ can be determined from $x(n)$. The function $X(\omega)$ is called the discrete time Fourier transform, or Fourier transform (FT) for short, of $x(n)$. Equation (2.2) shows how complex exponentials $X(\omega)\cdot e^{j\omega n}$ are specifically combined to form $x(n)$. The sequence $x(n)$ is called the inverse discrete time Fourier transform, or inverse FT for short, of $X(\omega)$. The consistency of (2.1) and (2.2) for stable $x(n)$ can be easily shown by combining them.

From (2.1) it can be seen that $X(\omega)$ is in general complex, even though $x(n)$ may be real, and that $X(\omega)$ is a function of a continuous variable ω, even though $x(n)$ is a function of a discrete variable n. In addition, $X(\omega)$ is always periodic with period of 2π, e.g. $X(\omega) = X(\omega+2\pi)$ for all ω. It can also be shown that $X(\omega)$ uniformly converges only for stable sequences. The function $X(\omega)$ is said to converge uniformly when $X(\omega)$ is finite and

$$\lim_{N\to\infty} \sum_{n=-N}^{N} x(n)\cdot e^{-j\omega n} \to X(\omega) \quad \text{for all } \omega \tag{2.3}$$

When $X(\omega)$ converges uniformly, it is an analytic function (infinitely differentiable with respect to ω).

We define a sequence $x(n)$ to be an eigenfunction of a system T if $T[x(n)] = k\cdot x(n)$ for some scalar k. Suppose we use a complex exponential $e^{j\omega n}$ as an input $x(n)$ to an LSI system with unit sample response $h(n)$. The output of the system, $y(n)$, can be obtained as follows:

$$\begin{aligned} y(n) &= \sum_{k=-\infty}^{\infty} h(k)\cdot x(n-k) = \sum_{k=-\infty}^{\infty} h(k)\cdot e^{j\omega(n-k)} \\ &= \sum_{k=-\infty}^{\infty} h(k)\cdot e^{-j\omega k}\cdot e^{j\omega n} \\ &= H(\omega)\cdot e^{j\omega n} \end{aligned} \tag{2.4}$$

From (2.4), $e^{j\omega n}$ is an eigenfunction of any LSI system and $H(\omega)$ is the FT of $h(n)$. The function $H(\omega)$ is called the frequency response of the LSI system. The notion that $e^{j\omega n}$ is an eigenfunction of an LSI system and $H(\omega)$ is the scaling factor that scales $e^{j\omega n}$ when subjected to the LSI system simplifies the system analysis when the input to the LSI system is a sinusoidal input. For example, the response of an LSI system with frequency response $H(\omega)$ when the input is $\cos\omega_0 n$ can be obtained as follows:

$$T[\cos\omega_0 n] = T\left[\frac{e^{j\omega_0 n}}{2} + \frac{e^{-j\omega_0 n}}{2}\right]$$

$$= \frac{1}{2}T[e^{j\omega_0 n}] + \frac{1}{2}T\,[e^{-j\omega_0 n}]$$

$$= \frac{1}{2}\cdot H(\omega_0)\cdot e^{j\omega_0 n} + \frac{1}{2}\cdot H(-\omega_0)\cdot e^{-j\omega_0 n} \qquad (2.5)$$

2.2 Examples of Fourier Transform

Some examples of $x(n)$ and $X(\omega)$ are shown below. These results can be easily obtained from (2.1) and (2.2).

$$x(n) \leftrightarrow X(\omega) \qquad (2.6)$$

$$\delta(n) \leftrightarrow 1 \qquad (2.7)$$

$$\alpha^n\cdot u(n) \text{ for } |\alpha|<1 \leftrightarrow \frac{1}{1-\alpha\cdot e^{-j\omega}} \qquad (2.8)$$

$$\alpha^n\cdot u(n) \text{ for } |\alpha|\geq 1 \leftrightarrow \text{Does Not Exist} \qquad (2.9)$$

$$\frac{\sin\omega_c n}{\pi n} \quad \leftrightarrow$$

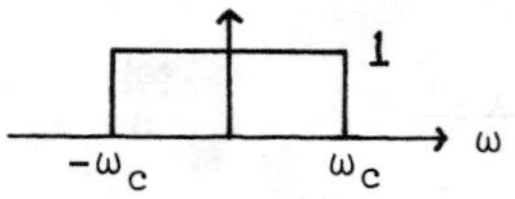

(2.10)

The sequence $\alpha^n\cdot u(n)$ for $|\alpha|\geq 1$ in (2.9) is an unstable sequence and therefore its FT does not exist (that is, it does not converge uniformly).

Strictly speaking, $\frac{\sin\omega_c n}{\pi n}$ in (2.10) is not absolutely summable, and therefore its FT does not uniformly converge. In fact, even though $\frac{\sin\omega_c n}{\pi n}$ was obtained from (2.2) by inverse Fourier transforming $X(\omega)$ given by

$$X(\omega) = 1 \quad \text{for } |\omega|\leq\omega_c$$

$$0 \quad \text{for } \omega_c\leq|\omega|\leq\pi \quad ,$$

$x(n)$ and $X(\omega)$ do not satisfy (2.3). Even though $X(\omega)$ in (2.10) does not uniformly converge, it converges in the mean square sense, e.g.

$$\lim_{N\to\infty} \int_{\omega=-\pi}^{\pi} |X(\omega) - \sum_{n=-N}^{N} x(n)\cdot e^{-j\omega n}|^2 \cdot d\omega = 0 \qquad (2.11)$$

Because of the important role that (2.10) plays in digital filtering and because of (2.11), we will accept (2.10) as a valid FT pair.

2.3 Properties

We can derive a number of useful properties from the FT pair in (2.1) and (2.2). Some of the more important properties often useful in practice, are listed below.

1. Linearity

$$a \cdot x_1(n) + b \cdot x_2(n) \leftrightarrow a \cdot X_1(\omega) + b \cdot X_2(\omega) \qquad (2.12)$$

2. Convolution

$$x(n) * h(n) \leftrightarrow X(\omega) \cdot H(\omega) \qquad (2.13)$$

3. Modulation

$$x(n) \cdot y(n) \leftrightarrow \frac{1}{2\pi} \int_{\theta=-\pi}^{\pi} X(\theta) \cdot Y(\omega - \theta) \cdot d\theta \qquad (2.14)$$

4. Shift of a Sequence

a. $x(n - n_o) \leftrightarrow X(\omega) \cdot e^{-j\omega n_o}$ (2.15)

b. $e^{j\omega_o n} \cdot x(n) \leftrightarrow X(\omega - \omega_o)$

5. Differentiation

$$-jn \cdot x(n) \leftrightarrow \frac{dX(\omega)}{d\omega} \qquad (2.16)$$

6. a. $x(0) = \frac{1}{2\pi} \cdot \int_{\omega=-\pi}^{\pi} X(\omega) \cdot d\omega$ (2.17a)

b. $X(0) = \sum_{n=-\infty}^{\infty} x(n)$ (2.17b)

7. Parseval's Theorem

a. $\sum_{n=-\infty}^{\infty} x(n) \cdot y^*(n) = \frac{1}{2\pi} \int_{\omega=-\pi}^{\pi} X(\omega) \cdot Y^*(\omega) \cdot d\omega$ (2.18a)

b. $\sum_{n=-\infty}^{\infty} |x(n)|^2 = \frac{1}{2\pi} \int_{\omega=-\pi}^{\pi} |X(\omega)|^2 \cdot d\omega$ (2.18b)

8. Symmetry Properties

a. $x^*(n) \leftrightarrow X^*(-\omega)$ (2.19)

b. $x(n)$: real $\leftrightarrow X(\omega) = X^*(-\omega)$

$X_R(\omega)$, $|X(\omega)|$: even

$X_I(\omega), \theta_x(\omega)$: odd

c. $x(-n) \leftrightarrow X(-\omega)$

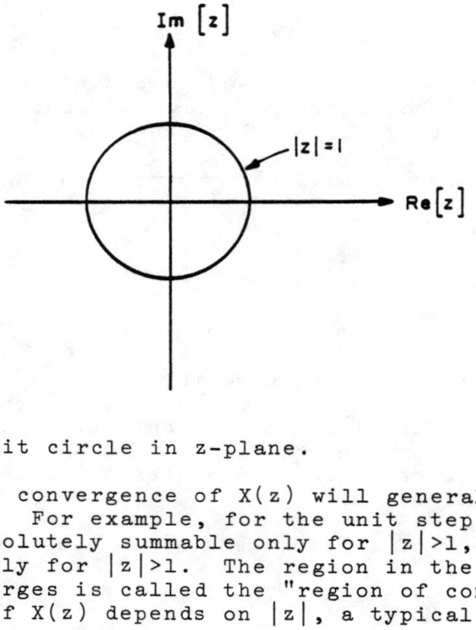

d. $x(n)$: real and even $\leftrightarrow$ $X(\omega)$: real and even

e. $x(n)$: real and odd $\leftrightarrow$ $X(\omega)$: pure imaginary and odd

9. Uniform Convergence

stable $x(n) \leftrightarrow X(\omega)$ uniformly converges (2.20)

Equation (2.13) states that convolution of two sequences $x_1(n)$ and $x_2(n)$ in the time domain corresponds to the multiplication of $X_1(\omega)$ and $X_2(\omega)$ in the frequency domain. This result simplifies the system analysis considerably in some cases. It is sometimes much easier to multiply two functions than to convolve two sequences. For example, $\frac{\sin\frac{\pi}{2}n}{\pi \cdot n} * \frac{\sin\frac{\pi}{3}n}{\pi \cdot n}$ would be quite tedious to compute directly in the time domain. If, however, we compute the FT of $\frac{\sin\frac{\pi}{2}n}{\pi \cdot n}$ and $\frac{\sin\frac{\pi}{3}n}{\pi \cdot n}$ from (2.10), multiply the two, and inverse FT the product, again using (2.10), the result immediately leads to $\frac{\sin\frac{\pi}{3}n}{\pi \cdot n}$.

The properties of FT can also be used to compute the FT of a complicated sequence or the inverse FT of a complicated function without explicitly evaluating (2.1). For example, the FT of $n \cdot [\frac{1}{2}]^n \cdot u(n)$ can be determined to be $\frac{\frac{1}{2}e^{-j\omega}}{(1-\frac{1}{2}e^{-j\omega})^2}$ using (2.8), (2.16) and (2.12)

3. z-TRANSFORM

3.1 z-Transform

The Fourier transform (FT) discussed in Section 2 uniformly converges only for stable sequences. As a result, many intersting classes of sequences, for instance the unit step sequence $u(n)$, cannot be represented by their FTs. In this section, we discuss the z-transform representation of a sequence, which converges for a much wider class of signals.

The z-transform (z-T) of a sequence $x(n)$ is denoted by $X(z)$ and is defined by

$$X(z) = \sum_{n=-\infty}^{\infty} x(n) \cdot z^{-n} \tag{3.1}$$

where z is a complex variable. From (3.1), $X(z)$ is related to $X(\omega)$ by

$$X(z)\Big|_{z=e^{j\omega}} = \sum_{n=-\infty}^{\infty} x(n) \cdot e^{-j\omega n} = X(\omega) \tag{3.2}$$

Equation (3.2) states that $X(\omega)$ is $X(z)$ evaluated at $z=e^{j\omega}$. The contour represented by $z=e^{j\omega}$ is the unit circle in the z-plane. as shown in Figure 3-1. For this reason, the z-T is considered a generalization of the FT.

Suppose X(z) in (3.1) is evaluated along $z=r\cdot e^{j\omega}$ where r is the radius and ω is the argument in the z-plane. The function X(z) can then be expressed as

$$X(z)\Big|_{z=r\cdot e^{j\omega}} = \sum_{n=-\infty}^{\infty} x(n)\cdot r^{-n}\cdot e^{-j\omega n} = F[x(n)\cdot r^{-n}] \quad (3.3)$$

where $F[x(n)\cdot r^{-n}]$ represents the FT of $x(n)\cdot r^{-n}$. Since the necessary and sufficient condition for the uniform convergence of X(ω) is the absolute summability (stability) of x(n), from (3.3), the necessary and sufficient condition for the uniform convergence of X(z) is the absolute summability of $r^{-n}\cdot x(n)$:

Uniform Convergence of X(z) $\leftrightarrow$

$$\sum_{n=-\infty}^{\infty} |r^{-n}\cdot x(n)| < \infty \quad \text{where } r=|z| \quad (3.4)$$

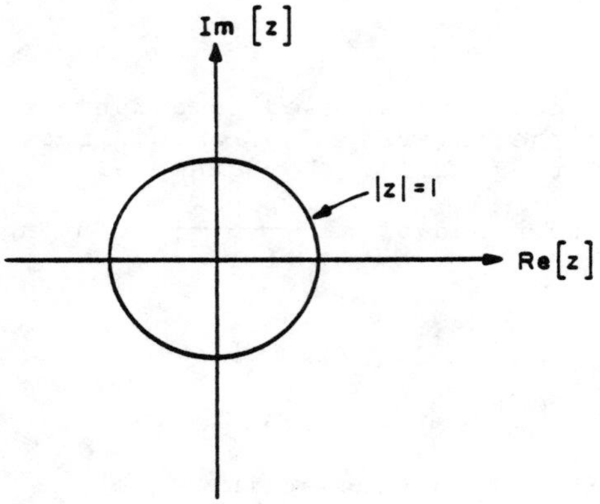

Figure 3-1.--Unit circle in z-plane.

From (3.4), the convergence of X(z) will generally depend on the value of $r=|z|$. For example, for the unit step sequence u(n), $r^{-n}\cdot u(n)$ is absolutely summable only for $|z|>1$, and therefore its z-T converges only for $|z|>1$. The region in the z-plane where X(z) uniformly converges is called the "region of convergence" (ROC). Since the ROC of X(z) depends on |z|, a typical ROC can be expressed as

$$\text{ROC:} \quad R_{x-} < |z| < R_{x+} \quad (3.5)$$

Outside the ROC, X(z) does not converge and does not exist. The ROC plays an important role in the z-T representation of a sequence, as we shall see shortly.

We can determine X(z) and its ROC from (3.1). It can be shown that x(n) can be determined from X(z) and its ROC using the following relationship:

$$x(n) = \frac{1}{2\pi j} \oint_C X(z) \cdot z^{n-1} \cdot dz \qquad (3.6)$$

where C is a closed contour which is in the ROC of $X(z) \cdot z^{n-1}$ and which encircles the origin in a counterclockwise direction. Equations (3.1) and (3.6) are the z-transform pair:

<u>z-Transform Pair</u>

$$X(z) = \sum_{n=-\infty}^{\infty} x(n) \cdot z^{-n} \qquad (3.7)$$

$$x(n) = \frac{1}{2\pi j} \oint_C X(z) \cdot z^{n-1} \cdot dz \qquad (3.8)$$

where C is a closed contour which is in the ROC of $X(z) \cdot z^{n-1}$ and which encircles the origin in a counterclockwise direction.

From (3.8), the result of the contour integration differs depending on the contour. Since the contour C can be determined from the ROC of X(z), a sequence x(n) is not specified by X(z) alone, but is specified by X(z) and its ROC:

$$x(n) \leftrightarrow X(z), \text{ROC} \qquad (3.9)$$

3.2 Pole-zero Representation of X(z)

For a rational X(z), it is often convenient to represent X(z) by the pole-zero plot. The poles of X(z) are those values of z for which X(z) becomes infinite and the zeros of X(z) are those values of z for which X(z) is zero. For example, $X(z) = \frac{z-2}{(z-3)^2 \cdot (z-4)}$ has poles at z=3 (double pole) and 4, and zeroes at z=2,∞ (double zero). We note that X(z) can be determined from its poles and zeros within a scaling factor. The pole-zero representation is often useful in roughly sketching the frequency response of an LSI system.

3.3 z-Transform Examples

From (3.7), the z-transform of a sequence and its ROC can be determined. Some examples are shown below.

$$x(n) \leftrightarrow X(z), \text{ ROC} \qquad (3.10)$$

$$\delta(n) \leftrightarrow 1, \text{ ROC: all } |z| \qquad (3.11)$$

$$a^n \cdot u(n) \leftrightarrow \frac{1}{1-a \cdot z^{-1}}, \text{ ROC: } |z|>|a| \qquad (3.12)$$

$$-a^n \cdot u(-n-1) \leftrightarrow \frac{1}{1-a \cdot z^{-1}}, \text{ ROC: } |z|<|a| \qquad (3.13)$$

$$a^n \cdot u(n) - b^n \cdot u(-n-1) \leftrightarrow \frac{1}{1-a \cdot z^{-1}} + \frac{1}{1-b \cdot z^{-1}}, \text{ for } |b|>|a|, \quad \text{ROC: } |a|<|z|<|b| \qquad (3.14)$$

$$a^n \cdot u(n) - b^n \cdot u(-n-1) \leftrightarrow \text{No convergence for } |b|<|a| \qquad (3.15)$$

Equations (3.12) and (3.13) clearly show that a unique specification of x(n) requires both X(z) and its ROC. Even though the two sequences $a^n \cdot u(n)$ and $-a^n \cdot u(-n-1)$ are quite different, their z-Ts are identical and they can be distinguished from their ROCs. Equation

(3.15) shows that there are sequences for which the z-T does not exist. The results in (3.12) and (3.13) will be useful when we discuss the inverse z-T problem in Section 3.5.

3.4 Properties

Many properties can be obtained from the z-transform pair in (3.7) and (3.8). Some of the properties are listed below.

<u>Properties of z-transform</u>

$$x(n) \leftrightarrow X(z)\ ,\ R_{x-} < |z| < R_{x+}$$
$$y(n) \leftrightarrow Y(z)\ ,\ R_{y-} < |z| < R_{y+}$$
$$h(n) \leftrightarrow H(z)\ ,\ R_{h-} < |z| < R_{h+}$$

1. Linearity

$$a\cdot x(n) + b\cdot y(n) \leftrightarrow a\cdot X(z) + b\cdot Y(z) \qquad (3.16)$$
$$\text{ROC: } R_- < |z| < R_+$$
$$R_- = \max\,[R_{x-}, R_{y-}],$$
$$R_+ = \min\,[R_{x+}, R_{y+}]$$

2. Convolution

$$x(n)*h(n) \leftrightarrow X(z)\cdot H(z) \qquad (3.17)$$
$$\text{ROC: } R_- < |z| < R_+$$

3. Shift of Sequence

$$x(n-n_0) \leftrightarrow X(z)\cdot z^{-n_0} \qquad (3.18)$$

4. Differentiation

$$n\cdot x(n) \leftrightarrow -z\cdot \frac{dX(z)}{dz} \qquad (3.19)$$

5. Symmetry Properties

a. $x^*(n) \leftrightarrow X^*(z^*)$ (3.20)

b. $x(-n) \leftrightarrow X(z^{-1})$, ROC: $\frac{1}{R_{x+}} < |z| < \frac{1}{R_{x-}}$

6. If $x(n) = 0$ for $n < 0$, then $x(0) = \lim_{z\to\infty} X(z)$ (3.21)

If $x(n) = 0$ for $n > 0$, then $x(0) = X(z)|_{z=0}$

<u>Properties of ROC</u>. 1. ROC is bounded by poles, and is a connected region with no poles inside the ROC.

2. right sided $x(n) \leftrightarrow$ ROC: $R_{x-} < |z| < \infty$ (outside some circle)
may include $|z| = \infty$

3. left sided $x(n) \leftrightarrow$ ROC: $0 < |z| < R_{x+}$ (inside some circle)
may include $|z| = 0$

4. both sided $x(n) \leftrightarrow$ ROC: $R_{x-} < |z| < R_{x+}$

5. finite extent $x(n) \leftrightarrow$ ROC: $0<|z|<\infty$
may include 0 and/or ∞

6. stable $x(n) \leftrightarrow$ ROC: includes $|z| = 1$

3.5 Inverse z-Transform

The problem of determining x(n) from X(z) and ROC can be solved by using (3.8). Equation (3.8), however, requires contour integration with respect to complex variable z. Even though such an integral can be evaluated by using the complex residue theorem, such an evaluation is extremely tedious.

An alternative, much simpler approach to solving the inverse z-T problem is by using the partial fraction expansion method. In this method, the z-transform is converted to a sum of simpler z-transforms: for each of the simpler z-Ts, the inverse z-T is performed. We will illustrate the partial fraction expansion method with the following example. Suppose we wish to determine x(n) whose z-T is given by

$$X(z) = \frac{3-3\cdot z^{-1}}{(1-\frac{1}{2}z^{-1})(1-2\cdot z^{-1})}, \quad \text{ROC: } \tfrac{1}{2}<|z|<2 \tag{3.22}$$

The function X(z) can be expressed as

$$X(z) = \frac{1}{1-\frac{1}{2}z^{-1}} + \frac{2}{1-2z^{-1}}, \quad \text{ROC: } \tfrac{1}{2}<|z|<2 \tag{3.23}$$

We now express X(z) as

$$X(z) = X_1(z) + X_2(z) \tag{3.24}$$

where $X_1(z) = \dfrac{1}{1-\frac{1}{2}z^{-1}}$, ROC: $\frac{1}{2}<|z|$ (3.25)

and $X_2(z) = \dfrac{2}{1-2\cdot z^{-1}}$, ROC: $|z|<2$ (3.26)

The ROCs for $X_1(z)$ and $X_2(z)$ can be chosen by noting that the ROC is bounded by poles and the ROC of X(z) is at least the intersection of the ROC of $X_1(z)$ and $X_2(z)$. From the linearity property,

$$x(n) = x_1(n) + x_2(n) \tag{3.27}$$

The sequence $x_1(n)$ and $x_2(n)$ can be obtained from $X_1(z)$, $X_2(z)$, and their ROCs as follows:

$$x_1(n) = (\tfrac{1}{2})^n\cdot u(n) \quad \text{from (3.12) and (3.25)} \tag{3.28}$$

$$x_2(n) = -(2)^{n+1}\cdot u(-n-1) \quad \text{from (3.13) and (3.26).} \tag{3.29}$$

From (3.27), (3.28), and (3.29), $x(n)=(\frac{1}{2})^n\cdot u(n)-2^{(n+1)}\cdot u(-n-1)$ (3.30)

In the example above, the ROC was explicitly given. In some cases of practical interest, however, the ROC is given only implicitly. In such cases, the ROC has to be determined first before the partial fraction expansion method is used. For example, suppose the ROC in the above example is not explicitly given and x(n) is known to be stable. Since the ROC is bounded by poles and is a connected region, and the ROC of a stable sequence includes the unit circle, the ROC can be determined to be $\frac{1}{2}<|z|<2$.

4. DIFFERENCE EQUATION

4.1 Difference Equation with Initial Condition

Difference equations play a more important role for discrete time systems than differential equations do for analog systems. In addition to representing a wide class of discrete time systems, difference equations can be used to recursively generate their solutions. This can be exploited in realizing digital filters with infinite extent unit sample responses. We will discuss this further in Section 10.

In this section, we will consider the class of linear constant coefficient difference equations (LCCDEs) of the following form:

$$\sum_{k=0}^{N} a_k \cdot y(n-k) = \sum_{k=0}^{M} b_k \cdot x(n-k) \qquad (4.1)$$

where a_k and b_k are constants and a_0 is assumed to be non-zero. The LCCDE in (4.1) is said to be of the Nth order.

The LCCDE alone does not specify a system, since there are many solutions of y(n) in (4.1) for a given x(n). For example, if $y_1(n)$ is a solution to $y(n)-a\cdot y(n-1) = x(n)$, then so is $y_1(n)+k\cdot a^n$ for any constant k. Like differential equations, imposing initial conditions (ICs) guarantees a unique solution to the LCCDE. For the Nth order LCCDE in (4.1), the ICs needed are N values of y(n). For convenience, we'll assume that y(n) is known for $-N\leq n<0$.

The problem of solving an LCCDE with ICs can be stated as follows:

Given y(n) for $-N\leq n<0$, and x(n), find the solution to

$$\sum_{k=0}^{N} a_k \cdot y(n-k) = \sum_{k=0}^{M} b_k \cdot x(n-k) \qquad (4.2)$$

We will consider one approach to solving the above problem. In this approach, y(n) is computed recursively and is a natural way to compute y(n) using a computer.

Approach to solve (4.2)

Step 1. For $n \geq 0$, recursively solve for $y(n)$ using

$$a_0 \cdot y(n) = -\sum_{k=1}^{N} a_k \cdot y(n-k) + \sum_{k=0}^{M} b_k \cdot x(n-k)$$

Step 2. For $n \leq -N-1$, recursively solve for $y(n)$ using

$$a_N \cdot y(n-N) = -\sum_{k=0}^{N-1} a_k \cdot y(n-k) + \sum_{k=0}^{M} b_k \cdot x(n-k)$$

Step 3: Combine Steps 1 and 2 with ICs.

To illustrate the above method, consider the following example:

$$y(n)-a\cdot y(n-1) = x(n) \quad , \quad x(n) = b\cdot\delta(n) \quad , \quad y(-1) = y_0 \tag{4.3}$$

Step 1. For $n \geq 0, y(n) = a\cdot y(n-1)+b\cdot\delta(n)$.

$$y(0) = a\cdot y(-1)+b = a\cdot y_0+b$$
$$y(1) = a\cdot y(0)+0 = a\cdot(a\cdot y_0+b)$$
$$y(2) = a\cdot y(1)+0 = a^2\cdot(a\cdot y_0+b)$$
$$\vdots$$
$$y(n) = a^n\cdot(a\cdot y_0+b)$$

Step 2. For $n \leq -2, y(n-1)=a^{-1}\cdot y(n)-a^{-1}\cdot b\cdot\delta(n)=a^{-1}\cdot y(n)$

$$y(-2) = a^{-1}\cdot y_0$$
$$y(-3) = a^{-2}\cdot y_0$$
$$\vdots$$
$$y(n) = a^{n+1}\cdot y_0$$

Step 3. Combining the results of Step 1, Step 2, and IC,

$$y(n) = a^n\cdot(a\cdot y_0+b)\cdot u(n)+a^{n+1}\cdot y_0\cdot u(-n-2)+y_0\delta(n+1)$$
$$= a^{n+1}\cdot y_0+b\cdot a^n u(n) \quad \text{for all } n \tag{4.4}$$

If we consider the LCCDE with IC as a system that relates input $x(n)$ to the output $y(n)$, the system is not, in general, an LSI system. This can be seen by noting that if $x(n)=2\cdot b\cdot\delta(n)$ in (4.3), then the solution to (4.3) would have been

$$y(n)=a^{n+1}y_0+2\cdot b\cdot a^n\cdot u(n) \tag{4.5}$$

Comparing (4.4) and (4.5) doubling $x(n)$ by a factor of two does not double $y(n)$ by a factor of two, and therefore the system is not even linear, let alone LSI. By proper choice of IC, however, it is possible to make an LCCDE with IC become LSI. One such method is discussed in the next section. If the LCCDE with IC is an LSI system, we can use the convolution property of the z-transform to develop a method of obtaining a closed-form solution to the LCCDE with IC problem. This is also discussed in the next section.

4.2 LCCDE with initial rest condition

One method to force an LCCDE with IC to become LSI which is particularly useful in digital signal processing is by imposing an IC known as an initial rest condition (IRC).

An initial rest condition (IRC) is defined as follows:

$$\text{Initial Rest Condition} \leftrightarrow y(n)=0 \quad \text{for } n<n_0 \quad \text{whenever } x(n)=0 \text{ for } n<n_0. \tag{4.6}$$

To see that an LCCDE with IRC is an LSI system, consider $x(n)$ which is zero for $n<n_0$. Rewriting the LCCDE in (4.2) as

$$y(n) = -\frac{1}{a_0}\cdot\sum_{k=1}^{N} a_k\cdot y(n-k) + \frac{1}{a_0}\cdot\sum_{k=0}^{M} b_k\cdot x(n-k) \quad ,$$

we can solve for $y(n)$ recursively:

$$y(n) = 0 \quad \text{for } n<n_0$$

$$y(n_0) = \frac{1}{a_0}\cdot b_0\cdot x(n_0) = L[x(n_0)]$$

$$y(n_0+1) = -\frac{1}{a_0}\cdot a_1\cdot y(n_0)+\frac{b_0}{a_0}\cdot x(n_0+1)+\frac{b_1}{a_0}\cdot x(n_0)$$
$$=L[x(n_0),x(n_0+1)]$$

$$\vdots$$

$$y(n) = L[x(n_0),x(n_0+1),x(n_0+2), \cdots x(n)] \tag{4.7}$$

where L[ ] represents some linear combination of the arguments. From (4.7), the output is a linear combination of the input, and a shift in the input would only shift the output by the same amount without changing the shape of the output (consider $x(n)=0$ for $n<n_1$). In addition, the current output does not depend on any future value of the input. Therefore, we have the following results:

$$\text{LCCDE with IRC is a causal LSI system} \tag{4.8}$$

From (4.8), an LCCDE with IRC can be represented by the system function $H(z)$ and its ROC. To determine the system function, we take the z-transform of $\sum_{k=0}^{N} a_k\cdot y(n-k) = \sum_{k=0}^{M} b_k\cdot x(n-k)$, which results in

$$H(z) = \frac{Y(z)}{X(z)} = \frac{\sum_{k=0}^{M} b_k\cdot z^{-k}}{\sum_{k=0}^{N} a_k\cdot z^{-k}} \tag{4.9}$$

The ROC associated with H(z) in (4.9) is, due to the causality of the system, outside the pole which is farthest away from the origin.

Since the LCCDE with IRC has a system function which is a rational z-transform, the unit sample response of the system is in general of infinite duration. Finding the output of the system by directly convolving the input with an infinitely long unit sample response requires a large amount of computation. If the unit sample response is such that its z-transform is rational, an alternative way to convolve the input with the unit sample response is by using the LCCDE. If the output is computed recursively using the LCCDE, the number of arithmetic operations (multiplications and additions) required per output point is of the order of (N+M). This will play an important role in our discussion of the type of filter known as the infinite impulse response (IIR) digital filter.

To solve an LCCDE with IRC, the approach discussed in Section 4.1 can, of course, be used. In addition, we can exploit the notion that the system is LSI in solving the LCCDE with IRC. Specifically, from the LCCDE, we determine the system function H(z) and its ROC. We then determine the z-transform of the input x(n). By using the convolution property of the z-transform for LSI systems, the output (or solution) of the system can be obtained by inverse-z-transforming $Y(z)=X(z)\cdot H(z)$.

5. DISCRETE FOURIER SERIES

5.1 Discrete Fourier Series Pair

A sequence $\tilde{x}(n)$ is said to be periodic with period of N when $\tilde{x}(n)=\tilde{x}(n+N)$ for all n. Since $\tilde{x}(n)\cdot r^{-n}$ is not absolutely summable for any r, neither the Fourier transform nor the z-transform uniformly converges for a periodic sequence.

One convenient way to represent a periodic sequence in the frequency domain is the discrete Fourier series (DFS) representation. Specifically, a periodic sequence $\tilde{x}(n)$ with period of N can be obtained by appropriately combining complex exponentials of the form $\tilde{X}(k)\cdot e^{j\frac{2\pi}{N}k\cdot n}$. The exponential sequence $\tilde{X}(k)\cdot e^{j\frac{2\pi}{N}k\cdot n}$ for $0\leq k\leq N-1$ represents all complex exponential sequences which are periodic with period of N. The sequence $\tilde{X}(k)$, which represents the amplitude associated with the complex exponential, can be obtained from $\tilde{x}(n)$. The relationship between $\tilde{x}(n)$ and $\tilde{X}(k)$ is given by:

<u>Discrete Fourier Series Pair</u>

$$\tilde{X}(k) = \sum_{n=0}^{N-1} \tilde{x}(n)\cdot e^{-j\frac{2\pi}{N}k\cdot n} \tag{5.1}$$

$$\tilde{x}(n) = \frac{1}{N}\sum_{k=0}^{N-1} \tilde{X}(k)\cdot e^{j\frac{2\pi}{N}k\cdot n} \tag{5.2}$$

Equation (5.1) shows how the amplitude $\tilde{X}(k)$ associated with the exponential $e^{j\frac{2\pi}{N}k\cdot n}$ can be determined from $\tilde{x}(n)$. The sequence $\tilde{X}(k)$ is called the discrete Fourier series (DFS) coefficient of $\tilde{x}(n)$.

Equation (5.2) shows how complex exponentials $\tilde{X}(k)\cdot e^{j\frac{2\pi}{N}k\cdot n}$ are specifically combined to form $\tilde{x}(n)$. The sequence $\tilde{x}(n)$ is called the inverse DFS of $\tilde{X}(k)$. The consistency of (5.1) and (5.2) can be easily shown by combining them.

From (5.1) and (5.2), $\tilde{x}(n)$ is represented by $\tilde{X}(k)$ for $0\leq k\leq N-1$. The sequence $\tilde{X}(k)$, can, therefore, be defined arbitrarily for k outside $0\leq k\leq N-1$. For conceptual convenience and by convention, $\tilde{X}(k)$ is defined to be periodic with period of N, with one period of $\tilde{X}(k)$ given by $\tilde{X}(k)$ for $0\leq k\leq N-1$. Also by convention, (5.2) has a scaling factor of 1/N. Since $\tilde{x}(n)$ and $\tilde{X}(k)$ are periodic with period of N in the variables n and k respectively, and since $e^{\pm j\frac{2\pi}{N}k\cdot n}$ is periodic with period of N in both the variables k and n, the limit of summation in (5.1) and (5.2) can be over any one period.

5.2 Properties

We can derive a number of useful properties from the DFS pair of (5.1) and (5.2). Some of the more important properties, often useful in practice, are listed below. In all cases, $\tilde{x}(n)$ and $\tilde{y}(n)$ are assumed periodic with period of N.

1. Linearity

$$a\cdot\tilde{x}(n) + b\cdot\tilde{y}(n) \leftrightarrow a\cdot\tilde{X}(k) + b\cdot\tilde{Y}(k) \qquad (5.3)$$

2. Periodic Convolution

$$\tilde{x}(n) \circledast \tilde{y}(n) = \sum_{l=0}^{N-1} \tilde{x}(l)\cdot\tilde{y}(n-l) \leftrightarrow \tilde{X}(k)\cdot\tilde{Y}(k) \qquad (5.4)$$

3. Modulation

$$\tilde{x}(n)\cdot\tilde{y}(n) \leftrightarrow \frac{1}{N}\cdot\tilde{X}(k) \circledast \tilde{Y}(k) \qquad (5.5)$$

$$= \frac{1}{N}\cdot\sum_{l=0}^{N-1} \tilde{X}(l)\cdot\tilde{Y}(k-l)$$

4. Shift of a sequence

$$\tilde{x}(n-n_0) \leftrightarrow \tilde{X}(k)\cdot e^{-j\frac{2\pi}{N}k\cdot n_0} \qquad (5.6)$$

5. Symmetry Properties

a. $$\tilde{x}^*(n) \leftrightarrow \tilde{X}^*(-k) \qquad (5.7)$$

b. $$\tilde{x}(-n) \leftrightarrow \tilde{X}(-k) \qquad (5.8)$$

Equation (5.4) states that when two sequences $\tilde{x}(n)$ and $\tilde{y}(n)$ are periodically convolved, their DFS coefficients $\tilde{X}(k)$ and $\tilde{Y}(k)$ multiply.

periodic convolution, denoted by $\circledast$, is very similar in form to the linear convolution discussed in Section 1.3. The difference lies in the limits of the summation. Specifically, $\tilde{x}(l)\cdot\tilde{y}(n-l)$ is summed over only one period ($0\le l\le N-1$) in the periodic convolution $\tilde{x}(n) \circledast \tilde{y}(n)$, while $x(l)\cdot y(n-l)$ is summed over all values of l in the linear convolution $x(n) * y(n)$. An example of $\tilde{x}(n) \circledast \tilde{y}(n)$ is shown in Figure 5-1. In the figure, $\tilde{x}(n) \circledast \tilde{y}(n)$ is computed explicitly using the convolution sum. An alternative way to compute $\tilde{x}(n) \circledast \tilde{y}(n)$ is by performing the inverse DFS operation of $\tilde{X}(k)\cdot\tilde{Y}(k)$. This approach to compute $\tilde{x}(n) \circledast \tilde{y}(n)$ is useful under some conditions, as we will discuss in Section 6.4.

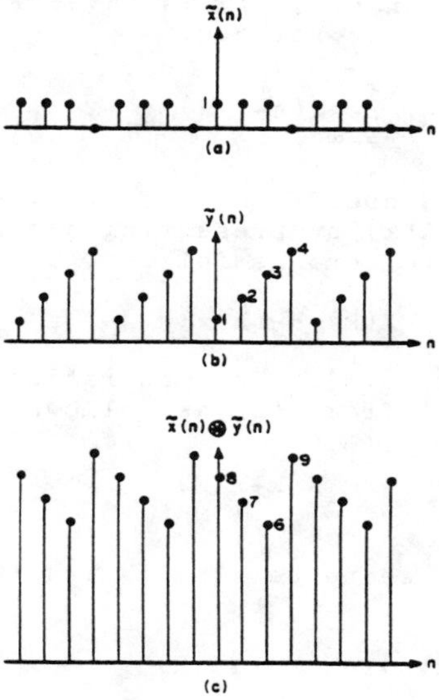

Figure 5-1.-- Example of periodic convolution.

6. DISCRETE FOURIER TRANSFORM

6.1 Discrete Fourier Transform Pair

In many signal processing applications, such as speech and image processing, the sequences we deal with are of finite duration. For such sequences, the Fourier transform and z-transform uniformly converge and are well defined. The FT and z-T representations $X(\omega)$ and $X(z)$ are functions of continuous variables ω and z. For finite duration sequences, which can be represented by a finite number of values, then, the FT and z-T are not computationally convenient frequency domain representations. In this section, we discuss the discrete Fourier transform (DFT) representation of finite duration sequences where a finite duration sequence is represented in the frequency domain by a finite number of values.

The DFT representation can be easily derived from the DFS representation of a periodic sequence discussed in Section 5. Consider a sequence $\tilde{x}(n)$ which is periodic with period of N. Suppose we form a finite duration sequence x(n) by preserving one period of $\tilde{x}(n)$ but setting all other values to zero as follows:

$$x(n) = \tilde{x}(n) \cdot R_N(n) \tag{6.1}$$

where

$$R_N(n) = \begin{cases} 1 & , 0 \le n \le N-1 \\ 0 & , \text{otherwise} \end{cases} \tag{6.2}$$

Clearly, the operation given by (6.1) is invertible, in that $\tilde{x}(n)$ can be determined from x(n) as follows:

$$\tilde{x}(n) = \sum_{r=-\infty}^{\infty} x(n-r \cdot N) \tag{6.3}$$

Now consider $\tilde{X}(k)$, the DFS coefficient of $\tilde{x}(n)$. Suppose we form a finite duration sequence X(k) by preserving one period of $\tilde{X}(k)$ but setting all other values to zero as follows:

$$X(k) = \tilde{X}(k) \cdot R_N(k) \tag{6.4}$$

where $R_N(k)$ is defined by (6.2). This operation is also invertible since $\tilde{X}(k)$ can be obtained from X(k) as follows:

$$\tilde{X}(k) = \sum_{r=-\infty}^{\infty} X(k-r \cdot N) \tag{6.5}$$

From the above, then, the sequence x(n) is related to X(k) as folfows:

$$x(n) \leftrightarrow \tilde{x}(n) \leftrightarrow \tilde{X}(k) \leftrightarrow X(k) \tag{6.6}$$

where "$\leftrightarrow$" denotes an invertible operation.

The relationship between x(n) and X(k) can be easily obtained from (6.6) and the DFS pair of (5.1) and (5.2) and is given by

<u>Discrete Fourier Transform Pair</u>

$$X(k) = \begin{cases} \sum_{n=0}^{N-1} x(n) \cdot e^{-j\frac{2\pi}{N}k \cdot n} & 0 \le k \le N-1 \\ 0 & \text{otherwise} \end{cases} \tag{6.7}$$

$$x(n) = \begin{cases} \frac{1}{N} \sum_{k=0}^{N-1} X(k) e^{j\frac{2\pi}{N}k \cdot n} & 0 \le n \le N-1 \\ 0 & \text{otherwise} \end{cases} \tag{6.8}$$

From (6.7) and (6.8), an N-point sequence x(n) is represented by an N-point sequence X(k) in the frequency domain. The sequence X(k) is called the DFT of x(n), and x(n) is called the inverse DFT (IDFT) of X(k).

The DFT pair above is defined only for a causal finite duration sequence. This is not a serious restriction in practice, since a finite duration sequence can always be shifted to make it causal, and this shift can be easily accounted for in most application problems.

6.2 Relation between X(k) and X(ω)

For a causal, finite-duration sequence x(n) which is zero outside $0 \le n \le N-1$, the DFT X(k) is very simply related to the discrete-time FT X(ω). From (2.1) and (6.7), it is very clear that

$$X(k) = X(\omega)\Big|_{\omega=\frac{2\pi}{N}k} \quad \text{for } 0 \le k \le N-1 \tag{6.9}$$

Equation (6.9) states that the DFT coefficients of x(n) are samples of X(ω) at N equally spaced points, beginning at ω=0. Since x(n) can be recovered from X(k), and X(ω) can be determined from x(n), X(ω) can be determined from X(k) as follows:

$$X(\omega) = \sum_{n=-\infty}^{\infty} x(n)\cdot e^{-j\omega n} = \frac{1}{N}\sum_{n=-\infty}^{\infty}\sum_{k=0}^{N-1} X(k)\cdot e^{j\frac{2\pi}{N}kn}e^{-j\omega n}$$

$$= \frac{1-e^{-j\omega N}}{N}\cdot\sum_{k=0}^{N-1}\frac{X(k)}{1-e^{j\frac{2\pi}{N}k}e^{-j\omega}} \tag{6.10}$$

Equation (6.10) shows that X(ω) contains considerable redundant information, e.g. N samples of X(ω) completely specify X(ω).

6.3 Properties

We can derive a number of useful properties from the DFT pair of (6.7) and (6.8). Alternatively, the relationship in (6.6) and the DFS properties discussed in Section 5.2 are often more convenient and easier to use in deriving the properties. Some of the more important properties, often useful in practice are listed below. In all cases, x(n) and y(n) are assumed to be zero outside $0 \le n \le N-1$, X(k) and Y(k) are N-point DFTs of x(n) and y(n), and $\tilde{x}(n)$, $\tilde{y}(n)$, $\tilde{X}(k)$, and $\tilde{Y}(k)$ are periodic sequences with period of N obtained from x(n), y(n), X(k), and Y(k).

1. Linearity

$$a\cdot x(n)+b\cdot y(n) \leftrightarrow a\cdot X(k)+b\cdot Y(k) \tag{6.11}$$

2. Circular convolution

$$x(n) \circledast y(n) = [\tilde{x}(n) \circledast \tilde{y}(n)]\cdot R_N(n) \leftrightarrow X(k)\cdot Y(k) \tag{6.12}$$

3. Relation between circular and linear convolution

$w_1(n) = 0$ outside $0 \le n \le N_1-1$

$w_2(n) = 0$ outside $0 \le n \le N_2-1$

$$w_1(n) * w_2(n) = w_1(n) \circledast w_2(n) \quad \text{with periodicity } N \ge N_1+N_2-1 \tag{6.13}$$

4. Circular shift of a sequence

$$\tilde{x}(n-n_0)\cdot R_N(n) \leftrightarrow X(k)\cdot e^{-j\frac{2\pi}{N}k\cdot n_0} \qquad (6.14)$$

5. Symmetry properties

a. $x^*(n) \leftrightarrow \tilde{X}^*(-k)\cdot R_N(k)$ (6.15)

b. $\tilde{x}(-n)\cdot R_N(n) \leftrightarrow \tilde{X}(-k)\cdot R_N(k)$ (6.16)

The result of circular convolution is identical to the result of linear convolution under some conditions. Suppose $w_1(n)$ is an N_1-point sequence which is zero outside $0\leq n\leq N_1-1$ and $w_2(n)$ is an N_2-point sequence which is zero outside $0\leq n\leq N_2-1$. Then (6.13) states that $w_1(n) * w_2(n)=w_1(n) \circledast w_2(n)$ if $w_1(n) \circledast w_2(n)$ is obtained with assumed periodicity $N\geq N_1+N_2-1$.

Equation (6.12) and (6.13) present an alternative way to perform linear convolution of two finite-duration sequences. To linearly convolve $w_1(n)$ and $w_2(n)$, we could assume the proper periodicity N, determine $W_1(k)$ and $W_2(k)$, multiply $W_1(k)$ and $W_2(k)$, and then perform the IDFT operation of $W_1(k)\cdot W_2(k)$. Even though this approach appears to be quite cumbersome, it often reduces the amount of computations involved in performing linear convolution in practical applications. This is discussed in the next section.

6.4 Overlap-add Method

Linear convolution of two sequences is desirable in a variety of contexts. In many cases, the input data x(n) is very long, while the unit sample response h(n) is short. In speech processing, for example, one minute of speech sampled at a rate of 10 kHz would have $6\text{x}10^5$ samples, while the unit sample response of a typical filter would be about 100 points long. If the speech signal x(n) were filtered using the convolution sum directly, we would need to perform about 100 multiplications and 100 additions per output sample.

An alternative to the direct convolution in the example above is to perform the inverse DFT of X(k)·H(k), where the DFT length is chosen to be greater than $6\text{x}10^5+100-1$. As we'll discuss in the next section, there are methods which compute the DFT and IDFT very efficiently. These methods are known as fast Fourier transform (FFT) algorithms. If these methods are used in the DFT and IDFT computations, performing the inverse DFT of X(k)·H(k) requires less computation than performing the convolution directly. One major problem with this approach, however, is the large amount of on-line storage and the long time delay required. This is because we need the entire x(n) before we can compute X(k), and X(k) and H(k) must be stored.

One way to exploit the computational efficiency of FFT algorithms while avoiding the large storage and delay requirements is by using the overlap-add method. In this method, x(n) is segmented into many segments $x_i(n)$. The sequence x(n), then can be represented by

$$x(n) = \sum_{i=1}^{M} x_i(n) \tag{6.17}$$

Convolving x(n) with h(n) and using the distributivity property of convolvution given by (1.16),

$$x(n)*h(n) = \left[\sum_{i=1}^{M} x_i(n) \right] * h(n) = \sum_{i=1}^{M} (x_i(n)*h(n)) \tag{6.18}$$

Since $x_i(n)$ is a much shorter sequence than x(n), $x_i(n)*h(n)$ can be computed by performing the inverse DFT operation of $X_i(k)\cdot H(k)$ with a much shorter DFT and IDFT size. Since the DFT and IDFT sizes used are much smaller, this needs much less memory and delay. Since $x_i(n)*h(n)$ is longer than $x_i(n)$, the results of $x_i(n)*h(n)$ and $x_{i+1}(n)*h(n)$ in (6.18)are overlapped and added. As a result, this method is called the "overlap-add" method. In typical speech processing applications, the overlap-add method reduces the number of arithmetic operations (multiplications and additions) by a factor of five to ten.

7. FAST FOURIER TRANSFORM

The discrete Fourier transform (DFT) discussed in Section 6 is used in a variety of digital signal processing applications, so it is of considerable interest to efficiently compute the DFT and inverse DFT. In this section, we discuss a set of algorithms collectively referred to as "fast Fourier transform" (FFT) algorithms. These FFT algorithms, when used to compute the DFT, offer considerable computational savings compared to directly computing the DFT in a straightforward manner. To appreciate the computational efficiency of FFT algorithms, we will first consider computing the DFT directly, and then discuss FFT algorithms.

7.1 Direct Computation

Consider an N-point complex seuquence x(n) which is zero outside $0 \le n \le N-1$. The DFT of x(n), X(k), is related to x(n) by the DFT pair as follows:

$$X(k) = \begin{cases} \sum_{n=0}^{N-1} x(n)\cdot e^{-j\frac{2\pi}{N}k\cdot n} & 0 \le k \le N-1 \\ 0 & \text{otherwise} \end{cases} \tag{7.1}$$

$$x(n) = \begin{cases} \frac{1}{N} \sum_{k=0}^{N-1} X(k)\cdot e^{j\frac{2\pi}{N}k\cdot n} & 0 \le n \le N-1 \\ 0 & \text{otherwise} \end{cases} \tag{7.2}$$

Since computing x(n) from X(k) is very similar to computing X(k) from x(n), we'll concentrate our discussion on computing X(k) from x(n).

From (7.1), directly computing X(k) for each k requires N complex multiplications and N-1 complex additions. Since there are N different values of k, the total number of arithmetic operations required in computing X(k) from x(n) is N^2 complex multiplications and N(N-1) complex additions.

It is possible to reduce the number of arithmetic operations by simple manipulations of the complex exponential $e^{-j\frac{2\pi}{N}k\cdot n}$ (such as periodicity with period of N in both variables k and n). Such an approach, however, typically results in the reduction of arithmetic operations by a factor of only two or four, and the number of arithmetic operations required remains proportional to N^2. To reduce the number of arithmetic operations for large N by orders of magnitude, a fundamentally new approach is needed. Those algorithms which require the number of arithmetic operations proportional to N·logN in computing an N-point DFT are known as FFT (fast Fourier transform) algorithms. They are discussed in the next section.

7.2 Fast Fourier Transform Algorithms

An FFT algorithm was formulated in the early part of the century, but went largely unnoticed at the time. In those days, it did not really matter whether computing the DFT of a sequence required ten thousand or a million multiplications. When an FFT algorithm was rediscovered by Cooley and Tukey [3] and introduced to the signal processing community in 1965, its significance was immediately recognized. Significant advances had been made in digital computer technology by the 1960s and performing ten thousand or a million multiplications was no longer an impossible task. The rediscovery of an FFT algorithm led to a flurry of activity in digital signal processing research, and is a landmark in the digital signal processing field.

Although there are many variations, all FFT algorithms are based on one simple principle: an N-point DFT can be computed by two $\frac{N}{2}$-point DFTs, or three $\frac{N}{3}$-point DFTs, etc. We'll develop one variation, known as the "decimation-in-time FFT algorithm."

To develop the decimation-in-time algorithm, we first assume $N=2^M$. Rewriting (7.1).

$$X(k) = \sum_{n=0}^{N-1} x(n)\cdot e^{-j\frac{2\pi}{N}k\cdot n}$$

$$= \sum_{n\ \text{even}} x(n)\cdot e^{-j\frac{2\pi}{N}k\cdot n} + \sum_{n\ \text{odd}} x(n)\cdot e^{-j\frac{2\pi}{N}k\cdot n}$$

$$= \sum_{r=0}^{\frac{N}{2}-1} x(2r)\cdot e^{-j\frac{2\pi}{\frac{N}{2}}\cdot k\cdot r} + e^{-j\frac{2\pi}{N}k} \sum_{r=0}^{\frac{N}{2}-1} x(2r+1)\cdot e^{-j\frac{2\pi}{\frac{N}{2}}k\cdot r}$$

$$\text{for } 0\le k\le N-1 \qquad (7.3)$$

Now we form $\frac{N}{2}$-point sequences g(n) and h(n) by the following relationship:

$$g(n) = x(2n)$$
$$h(n) = x(2n+1) \qquad (7.4)$$

Let the $\frac{N}{2}$-point DFTs of g(n) and h(n) be denoted by G(k) and H(k). Rewriting (7.3),

$$X(k) = \sum_{r=0}^{\frac{N}{2}-1} g(r)\cdot e^{-j\frac{2\pi}{\frac{N}{2}}k\cdot r} + e^{-j\frac{2\pi}{N}k} \cdot \sum_{r=0}^{\frac{N}{2}-1} h(r)\cdot e^{-j\frac{2\pi}{\frac{N}{2}}k\cdot r}$$

$$= G(k) + e^{-j\frac{2\pi}{N}k}\cdot H(k) \qquad \text{for } 0\leq k\leq\frac{N}{2}-1 \qquad (7.5)$$

$$\text{and } G(k-\tfrac{N}{2}) + e^{-j\frac{2\pi}{N}k}\cdot H(k-\tfrac{N}{2}) \qquad \text{for } \tfrac{N}{2} \leq k\leq N-1$$

The separation of k for $0\leq k\leq\frac{N}{2}-1$ and $\frac{N}{2}\leq k\leq N-1$ is due to the fact that G(k) and H(k) are $\frac{N}{2}$-point DFTs and therefore are zero for $\frac{N}{2}\leq k\leq N-1$. From (7.5), it is clear that one N-point DFT can be computed by linearly combining the appropriate coefficients of two $\frac{N}{2}$ -point DFTs.

The next step in the algorithm development is to compute each $\frac{N}{2}$-point DFT by computing two $\frac{N}{4}$-point DFTs using essentially the same approach as before. Each $\frac{N}{4}$-point DFT can now be computed by computing two $\frac{N}{8}$-point DFTs, and this procedure can be continued until we are left with one-point DFTs. A one-point DFT is itself. This method is called "decimation-in-time" because the division of an N-point sequence into two $\frac{N}{2}$-point sequences is made in the time domain.

It is convenient to sketch (7.5) using a signal flowgraph representation. Two elements needed in the signal flowgraph representation are shown in Figure 7-1. The signal flows from or toward a "node" represnted by a "o" in the flowgraph. The line that connects two nodes is called a "branch", and the scaler "c" which scales the signal is called "branch transmittance". It is customary to omit the branch transmittance when it is unity, as in Figure 7-1(b).

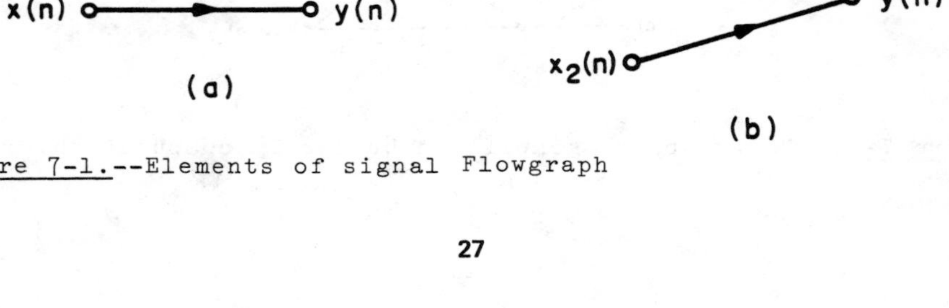

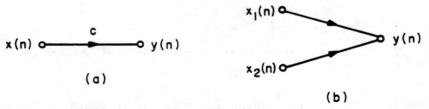

Figure 7-1.--Elements of signal Flowgraph

Using the elements in Figure 7-1 and denoting $e^{-j\frac{2\pi}{N}}$ by W_N, the decimation-in-time FFT algorithm can be represented for N=8 as in Figure 7-2. From Figure 7-2, the basic element is a "butterfly" computation, shown in Figure 7-3. With simple algebraic manipulation, it can be shown that the butterfly computation in Figure 7-3 can be replaced by the one shown in Figure 7-4.

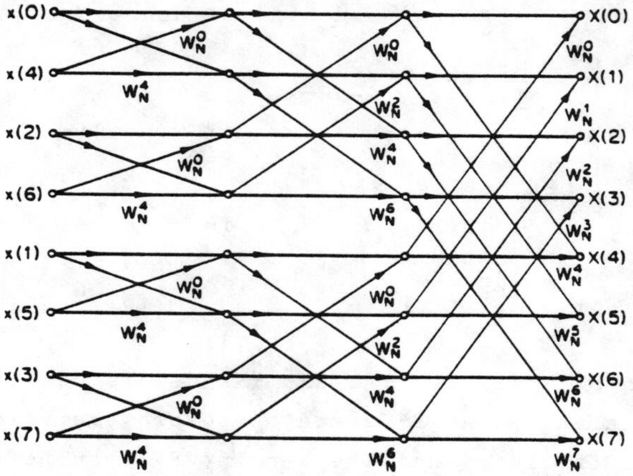

Figure 7-2.--Decimation-in-time FFT algorithm based on (7.5) (After [1]).

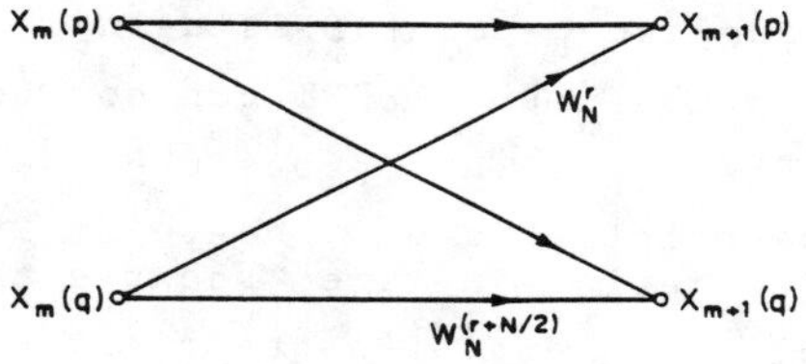

Figure 7-3.--Flowgraph of basic butterfly computation in Figure 7-2. (After [1]).

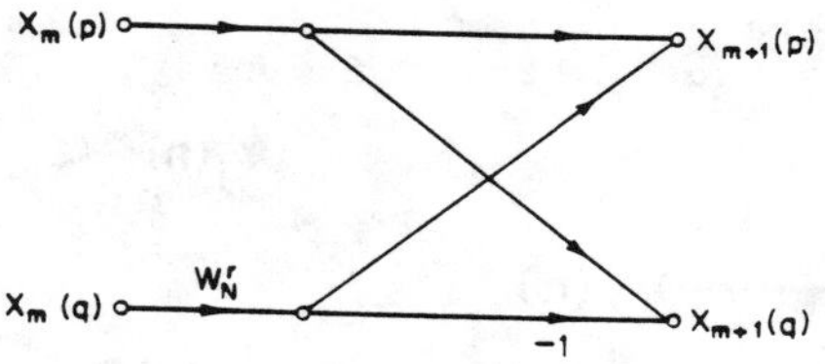

Figure 7-4.--Flowgraph of simplified butterfly computation. (After [1]).

The advantage of this new butterfly computation is that it requires only one complex multiplication, while the method in Figure 7-3 requires two. Replacing each butterfly computation in Figure 7-3 with that in Figure 7-4, we now have the signal flowgraph shown in Figure 7-5. We'll refer to the algorithm in Figure 7-5 as the "standard form of decimation-in-time FFT algorithm."

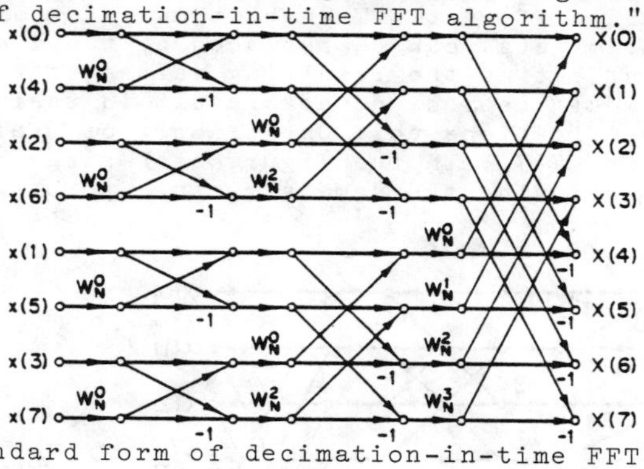

Figure 7-5.--Standard form of decimation-in-time FFT algorithm. (After [1])

From Figure 7-5, we can make a number of observations about the FFT algorithm. Perhaps the most important aspect is the number of arithmetic operations required. For an N-point DFT there are $\log_2 N$ stages. For example, when N=8, the FFT algorithm has $\log_2 8=3$ stages, as shown in Figure 7-5. In each stage, there are $\frac{N}{2}$ butterfly computations. Each butterfly computation requires one complex multiplication (multiplication by -1 is not considered a multiplication) and two complex additions. The total number of arithmetic operations required is $\frac{N}{2}\log_2 N$ complex multiplications and $N\log_2 N$ complex additions. For small values of N, the advantage of $N\log_2 N$ over N^2 is not significant. For large N (N=1024, for example), the number of arithmetic operations is reduced by orders of magnitude.

The FFT algorithm in Figure 7-5 naturally leads to "in-place" computation. If the DFT is computed directly, we need 2N storage locations: N storage locations for x(n), and N for X(k). In the FFT algorithm in Figure 7-5, the butterfly computation is a basic computational element, and each butterfly computation in a given stage does not depend on other butterfly computations in the same stage. Therefore, the result of each butterfly computation can be stored in the same storage locations that contained the input to the butterfly computation. In this way, if we do not need the data x(n), the DFT X(k) can be stored in the same storage locations that originally contained the data, thus requiring only N storage locations. When N is large (in image processing applications, the data size can reach a million points), in-place computation offers significant saving in amount of storage required.

From Figure 7-5, the output X(k) is in the normal order. The input x(n), however, is not. Because of the way in which the data have been divided into two parts in each stage, the order of the input data is in bit-reversed order. When N=8, for example, the input x(n) is in the order n=0,4,2,6,1,5,3,7 ; in binary representation,

000,100,010,110,001,101,011,111. The above binary representation is in the bit-reversed form of the normal order, which is 000,001,010,011,100,101,110,111.

The signal flowgraph in Figure 7-5 can be modified in a variety of ways to obtain different FFT algorithms. For example, the dotted portion in Figure 7-6(a) can be replaced by the flowgraph in Figure 7-6(b) without affecting the resulting values of X(k). Even though such arbitrary modifications do not result in useful FFT algorithms, more planned modifications do. One example of useful modifications is shown in Figure 7-7. In the figure each stage has exactly the same form, thus allowing the same structure to be used for each stage.

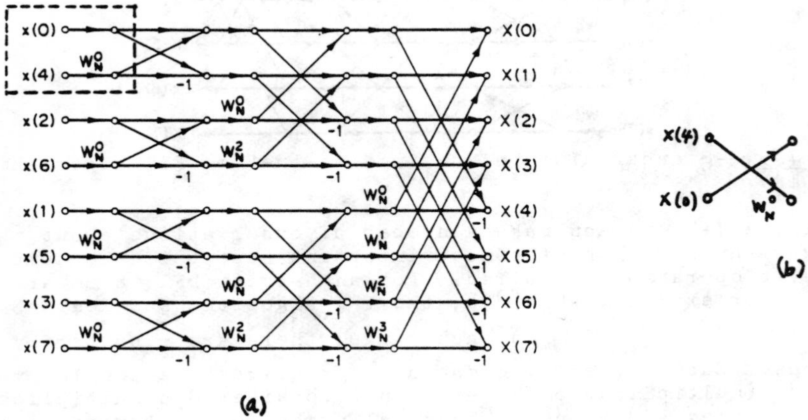

Figure 7-6.--Simple modification of signal flowgraph.

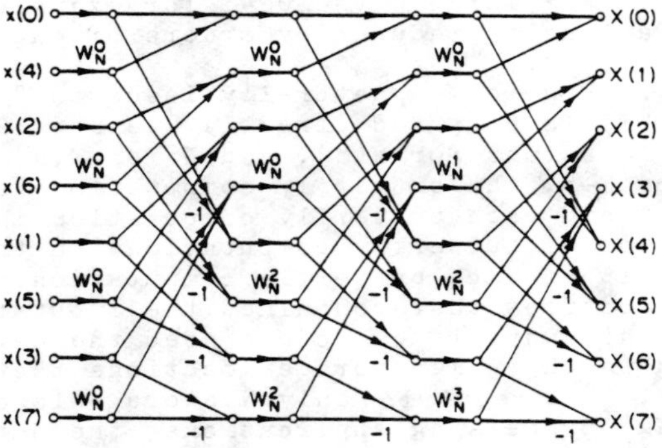

Figure 7-7.--Example of useful modifications of the standard form of decimation-in-time FFT algorithm. (After [1])

In the discussion above, we have considered the number of arithmetic operations and the amount of storage required. Another important consideration in comparing different methods of computing the DFT is the complexity of the algorithm. For example, if an FFT algorithm is quite complex to implement, its advantage in number of required arithmetic operations may be lost, since computing the DFT directly is very simple. Fortunately, the FFT algorithm in Figure 7-5 is simple to implement. A typical FORTRAN subroutine that implements an FFT algorithm has twenty to thirty lines of statements.

We considered above the case where $N=2^M$. The same general approach, however, can be applied in the case where N is a highly composite number, i.e. $N=p_1 \cdot p_2 \cdot p_3 \cdots p_M$ where p_i may not be 2. For example, when $p_M=3$, we can compute three $\frac{N}{3}$-point DFTs rather than one N-point DFT. The number of arithmetic operations required for $N=p_1 \cdot p_2 \cdots p_M$ is on the order of $N \cdot \sum_{i=1}^{M} (p_i-1)$ complex multiplications and $N \cdot \sum_{i=1}^{M} (p_i-1)$ complex additions. The reason there is a factor of two difference in the number of multiplications between $\frac{N}{2}\log_2 N$ and $N \cdot \sum_{i=1}^{M} (p_i-1)$ when $p_i=2$ is that the number of multiplications is reduced by a factor of 2 by manipulating the butterfly computation. This can be achieved when $p_i=2$, but is not generally possible for other values of p_i.

Although the FFT algorithm has a clear advantage over computing the DFT directly, there are cases when it is more advantageous to compute the DFT directly. For example, when we need to compute only a few DFT coefficients, it may be more advantageous to directly compute only those coefficients we need. In addition, in hardware environments with a large amount of parallelism, direct computation of the DFT can exploit the parallelism better than an FFT algorithm which has many stages through which the data flows.

7.3 Recent Developments in the DFT Computation

By the early 1970s, extensions, modifications, and interpretations had been made to Cooley and Tukey's 1965 algorithm, and the FFT was a standard tool of the signal processing community. In 1975, however, S. Winograd of IBM presented a new approach to computing the DFT, the Winograd Fourier transform (WFT). The WFT is based on the algebraic complexity theory of computation which is concerned with determining the number of multiplications required to perform a computation and systematic development of minimal algorithms. For certain cases the number of multiplications required in the WFT is proportional to N for an N-point DFT computation. Despite this advantage the WFT does not compete favorably with the FFT discussed in Section 7.2 in most applications. With available digital hardware, minimizing the number of multiplications alone does not necessarily maximize an algorithm's efficiency. This reduces the relative advantage of the WFT algorithm, which is quite complex to program and is more sensitive to finite precision arithmetic. A typical FORTRAN subroutine that implements the WFT requires something on the order of 500 statements.

Even though the WFT itself has not proved very useful in practice, parts of the WFT have been successfully used in developing other methods to compute the DFT. One such example is the prime factor algorithm (PFA), which uses the prime-factor index maps associated with the Winograd short DFT algorithms. Further development of these algorithms is being actively investigated. For further reading on this topic, see [4,5].

8. DIGITAL PROCESSING OF ANALOG SIGNALS

Processing analog signals, such as speech, images, radar signals, and geophysical data, is an important application area for digital signal processing. In such applications, an analog signal is first sampled, next processed by a digital signal processing algorithm, and then converted to an analog signal. In this section, we discuss issues related to this process.

8.1 Notation for Analog Signal Representation

To differentiate an analog signal from a sequence, we denote the analog signal by $x_a(t)$. The continuous time Fourier transform (FT) of $x_a(t)$, $X_a(\Omega)$, is related to $x_a(t)$ as follows:

$$X_a(\Omega) = \int_{t=-\infty}^{\infty} x_a(t) \cdot e^{-j\Omega t} \cdot dt$$

$$x_a(t) = \frac{1}{2\pi} \int_{\Omega=-\infty}^{\infty} X_a(\Omega) \cdot e^{j\Omega t} \cdot d\Omega \tag{8.1}$$

Equation (8.1) is called the FT pair for analog signals.

The Laplace transform of $x_a(t)$,$X_a(s)$, is related to $x_a(t)$ as follows:

$$X_a(s) = \int_{t=-\infty}^{\infty} x_a(t) \cdot e^{-st} \cdot dt$$

$$x_a(t) = \frac{1}{2\pi j} \cdot \int_{\sigma-j\infty}^{\sigma+\infty} X_a(s) \cdot e^{st} \cdot ds \tag{8.2}$$

where σ=Re[s] is in the region of convergence (ROC) of $X_a(s)$. Equation (8.2) is called the Laplace transform pair.

If $X_a(s)$ has an ROC that includes the $j\Omega$ axis ($s=j\Omega$), then from (8.1) and (8.2), $X_a(\Omega)$ is related to $X_a(s)$ by the following expression:

$$X_a(\Omega) = X_a(s)\big|_{s=j\Omega} \tag{8.3}$$

From (8.3), the FT of an analog signal is often considered to be a special case of the Laplace transform.

8.2 Analog to Digital Conversion

Suppose we obtain a discrete time signal $x(n)$ by sampling an analog signal $x_a(t)$ with sampling period of T as follows:

$$x(n) = x_a(t)\big|_{t=n\cdot T} \qquad (8.4)$$

Equation (8.4) represents the input-output relationship of an ideal A/D (analog to digital) converter. We now wish to determine the relationship between $X(\omega)$, the discrete time FT of $x(n)$, and $X_a(\Omega)$, the continuous time FT of $x_a(t)$.

To establish the relation between $X(\omega)$ and $X_a(\Omega)$, it is useful to represent (8.4) by the system shown in Figure 8-1. In the figure, $p_a(t)$ denotes a train of impulses periodic with period of T. The system G converts an analog signal $x_s(t)$ to a sequence $x(n)$ by measuring the area under each impulse and using it as the amplitude of the sequence. An example of $x_a(t)$, $p_a(t)$, $x_s(t)$, and $x(n)$ is shown in Figure 8-2. The relationships among $x_a(t)$, $x_s(t)$, and $x(n)$

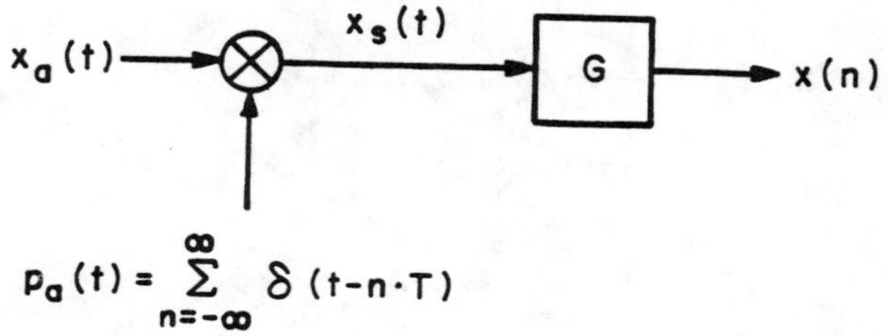

Figure 8-1.--System that represents $x(n)=x_a(t)\big|_{t=n\cdot T}$

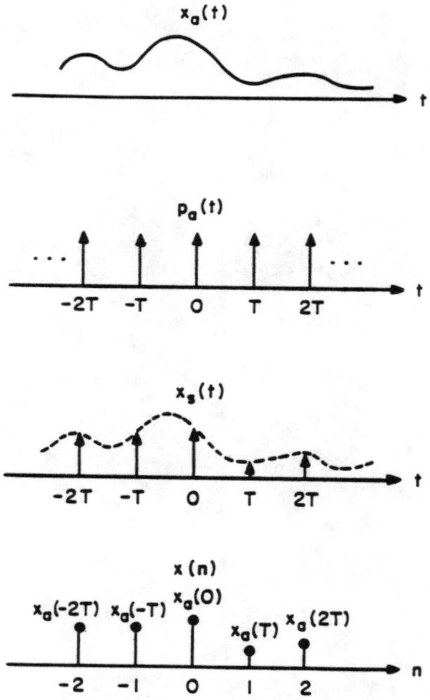

Figure 8-2.--Example of $x_a(t)$, $p_a(t)$, $x_s(t)$, and $x(n)$.

are as follows:

$$x_s(t) = \sum_{n=-\infty}^{\infty} x_a(n\cdot T)\cdot\delta(t-n\cdot T)$$

$$x(n) = x_a(n\cdot T) \tag{8.5}$$

To represent $x_a(t)$, $p_a(t)$, $x_s(t)$, and $x(n)$ in the frequency domain, we first note that $P_a(\Omega)$, the continuous time FT of $p_a(t)$, is given by

$$P_a(\Omega) = \frac{2\pi}{T}\cdot \sum_{m=-\infty}^{\infty} \delta[\Omega-m\cdot\frac{2\pi}{T}] \tag{8.6}$$

Since $x_s(t)=x_a(t)\cdot p_a(t)$, $X_s(\Omega)$ is related to $X_a(\Omega)$ by

$$X_s(\Omega) = \frac{1}{2\pi}\cdot X_a(\Omega)*P_a(\Omega)$$

$$= \frac{1}{2\pi}\cdot X_a(\Omega)*\frac{2\pi}{T} \sum_{m=-\infty}^{\infty} \delta[\Omega-m\cdot\frac{2\pi}{T}]$$

$$= \frac{1}{T}\cdot \sum_{m=-\infty}^{\infty} X_a[\Omega-m\cdot\frac{2\pi}{T}] \tag{8.7}$$

Equation (8.7) relates $X_s(\Omega)$ to $X_a(\Omega)$. To relate $X_s(\Omega)$ to $X(\omega)$, from (8.1) and (8.5),

$$X_s(\Omega) = \int_{t=-\infty}^{\infty} x_s(t)\cdot e^{-j\Omega t}\cdot dt$$

$$= \int_{t=-\infty}^{\infty} \sum_{n=-\infty}^{\infty} x_a(n\cdot T)\cdot\delta(t-n\cdot T)\cdot e^{-j\Omega t}\cdot dt$$

$$= \sum_{n=-\infty}^{\infty} x_a(n\cdot T)\cdot e^{-j\Omega\cdot nT} \tag{8.8}$$

From (8.5),

$$X(\omega) = \sum_{n=-\infty}^{\infty} x(n)\cdot e^{-j\omega n}$$

$$= \sum_{n=-\infty}^{\infty} x_a(n\cdot T)\cdot e^{-j\omega n} \tag{8.9}$$

Comparing (8.8) and (8.9),

$$X(\omega) = X_s(\Omega)\Big|_{\Omega=\frac{\omega}{T}} \tag{8.10}$$

Combining (8.7) and (8.10),

$$X(\omega) = \frac{1}{T}\cdot \sum_{m=-\infty}^{\infty} X_a[\frac{\omega}{T}-m\cdot\frac{2\pi}{T}] \tag{8.11}$$

Two examples of $X_a(\Omega)$, $P_a(\Omega)$, $X_s(\Omega)$, and $X(\omega)$ are shown in Figure 8-3. Figure 8-3(a) shows the case when Ω_c, the cut-cff frequency of $x_a(t)$, is less than $\frac{\pi}{T}$. Figure 8-3(b) shows the case when Ω_c is greater than $\frac{\pi}{T}$.

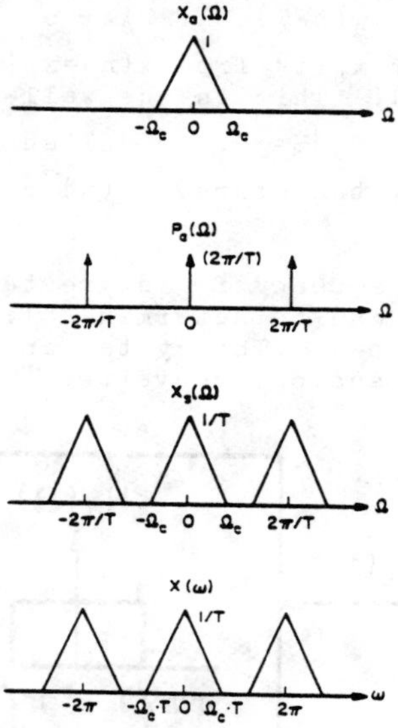

Figure 8-3(a).--Example of $X_a(\Omega)$, $P_a(\Omega)$, and $X(\omega)$ for the case when $\Omega_c < \frac{\pi}{T}$.

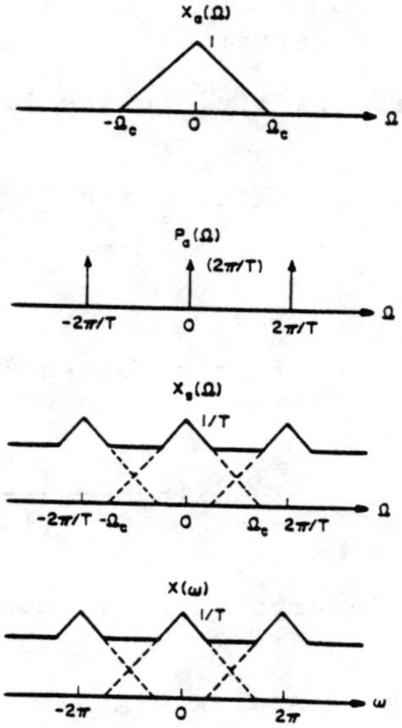

Figure 8-3(b).--Example of $X_a(\Omega)$, $P_a(\Omega)$, $X_s(\Omega)$, and $X(\omega)$ for the case when $\Omega_c > \frac{\pi}{T}$.

8.3 Digital to Analog Conversion

From Figure 8-3, it is clear that if $\Omega_c<\frac{\pi}{T}$ or $\frac{1}{T}>\frac{\Omega_c}{\pi}$, then $x_a(t)$ can be recovered from $x(n)=x_a(n\cdot T)$. If $\Omega_c>\frac{\pi}{T}$ or $\frac{1}{T}<\frac{\Omega_c}{\pi}$, however, it is not possible to recover $x_a(t)$ from $x(n)=x_a(n\cdot T)$ without additional information about $x_a(t)$. This is the well-known "sampling theorem." The sampling frequency, $f_s=\frac{1}{T}=\frac{\Omega_c}{\pi}$, is called the "Nyquist rate." Above the Nyquist rate, therefore, $x_a(t)$ can be obtained from $x(n)=x_a(n\cdot T)$.

Suppose the sampling frequency f_s is greater than the Nyquist rate. From Figure 8-3, $x_a(t)$ can be determined from $x(n)=x_a(n\cdot T)$ by the system shown in Figure 8-4. The system in the figure is called the "ideal D/A (digital to analog) converter."

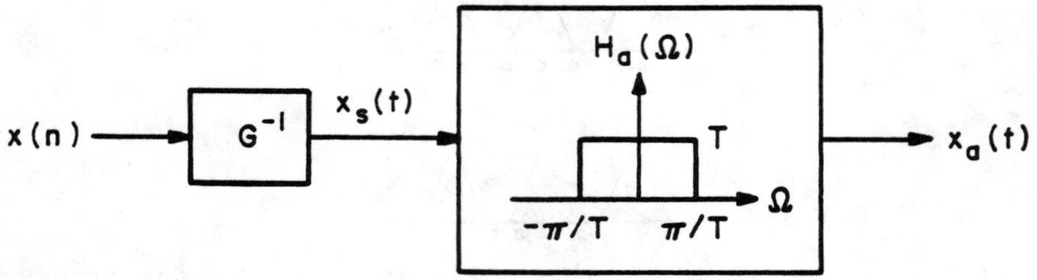

Figure 8-4.--Ideal D/A converter.

To express $x_a(t)$ in terms of $x(n)$ in the ideal D/A converter, we first note that

$$x_s(t) = \sum_{n=-\infty}^{\infty} x(n)\cdot\delta(t-n\cdot T) \tag{8.12}$$

From Figure 8-4, then,

$$x_a(t) = x_s(t)*h_a(t) = \sum_{n=-\infty}^{\infty} x(n)\cdot\delta(t-n\cdot T)* \frac{\sin\frac{\pi}{T}\cdot t}{\frac{\pi}{T}\cdot t}$$

$$= \sum_{n=-\infty}^{\infty} x(n)\cdot\frac{\sin\frac{\pi}{T}(t-n\cdot T)}{\frac{\pi}{T}(t-n\cdot T)} \tag{8.13}$$

Equation (8.13) is the interpolation formula that determines $x_a(t)$ from its samples $x(n) = x_a(n\cdot T)$.

8.4 Digital Processing of Analog Signals

An analog signal can often be processed by digital signal processing techniques using the A/D and D/A converters discussed in Sections 8.2 and 8.3. Digital processing of analog signals can, in general, be represented by the system in Figure 8-5. The analog low-pass filter limits the bandwidth of the analog signal to reduce the effect of aliasing.

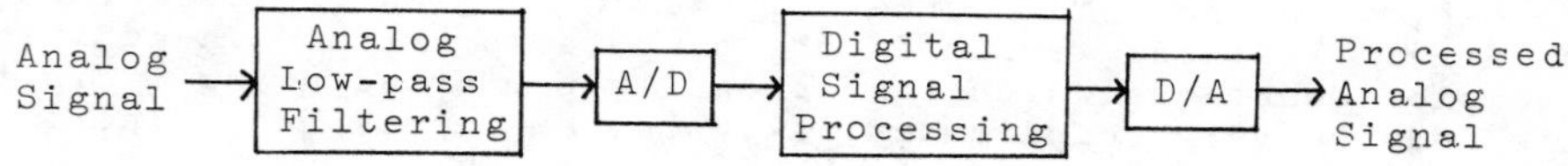

Figure 8-5.--System for digital processing of analog signals.

The following very simple example illustrates the approach in Figure 8-5. Consider a signal restoration problem, where the noisy observation $y_a(t)$ is given by

$$y_a(t) = s_a(t)+w_a(t) \tag{8.14}$$

where $s_a(t)$ and $w_a(t)$ denote the signal and the background noise, respectively. The spectra of $s_a(t)$, $w_a(t)$, and $y_a(t)$ in this example are shown in Figure 8-6.

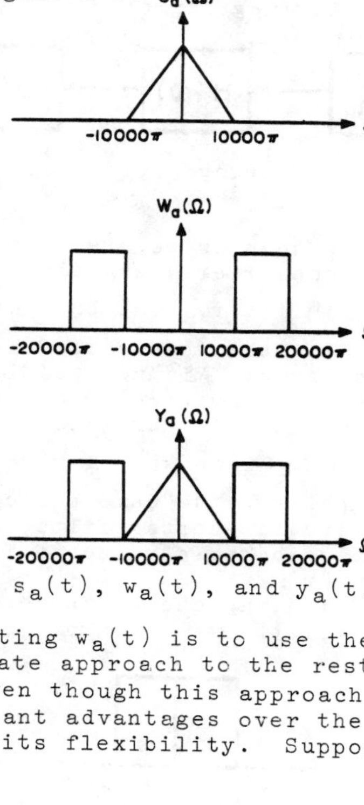

Figure 8-6.--Spectra of $s_a(t)$, $w_a(t)$, and $y_a(t)$.

One approach to eliminating $w_a(t)$ is to use the system shown in Figure 8-7.* An alternate approach to the restoration problem is shown in Figure 8-8. Even though this approach appears more complicated, it has important advantages over the method in Figure 8-7. One major advantage is its flexibility. Suppose we wish to change

*The analog filter, of course, cannot be realized in practice. We use it here for illustration purposes.

the analog filter to have a different $H_a(\Omega)$ because of change in signal and noise characteristics. To obtain an analog filter with different characteristics, we would have to change the hardware components: capacitors, resistors, etc. Using the approach in Figure 8-8, we do not change the computer or digital processor to change $H(\omega)$. Instead, we modify the computer program. In addition, many complex functions which can be easily performed using a digital computer may be very complicated, awkward, or impossible to perform using an analog system.

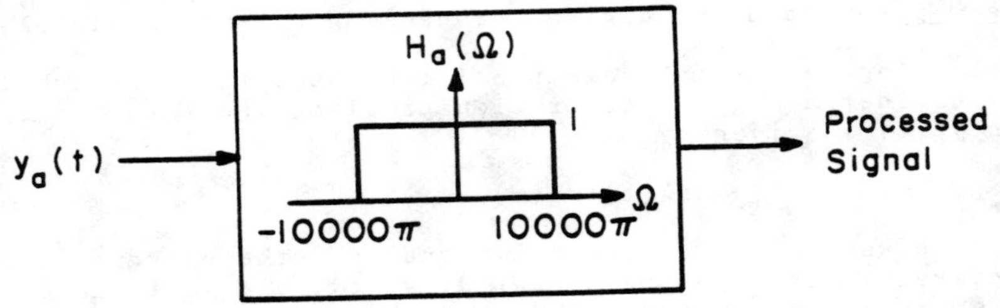

Figure 8-7.--Analog filtering approach to background noise reduction.

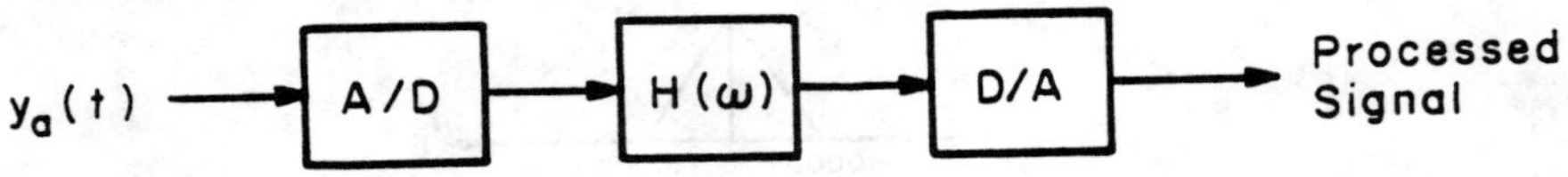

Figure 8-8.-- Digital filtering approach to background noise reduction.

In the above system, we need to determine both the sampling period T for the A/D and D/A converters, and also $H(\omega)$, the frequency response of the digital filter. One choice of $H(\omega)$ for $T = \frac{1}{20000}$ is shown in Figure 8-9. Other choices of T and $H(\omega)$ are also possible. In general, a larger T is more efficient. With a larger T, the amount of data which must be processed is smaller. For example, speech of one second's duration corresponds to 10000 points when $T = \frac{1}{10000}$, and 20000 points when $T = \frac{1}{20000}$. In addition, the digital filter that achieves the same performance in the analog domain typically requires a shorter filter length with a larger T. In the specific example above, a value of T greater than $\frac{1}{20000}$ can be chosen, since aliasing due to large T will begin affecting the background noise first.

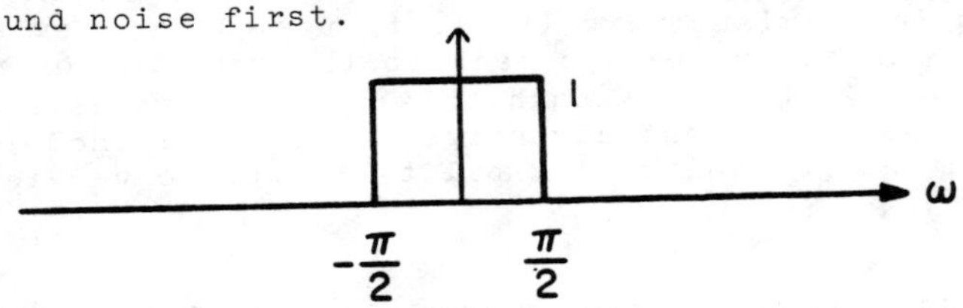

Figure 8-9.--Frequency response of digital filter in Figure 8-8 for the case when $T = \frac{1}{20000}$.

In practice, the input data is not band-limited in most cases. To avoid aliasing due to the A/D converter, the input is first low-pass filtered by an analog low-pass filter to limit the bandwidth of the input signal.

In the above discussions, we have assumed that the A/D and D/A converters are ideal. For example, it was assumed that the analog signal $x_a(t)$ could be sampled at any desired sampling rate with infinite precision in the amplitude representation. In practice, this is neither possible nor necessary. For most audio processing applications, for example, converters of 14- to 16-bit accuracy and a sampling rate of 50khz is sufficient. A variety of A/D and D/A converters that trade off accuracy and speed have been developed. Details of the basic principles behind different converters and the impact that recent advances in hardware technology have on converters can be found in [6,7,8].

9. FINITE IMPULSE RESPONSE DIGITAL FILTER

9.1 Introduction

Three steps are generally followed in using digital filters. The first is the filter specification, in which we specify the characteristics required of the filter. The filter specification depends, of course, on the application for which the filter is needed. For example, in a case where we wish to restore a signal which has been degraded by background noise, the filter characteristics we require depend on the spectral characteristics of the signal and background noise. The second step in using digital filters is design. In the filter design step, we determine $h(n)$, the unit sample response of the filter, or its system function $H(z)$, that meets the design specification. The third is the filter implementation step, in which we realize a discrete time system with the given $h(n)$ or $H(z)$.

The three steps above are closely related to each other. For example, it does not make much sense to specify a filter that cannot be designed. Neither does it make much sense to design a filter which cannot be implemented. Despite the close relationship among the three steps, we will discuss them separately for convenience. We will, however, point out the interrelationships at various points in our discussion.

We'll restrict ourselves to a certain class of digital filters for practical reasons. One restriction is that $h(n)$ is real. In practice, we typically deal with real data, so to ensure that the processed signal is real, we require $h(n)$ to be real. Another restriction is the causality of $h(n)$, i.e. $h(n)=0$ for $n<0$. In practice, we cannot generally anticipate the future input for the current output, and a non-causal $h(n)$ corresponds to more delay than a causal $h(n)$, which is not desirable in real-time applications. Another restriction is the stability of $h(n)$, i.e. $\sum_{n=-\infty}^{\infty} |h(n)|<\infty$. In practice, an unbounded output can cause many difficulties such as system overload. For these practical reasons, we will restrict our discussion to the class of digital filters, whose unit sample response h(n) is real, causal, and stable.

Digital filters can be classified into two groups. In the first group h(n) is a finite duration sequence and the filters in this group are called finite impulse response (FIR) filters. In the second group, h(n) is of infinite extent and the filters in this group are called infinite impulse response (IIR) filters. In this section, we'll concentrate on FIR filters. The IIR filters will be discussed in Section 10. As we'll see, the design and implementation of FIR filters differ considerably from those of IIR filters.

9.2 Linear Phase Filters

A digital filter h(n) is said to have linear phase if $H(\omega)$ can be expressed in the following form:

$$H(\omega) = H_M(\omega)\cdot e^{-j\alpha\omega} \tag{9.1}$$

where $H_M(\omega)$ is a real function of ω and α is some constant. To illustrate the significance of a linear phase filter, suppose a signal s(n) is filtered by a linear phase filter with $H_M(\omega)=1$ and some integer α. The filtered signal will have a spectrum given by $S(\omega)\cdot e^{-j\omega\alpha}$, which is $s(n-\alpha)$ in the time domain. The linear phase, in this example, clearly preserves the shape of the signal. If the frequency response of the filter has non-linear phase with $|H(\omega)|=1$, the shape of the signal will not in general be preserved. The example demonstrates one characteristic of a linear phase filter, its tendency to preserve the shape of the signal component in the passband region of the filter. This characteristic is quite useful in applications such as image processing, where the shape of the signal is very important.

It is useful to generalize the notion of linear phase to include the case when $H(\omega)$ can be expressed as

$$H(\omega) = H_M(\omega)\cdot e^{-j\alpha\omega}\cdot e^{j\beta} \tag{9.2}$$

where $H_M(\omega)$ is a real function of ω, and α and β are some constants. The case where $\beta=0$ corresponds to the more strict definition of linear phase.

For a real, causal, and stable FIR filter, h(n)=0 outside $0\leq n\leq N-1$, where N denotes the filter length. By algebraic manipulation, it can be shown that an FIR filter has generalized linear phase if and only if $h(n) = \pm h(N-1-n)$:

$$\text{Generalized linear phase for FIR filter} \leftrightarrow h(n)= \pm h(N-1-n) \tag{9.3}$$

From (9.3), it is quite easy to obtain linear phase FIR filters. In addition, as we'll see later, requiring linear phase simplifies the filter design problem. For these reasons, and because of the signal shape preservation characteristic of linear phase filters, we'll restrict our discussion of FIR filters to linear phase filters.

Depending on whether h(n) is symmetric (h(n)=h(N-1-n)) or anti-symmetric (h(n)=-h(N-1-n)) with respect to the point of symmetry, and depending on whether the length of the filter is odd or even, all linear phase FIR filters can be classified into four different types. The characteristics of these four different types are tabulated in Table 9-1. Examples of the different types are shown in Figure 9-1. Table 9-1 will be useful in our discussion on the design of FIR filters.

9.3 Filter Specification

Like analog filters, digital filters are generally specified in the frequency domain. Since $H(\omega)=H(\omega+2\pi)$ for all ω, $H(\omega)$ for $0\leq\omega\leq 2\pi$ completely specifies $H(\omega)$. In addition, since $h(n)$ is assumed real, $H(\omega)=H^*(-\omega)$. Specifying $H(\omega)$ for $0\leq\omega\leq\pi$ therefore completely specifies $H(\omega)$ for all ω. Examples of $H(\omega)$ for ideal low-pass, high-pass, band-pass, and band-stop filters are shown in Figure 9-2. The frequency response in Figure 9-2(a) is called a low-pass filter, since an analog signal is low-pass filtered when it is used for digital filtering of analog signals. Other filters are named in an analogous manner.

Since $H(\omega)$ is a complex function of ω, we need to specify both the magnitude and the phase of $H(\omega)$. For FIR filters, we require linear phase and therefore need to specify only the magnitude response. The scheme that is used for the magnitude specification is called a "tolerance" scheme. To illustrate this scheme, let us consider the specification of a low-pass filter. Ideally, a low-pass filter has only a passband region and a stopband region. In practice, a sharp transition between the two regions cannot be achieved, and the passband region corresponds to $0\leq\omega\leq\omega_p$ and the stopband region corresponds to $\omega_s\leq\omega\leq\pi$. The frequencies ω_p and ω_s are called, respectively, the "passband frequency" and the "stop-band frequency." The frequency range $\omega_p<\omega<\omega_s$ is called the "transition band." Ideally, the magnitude response $|H(\omega)|$ is unity in the passband region and is zero in the stopband region. In practice, we require $1-\delta_1\leq|H(\omega)|\leq 1+\delta_1$ for $0\leq\omega\leq\omega_p$ and $|H(\omega)|<\delta_2$ for $\omega_s\leq\omega\leq\pi$. The variables δ_1 and δ_2 are called "passband tolerance" and "stopband tolerance" respectively. A low-pass filter is, then, completely specified by four parameters δ_1, δ_2, ω_p, and ω_s. Other filters can also be specified in an analogous manner.

9.4 FIR Filter Design

The problem of designing a filter is basically that of determining an $h(n)$ or $H(z)$ that meets the design specification. In designing FIR filters, the first step is to decide on the type of filter. In Section 9.2, we discussed the characteristics of four different types of linear phase FIR filters. Some types are inherently not suitable for designing certain classes of filters. For example, from Table 9-1, Type II filters will always have $H(\omega)=0$ at $\omega=\pi$, and therefore are not suitable for the design of high-pass filters. Table 9-2 shows what types of filters are not suitable for which classes of digital filters.

The three standard approaches to designing FIR filters are the window method, the frequency sampling method, and the optimal filter design method. They are discussed in this section.

TABLE 9-1. PROPERTIES OF LINEAR PHASE FIR FILTERS

Type	Symmetry of h(n)	N (Filter Length)	α (Point of Symmetry)	β	Form of $H_M(\omega)$	Constraints on $H_M(\omega)\cdot e^{j\beta}$
I	h(n)=h(N-1-n) symmetric	odd	$\frac{N-1}{2}$	0	$\sum_{n=0}^{\frac{N-1}{2}} a(n)\cdot \cos\omega n$	real
II	h(n)=h(N-1-n) symmetric	even	$\frac{N-1}{2}$	0	$\sum_{n=1}^{\frac{N}{2}} b(n)\cdot \cos\omega(n-\frac{1}{2})$	real $H_M(\pi)=0$
III	h(n)=-h(N-1-n) anti-symmetric	odd	$\frac{N-1}{2}$	$\frac{\pi}{2}$	$\sum_{n=1}^{\frac{N-1}{2}} c(n)\cdot \sin\omega n$	pure imaginary $H_M(0)=H_M(\pi)=0$
IV	h(n)=-h(N-1-n) anti-symmetric	even	$\frac{N-1}{2}$	$\frac{\pi}{2}$	$\sum_{n=1}^{\frac{N}{2}} d(n)\cdot \sin\omega(n-\frac{1}{2})$	pure imaginary $H_M(0)=0$

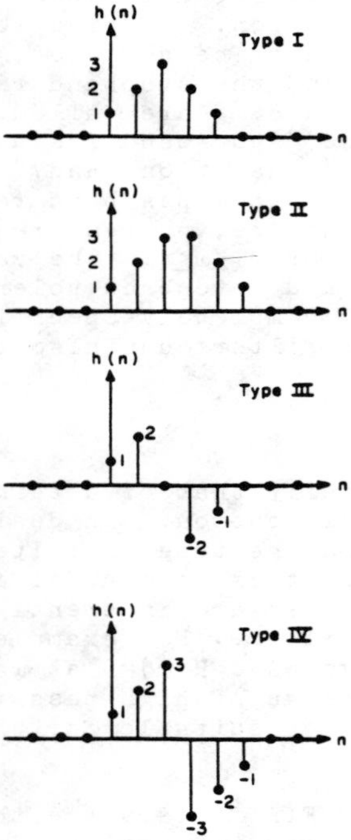

Figure 9-1.--Example of four different types of linear phase FIR filter.

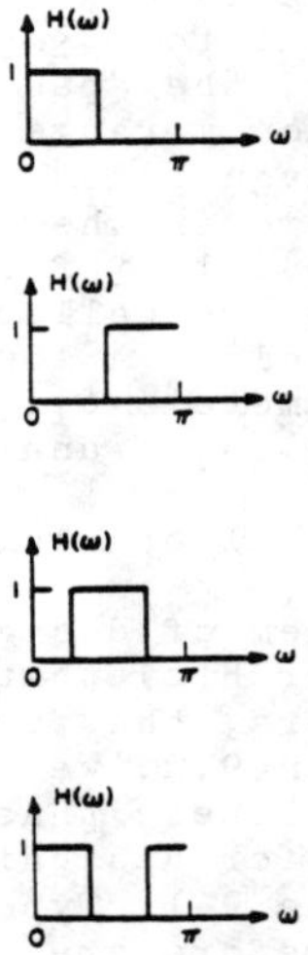

Figure 9-2.--Frequency response H(ω) for ideal digital low-pass, high-pass, band-pass, and band-stop filters.

Table 9-2. Filter Types Not Suitable For Design of Certain Classes Of Filter

Type	Low-pass	High-pass	Band-pass	Band-stop	Differentiator
I					
II		no		no	
III	no	no		no	
IV	no			no	

9.4.1 Window method

In the window method, the desired frequency response $H_d(\omega)$ is assumed to be known. By inverse Fourier transforming $H_d(\omega)$, we can determine the desired unit sample response of the filter, $h_d(n)$. In general, $h_d(n)$ is an infinite duration sequence. In the window method, an FIR filter is obtained by multiplying a window $w(n)$ to $h_d(n)$. If $h_d(n)$ is symmetric (or anti symmetric) with respect to a certain point of symmetry (or antisymmetry) and $w(n)$ is also symmetric with respect to the same point of symmetry, then $h_d(n) \cdot w(n)$ is a linear phase filter.

Two important design parameters in the window method are shape and length of the window. To gain some insight into how these design parameters are chosen, we express $h(n)=h_d(n)\cdot w(n)$ in the frequency domain as follows:

$$H(\omega) = H_d(\omega) \circledast W(\omega) = \frac{1}{2\pi} \int_{\theta=-\pi}^{\pi} H_d(\theta)\cdot W(\omega-\theta)\cdot d\theta \qquad (9.4)$$

An example of $H_d(\omega)$, $W(\omega)$, and $H(\omega)$ is shown in Figure 9-3.

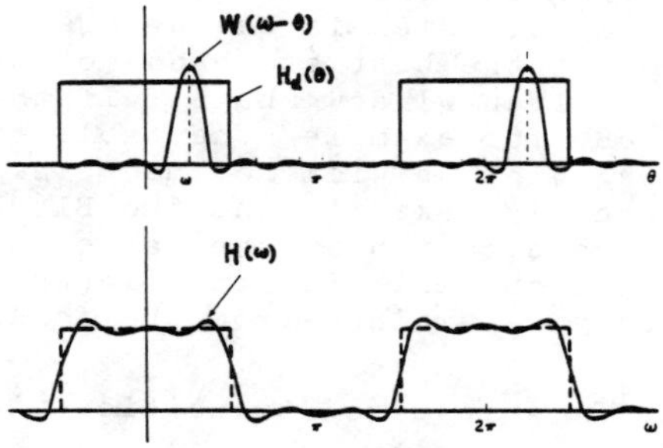

Figure 9-3.--Example of $H_d(\omega)$, $W(\omega)$, and $H(\omega)$. (After [1]).

From Figure 9-3, the mainlobe width of $W(\omega)$ affects the transition width and the sidelobe of $W(\omega)$ affects the passband and stopband tolerance. A good choice of the window, therefore, corresponds to choosing a window which has good mainlobe and sidelobe behaviors (small mainlobe width and small sidelobe amplitude).

Five commonly used windows are rectangular, Bartlett (triangular), Hanning, Hamming, and Blackman. The mainlobe and sidelobe characteristics of these windows are tabulated in Table 9-3. Characteristics of additional windows can be found in [9]. From the first three columns of Table 9-3, the shape of the window affects both mainlobe and sidelobe behavior. The length of the window (N in Table 9-3) affects primarily the mainlobe behavior. This can be seen from the second column of Table 9-3, where peak amplitude of sidelobe is not a function of N.

Table 9-3. Properties of Different Windows

Window	*Peak Amplitude of Side Lobe (dB)*	*Transition Width of Main Lobe*	*Minimum Stopband Attenuation (dB)*
Rectangular	−13	$4\pi/N$	−21
Bartlett	−25	$8\pi/N$	−25
Hanning	−31	$8\pi/N$	−44
Hamming	−41	$8\pi/N$	−53
Blackman	−57	$12\pi/N$	−74

The effect of the shape and size of the window can be summarized as follows:

Window shape → sidelobe behavior
mainlobe behavior

Window size → mainlobe behavior (9.5)

Since the sidelobe behavior (and therefore the passband and stopband tolerance) is affected by only the shape of the window, the window shape is chosen first based on the passband and stopband tolerance requirements, and then the window size is determined based on the transition width requirements. Table 9-3 contains information useful in choosing the shape and length of the window. The last column in the table, which is based on empirical observations, contains information on the best stopband (or passband) attenuation we can expect from a given window shape. If the required stopband attenuation is -50 dB, for example, the table shows that rectangular, Bartlett, and Hanning windows will not meet the specification, the Hamming window may barely make it, and the Blackman window most likely will. The second column in the table shows the appropriate length of the window required. For the Blackman window, for example, N can be approximately determined from the following relationship:

$$\frac{12\pi}{N} = \text{required transition width} \qquad (9.6)$$

A digital filter designed by the window method is shown in Figure 9-4.

The window method is not optimal, in the sense that there exists in general a filter which meets the given design specification and whose length is shorter than the filter designed by the window method. For an arbitrary $H_d(\omega)$, determining $h_d(n)$ from $H_d(\omega)$ may require a large inverse DFT computation. In addition, due to a lack of control over the frequency domain specification parameters, it is sometimes necessary to design several filters to meet the given design specification. Despite these disadvantages, the window method is sometimes used in practice because of its conceptual and computational simplicity.

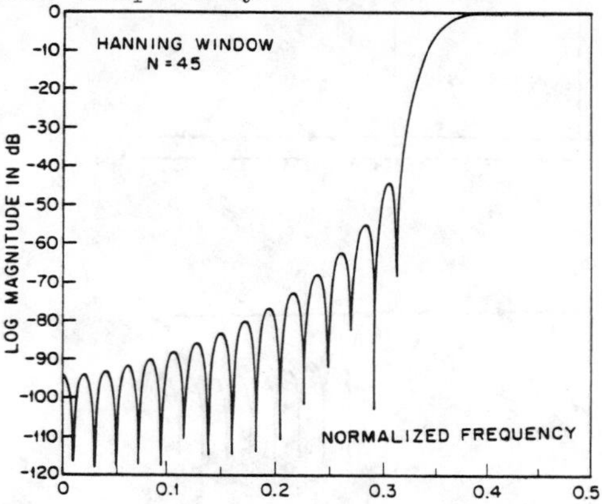

Figure 9-4.--Hanning window highpass filter frequency response. Note that normalized frequency = $\frac{\omega}{2\pi}$. (After [2])

9.4.2. Frequency sampling method

In the frequency sampling method, the desired frequency response $H_d(\omega)$ is sampled at equally spaced points and the result is inverse discrete Fourier transformed. Specifically, let H(k) be obtained by the following relationship:

$$H(k) = H_d(\omega)\Big|_{\omega=\frac{2\pi}{N}k} \tag{9.7}$$

The unit sample response of the filter, h(n), is obtained from

$$h(n) = \mathrm{IDFT}[H(k)] \tag{9.8}$$

From Section 6.2, the resulting $H(\omega)$ is given by

$$H(\omega) = \frac{1-e^{-j\omega N}}{N} \cdot \sum_{k=0}^{N-1} \frac{H(k)}{1-e^{j\frac{2\pi}{N}k} \cdot e^{-j\omega}} \tag{9.9}$$

An example of $H(\omega)$ obtained by the frequency sampling method is shown in Figure 9-5.

When $H_d(\omega)$ is sampled exactly, as in Figure 9-5, it has been observed that the resulting filter has a poor stopband and passband tolerance. It has been empirically observed that the stopband and passband behavior can be improved considerably if some transition samples are taken in the frequency region where $H_d(\omega)$ has a sharp transition. An example allowing one transition sample is shown in Figure 9-6. It has been empirically observed that stopband attenuation of -100 dB can be achieved with three transition samples.

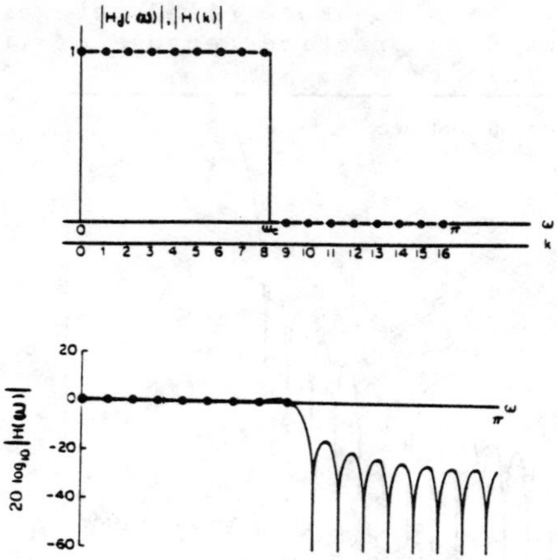

<u>Figure 9-5.</u>-- Example of frequency sampling method with no transition samples. (After [1])

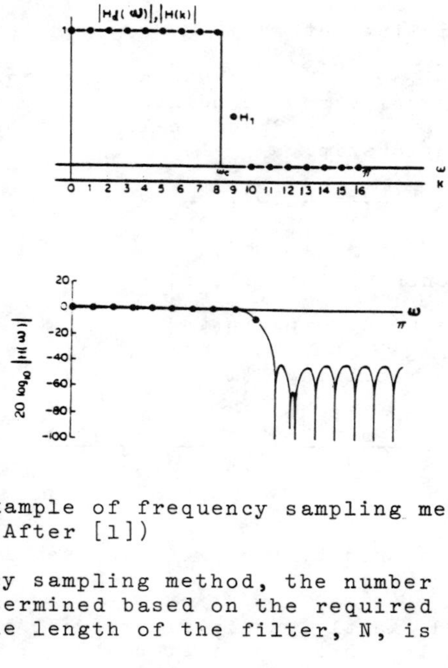

<u>Figure 9-6.</u>--Example of frequency sampling method with one transition sample. (After [1])

In the frequency sampling method, the number of transition samples, M, is first determined based on the required stopband and passband tolerances. The length of the filter, N, is then estimated by

$$(M+1) \cdot \frac{2\pi}{N} = \text{required transition width} \qquad (9.10)$$

The specific values of the transition samples are then determined in some reasonable fashion [2]. A digital filter designed by the frequency sampling method is shown in Fiqure 9-7.

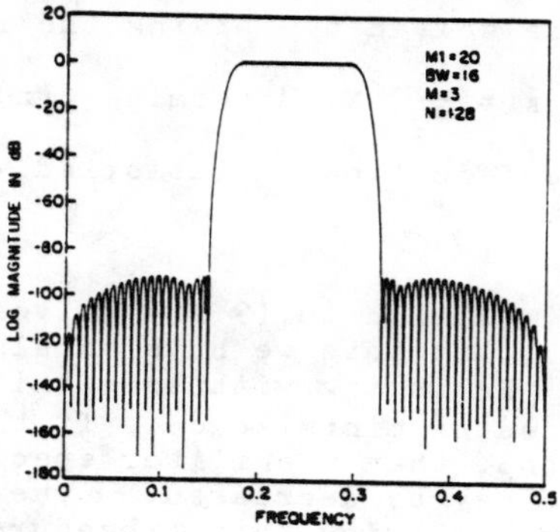

Figure 9-7.--Frequency response of a frequency sampling bandpass filter. Note that frequency = $\frac{\omega}{2\pi}$. (After [2])

As with the window method, the filter designed by the frequency sampling method is not optimal, in the sense that there exists in general a filter which meets the given design specification and whose length is shorter than the filter designed by the frequency sampling method. In addition, due to a lack of control over the frequency domain specification parameters, we may have to design several filters to meet the given design specification. Despite these disadvantages, the frequency sampling method is sometimes used in practice because of its conceptual and computational simplicity. Determining specific values for the transition samples is cumbersome compared to using the window method, but the inverse transformation of $H_d(\omega)$ is not needed in the frequency sampling method. Performance, measured in terms of the filter length needed to meet a given design specification, appears to be similar for both the window and frequency sampling method.

9.4.3 Optimal filter design

The major disadvantage of both the window method and the frequency sampling method is their non-optimality. In this section, we will discuss the optimal filter design method.

In our discussion, we'll concentrate on the design of a low-pass filter with design specification parameters δ_1, δ_2, ω_p and ω_s. In addition, we'll use Type I filters. Our discussion will extend in a straightforward way to the design of a more general filter and other types of filters [10].

The problem of designing an optimal filter can be stated as follows:

> Given δ_1, δ_2, ω_p and ω_s, determine h(n) such that the design specification is satisfied with the smallest possible N. (9.11)

The above problem can be solved by solving the following problem:

Given ω_p, ω_s, $\kappa = \frac{\delta_1}{\delta_2}$, N, determine h(n) such that the design specification is satisfied with the smallest possible δ_2 (9.12)

To illustrate that the solution to (9.12) solves the optimal filter design problem of (9.11), suppose we have an algorithm that solves the problem of (9.12). Beginning with some initial estimate of N, the algorithm can be used to minimize δ_2. If the resulting δ_2 is smaller than the given δ_2, then the filter specification has been met and the filter length N may be decreased. Otherwise, the filter length N can be increased. By following this procedure, the minimum possible value of N that meets the design specification can be found. The corresponding h(n) is a solution to the optimal filter design problem.

The problem stated in (9.12) can be shown to be a special case of a weighted Chebyshev approximation problem. The weighted Chebyshev approximation problem is a functional approximation problem. In the problem, the coefficients a(n) are chosen so that $H_M(\omega) = \sum_{n=0}^{\frac{N-1}{2}} a(n)\cdot\cos\omega n$ best approximates the desired frequency response in the frequency range of interest (passband and stopband region, in the case of a low-pass filter). The criterion used is the minimization of the maximum error between the desired frequency response and the resulting filter frequency response. For this reason, the problem is sometimes called a "min-max" problem. The weighted Chebyshev approximation problem has been studied extensively in mathematics. One theorem, the "alternation theorem," states that the problem has a unique solution. The theorem also provides a necessary and sufficient condition for the unique solution. The Remez multiple exchange algorithm exploits this necessary and sufficient condition to solve the weighted Chebyshev approximation problem. The Remez exchange algorithm was first used by Parks and McClellan to solve the optimal filter design problem of (9.12). The optimal filter design algorithm based on the Remez exchange algorithm is sometimes called the "Parks-McClellan" algorithm. The algorithm is an iterative procedure which is guaranteed to converge to the desired solution. Experience has shown that the algorithm converges very fast, and it is widely used in practice to design optimal filters. A computer program that implements the algorithm exists in the open literature [11].

One characteristic of optimal filters is that the error between the desired frequency response and the designed filter frequency response achieves its maximum and minimum consecutively at a finite number of frequencies in the frequency range of interest (passband and stopband regions). The frequencies at which the error achieves its maximum and minimum are called alternation frequencies, and the alternation theorem states the minimum number of alternation frequencies all optimal filters should have. This number is $\frac{N+3}{2}$ for Type I filters. Because of this characteristic, optimal filters are called "equi-ripple" filters. Some examples of optimal filters are shown in Figure 9-8.

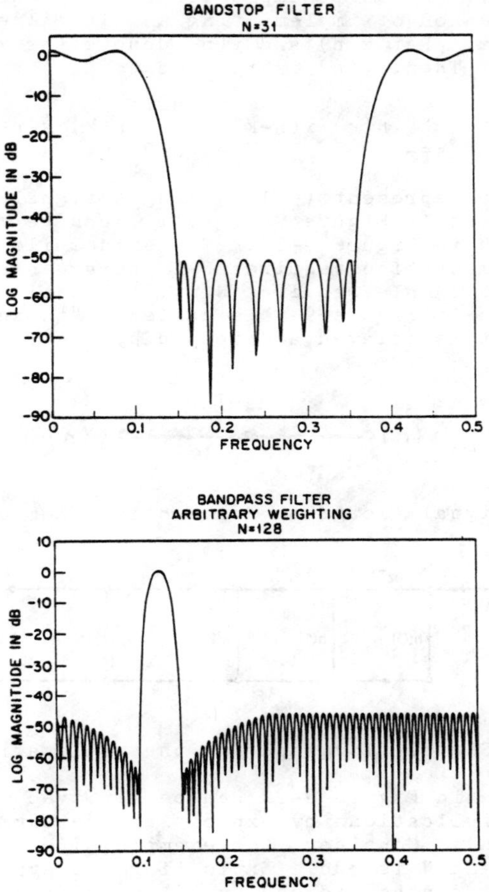

Figure 9-8.--Two examples of optimal FIR filters. Note that frequency = $\frac{\omega}{2\pi}$. (After [2])

Designing an optimal filter requires considerably more computation than the window method or the frequency sampling method. When one uses a computer, however, the difference between the optimal filter design method and the window or frequency sampling method is minor. In typical applications, it has been empirically observed that the length of the filter designed by the window or frequency sampling method is longer than the length of the optimal filter by 20% to 50% to meet a given design specification. For this reason, the optimal filter design method is the most widely used FIR filter design method.

9.5 Implementation of FIR Filters

In implementing a filter, the object is to realize a discrete time sytem with h(n), the unit sample response of the filter designed. The simplest method of implementing an FIR filter is to use the convolution sum. Let x(n) and y(n) denote the input and output of the filter. Then, y(n) is related to x(n) by

$$y(n) = \sum_{k=-\infty}^{\infty} h(k)\cdot x(n-k) = \sum_{k=0}^{N-1} h(k)\cdot x(n-k) \qquad (9.13)$$

It is useful to represent (9.13) using a signal flowgraph. Using the delay element in Figure 9-9, (9.13) can be represented by the signal flowgrph in Figure 9-10. The signal flowgraph clearly shows that the implementation requires N-1 storage elements, and computing each output sample requires N multiplications and N-1 additions. The realization in Figure 9-9 is called a "tapped delay line" filter, since the input is delayed and tapped by the filter coefficients.

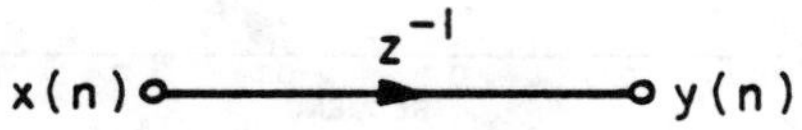

Figure 9-9.--Signal flowgraph for a delay element.

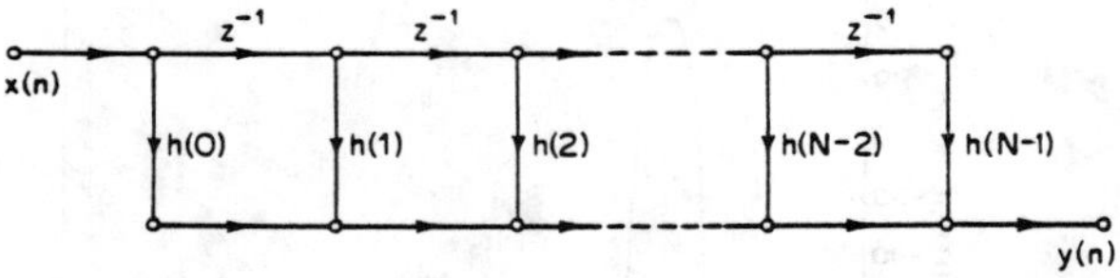

Figure 9-10.--Signal flowgraph for the realization of (9.13).

The realization in Figure 9-10 can be improved in the required number of multiplications by exploiting the symmetry or anti-symmetry of h(n). Consider, for example, Type I filters, for which h(n)=h(N-1-n) and N is odd. Exploiting the symmetry, the number of multiplications can be reduced by 50% without affecting the number of additions or storage elements required.

When N is 20 or longer, it is computationally more efficient to use an FFT algroithm. As we discussed in Section 6.4, the overlap-add method can be used to perform the filtering operation. For N=100, the reduction in the number of multiplications is approximately on the order of 10.

10. INFINITE IMPULSE RESPONSE FILTERS

10.1 Introduction

An IIR (infinite impulse response) filter has a unit sample response which is of infinite extent. As a result, an IIR filter differs from an FIR filter in some major ways. In this section, we discuss issues related to IIR digital filters.

An IIR filter with an arbitrary unit sample response h(n) cannot be realized, since computing each output sample requires a very large amount of arithmetic operations. As a result, in addition to requiring h(n) to be real, causal, and stable, we require h(n) to have a rational z-transform. Specifically, we require H(z), the z-transform of h(n), to be a rational function of the following form:

$$H(z) = \frac{\sum_{k=0}^{q} b_k \cdot z^{-k}}{1 - \sum_{k=1}^{p} a_k \cdot z^{-k}} \qquad (10.1)$$

As we discussed in Section 4.2, a causal h(n) with a rational z-transform can be realized by an LCCDE (linear constant coefficient difference equation) with IRC (initial rest condition), where the output is computed recursively. For example, suppose h(n) is given by

$$h(n) = [\tfrac{1}{2}]^n \cdot u(n) \qquad (10.2)$$

Then h(n) is causal and H(z) is given by

$$H(z) = \frac{1}{1-\frac{1}{2}z^{-1}} \qquad (10.3)$$

The above IIR filter can be realized by

$$y(n) = \tfrac{1}{2}y(n-1)+x(n) \quad \text{with IRC} \qquad (10.4)$$

The number of arithmetic operations required per output point is one multiplication and one addition when the filter is realized by (10.4), even though h(n) is of infinite extent.

Another major difference between IIR and FIR filters is in stability. An FIR filter is always stable as long as h(n) is bounded (finite for all n), so stability is not an issue in design or implementation. With an IIR filter, however, all the poles of H(z) must lie inside the unit circle to ensure filter stability. This imposes some restrictions on the design and implementation of IIR filters.

Linear phase is very easy to achieve for an FIR filter, and we require that all FIR filters have it. With an IIR filter, however, controlling the phase characteristic is very difficult. As a result, only the magnitude response is typically specified when an IIR filter is designed. The phase characteristic of the resulting filter is then regarded as acceptable phase. This lack of control over the phase characteristics limits the usefulness of IIR digital filters.

10.2 Design of IIR Filters

In designing an IIR filter, the problem is determining a rational

H(z) that meets a given magnitude specification. The magnitude specification used is the tolerance scheme discussed in Section 9.3. There are two standard approaches to designing IIR filters. One is to design the filter from an analog filter system function, and the other is to design it directly. We'll concentrate our discussion on the first approach, which is much more useful in designing a filter that meets a given magnitude specification.

The four steps in designing a digital filter from an analog filter are:

Steps for Designing an IIR Filter

Step 1. Specification of digital filter.

Step 2. Translation to a specification of an analog filter.

Step 3. Determination of analog filter system function $H_a(s)$.

Step 4. Tranformation of $H_a(s)$ to a digital filter system function H(z).

Step 1 depends on the specific application being considered. Step 2 depends on Step 4, since translation of the digital filter specification to an analog filter specification depends on how the analog filter is transformed back to the digital domain. Step 3 is a well-known analog filter design problem. The analog filters typically used are the Butterworth, Chebyshev, and elliptic filters. Step 4 is the transformation of an analog to a digital filter system function, and is briefly discussed below.

There are several properties which it is desirable for the transformation in Step 4 to satisfy. First, it is desirable for Step 4 to always transform a causal and stable $H_a(S)$ to a causal and stable H(z). This property guarantees that the resulting digital filter will be causal and stable if we design a causal and stable $H_a(s)$ in Step 3. Second, it is desirable that the transformation result in a digital filter with a rational H(z) whenever the analog filter has a rational $H_a(s)$. Since the analog filters used have rational $H_a(s)$, this property guarantees that the resulting digital filter will have a rational H(z), so can be realized by a difference equation. Third, the analog filter frequency response $H_a(\Omega)=H_a(s)|s=j\Omega$ should map to the digital filter frequency response $H(\omega)=H(z)|z=e^{j\omega}$. This property simplifies Step 2 considerably. If the $j\Omega$ axis in the s-plane does not map to the unit circle in the z-plane, translating the digital filter frequency response to the analog domain could be a complex task.

Two transformations satisfy all the above properties and are often used in designing IIR filters: impulse invariance and bilinear transformation, discussed below.

Impulse invariance method. In the impulse invriance method, the unit sample response of the digital filter is obtained by sampling the impulse response of the analog filter. Specifically, from $H_a(s)$, $h_a(t)$ is determined. The sequence $h(n)$ is then obtained by

$$h(n) = h_a(t)\big|_{t=n\cdot T} \tag{10.5}$$

From (10.5), $H(z)$ is obtained. Since T has no effect on the design or implementation of an IIR filter, T is assumed to be 1 for simplicity.

From (8.11), $H(\omega)$ is related to $H_a(\Omega)$ with T=1 by

$$H(\omega) = \sum_{r=-\infty}^{\infty} H_a(\omega-2\pi r) \tag{10.6}$$

It can be seen from (10.6) that $H(\omega)$ is not simply related to $H_a(\Omega)$ because of aliasing. If there is no aliasing effect, then

$$H(\omega) = H_a(\omega) \quad \text{for } -\pi\leq\omega\leq\pi \tag{10.7}$$

In this case, the shape of $H(\omega)$ is the same as $H_a(\Omega)$.

Since $H_a(\Omega)$ is not band-limited in practice, aliasing will be present. This causes some difficulty in designing a digital filter using the impulse invariance method. Specifically, in Step 2, in order to translate the digital filter specification to an analog specification, we need to determine $H_a(\Omega)$ from $H(\omega)$ using (10.6). This is a very difficult task. In practice, what is generally done is to use (10.7) for step 2, then carry out the remaining steps. Because of this approximation, the resulting filter is not guaranteed to satisfy the digital filter specification, so some iteration may be needed. Another difficulty which arises due to aliasing is that without an additional transformation, the method cannot be used in designing filters which have significant aliasing, such as high-pass or band-stop filters.

<u>Bilinear transformation.</u> In bilinear transformation, $H(z)$ is obtained from $H_a(s)$ as follows:

$$H(z) = H_a(s)\Big|_{s=\frac{1-z^{-1}}{1+z^{-1}}} \tag{10.8}$$

From (10.8), $H(\omega)$ is related to $H_a(\Omega)$ as follows:

$$H(\omega) = H_a(\Omega)\Big|_{\Omega=\tan\frac{\omega}{2}} \tag{10.9}$$

It can be seen, from (10.9) that there is no aliasing effect. From (10.9), the frequencies ω and Ω are nonlinearly related, so the detailed shape of $H_a(\Omega)$ is not preserved in $H(\omega)$. Although this is not a problem in designing piece-wise constant filters such as low-pass or high-pass filters, it can be a problem in designing filters such as differentiators.

Because there is no aliasing effect, Step 2 is quite simple. If the analog filter designed in Step 3 meets the analog filter specification that results from Step 2, then the resulting digital filter is guaranteed to meet the digital filter specification and therefore no iterations are necessary. In addition, the bilinear transformation method can be used to design high-pass or band-stop filters without additional transformation. A computer program that designs an IIR filter from an analog filter using the bilinear transformation can be found in [12].

Design examples based on the above approach can be found in a later chapter.

10.3 Implementation of IIR Filters

An IIR filter has an infinite extent unit sample response, and it cannot be realized in practice by direct convolution. For this reason, we have required the IIR filter to have a rational z-transfrom of the form given by (10.1). As we discussed in Section 10.1, (10.1) can be realized by using a difference equation. The difference equation can be realized in several different ways, discussed in this section.

Direct Form I. The difference equation that realizes H(z) in (10.1) is given by:

$$y(n) = \sum_{k=1}^{p} a_k \cdot y(n-k) + \sum_{k=0}^{q} b_k \cdot x(n-k) \qquad (10.10)$$

The difference equation can be realized in a straightforward manner by the signal flowgraph in Figure 10-1. The realization in Figure 10-1 is called Direct Form I, since it is straightforward realization of the difference equation.

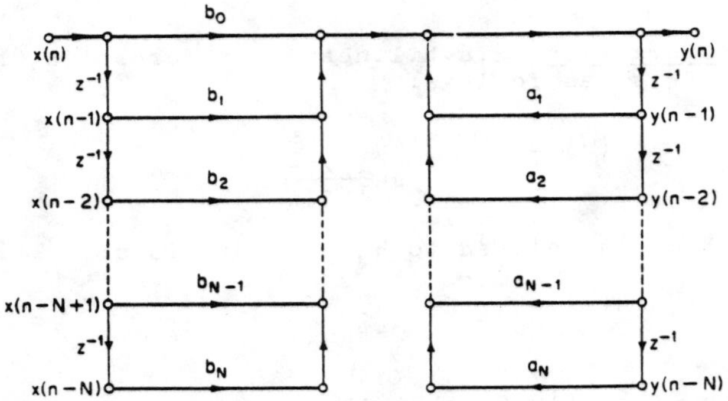

Figure 10-1.--Direct Form I for the realization of (10.10) with p=q=N. (After [1]).

Direct Form II. The realization in Figure 10-1 can be viewed as a cascade of two systems whose system functions are given by $H_1(z)$ and $H_2(z)$. By changing the order of $H_1(z)$ and $H_2(z)$ and eliminating the redundant delay elements, the realization in Figure 10-1 leads to the realization (called Direct Form II) in Figure 10-2. The advantage of Direct Form II over Direct Form I is the reduction in the number of storage elements required while the number of arithmetic operations required is constant.

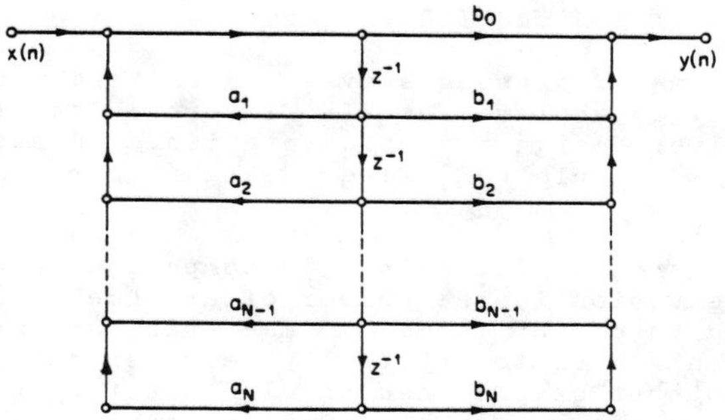

Figure 10.2.--Direct Form II for the realization of (10.10) with p=q=N. (After [1]).

Cascade Form. The system function in (10.1) can be expressed in the following form:

$$H(z) = A \cdot \prod_{k} H_k(z) \tag{10.11}$$

where $H_k(z)$ is given by

$$H_k(z) = \frac{1+\beta_{1k}\cdot z^{-1}+\beta_{2k}\cdot z^{-2}}{1-\alpha_{1k}\cdot z^{-1}-\alpha_{2k}\cdot z^{-2}} \tag{10.12}$$

The constants A, β_{1k}, β_{2k}, α_{1k}, and α_{2k} are real, since h(n) is real. From (10.11) and (10.12), H(z) is represented by a cascade of second-order sections, $H_k(z)$. The signal flowgraph based on (10.11) and (10.12) with each second-order section $H_k(z)$ realized by Direct Form II is shown in Figure 10-3.

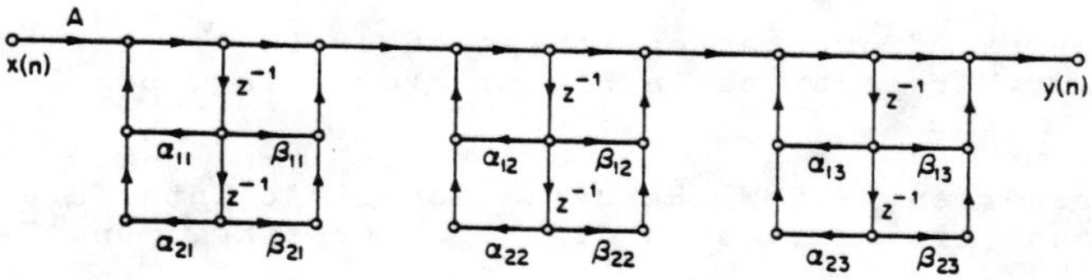

Figure 10-3.--Cascade structure with a Direct Form II realization of each second-order subsystem. (After [1]).

The cascade form is known to be less sensitive to coefficient quantization. One qualitative way to see this is to note that the change in one branch transmittance in the two direct forms will affect all the poles or zeros of the system function. In the cascade form, however, the change in one branch transmittance affects only the poles or zeros in the second-order section where the branch transmittance is affected. Further details on the effects of coefficient quantization can be found in [1,2].

10.4 Comparison of FIR and IIR Filters

FIR filters have many advantages over IIR filters. Linear phase is extremely easy to achieve for FIR filters. Stability is not an issue in FIR filter design or implementation. In addition, convenient and very flexible FIR filter design algorithms have been developed.

The main advantage of an IIR filter in comparison with an FIR filter is the reduction in the number of arithmetic operations when implemented in direct form. To meet the same magnitude specification, an FIR filter typically requires more arithmetic operations per output sample than an IIR filter by a factor of more than 3-5. This is based on the assumption that an FIR filter is implemented by direct convolution. If an FIR filter is implemented exploiting the computational efficiency of FFT algorithms, this advantage of an IIR filter disappears.

Because of the advantages of FIR filters over IIR filters, FIR filters are more common in practice.

REFERENCES

1. A. V. Oppenheim and R. W. Schafter, Digital Signal Processing, Prentice-Hall, 1975.

2. L. R. Rabiner and B. Gold, Theory and Applications of Digital Signal Processing, Prentice-Hall, 1975.

3. J. W. Cooley and J. W. Tukey, "An algorithm for the machine calculation of Complex Fourier Series", Math. Computation, Vol. 19, pp. 297-301, 1965.

4. J. H. McClellan and C. Rader, Number Theory in Digital Signal Processing, Prentice Hall, 1979.

5. C. S. Burrus, "Computation of the discrete Fourier transform", Trends and Perspectives in Signal Processing, pp. 1-4, April, 1982.

6. R. W. Brodersen, Max W. Hauser, "Monolitic Interface Circuits", Trends and Perspectives in Signal Processing, pp. 7-11, April, 1982.

7. G. B. Clayton, Data Converters, John Wiley and Sons, New York, 1982.

8. Analog-Digital Conversion Notes, Analog Devices, 1977.

9. F. J. Harris, "On the Use of Windows for harmonic Analysis with the Discrete Fourier Transform", IEEE Proc., vol. 66, pp. 51-83, Jan. 1978.

10. L. R. Rabiner, J. H. McClellan, and T. W. Parks, "FIR Digital Filter Design Techniques Using Weighted Chebyshev Approximation" Proc IEEE, Vol. 63, pp. 595-608, April, 1975.

11. J. H. McClellan, T. W. Parks, and L. R. Rabiner, "A Computer Program for Designing Optimum FIR Linear Phase Digital Filters" IEEE Trans. on Audio and Electroacoustics, Vol. AU-21, Dec., 1973.

12. A. H. Gray, Jr. and J. R. Markel, "A Computer Program for Designing Digital Elliptic Filters", IEEE Trans. on Acoust. Speech, Sig. Proc., pp. 529-538, Dec., 1976.

ACKNOWLEDGMENTS

The author would like to thank A. V. Oppenheim and R. Schafer for permitting use of figures from Digital Signal Processing (Prentice-Hall, 1975), and L. R. Rabiner and B. Gold for permitting use of figures from Theory and Applications of Digital Signal Processing (Prentice-Hall, 1975).

2

Linear Estimation for Stationary and Near-Stationary Processes

THOMAS KAILATH

In recent years, linear estimation theory and application have focused on processes with minimal state-space models. Such models are not always readily available or appropriate for signal processing problems and therefore attention has turned to other assumptions, especially stationarity and near-stationarity. In this chapter we shall present a fairly detailed account of the stationary theory and then more briefly describe the nature of some of the modifications for nonstationary and near-stationary processes. Particular attention will be paid to lattice filter models and recursions.

TABLE OF CONTENTS

4.3 Simplifications in the Near-Stationary Case

References

INTRODUCTION

The estimation of random signals is an important part of modern signal processing. Besides problems where signal estimation is the main issue, the techniques of statistical estimation theory are important for a variety of related problems, e.g., in spectral estimation, in signal detection and in digital filter design.

Several optimization criteria can be used for statistical estimation problems, but for signal processing one of the more important, at least in in the sense of having had the most applications, is the ***linear least squares criterion***. The study of such problems was introduced into engineering by Norbert Wiener in the early 1940s, in the context of stationary processes observed over infinite or semi-infinite intervals. In the late 1950s interest shifted to nonstationary processes with known finite dimensional state-space models, where the availability of the computationally efficient recursive Kalman-Bucy filter has essentially monopolized the field in the last two decades. In the mid-seventies, however, attention (in electrical engineering) began to shift back to general stationary processes, although now observed only over finite intervals. In particular, an algorithm due to Levinson (1947) was reintroduced to the speech processing community in connection with the so-called LPC (linear predictive coding) technique (see Makhoul (this volume)). Independently, there was increasing interest in high resolution spectral estimation methods, especially the so-called Maximum Entropy Method and its variants (see Marple (this volume)), which were closely related to the Levinson algorithm. More recently, there have been applications to digital filter design (see Dewilde (this volume)). Extensions of the Levinson algorithm to nonstationary processes have been used in communications (see Proakis (this volume). Finally we mention that many of these new algorithms are capable of VLSI implementation, and these features will be briefly discussed by Kung and by Whitehouse in their papers in this volume.

In view of these various applications, and others that we have not mentioned here, in this chapter we shall cover the theoretical basis of these algorithms in some detail. We shall begin with the basic problem of computing the innovations for a multichannel (or vector) stationary discrete-time process. After elaborating several interesting properties of the stationary case, we shall demonstrate the desirability of extending the results to nonstationary processes and briefly indicate how this can be done.

It is assumed that the reader is acquainted with the basic facts about linear least-squares estimation, as can be found in the introductory chapters of several textbooks, e.g., Kailath (1981).

1. Innovations For Scalar Stationary Processes

The study of estimation and modeling problems for second-order random processes can be approached in several ways. We have found it useful to begin by studying how to determine the innovations process associated with the given process, because once the innovations are known the solution of several other problems becomes much easier.

Therefore we shall start in Section 1.1 with formulating the basic so-called Yule-Walker linear equations with a symmetric Toeplitz coefficient matrix. These equations first arose,as we shall show, in the problem of autoregressive model fitting. In Section 1.2, we shall show how the Toeplitz structure can be exploited to derive the fast Levinson-Durbin algorithm for solving the equations with $O(N^2)$ elementary computations, as compared to $O(N^3)$ for a general set of N-th order linear equations.

1.1 The Yule-Walker Equations

Given a stationary stochastic process $\{y_0, y_1, \cdots\}$ with

$$Ey_i = 0, \quad Ey_i y_j = R_{i-j} \tag{1.1}$$

the goal is to find a recursive procedure for determining the *innovations* $\{e_0, e_1, \ldots\}$, defined as (see, e.g., Kailath (1981))

$$e_m := y_m - \hat{y}_{m|m-1}, \quad m = 0,1,\ldots \tag{1.2}$$

where

$$e_0 := y_0$$

and

$$\begin{aligned}\hat{y}_{m|m-1} &:= \text{the linear least-squares estimate of } y_m \text{ given } \{y_0, \ldots, y_{m-1}\} \\ &= -A_{m,1}y_{m-1} \cdots -A_{m,m}y_0, \text{ say} .\end{aligned} \tag{1.3}$$

More compactly, we shall write

$$e_m := y_m + A_{m,1}y_{m-1} + \cdots + A_{m,m}y_0 = \mathbf{A}_m\mathbf{Y}_m \tag{1.4}$$

where

$$\mathbf{A}_m := [A_{m,m} \quad \cdots \quad A_{m,1} \quad 1], \quad \mathbf{Y}_m := [y_0 \quad \cdots \quad y_m] \tag{1.5}$$

It will be useful to note that the quantity

$$\hat{y}_{t|t-m:t-1} := \text{the l.l.s.e. of } y_t \text{ given } \{y_{t-m}, \ldots, y_{t-1}\}$$

will, because of our assumption of stationarity, have the form

$$\hat{y}_{t|t-m:t-1} = -A_{m,1}y_{t-1} - \cdots - A_{m,m}y_{t-m} \tag{1.7}$$

where the coefficients $\{A_{m,i}\}$ are exactly those above. Therefore, we shall say that

$\mathbf{A}_m$:= the $m-th$ *order* prediction-error-filter.

We shall define

$$\begin{aligned} e_{m,t} &:= y_t + A_{m,1}y_{t-1} + \cdots + A_{m,m}y_{t-m} \\ &= \text{the } m-th \text{ order prediction error residual at time } t \end{aligned} \tag{1.8}$$

Note that the residuals will be equal to the innovations only when $t = m$, i.e., $e_{m,m} = e_m$; it turns out to be useful, especially in our later analysis of nonstationary processes, to imbed the problem of finding the innovations $\{e_m, e \geq 0\}$ in the larger problem of finding the residuals $\{e_{m,t}, m \geq 0, t \geq 0\}$.

The vector $\mathbf{A}_m$ can be determined by using the well known orthogonality properties that characterize linear least-squares estimates (see, e.g., Kailath (1981)),

$$e_{m,t} \perp y_{t-1}, \ldots, y_{t-m} \tag{1.9}$$

where

$a \perp b$ denotes $Eab = 0$

It will be convenient to add to the m equations resulting from (1.9), the equation

$$Ee_{m,t}y_t = Ee_{m,t}e_{m,t} := R_m^e \ , \quad R_0^e = R_0 \tag{1.10}$$

Combining (1.4), (1.9) and (1.10) yields

$$E[(\mathbf{A}_m \mathbf{Y}_{t-m:t})\mathbf{Y}_{t-m:t}] = \mathbf{A}_m \mathbf{R}_m = [0 \ \cdots \ 0 \ R_m^e] \ , \tag{1.11}$$

where we have defined

$$\mathbf{Y}_{t-m:t} := [y_{t-m} \quad \cdots \quad y_t]$$

and noted that by stationarity

$$E\mathbf{Y}_{t-m:t}\mathbf{Y}_{t-m:t}^T = E\mathbf{Y}_m\mathbf{Y}_m = [R_{i-j}] = \mathbf{R}_m$$

The equations (1.11) are often called the *Yule-Walker equations*, because they were first derived by Yule (1927) and Walker (1931) in studying the properties of autoregressive (or AR) processes. An AR process is one of the form

$$y_t + a_1 y_{t-1} + \cdots + a_m y_{t-m} = e_t$$

where $\{e_t\}$ is a white noise innovations sequence such that

$$Ee_t e_s = 0 \ , \quad t \neq s \ , \qquad Ee_t y_s \equiv 0 \ , \quad s < t \ .$$

and n is called the order of the autoregression. For such a process, the $\{A_{m,i}\}$ defined by the above construction, i.e., by (1.11), would coincide with the $\{a_i\}$ when $n = m$; then $e_{n,t} = e_t$, and R_n^e would be equal to the variance of the process $\{e_t\}$. Therefore, our analysis of a stationary process in terms of the residuals $e_{m,t};\ m \geq 0,\ t \leq m\}$ is really one of studying a general stationary process by fitting a succession of AR models to it; there will be an exact fit only when the process is itself AR of some finite order.

In any case, the issue now is the (efficient) solution of the Yule-Walker equations (1.11). A unique solution requires that the $(m + 1)$-dimensional matrix $\mathbf{R}_m$ be strictly positive definite, which we shall write as

$$\mathbf{R}_m > 0 \ , \quad m = 0,1,\ldots \tag{1.12}$$

This is a *nondegeneracy assumption* on the stochastic process $\{y_i\}$, because $\mathbf{R}_m$ singular implies a linear dependence between the random variables $\{y_0,\ldots,y_m\}$ that would allow perfect prediction of some of them. It follows from this interpretation that

$$\mathbf{R}_m > 0 \ \text{ implies } \ R_m^e > 0 \ , \quad m = 0,1,\ldots \tag{1.13}$$

However we can also give a direct algebraic proof of (1.13) that will be useful later. Note that

$$\begin{bmatrix} 1 & & & & 0 \\ & \cdot & & & \cdot \\ & & \cdot & & \cdot \\ & & & \cdot & \cdot \\ & & & 1 & 0 \\ A_{m,m} & \cdot & \cdot & A_{m,1} & 1 \end{bmatrix} \mathbf{R}_m = \begin{bmatrix} & & R_{-m} \\ \mathbf{R}_{m-1} & & \cdot \\ & & \cdot \\ & & \cdot \\ & & R_{-1} \\ 0 \ \ \cdot & 0 & R_m^e \end{bmatrix} \tag{1.14}$$

so that

$$\det \mathbf{R}_m = \det \mathbf{R}_{m-1} \cdot \det R_m^e$$

which yields (cf. (1.12))

$$\det R_m^e = \det \mathbf{R}_m / \det \mathbf{R}_{m-1} \ , \quad m = 1,\ldots \tag{1.15}$$

The desired result (1.13) now follows easily

Equation (1.11) gives us $m + 1$ linear equations for the $m + 1$ unknowns, which are R_m^e and the m nonunity coefficients of $\mathbf{A}_m$. For a general coefficient matrix, the number of elementary computations (multiplications and additions) required to solve a set of m linear equations is well-known [see, e.g., Stewart (1973)] to be of the order of m^3, written $O(m^3)$, unless the coefficient matrix has special structure, e.g., sparseness or bandedness. In our case, the coefficient matrix is Toeplitz and this fact was exploited by Levinson (1947), and independently by Durbin (1960), to give a solution of (1.11) that needed only $O(m^2)$ elementary computations. We shall first give an algebraic derivation; later in Section 4.2 we shall give a purely stochastic derivation.

1.2 The Levinson-Durbin Algorithm

The Levinson-Durbin algorithm proceeds via a simple updating procedure for going from $\{\mathbf{A}_m, R_m^e\}$ to $\{\mathbf{A}_{m+1}, R_{m+1}^e\}$. To motivate this algorithm, note that if we have found the solution $\mathbf{A}_m$ to

$$\mathbf{A}_m \mathbf{R}_m = [0 \cdots 0 \;\; R_m^e], \tag{1.16}$$

then the "trial solution" vector $[0 \;\; \mathbf{A}_m]$ would give the result

$$[0 \;\; \mathbf{A}_m] \mathbf{R}_{m+1} = [\Delta_{m+1} \cdots 0 \;\; R_m^e] \tag{1.17}$$

where

$$\Delta_{m+1} := A_{m,m} R_1 + \cdots + A_{m,1} R_m + R_{m+1}. \tag{1.18}$$

If it happened that Δ_{m+1} was zero, then we could immediately claim that

$$\mathbf{A}_{m+1} = [0 \;\; \mathbf{A}_m] \text{ and } R_{m+1}^e = R_m^e$$

This, however, will not happen in general, and therefore one must look for some way of modifying $[0 \;\; \mathbf{A}_m]$ so as to get rid of the Δ_{m+1} on the RHS of (1.17). Some reflection might suggest that we explore equations of the form

$$\mathbf{B}_m \mathbf{R}_m = [R_m^r \;\; 0 \cdots 0] \tag{1.19}$$

for the unknowns R_m^r and

$$\mathbf{B}_m := [1 \;\; B_{m,1} \cdots B_{m,m}]. \tag{1.20}$$

As with R_m^e, it is easy to check (cf. (1.14)-(1.15)) that

$$\det \mathbf{R}_m = \det \mathbf{R}_{m-1} \cdot R_m^r \tag{1.21}$$

so that our nondegeneracy assumption (1.12) also ensures that

$$R_m^r > 0, \quad m = 0, 1, \ldots. \tag{1.22}$$

We also note that

$$R_0^r = R_0 \tag{1.23}$$

We shall now show that knowledge of $\{\mathbf{A}_m, R_m^e\}$ and $\{\mathbf{B}_m, R_m^r\}$ will allow us to easily determine the corresponding quantities with subscript $m+1$. To do so, we first check that, as in (1.17)-(1.18), we shall have

$$[\,\mathbf{B}_m \quad 0]\,\mathbf{R}_{m+1} = [R_m^r \quad 0 \cdots 0 \quad \tilde{\Delta}_{m+1}] \tag{1.24}$$

where

$$\tilde{\Delta}_{m+1} := R_{-m-1} + B_{m,1}R_{-m} + \cdots + B_{m,m}R_{-1} \tag{1.25}$$

Let us write (1.17) and (1.24) together as

$$\begin{bmatrix} 0 & \mathbf{A}_m \\ \mathbf{B}_m & 0 \end{bmatrix} \mathbf{R}_{m+1} = \begin{bmatrix} \Delta_{m+1} & 0 & . & . & . & 0 & R_m^e \\ R_m^r & 0 & . & . & . & 0 & \tilde{\Delta}_{m+1} \end{bmatrix} \tag{1.26}$$

Then the solutions to the equations

$$\begin{bmatrix} \mathbf{A}_{m+1} \\ \mathbf{B}_{m+1} \end{bmatrix} \mathbf{R}_{m+1} = \begin{bmatrix} 0 & . & . & 0 & R_{m+1}^e \\ R_{m+1}^r & 0 & . & . & 0 \end{bmatrix} \tag{1.27}$$

can be determined by forming suitable linear combinations of the rows of (1.26). For example, we can obtain $\mathbf{A}_{m+1}$, and simultaneously R_{m+1}^e, as

$$\mathbf{A}_{m+1} = [0 \quad \mathbf{A}_m] - \Delta_{m+1}R_m^{-r}[\,\mathbf{B}_m \quad 0] \tag{1.28a}$$
$$R_{m+1}^e = R_m^e - \Delta_{m+1}R_m^{-r}\tilde{\Delta}_{m+1} \tag{1.28b}$$

Similarly, we can find

$$\mathbf{B}_{m+1} = [\,\mathbf{B}_m \quad 0] - \tilde{\Delta}_{m+1}R_m^{-e}[0 \quad \mathbf{A}_m] \tag{1.29a}$$
$$R_{m+1}^r = R_m^r - \tilde{\Delta}_{m+1}R_m^{-e}\Delta_{m+1}\,. \tag{1.29b}$$

where we have used the convenient notations

$$R_m^{-r} = [R_m^r]^{-1}\,, \qquad R_m^{-e} := [R_m^e]^{-1}\,.$$

It is useful to write these relations in the compact forms

$$\begin{bmatrix} \mathbf{A}_{m+1} \\ \mathbf{B}_{m+1} \end{bmatrix} = \Theta_{m+1} \begin{bmatrix} 0 & \mathbf{A}_m \\ \mathbf{B}_m & 0 \end{bmatrix} \tag{1.30a}$$

and

$$\begin{bmatrix} R^e_{m+1} & 0 \\ 0 & R^r_{m+1} \end{bmatrix} = \Theta_{m+1} \begin{bmatrix} R^e_m & \Delta_{m+1} \\ \tilde{\Delta}_{m+1} & R^r_m \end{bmatrix} \tag{1.30b}$$

where Θ_{m+1} is the (2×2) matrix

$$\Theta_{m+1} := \begin{bmatrix} 1 & -\Delta_{m+1} R_m^{-r} \\ -\tilde{\Delta}_{m+1} R_m^{-e} & 1 \end{bmatrix} \tag{1.30c}$$

Even more compactness may be achieved by associating polynomials with vectors in the natural way: define

$$A_m(z) := z^m + A_{m,1} z^{m-1} + \cdots + A_{m,m} = \mathbf{A}_m \; V_m^T(z) \text{, say} \tag{1.31}$$

$$B_m(z) := B_{m,m} z^m + \cdots B_{m,1} z + 1 = \mathbf{B}_m \; V_m^T(z) \text{, say} \tag{1.32}$$

where

$$V_m(z) := [1 \;\; z \;\; \cdots \;\; z^m]$$

Note that

$$[0 \;\; \mathbf{A}_m] \; V_{m+1}^T(z) = z \, A_{m,m} + \cdots + z^{m+1} = z \, A_m(z)$$

and

$$[\, \mathbf{B}_m \;\; 0] \; V_{m+1}^T(z) = I + \cdots + B_{m,m} z^m = B_m(z)$$

Therefore the recursions (1.30a) can be rewritten as

$$\begin{bmatrix} A_{m+1}(z) \\ B_{m+1}(z) \end{bmatrix} = \Theta_{m+1} \begin{bmatrix} z \, A_m(z) \\ B_m(z) \end{bmatrix}, \quad m = 0, 1, \ldots \tag{1.33a}$$

$$= \Theta_{m+1}(z) \begin{bmatrix} A_m(z) \\ B_m(z) \end{bmatrix}, \text{ say} \tag{1.33b}$$

where

$$\Theta_{m+1}(z) = \Theta_{m+1} \begin{bmatrix} z & 0 \\ 0 & 1 \end{bmatrix} = \begin{bmatrix} z & -\Delta_{m+1}R_m^{-r} \\ -z\,\tilde{\Delta}_{m+1}R_m^{-e} & 1 \end{bmatrix} \tag{1.33c}$$

We note that

$$R_0^e = R_0 = R_0^r \;, \quad A_0(z) = 1 = B_0(z)\,. \tag{1.33d}$$

Starting with these, we can use in succession (1.18), (1.25) and (1.33a) or (1.33b) to compute all the polynomials (and corresponding row vectors) $\{A_m(z), B_m(z);\ m \geq 1\}$.

Stochastic Meaning of B_m. The defining relation (1.19)-(1.20)

$$[1 \;\; B_{m,1} \;\; \cdots \;\; B_{m,m}]\mathbf{R}_m = [R_m^r \;\; 0 \;\; \cdots \;\; 0]$$

shows that (using $\mathbf{R}_m = E\mathbf{Y}_{t-m:t}\mathbf{Y}_{t-m:t}^T$)

$$r_{m,t} := y_{t-m} + B_{m,1}y_{t-m+1} + \cdots + B_{m,m}y_{t-m+i} \perp y_{t-m+i} \;, \quad 1 \leq i \leq m \tag{1.34}$$

Therefore we can identify $r_{m,t}$ as

$$\begin{aligned} r_{m,t} = \mathbf{B}_m\mathbf{Y}_{t-m:t} &= \text{the error in estimating } y_t \text{ from } \{y_{t+1},\ldots,y_{t+m}\} \quad (1.35) \\ &= \text{the } m\text{-}th \text{ order backwards prediction-error residual} \end{aligned}$$

and

R_m^r = the corresponding mean-square error.

We shall call $\mathbf{B}_m$ the m-th order *backwards prediction-error filter*.

Symmetric $\mathbf{R}_m$. Actually we have not so far really used the assumption that the Toeplitz matrix is a covariance matrix, so that

$$R_{-i} = R_i \tag{1.36}$$

and $\mathbf{R}_m$ will be symmetric. With (1.36) we shall find by comparing (1.18) and (1.25) that

$$\Delta_{m+1} = \tilde{\Delta}_{m+1} \quad , \quad m = 0,1,... \tag{1.37a}$$

and from (1.15) and (1.21) that

$$R_m^e = R_m^r \ . \tag{1.37b}$$

Furthermore, we see from (1.16) and (1.19) that we can take $\mathbf{B}_m$ to be the "reverse" of $\mathbf{A}_m$, namely that

$$[1 \ B_{m,1} \ \cdots \ B_{m,m}] = [1 \ A_{m,1} \ \cdots A_{m,m}] \tag{1.38}$$

In terms of the associated polynomials, this means that

$$\begin{aligned} B_m(z) &:= B_{m,m} z^m + \cdots + B_{m,1} z + 1 \\ &= A_{m,m} z^m + \cdots + A_{m,1} z + 1 \\ &= z^m A_m(z^{-1}) := A_{m*}(z), \ say, \end{aligned} \tag{1.39}$$

the so-called "reverse" (polynomial) of $A_m(z)$. Therefore in the symmetric scalar case, we need only be concerned with a single set of polynomials $\{A_0(z) = I, A_1(z),...\}$; the more general formulation with two sets of polynomials will be needed in the multichannel case, where $\mathbf{R}$ will be a block Toeplitz covariance matrix (see Section 3).

Of course the recursions (1.33) will still call upon the reverse polynomials. Let us define

$$k_{m+1} = \Delta_{m+1} / R_m^e = \Delta_{m+1} / R_m^{e/2} R_m^{r/2} \tag{1.40a}$$

where we recall that (cf. (1.18))

$$\Delta_{m+1} = R_{m+1} + A_{m,1} R_m + \cdots + A_{m,m} R_1 \tag{1.40b}$$

Then the recursions (1.33) take the form

$$\begin{bmatrix} A_{m+1}(z) \\ A_{m+1*}(z) \end{bmatrix} = \begin{bmatrix} 1 & -k_{m+1} \\ -k_{m+1} & 1 \end{bmatrix} \begin{bmatrix} zA_m(z) \\ A_{m*}(z) \end{bmatrix} \ , \quad \begin{bmatrix} A_0(z) \\ A_{0*}(z) \end{bmatrix} = \begin{bmatrix} 1 \\ 1 \end{bmatrix} \tag{1.41a}$$

We note also that

$$R^e_{m+1} = (1 - k^2_{m+1})R^e_m = \prod_{i=1}^{m+1} (1 - k_i^2)R_0 \tag{1.41b}$$

and

$$k_{m+1} = - A_{m+1,m+1} \tag{1.41c}$$

The algorithm provided by (1.41) is a streamlined form (perhaps first due to Burg, 1958, unpublished) of one first described by Levinson in (1947). It has been rediscovered often, notably by Durbin (1960), in the problem of autoregressive model fitting. Therefore we shall call it the Levinson-Durbin recursion.

We should note that it is not essential to bring in the reverse polynomial to have a recursion for the $\{A_i(z)\}$. We could eliminate $A_{m*}(z)$ in (1.41a) to obtain the "three-term recurrence formula"

$$k_m A_{m+1}(z) = (zk_m + k_{m+1})A_m(z) - k_{m+1}z(1 - k^2_m)A_{m-1}(z) \tag{1.42a}$$

which can be initialized with

$$A_0(z) = 1, \ A_1(z) = z - k_1 \tag{1.42b}$$

This representation has interesting implications, because three-term recurrences are closely associated with orthogonal polynomials (see, e.g., Szegö (1939)). We shall explore this connection in Section 2.2.

The nondegeneracy assumption (1.12) ensures that $R^e_i \neq 0$ (cf. (1.13)) and therefore we see from (1.41b) that we must have

$$|k_i| < 1 \tag{1.43}$$

This is a partial justification for the widely used designation of the $\{k_i\}$ as ***reflection coefficients***; a fuller justification will be given in Section 2.6.

The $\{k_i\}$ also have an important stochastic interpretation, which will be encountered again in Sect. 4.2 in connection with general nonstationary processes. This interpretation follows by combining (1.40b) with the definitions (1.8) and (1.35) of $e_{m,t}$ and $r_{m,t}$ to obtain

$$\Delta_{m+1} = \mathbf{A}_m (E\mathbf{Y}_{t-m:t} y_{t-m-1})$$
$$= Ee_{m,t} y_{t-m-1} = Ee_{m,t} r_{m,t-1} + 0 \tag{1.44}$$

so that we can interpret

$$k_{m+1} = \Delta_{m+1} / R_m^{e/2} R_m^{r/2}$$
$$= \text{the correlation coefficient of } e_{m,t} \text{ and } r_{m,t-1}, \tag{1.45}$$

an interpretation again consistent with the fact that $|k_{m+1}| \le 1$.

Moreover since

$e_{m,t}$ = the residual in estimating y_t given $\{y_{t-1}, \ldots, y_{t-m}\}$

and

$r_{m,t-1}$ = the residual in estimating y_{t-m-1} given $\{y_{t-m}, \ldots, y_{t-1}\}$

we see that, in statistical terminology (see, e.g., Anderson (1958),

k_{m+1} = the partial correlation coefficient of y_t and y_{t-m-1} given the random variables $\{y_{t-1}, \ldots, y_{t-m}\}$. (1.46)

This fact has led to the $\{k_i\}$ sometimes being known as PARCOR coefficients.

Note that when $k_m = 1$, the prediction problem becomes degenerate, i.e., $R_m^e = 0$; we shall examine the significance of this limiting case in Problem 2.6.

Returning to the recursions (1.41), note that the matrix in (1.41a) will be invertible as long as $k_m \ne 1$. Therefore we can also write the *backwards recursions*: given $A_m(z)$,

$$\begin{bmatrix} A_{m-1}(z) \\ A_{m-1*}(z) \end{bmatrix} = \frac{1}{1-k_m^2} \begin{bmatrix} z^{-1} & z^{-1}k_m \\ k_m & 1 \end{bmatrix} \begin{bmatrix} A_m(z) \\ A_{m*}(z) \end{bmatrix} \tag{1.47}$$

The forward and backward recursions (1.41) and (1.47) have many important mathematical and physical consequences, arising from the interplay between the covariance sequence $\{R_0, R_1, \ldots, R_m\}$, the polynomials $\{A_0(z), \ldots, A_m(z)\}$ and the reflection coefficients $\{k_1, \ldots, k_m\}$.

Here let us just note that there is a unique correspondence between the sets

$$\{1, A_{m,1}, \ldots, A_{m,m}\} \longleftrightarrow \{k_1, \ldots, k_m\} \tag{1.48}$$

Given $\{k_1, \ldots, k_m\}$, $A_m(z)$ is uniquely determined by the recursions (1.41a). Conversely given $A_m(z)$, we can find k_m as $-A_{m,m}$, then use the backward recursions (1.47) to find $A_{m-1}(z)$, $k_{m-1} = -A_{m-1,m-1}$ and so on.

A natural question to ask is how we know that a given $A_m(z)$ will yield a set $\{k_1, \ldots, k_m\}$ with $|k_i| < 1$? The answer, which will arise from the interpretation of the polynomials $\{A_m(z)\}$ as orthogonal polynomials in Section 2.2, is that if and only if $A_m(z)$ is stable (i.e., has its roots within the unit circle) will the derived $\{k_i\}$ have magnitude less than unity. This is in fact the celebrated Schur-Cohn Test for stability of a polynomial (see, e.g., Jury (1974) and also Vieira and Kailath (1977)).

We go now to Section 2.0 for a fuller exploration of this and many related properties.

2. More on Scalar Stationary Processes

The Levinson-Durbin algorithm has many implications for the study and the analysis of stationary processes, some of which we shall explore in this section. In Section 2.1, we shall show that the reflection coefficients provide a nice characterization of a stationary covariance sequence, which enables us to study certain useful (autoregressive and maximum entropy) covariance extension problems. The Levinson-Durbin algorithm is also closely connected with the Szegö unit-circle orthogonal polynomials, a fact that immediately makes available a lot of information about the solution to the prediction problem (see Section 2.2). Next, we shall examine some computational aspects, and in particular the topical question of how the algorithm can be speeded up by using parallel computation. It turns out that this requires an alternative algorithm, traceable to Schur (1917), which has several other interesting properties (Section 2.3). In fact, it is further analysis of the Schur algorithm that will lead us to fast algorithms for nonstationary processes. In Section 2.4 we shall describe the important lattice filter implementation of the prediction error equation. In

Section 2.5 we shall discuss its use for approximate autoregressive modelling of the given process, following up some of the discussion at the end of Section 2.1. Several alternative feedback lattice and transmission line models, using sections with four to zero multipliers each, are presented in Section 2.6; the multiplier-free sections are the so-called CORDIC (Coordinate Rotation Digital Computers) blocks, which are increasingly being studied for VLSI implementation.

2.1 The Reflection Coefficients, Autoregressive Extensions and Related Properties

The Levinson algorithm shows that the covariance sequence $\{R_0,\dots,R_m\}$ completely determines the m-th order prediction polynomial $A_m(z)$ and a set of reflection coefficients $\{k_1,\dots,k_m\}$. Conversely, we have the following interesting and important result.

Lemma 2.1. Any set of numbers

$$\{R_0,k_1,\dots,k_m : R_0 > 0\,, \quad |k_i| < 1\}$$

uniquely determines a covariance sequence $\{R_0,R_1,\dots,R_m\}$.

Proof. Given $\{R_0,k_1,\dots,k_m\}$, we can construct $\{A_m(z),R_m^e\}$ by using (1.41). Then we can generate $\{R_1,\dots,R_m\}$ recursively via the formulas (cf. (1.40))

$$\Delta_{j+1} = k_{j+1}R_j^e \tag{2.1a}$$

$$R_{j+1} + A_{j,1}R_j + \cdots + A_{j,j}R_1 = \Delta_{j+1} \tag{2.1b}$$

for $j = 0,1,\dots,m-1$.

A noteworthy application of this result is following.

The Covariance Extension Problem. Given a nondegenerate covariance sequence $\{R_0,\dots,R_m\}$, determine an additional term R_{m+1} so that $\{R_0,\dots,R_m,R_{m+1}\}$ is also a covariance sequence.

Of course a necessary and sufficient condition is that the Toeplitz matrix $\mathbf{R}_{m+1}$ be nonnegative definite, which since $\mathbf{R}_m$ is already known to be so, will be true if and only if

$$\det \mathbf{R}_{m+1} > 0\ .$$

But this yields a quite complicated condition on the allowable values of $\mathbf{R}_{m+1}$, especially for large m (see Problem 2.1). However a simple solution is available via the reflection coefficients:

Form the reflection coefficients $\{k_1,\ldots,k_m\}$ associated
with the given covariance sequence $\{R_0,\ldots,R_m\}$.
Let k_{m+1} be any number of magnitude less than 1 .
Then the sequence $\{R_0,R_1,\ldots,R_m,R_{m+1}\}$ associated (2.2)
with $\{R_0,k_1,\ldots,k_m,k_{m+1}\}$ will yield an allowable R_{m+1} .

This procedure can obviously be used to obtain an extension of arbitrary length $\{R_0,\ldots,R_m,R_{m+1},R_{m+2},\ldots,R_{m+l}\}$, $l \geq 0$. If we extend the sequence out to infinity, then we could form a *power-spectral density function*

$$W(e^{i\vartheta}) = \sum_{-\infty}^{\infty} R_n e^{jn\vartheta} \quad , \quad R_{-n} := R_n^* \; . \tag{2.3}$$

of a stationary process $\{y_i\}$, the first m terms of whose covariance sequence are equal to the given values $\{R_0,R_1,\ldots,R_m\}$. The superscript denotes complex conjugate.

A particularly natural extension would seem to be achieved by choosing the additional reflection coefficients as zero, i.e.,

$$k_{m+l} = 0, \quad l = 1,2,\ldots \quad , \tag{2.4a}$$

It turns out that the corresponding power spectral density function is

$$W(e^{i\vartheta}) = R_m^e / \, |A_m(e^{i\vartheta})|^2 \tag{2.4b}$$

This is often called the *maximum entropy spectrum* associated with $\{R_0,\ldots,R_m\}$, for reasons explained in Problem 2.3. The choice (2.4) is clearly only one way of associating a power spectral density function with a given covariance sequence $\{R_0,\ldots,R_m\}$, though it does correspond to an "obvious" extension. Any other extension would seem more constrained, thus providing perhaps a subjective justification of the name maximum entropy.

A Proof of (2.4). Suppose we have $\{R_0,R_1,\ldots,R_N\}$, have computed $\{k_1,\ldots,k_m\}$ and now choose $k_{m+1} = 0$. Then the new covariance term,

$\hat{R}_{m+1}$ say, can be found via (2.1) as

$$\hat{R}_{m+1} + A_{m,1}R_m + .. + A_{m,m}R_1 = 0 \tag{2.5}$$

If we choose $k_{m+2} = 0$, then

$$\hat{R}_{m+2} + A_{m+1,1}\hat{R}_{m+1} + \cdots + A_{m+1,m+1}R_1 = 0 \tag{2.6}$$

But since $k_{m+1} = 0$, the Levinson-Durbin algorithm (cf. (1.41)) shows that

$$A_{m+1}(z) = zA_m(z) \tag{2.7}$$

so that

$$A_{m+1,m+1} = 0 \tag{2.8}$$

and

$$A_{m+1,i} = A_{m,i}\ , \quad i = 1,\ldots,m \tag{2.9}$$

and we have

$$\hat{R}_{m+2} + A_{m,1}\hat{R}_{m+1} + \cdots + A_{m,m}R_2 = 0 \tag{2.10}$$

Proceeding in this way shows that

$$A_{m+j}(z) = A_m(z)\ , \quad j \geq 0 \tag{2.11}$$

and

$$[I \;\; A_{m,1} \;\cdots\; A_{m,m} \;\; 0 \;\cdots\;]\mathbf{R}_\infty = [R_m^e \;\; 0 \;\cdots\; 0 \;\cdots\;] \tag{2.12}$$

where $\mathbf{R}_\infty$ is a Toeplitz covariance matrix with first column $\{R_0, R_1, \ldots, R_m, \hat{R}_{m+1}, \hat{R}_{m+2}, \ldots\}$. We also have

$$[A_{m,m} \;\cdots\; A_{m,1} \;\; I \;\; 0 \;\cdots\;]\mathbf{R}_\infty = [0 \;\cdots\; R_m^e \;\; 0 \;\cdots\;] \tag{2.13}$$

When we multiply out the product

$$A(e^{i\vartheta})W(e^{i\vartheta})A^*(e^{i\vartheta}) \tag{2.14}$$

and use the above relations, we shall find that the product is R_m^e, thus establishing (2.4).

Autoregressive Modeling. The formula (2.4b) shows that the maximum entropy extension of the covariance sequence, corresponding to choosing $k_{m+i} = 0$, $i > 0$, is equivalent to approximating the stochastic process $\{y_t\}$ by a so-called *autoregressive process* of order m; the approximation has the property that the first m terms of its covariance function are exactly the first m terms of the covariance function of the original process. Parzen (1969) was one of the first to suggest this result as the basis of a method of spectral estimation: Burg (1967) made the same suggestion based on the maximum entropy interpretation. Furthermore, Burg (1968) suggested a specific method of estimation from a data sample, since in practice the covariance sequence will not be available. Burg's method hypothesizes a lattice structure and estimates the reflection coefficients by minimizing a sum of forward and backward squared residuals; see Childers (1978) for reprints of several papers on the Burg technique and later variants, due to Ulrych and Clayton (1976) and others (see also Marple (1980)).

2.2 Szegö Orthogonal Polynomials

Given a scalar covariance sequence $\{R_0, R_1, \ldots\}$, its spectral density function is defined as

$$W(e^{i\vartheta}) = \sum_{n=-\infty}^{\infty} R_n e^{in\vartheta}, \quad R_{-n} = R_n^* \tag{2.15}$$

We assume that the series converges, perhaps in a distributional sense.

The Szegö orthogonal polynomials on the unit circle are polynomials in z, $\varphi_m(z)$ being of degree m, such that

$$\langle \varphi_m(z), \varphi_n(z) \rangle := \int_{-\pi}^{\pi} \varphi_m(e^{i\vartheta}) \varphi_r^*(e^{i\vartheta}) W(e^{i\vartheta}) d\vartheta = \delta_{nm} \tag{2.16}$$

They were first studied by Szegö in 1918. We shall see that we can identify

$$\varphi_m(z) = R_m^{-e/2} A_m(z) \tag{2.17}$$

The reason is that by direct substitution from the definitions, we have

$$\langle A_m(z), A_n(z) \rangle = \int_{-\pi}^{\pi} A_m(e^{i\vartheta}) A_n^*(e^{i\vartheta}) W(e^{i\vartheta}) d\vartheta$$

$$= \sum_{k,l} A_{m,k} \langle e^{ik\vartheta}, e^{il\vartheta} \rangle A^*_{n,l} = \sum_{k,l} A_{m,k} R_{i-j} A^*_{n,l} \tag{2.18}$$

$$= \mathbf{A}_m \mathbf{R}_m \mathbf{A}^*_n = R^e_m \delta_{mn} \tag{2.19}$$

where the last equality follows from the defining equation (1.11) for A_m.

With the identification (2.17) we can refer directly to the extensive literature on the Szegö polynomials to discover several properties of the $\{A_m(z)\}$; two basic references are Szegö [(1939), chapter 11] and Grenander and Szegö (1958). Geronimus (1948), (1958) presents many properties of these polynomials; the books of Akhiezer and Krein (1938) and Akhiezer (1961) also contain much relevant information. For example, we find that the basic Levinson-Durbin recursions (1.41) were derived for the $\{\varphi_m(z)\}$ by Szegö in 1938 and published in the first edition of his book cited above; they were also independently discovered by Geronimus in 1940.

The connection (2.17) can also be established in a stochastic fashion by using the so-called Kolmogorov isomorphism between the linear (Hilbert) space spanned by the random variables $\{y_i\}$ and the (deterministic) space sparned by the powers $\{A_m(z)\}$ are isomorphic to the innovations $\{e_m\}$, thus easily yielding (2.19). The interested reader can find a tutorial exposition in Kailath, Vierra and Morf (1978).

With orthogonal polynomials, there is a classical formula known as the *Christoffel-Darboux formula*:

$$K_m(z,w) = \sum_{i,j=0}^{m} z^i w^{*j} (R_m)^{-1}_{i,j} \tag{2.20}$$

$$= \sum_{k=0}^{m} A^*_k(z) R_k^{-e} A_k(w) = \frac{A^*_{m*}(z) A_{m*}(w) - zw^* A^*_m(z) A_m(w)}{R^e_m(1 - zw^*)} \tag{2.21}$$

$$= \frac{A^*_{m+1*}(z) A_{m+1*}(w) - A^*_{m+1}(z) A_{m+1}(w)}{R^e_{m+1}(1 - zw^*)} \tag{2.22}$$

The proof of this formula will be deferred to Section 3.2, where it will be established in the matrix (or multichannel) case.

A nice application of the Christoffel-Darboux formula is to establish a previously mentioned result that

$$A_m(z) \neq 0 \quad , \quad |z| \geq 1 \tag{2.23}$$

Proof. Suppose $A_m(z_0) = 0$ for some z_0 such that $|z_0| > 1$. Then if we put $z = w = z_0$ in (2.21) we shall have

$$\sum_{k=0}^{m} |A_k(z_0)|^2 / R_k^e = \frac{|A_{m*}(z_0)|^2}{R_m^e(1-|z_0|^2)} \leq 0$$

which is a contradiction.

Now suppose $A_m(z_0) = 0$ for some z_0 with $|z_0| = 1$. Then $A_{m*}(z_0) = 0$ as well. But then we can conclude from the backwards recursions (1.47) that $A_{m-1}(z_0) = 0 = A_{m-1*}(z_0)$, and so on till we get $1 = A_0(z_0) = 0$, which is impossible. Hence...

The Christoffel-Darboux formulas also yield certain useful factorizations and representations of $\mathbf{R}_m^{-1}$, which we shall present in Section 3.2 (in the more general multichannel case).

There are several other applications of the connection we have made with the Szegö polynomials, especially in studying the asymptotic properties of the prediction filters; the latter will be briefly noted in Section 2.5.

2.3 Serial and Parallel Computation; The Schur Algorithm for the $\{k_i\}$

For simplicity, we shall continue for a while with the assumption of a scalar process.

The recursions (1.41) show that it takes $O(m)$ multiplications and additions to go from stage m to stage $m+1$. Therefore for N stages it will take of the order of

$$1 + 2 + \cdots + N = N(N+1)/2$$

or $O(N^2)$ elementary computations to obtain $\mathbf{A}_N$ via the Levinson algorithm, as compared to $O(N^3)$ if we had directly solved the defining equation (1.16) without exploiting its Toeplitz structure.

This order of magnitude reduction can be quite significant for large N, and the Levinson algorithm has therefore been widely used in certain applications, in particular in seismic oil exploration. In 1976 it was estimated that "at least 5 million sets of normal equations, many involving up to 100 unknowns or more, are solved each day by the seismic industry" (see Robinson [1978, p. 474]).

At present, with the increasing progress in microelectronics, attention is turning to the use of parallel computing architectures. Therefore it is natural to ask if the computation of $\mathbf{A}_N$ could be speeded up by using parallel computation.

If an elementary computation is assumed to take one time unit, then with a single serial processor, we would require $O(N^2)$ time units to compute $\mathbf{A}_N$ by the Levinson algorithm, which we recall is based on the recursion (cf. (1.28))

$$\mathbf{A}_{N+1} = [0 \quad \mathbf{A}_N\,] - k_{N+1}[\mathbf{A}_N \quad 0\,]$$

Now with parallel computers, the multiplications required to form $k_{N+1}\mathbf{A}_N$ could be carried out simultaneously and so could the componentwise additions. Therefore we would expect that the time required to compute $\mathbf{A}_N$ would be $O(N)$ if we had $O(N)$ parallel computing devices. Unfortunately this is too hasty a conclusion. We have neglected to examine the time required to form the reflection coefficients. Thus recall from (1.40) that

$$k_{N+1} := \Delta_{N+1}/\,R_N^e\ ,$$

where

$$\Delta_{N+1} := R_{N+1} + A_{N,1}R_N + \cdots + A_{N,N}R_1$$

The point is that while the subproducts in Δ_{N+1} can be computed simultaneously if we had N parallel computing devices, unfortunately the sum of these products cannot be done immediately: it will take at least $O(\log N)$ time units to form the sum in Δ_{N+1}.

Therefore, even with the availability of N parallel computing devices, the Levinson algorithm will require $O(N \log N)$ time units to compute $\mathbf{A}_N$. This is much less than the $O(N^2)$ time units needed with a single processor, but it is still disappointing that we cannot reduce the time to $O(N)$ with $O(N)$ processors. It turns out that in fact this can be achieved by using a different scheme, the Schur algorithm, for computing the $\{k_i\}$.

The Schur Algorithm. The Schur algorithm was originally presented in 1917 as a test to see if a power series in z was analytic and bounded in the unit disc. In our problem it reduces to a simple three step procedure (see Dewilde, Vieira and Kailath (1978)), Lev-Ari and Kailath (1984)) for the computation of the $\{k_i\}$ associated with a covariance sequence $\{R_0, R_1, ..., R_N\}$.

Step 1. Start with a so-called generator matrix (superscript T denotes matrix transpose)

$$G_0^T := \begin{bmatrix} R_0 & R_1 & \cdots & R_N \\ 0 & R_1 & \cdots & R_N \end{bmatrix}$$

and shift the first column down one row to get

$$\tilde{G}_1^T = \begin{bmatrix} 0 & R_0 & \cdots & R_{N-1} \\ 0 & R_1 & \cdots & R_N \end{bmatrix}$$

Step 2. Compute k_1 as the ratio of the (2,2) and (2,1) entries of $\tilde{G}_1$. The reason for this special choice is made evident in the next step.

Step 3. Form a matrix

$$\overline{\Theta}(k_1) = \frac{1}{\sqrt{1 - k_1^2}} \begin{bmatrix} 1 & -k_1 \\ -k_1 & 1 \end{bmatrix}$$

and apply it to $\tilde{G}_1$. The special choice of k_1 then yields a new Schur-reduced generator of the form

$$G_1^T = \overline{\Theta}(k_1)\tilde{G}_1^T = \begin{bmatrix} 0 & x & x & \cdots & x \\ 0 & 0 & x & \cdots & x \end{bmatrix}$$

where the x denotes elements whose exact value is not relevant at the moment; $\overline{\Theta}$ is known as a J-rotation matrix because

$$\overline{\Theta}(k)\, J\, \overline{\Theta}^T(k) = J \quad , \qquad J = \begin{bmatrix} 1 & 0 \\ 0 & -1 \end{bmatrix}$$

Its role is to rotate in the J-metric (i.e.., to hyperbolate) row of $\tilde{G}_1$ to lie along the first coordinate direction.

Now we can repeat Steps 1, 2, 3 to obtain k_2 and a new reduced generator matrix G_2. And so on.

It can be shown that the $\{k_i\}$ computed in the Schur algorithm are exactly the same quantities as defined before in the Levinson algorithm; we shall give a proof in Section 2.4 (see also Problem 2.5). However, note that the Schur algorithm only requires sets of 2×1 row vector by 2×2 matrix multiplications, which can be carried out in parallel at each stage. Therefore, we shall need only $O(N)$ time units to carry out the Schur algorithm with N processors, as compared to $O(N \log N)$ for the Levinson algorithm.

In fact, a parallel (and pipelined) lattice VLSI computing structure based on the Schur algorithm has already been designed and built (see Kung and Hu (1983)).

The Schur algorithm has many other interesting properties (see, e.g., Lev-Ari (1983)). Here we content ourselves with the following remarks.

We note that the Schur algorithm gives us the $\{k_i\}$ and not the $\{A_i(z)\}$; nevertheless, once we have the $\{k_i\}$, we can readily obtain the $\{A_i(z)\}$ by using the recursion (1.41),

$$\begin{bmatrix} A_{m+1}(z) \\ A_{m+1*}(z) \end{bmatrix} = \Theta(k_i) \begin{bmatrix} zA_m(z) \\ A_{m*}(z) \end{bmatrix}, \quad \begin{bmatrix} A_0(z) \\ A_{0*}(z) \end{bmatrix} = \begin{bmatrix} 1 \\ 1 \end{bmatrix}.$$

Thus the Schur algorithm provides a complete and useful alternative to the Levinson algorithm; this is a fact that not yet fully appreciated in the signal processing literature. Moreover, it turns out that for a physical implementation of the prediction-error filters $\{A_i(z)\}$ it is sufficient to have the reflection coefficients, as will be shown in the next section.

2.4 Transversal and Lattice Filter Implementations

Perhaps the most straightforward way of implementing the prediction-error-order filters is via tapped-delay-line structures (also called transversal filters). Thus consider the m-th order filter

$$A_m(z) = z^m + A_{m,1}z^{m-1} + \cdots + A_{m,m} \ .$$

Its response to the input $\{y_0, y_1, \ldots, y_t\}$ will be

$$e_{m,t} = y_t + A_{m,1}y_{t-1} + \cdots + y_{t-m} \ , \quad t \geq 0 \tag{2.24}$$

(which will be the innovation e_t only when $m = t$). We assume that $y_j = 0, j < 0$. A tapped-delay-line structure for implementing this calculation is shown in Figure 1. If we wish to use a higher order filter, or if we wished to compute the innovations for all $t \geq 0$, we would need to have filters of several orders. Equivalently we could have a single, "growing memory" tapped-delay-line filter, with coefficients that varied with the order:

$$\begin{bmatrix} I & & & & \\ A_{1,1} & 1 & & & \\ A_{2,2} & A_{2,1} & 1 & & \\ \cdot & & & \cdot & \\ \cdot & & & & \cdot \\ \cdot & & & & \cdot \\ A_{m,m} & A_{m,m-1} & \cdots & A_{m,1} & 1 \end{bmatrix} \tag{2.25}$$

This growing memory, tapped delay implementation is, to quote Willsky [1979, p. 31], "hardly appealing", and it was to avoid this prospect that the assumption that $\{y_t\}$ had a Markovian state-space model was introduced. When say an n-dimensional state vector essentially determines the process $\{y_t\}$, estimates of $\{y_t\}$ can be obtained in terms of estimates of the n state variables and therefore it should not be surprising that the innovations of $\{y_t\}$ can be determined with a fixed memory (equal to n) filter, though one that will be time-variant. In fact, the solution will be the Kalman filter, which has been widely studied and even more widely used. Perhaps even too much so because in many signal processing problems, fixed memory state-space models may not be readily available or even appropriate. Moreover, with present technology, memory is not really a limiting factor and it would be more desirable to have a tradeoff in the opposite direction: constant gains but growing memory.

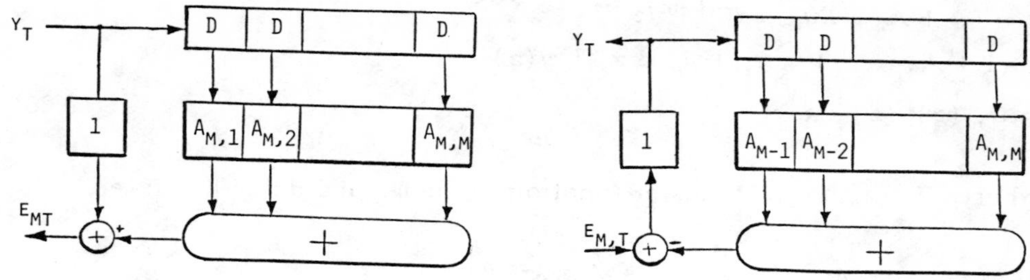

(a) The Whitening Filter

(b) The Modeling Filter

FIGURE 1: Tapped-delay-line realizations.

To show this, it will be convenient to introduce z-transform descriptions (see Lim (this volume)). that the innovations of $\{y_t\}$ can be determined with a fixed memory (equal to n) filter, though one that will be time-variant. In fact, the solution will be the Kalman filter, which has been widely studied and even more widely used. Perhaps even too much so because in many signal processing problems, fixed memory state-space models may not be readily available or even appropriate. Moreover, with present technology, memory is not really a limiting factor and it would be more desirable to have a tradeoff in the opposite direction: constant gains but growing memory.

It turns out that the stationarity assumption and in particular the Levinson and Schur algorithm do enable essentially a time-invariant growing memory filter.

To show this, it will be convenient to introduce z-transform descriptions (see Lim (this volume)).

From

$$e_{m,t} = y_t + A_{m,1}y_{t-1} + \cdots + A_{m,m}y_{t-m}$$

we can write (recall $y_t = 0, t < 0$)

$$e_m(z) = \sum_{-\infty}^{\infty} e_{m,t}z^{-t} = \Sigma y_t z^{-t} + A_{m,1}\Sigma y_{t-1}z^{-(t-1)} \cdot z^{-1}$$

$$+ \cdots + A_{m,m} \Sigma y_{t-m} z^{-(t-m)} z^{-m}$$
$$= (1 + A_{m,1} z^{-1} + \cdots + A_{m,m} z^{-m}) y(z)$$
$$= \mathcal{a}_m(z) y(z) \text{, say} \tag{2.26}$$

where $\mathcal{a}_m(z)$ is the transfer function of the tapped-delay-line filter,

$$\mathcal{a}_m(z) = 1 + A_{m,1} z^{-1} + \cdots + A_{m,m} z^{-m}$$
$$= z^{-m} A_m(z) = A_{m*}(z^{-1}) \tag{2.27}$$

Since we have a recursion for the $\{A_m(z)\}$, we should be able to deduce one for the $\{\mathcal{a}_m(z)\}$. First define

$$\mathcal{a}_{m*}(z) = z^{-m} + A_{m,1} z^{-(m-1)} + \cdots + A_{m,m}$$
$$= A_m(z^{-1}) = z^{-m} \mathcal{a}_m(z^{-1}) \ . \tag{2.28}$$

Then from the Levinson recursions (1.41)

$$\begin{bmatrix} A_m(z) \\ A_{m*}(z) \end{bmatrix} = \begin{bmatrix} z & -k_m \\ -zk_m & 1 \end{bmatrix} \begin{bmatrix} A_{m-1}(z) \\ A_{m-1*}(z) \end{bmatrix}$$

we can conclude that

$$\begin{bmatrix} \mathcal{a}_{m*}(z) \\ \mathcal{a}_m(z) \end{bmatrix} = \begin{bmatrix} z^{-1} & -k_m \\ -z^{-1}k_m & 1 \end{bmatrix} \begin{bmatrix} \mathcal{a}_{m-1*}(z) \\ \mathcal{a}_{m-1}(z) \end{bmatrix} ,$$

or

$$\begin{bmatrix} \mathcal{a}_m(z) \\ \mathcal{a}_{m*}(z) \end{bmatrix} = \begin{bmatrix} 1 & -z^{-1}k_m \\ -k_m & z^{-1} \end{bmatrix} \begin{bmatrix} \mathcal{a}_{m-1}(z) \\ \mathcal{a}_{m-1*}(z) \end{bmatrix} , \quad \begin{bmatrix} \mathcal{a}_0(z) \\ \mathcal{a}_{0*}(z) \end{bmatrix} = \begin{bmatrix} 1 \\ 1 \end{bmatrix} \tag{2.29}$$

This suggests that we can realize $\mathcal{a}_m(z)$ (and at the same time, $\mathcal{a}_{m*}(z)$)

by using the cascade lattice structure shown in Figure 2.

Moreover, to increase the order of the filter, say to $m + 1$, we only need to switch in one more section, with reflection coefficients k_{m+1}, without affecting the previous section. Or to describe it differently, we have a lattice filter with reflection coefficients of the simple time-variant form

$$k_i(t) = \begin{cases} 0 , & t < i \\ k_i , & t \geq i \end{cases} \qquad (2.30)$$

This should however be a much more manageable form of time variation than in the transversal filter. The fact that the $\{|k_i|\}$ are smaller than unity also contrasts favorably with the fact that the coefficients $\{A_{m,i}\}$ of the transversal filter can vary over a large range -- a fact that is significant from the point of view of digital implementation with fixed point arithmetic.

Relation to the Schur Algorithm. The lattice filter is completely parametrized by the reflection coefficients $\{k_i\}$. Therefore the Schur algorithm of Section 2.3 suffices to determine it, and there is no need to use the Levinson algorithm, or to compute the $\{A_m(z)\}$ from the Schur coefficients.

The intimate relation between the lattice filter and the Schur algorithm can be made evident by using the covariance sequence $\{R_0, R_1, \ldots\}$ as input to the lattice filter, and tracing its flow through it. At time 0, we feed in R_0; at time 1, R_1; and so on. Now at time 1, the input to the top line of the first lattice section will be R_1, while because of the unit delay, the input at the bottom line will be R_0. Their ratio is exactly k_1, as defined in the Levinson algorithms.

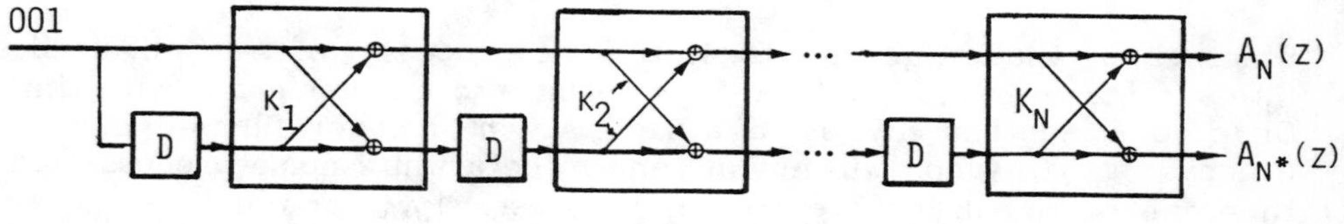

FIGURE 2: Cascade feedforward lattice filter, when excited by an impulse the outputs are $A_n(z)$ and $A_{n*}(z)$. When excited by $\{y_t\}$, the outputs will be $e_{n,t}$ and $r_{n,t}$.

Now at time 2, we can compute from the figure that $R_2 - k_1 R_1$ will be the input to the second lattice section and by referring to the Levinson construction, we see that $R_2 - k_1 R_1 = \Delta_2$ (note that $k_1 = -A_{1,1}$). On the other hand the input to the second lattice section will, because of the unit delay, be the output of the first section at time 1. But this output is $R_0 - k_1 R_1$, which can be seen to be equal to $R_1^r = R_1^e$. Therefore the ratio, at time 2, of the inputs to the second lattice section, will be exactly k_2. Continuing in this way, we can show that the ratio, at time m, of the inputs to the m-th lattice section will be equal to k_m (see Problem 2.6).

We should remark that this discussion essentially provides a proof of the claim made in Section 2.3 that the coefficients $\{k_i\}$ found in the Schur algorithm are exactly those of the Levinson algorithm.

Therefore the lattice filter gives an online method of computing the reflection coefficients. An equivalent result, but hidden in some algebraic formulas was noted by LeRoux and Gueguen (1977); the nice graphical representation above was first explicitly noted by Morf et al. (1977). As we might expect, the above construction is very closely related to the Schur algorithm; this will become explicit when one translates the Schur algorithm as described in Section 2.3 into flow graph (block diagram) form, although we shall not pursue this argument here.

2.5 Approximate Autoregressive Models and Asymptotic Behaviour

Given a stationary stochastic process $\{y_0, \ldots, y_t, \ldots\}$, we have defined the m-th order prediction residual as

$$e_{m,t} = y_t + A_{m,1} y_{t-1} + \cdots + A_{m,m} y_{t-n} .$$

When $m = t$,

$$e_{t,t} = y_t - \hat{y}_{t|t-1} = e_t, \text{ the innovation at time } t .$$

which shows that in general to whiten the process $\{y_t\}$, we will need a filter, in transversal- or lattice-filter form, whose order grows with time. Or to put it another way, use of a fixed, say n-th order, filer will give the process $\{e_{n,t}\}$, which will not in general be a white noise process. It is then natural to ask if there are any processes $\{y_t\}$ for which $\{e_{n,t}\}$ will be white for a finite n? Otherwise, we should try to obtain a measure of how the n-th order approximation approaches a complete process model as $n \to \infty$.

Autoregressive (AR) Processes. The Levinson and Schur algorithms will effectively terminate at step n if the process is autoregressive (AR), i.e., it has the form

$$y_t + a_1 y_{t-1} + \cdots + a_n y_{t-n} = u_t \ , \quad t \geq 0$$

where u_t is a white noise sequence with

$$Eu_t = 0 \ , \quad Eu_t u_s^T = Q\delta_{ts} \ ,$$

and

$$Eu_t y_s^T \equiv 0 \ , \quad t \geq s \ .$$

If $\{y_t\}$ is to be stationary, we also have to assume that the covariance matrix of the "initial" random variables $\{y_{-n},\ldots,y_{-1}\}$ equals the covariance matrix of any other consecutive set of process variables $\{y_t,\ldots,y_{t+n-1}\}$. We shall not pursue the question of what this matrix should be, except to say that the condition can always be met if (and only if) the polynomial

$$a(z) = z^n + a_1 z^{n-1} + \cdots + a_n$$

has all its roots within the unit circle.

The definition of an AR process shows that an n-th order filter gives a white noise sequence, viz., u_t, and therefore we should expect that the Levinson algorithm should terminate after n steps. To see this, we must study the covariance sequence of the AR process. By multiplying both sides of the defining equation by y_{t-i} and averaging, we find the relations

$$R_i + a_1 R_{i-1} + \cdots + a_n R_{i-n} = 0 \ , \quad i > 0$$

and

$$R_0 + a_1 R_{-1} + \cdots + a_n R_{-n} = Q$$

We can write this in matrix form as

$$[\ldots 0 \ \cdots \ 0 \ a_n \ \cdots \ a_1 \ 1]\mathbf{R} = [\ldots 0 \ \cdots \ 0 \ Q]$$

where $\mathbf{R}$ is the covariance matrix of $\{y_0, y_1, \ldots, \}$. Then comparing with Eq. (1.16) we can identify the n-th order prediction error filter as

$$\mathbf{A}_n = [a_n \quad \cdots \quad a_1 \quad 1]$$

so that

$$A_n(z) = a(z)$$

while all higher order prediction error polynomials will be of the form

$$A_{n+i}(z) = z^i a(z) \ .$$

From these formulas we can deduce that

$$k_{n+i} = 0 \ , \quad i > 0$$

so that the lattice filter will have n nontrivial sections (assuming $\mathbf{R}_n > 0$).

Asymptotics as $n \to \infty$. If the process $\{y_t\}$ is not AR, and if $\mathbf{R}_n > 0$ for all n, then the reflection coefficients will never be zero and we shall have an lattice filter with an infinite number of sections.

The asymptotic behaviour of the $\{A_n(z), R_n^e\}$ as $n \to \infty$ has been studied in some detail by Szegö and by Geronimus (1958). In Chapter 2 of the latter reference it is shown for example that if the process is "purely" nondeterministic" in the sense that its spectral density $W(e^{i\vartheta})$ is such that

$$\int_0^{2\pi} \ln W(e^{i\vartheta}) \, d\vartheta > -\infty$$

then

$$\lim_{n \to \infty} R_n^e = \exp \frac{1}{2\pi} \int_0^{2\pi} \ln W(e^{i\vartheta}) \, d\vartheta$$

This is known as Szegö's formula and is a celebrated result in the prediction theory of stationary processes, (see, e.g., Doob (1953)). In prediction theory it is known that under the nondeterminism condition the spectral density $W(e^{i\vartheta})$ has a unique canonical factorization

$$W(e^{i\vartheta}) = \Phi(e^{i\vartheta})\Phi^*(e^{i\vartheta})$$

where $\Phi(e^{i\vartheta})$ is the (a.e.) limit of a so-called outer function $\Phi(z)$. Roughly speaking an outer function is one for which the inverse transforms of $\Phi(z)$ and $\Phi^{-1}(z)$ are both causal; when $W(e^{i\vartheta})$ is rational, then $\Phi(z)$ is the minimum-phase factor of $W(z)$, i.e., $\Phi(z)$ has no poles or zeros outside the unit circle.

Now it can be shown that under the nondeterminism condition, the function

$$a_n(z) = R_n^{-e/2} z^{-n} A_n(z) \to \pi(z) \ , \quad z > 1$$

where $1/\pi(z)$ is the canonical outer factor of the spectral density function, i.e.,

$$\frac{1}{|\pi(e^{i\vartheta})|^2} = W(e^{i\vartheta})$$

Note that $a_n(z)$ is (up to the scale factor $R_n^{-e/2}$) the transfer function of the n-th order approximate whitening filter, while $1/a_n(z)$ is the transfer function of the n-th order approximate autoregressive model of the process. It is interesting to remark that $W(z)$ is rational, then $1/a_n(z)$, which has no zeros for any n, will converge to a rational function with poles and zeros! Note also that it is the filter transfer function $a_n(z)$ and not the orthogonal polynomial $A_n(z)$ that converges as $n \to \infty$.

We shall not pursue these interesting avenues any further here; the interested reader can refer to Chapters 2 and 8 of Geronimus (1958) for more information, e.g., estimates of the rate of convergence.

We return instead to further examination of the finite-order models.

2.6 Feedback Lattice and Transmission-Line Models

Once we have chosen an approximation of a certain order, say

$$y_t + A_{n,1}y_{t-1} + \cdots + A_{n,n}y_{t-n} = e_{n,t}$$

we can approximately model the process by passing white noise through a linear filter with transfer function (cf. Section 4)

$$\frac{1}{a_n(z)} = \frac{1}{1 + A_{n,1}z^{-1} + \cdots + A_{n,n}z^{-n}}$$

$$= \frac{z^n}{z^n + A_{n,1}z^{n-1} + \cdots + A_{n,n}}$$

This filter can be realized in any of several different forms, e.g., the so-called direct forms of digital filter theory (see, e.g., Oppenheim and Schafer (1975)), which are known as controller and observer canonical forms in linear system theory (see, e.g., Kailath (1980)). Now it is well known that these direct forms do not have good numerical properties, in that they are very sensitive to (roundoff) perturbations in the parameter values, and therefore alternative structures are often sought.

A nice structure can be obtained by using simple signal flow graph rules (see, e.g., Mason and Zimmerman (1960)) to form the inverse of the lattice-form approximate whitening filter of Figure 2. The general flow-graph recipe for forming the inverse is: form a delay-free path from output back to input, replace all gains on this path by their reciprocals and change the sign of all inputs entering any summation node along this path.

Using this rule we can obtain the modeling filters shown in Figure 3. This structure is known as a feedback lattice form. It is a cascade of essentially identical sections, each characterized by a scattering matrix of the form

$$\begin{bmatrix} e_{n-1}(z) \\ r_n(z) \end{bmatrix} = \begin{bmatrix} 1 & k_n \\ -k_n & 1-k_n^2 \end{bmatrix} \begin{bmatrix} e_n(z) \\ z^{-1}r_{n-1}(z) \end{bmatrix}$$

where $\{e_n(z), z^{-1}r_{n-1}(z)\}$ are the "incident waves" on the n-th section and $\{e_{n-1}(z), r_n(z)\}$ are the "reflected waves" from the n-th section (and $e_n(z) = \sum e_{n,t}z^{-t}$, $r_n(z) = \sum r_{n,t}z^{-t}$).

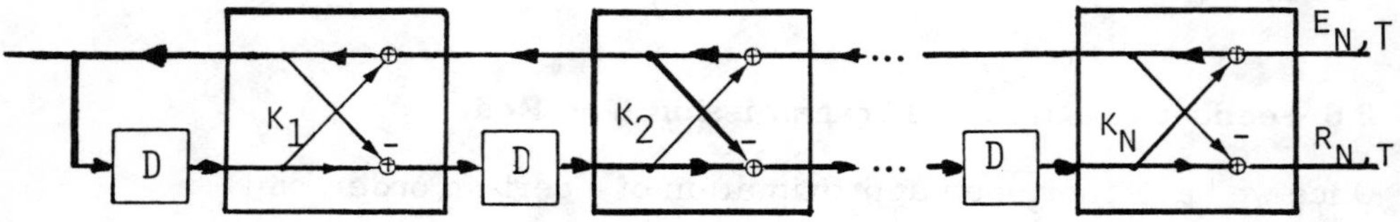

FIGURE 3: Cascade feedback lattice structure. When excited by $e_{n,t}$ the outputs will be y_t and $r_{n,t}$; when excited by an impulse the outputs will be $1/A_n(z)$ and $A_{n*}(z)$.

The above description suggests that the feedback lattice filter can be redrawn in a "transmission-line type" form as in Figure 4 This looks interesting, but requires 3 multipliers per section, while the lattice section requires only two. In fact, we can by a suitable normalization, obtain a configuration that requires only one multiplier per section (Itakura and Saito (1971)). The technique is the following: use scaling parameters, δ_n and δ_{n-1}, at the input and output of the n-th section and define

$$\bar{\bar{e}}_n(z) = e_n(z)/\delta_n \ , \quad \bar{\bar{r}}_n(z) = r_n(z)/\delta_n$$

Then the scattering matrix relations can be rewritten as

$$\bar{\bar{e}}_{n-1}(z) = (\delta_n/\delta_{n-1})\bar{\bar{e}}_n(z) + k_n z^{-1}\bar{\bar{r}}_{n-1}(z)$$
$$\bar{\bar{r}}_n(z) = -k_n\bar{\bar{e}}_n(z) + (1-k_n^2)(\delta_{n-1}/\delta_n)z^{-1}\bar{\bar{r}}_{n-1}(z)$$

If we now choose

$$\delta_n/\delta_{n-1} = 1 + k_n$$

then the scattering relation will be

$$\begin{bmatrix}\bar{\bar{e}}_{n-1}(z)\\ \bar{\bar{r}}_n(z)\end{bmatrix} = \begin{bmatrix}1+k_n & k_n\\ -k_n & 1-k_n\end{bmatrix}\begin{bmatrix}\bar{\bar{e}}_n(z)\\ z^{-1}\bar{\bar{r}}_{n-1}(z)\end{bmatrix}$$

which was a form derived by Kelly and Lochbaum (1962) in making a discrete model of the vocal tract (see Flanagan [1972, sec. 6. p 262]) and Markel and Gray [1975, ch 4]). The same structure was obtained by Goupillaud (1961) as a model for seismic exploration (see Problem 2.7). This structure appears to need 4 multipliers per section, but as noted by Itakura and Saito (1971), by combining terms we can obtain an equivalent

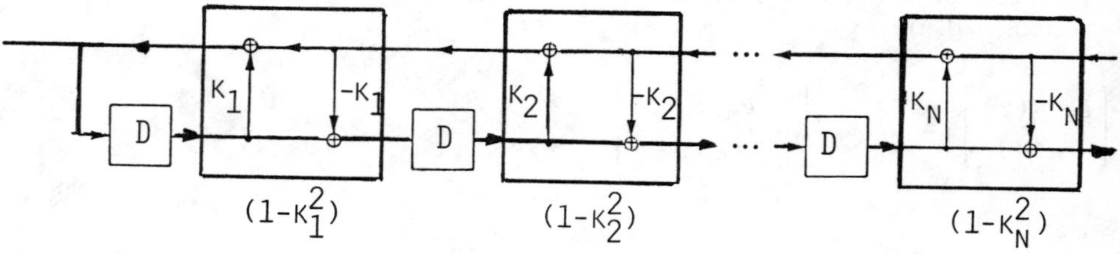

FIGURE 4: Transmission line representation of the unnormalized feedback lattice filter of figure 3.

form that needs only one multiplier:

$$\bar{e}_{n-1}(z) = \bar{e}_n(z) + k_n(\bar{e}_n(z) + z^{-1}\bar{r}_{n-1}(z))$$
$$\bar{r}_n(z) = z^{-1}\bar{r}_{n-1}(z) - k_n(\bar{e}_n(z) + z^{-1}\bar{r}_{n-1}(z))$$

We can implement these equations with just one multiplier, as we shall leave to the reader to show.

We may note that alternative but very similar one-multiplier forms may be obtained by using the normalizations

$$\delta_n / \delta_{n-1} = 1 - k_n$$

These one-multiplier forms are attractive in one way, but we should note that often the numerical properties of the implementation, in particular, the sensitivity to finite precision arithmetic, may be more significant to the user than the number of multipliers. In fact it turns out that the best form from a numerical point of view is a normalized form, which turns out to have 4 multipliers per section. We hasten to add however that there are direct methods, using what are called CORDICs for implementation of the normalized forms without any multipliers (see below).

This normalized form is naturally suggested by the stochastic interpretation. We normalize the variables to have unit variance

$$\bar{e}_n(z) = e_n(z)/ R_n^{e/2} \quad , \quad \bar{r}_n(z) = r_n(z)/ R_n^{e/2}$$

(since $R_n^r = R_n^e$ in the scalar case). Therefore, in our earlier notation

$$\delta_n / \delta_{n-1} = (R_n^e / R_{n-1}^e)^{1/2} = (1-k_n^2)^{1/2}$$

by virtue of the basic formula (1.41b). Consequently, the new scattering relation will be

$$\begin{bmatrix} \bar{e}_{n-1}(z) \\ \bar{r}_n(z) \end{bmatrix} = Q(k_n) \begin{bmatrix} \bar{e}_n(z) \\ z^{-1}\bar{r}_{n-1}(z) \end{bmatrix}$$

where

$$Q(k_n) := \begin{bmatrix} (1-k_n^2)^{1/2} & k_n \\ -k_n & (1-k_n^2)^{1/2} \end{bmatrix}$$

We note that $Q(k_n)$ is *orthogonal*, i.e.,

$$Q(k_n)\, Q^T(k_n) = I$$

which is the source of the good numerical properties of the filter (no overflow oscillations, no limit cycles). The corresponding transmission-line structure (Figure 5) is that of a lossless discrete transmission-line, a fact that also has several important implications, though for reasons of space we shall not enter into them here. For the active reader we may mention that this structure, or rather its inverse, provides a graphical description of the Schur algorithm, and is closely related to the so-called inverse scattering problems of physics (see Bruckstein and Kailath (1984)).

It is interesting to calculate the corresponding normalized whitening filter section, which turns out to be

$$\begin{bmatrix} \bar{e}_n(z) \\ \bar{r}_n(z) \end{bmatrix} = \bar{\Theta}(k_n) \begin{bmatrix} \bar{e}_{n-1}(z) \\ z^{-1}\bar{r}_{n-1}(z) \end{bmatrix}$$

where

$$\bar{\Theta}(k_n) \equiv (1 - k_n^2)^{-1/2} \begin{bmatrix} 1 & -k_n \\ -k_n & 1 \end{bmatrix}$$

CORDIC Implementations. The normalized sections again require 4 multipliers for a direct implementation. However, note that the matrix $Q(k_n)$ is orthogonal and therefore represents a circular rotation, while $\bar{\Theta}(k_n)$ is a J-rotation, $J = 1 \oplus -1$, i.e., it represents a hyperbolic rotation. It turns out that in 1959, Volder proposed a simple scheme for implementing rotations (hyperbolic, circular and linear) by using what he called

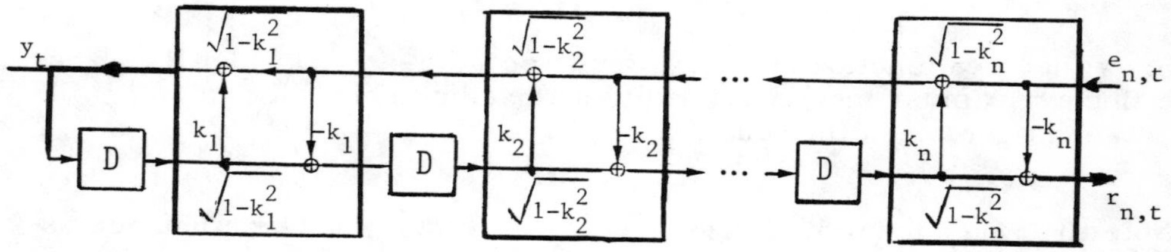

FIGURE 5: Normalized transmission line model.

Coordinate Digital Computers or CORDICs. CORDICs use a circuit built only of registers, shifters and adders to iteratively compute functions such as $\sqrt{x^2 + y^2}$, $\tan^{-1} y/x$, $\sqrt{x^2 - y^2}$, $\tanh^{-1} y/x$, $x \cos\alpha - y \sin\alpha$, etc.; they have been employed for this purpose in several pocket calculators, e.g., the HP-35.

In recent work CORDICs have been suggested for a variety of one- and two-dimensional VLSI computing structures of the systolic and wavefront array types -- see, e.g., the papers of Ahmed, Delosme and Morf (1982), Rao and Kailath (1983a). Despain (1974), (1979) was one of the first to draw attention to CORDICs for signal processing of the FFT type, and more recently Haviland and Tuszynski (1980) have described a VLSI CORDIC chip for arithmetic calculations. We should note that there are several factors (speed, accuracy, design overhead, etc.) that have to be considered before settling on an actual implementation via CORDICs or via some other schemes, e.g., array multipliers, perhaps coupled with stored trigonometric tables, etc.

Problem 2.1. Extending a Covariance Sequence. Let $\{R_0, R_1, \ldots, R_m\}$ be a scalar positive-definite covariance sequence. Show that any real number R_{m+1} such that $\{R_0, R_1, \ldots, R_{m+1}\}$ is also a covariance sequence must be such that

$$-\left[\sum_1^m A_{m,i} R_{m+1-i}\right] - R_m^e \le R_{m+1} \le -\left[\sum_1^m A_{m,i} R_{m+1-i}\right] + R_m^e$$

Problem 2.2. Again assume that $\{R_0, \ldots, R_m\}$ is a scalar covariance sequence. Show that we can write $\det \mathbf{R}_{m+1}$ in the form

$$\det \mathbf{R}_{m+1} = -(det\ \mathbf{R}_{m-1}) R_{m+1}^2 + \beta R_{m+1} + \gamma$$

where β and γ are functions of $\det \mathbf{R}_{m-1}$ and of the quantity $(\Sigma A_{m,i} R_{m+1-i})$.

Show that as a function of R_{m+1}, for given $\{R_0, \ldots, R_m\}$, $\det \mathbf{R}_{m+1}$ has a single maximum and (therefore) that the admissible values of R_{m+1} lie between the roots of the equation $\det \mathbf{R}_{m+1} = 0$.

Note that the admissible range of R_{m+1} is $2R_m^e$, which is a nonincreasing function of m.

Problem 2.3. The Maximum Entropy Extension. Refer to Problem 2.1.

Show that choosing R_{m+1} as the midpoint of the admissible range, i.e.,

$$R_{m+1} = -\left[\sum_{1}^{m} A_{m,i} R_{m+1-i}\right]$$

yields

$$k_{m+1} = 0 \quad \text{and} \quad R_{m+1}^{e} = R_{m}^{e} .$$

Show that this choice of R_{m+1} also yields the maximum possible value of det $\mathbf{R}_{m+1}$.

Remark: Shannon (1948) showed that the entropy of an $(m+1)$-dimensional Gaussian vector $\{y_0, \ldots, y_m\}$ is given by $H = \frac{1}{2} log(2\pi e)^{m+1} \cdot \det \mathbf{R}_m$. Therefore the above extension is the one that yields maximum entropy for the $(m+2)$-dimensional Gaussian vector.

Problem 2.4. Find the maximum entropy extension of the covariance sequence $\{1, a, a^2, \ldots, a^m\}$.

Problem 2.5. Reflection Coefficients from the Lattice Filter. Consider an m-section feedforward lattice filter that realizes the polynomials $a_m(z)$ and $a_{m*}(z)$ -- see Figure 2. If we add in one more section, the outputs will be $a_{m+1}(z)$ and $a_{m+1*}(z)$. Feed the sequence $\{R_0, R_1, \ldots, R_{m+1}, \ldots\}$ into the filter.

i) Show that at time $m+1$, the input at the top line of the $(m+1)^{th}$ lattice section will be

$$R_{m+1} + A_{m,1} R_m + \cdots + A_{m,m} R_1 N = \Delta_{m+1}$$

ii) Show that at time m, the output at the bottom line of the mth lattice section will be

$$A_{m,m} R_m + \cdots + A_{m,1} R_1 + R_0 = R_m^r = R_m^e$$

iii) Hence show that k_{m+1} is the ratio of the inputs to the $(m+1)^{th}$ lattice section.

Problem 2.6. Singular Extensions. Refer to Problem 2.1.

We assume that $\mathbf{R}_m$ is nonsingular and choose R_{m+1} to be at one of the extremities of the admissible range, i.e.,

$$R_{m+1} = -[A_{m,i}R_{m+1-i}] \pm R_m^e \quad .$$

Show that now

i) $|k_{m+1}| = 1$
ii) $R_{m+1}^e = 0$, $\det \mathbf{R}_{m+1} = 0$, all $i > 0$, and that in fact, all R_{m+1}, $i > 0$, are now uniquely determined by $\{R_0, \ldots, R_m\}$.
iii) $A_{m*}(z) = A_m(z)$, so that the roots of $A_m(z)$ must be symmetrically located with respect to the unit circle. Because $A_m(z)$ is stable, this means that all the roots must lie *on* the unit circle.
iv) $\{R_k, |k| \le m\}$ can be expressed as a sum of harmonics (cf. Geronimus ((1948, Eq. (1.10)).

Problem 2.7 Layered-Earth Models. The scattering matrix of a section of the Kelly-Lochbaum modeling filter has the form

$$\begin{bmatrix} 1+k & k \\ -k & 1-k \end{bmatrix}$$

Show that the scattering matrix of a section of the corresponding whitening filter has the form

$$\frac{1}{1+k}\begin{bmatrix} 1 & k \\ k & 1 \end{bmatrix} .$$

Compare with the discussion in Claerbout (1976, Ch. 8).

3. MULTICHANNEL STATIONARY PROCESSES

In this section, we shall briefly note extensions of several of the results of the preceding sections to the case of vector or multichannel processes, where each random variable $y(i)$ is a $p \times 1$ vector. The covariance lags

$$R(i-j) = Ey(i)y^T(j)$$

will then be $p \times p$ matrices obeying the symmetry condition

$$R(i-j) = R^T(j-i) \text{ or } R_i = R_{-i}^T$$

where we note again that the superscript T denotes transpose (Hermitian) transpose for complex-valued variables). Therefore the matrix

$$\mathbf{R} = [R(i-j)]$$

will not be block symmetric, although as a covariance matrix it will, of course, be symmetric in the usual sense.

Now in Section 1.2 we deliberately postponed use of the assumption of symmetry as much as possible, so that most of that development will apply to the multichannel problem.

In fact, with the simple reinterpretation that $\{A_{m+1}(z), B_{m+1}(z)\}$ are polynomial vectors with $p \times p$ matrix coefficients, the recursions (1.33) will apply to the present problem. We repeat them here for convenience:

$$\begin{bmatrix} A_{m+1} \\ B_{m+1} \end{bmatrix} = \Theta_{m+1} \begin{bmatrix} zA_m(z) \\ B_m(z) \end{bmatrix} \tag{3.1a}$$

where

$$A_0(z) = I = B_0(z), \quad I = p \times p \text{ identity matrix}$$

and

$$\Theta_{m+1} := \begin{bmatrix} I & -\Delta_{m+1} R_m^{-r} \\ -\tilde{\Delta}_{m+1} R_m^{-e} & I \end{bmatrix} \tag{3.1b}$$

with (cf. (1.18) and (1.25))

$$\Delta_{m+1} = A_{m,m} R_1 + \cdots + A_{m,1} R_m + R_{m+1} \cdot \tag{3.1c}$$

$$\tilde{\Delta}_{m+1} = R_{-m-1} + B_{m,1} R_{-m} + \cdots + B_{m,m} R_{-1} \tag{3.1d}$$

Moreover, we also have the relations (cf. (1.28b) and (1.29b))

$$R_{m+1}^e = R_m^e - \Delta_{m+1} R_m^{-r} \tilde{\Delta}_{m+1} \tag{3.1e}$$

$$R^r_{m+1} = R^r_m - \tilde{\Delta}_{m+1} R_m^{-e} \Delta_{m+1} \tag{3.1f}$$

In the symmetric scalar case, it turned out that $B_m(z)$ was just $A_m(z)$ with its coefficients reversed and, therefore, the scalars Δ_{m+1} and Δ_{m+1} became equal. The simple relation between $A_m(z)$ and $B_m(z)$ does not carry forward to the multichannel case. But the block symmetry of $\mathbf{R}$, viz., the fact that $R_{-i} = R_i^T$, yields the result, first noted by Burg (as cited in Wiggins and Robinson (1965)), that

$$\tilde{\Delta}_{m+1} = \Delta^T_{m+1} \ . \tag{3.1g}$$

For a proof, we may just note that

$$\begin{aligned}&[0 \quad \mathbf{A}_m]\mathbf{R}_{m+1} \cdot [\mathbf{B}_m^T \quad 0] \\ &\quad = [\Delta_{m+1} \ 0 \ \cdots \ 0 \ R_m^e] \cdot [I \ \ B^T_{m,1} \ \cdots \ B^T_{m,m} \ \ 0]^T = \Delta_{m+1}\end{aligned}$$

while

$$\begin{aligned}&[\mathbf{B} \quad 0]\mathbf{R}_{m+1} \cdot [0 \quad \mathbf{A}_m^T]^T \\ &\quad = [R_m^r \ 0 \ \cdots \ 0 \ \tilde{\Delta}_{m+1}] \cdot [0 \ \ A^T_{m,m} \ \cdots \ A^T_{m,1} \ \ I]^T = \tilde{\Delta}_{m+1} \ .\end{aligned}$$

But the left hand sides of these two expressions are transposes of each other; hence so are Δ_{m+1} and Δ_{m+1}.

The recursions (3.1) will be called the LWR recursions, after Levinson (1947) who essentially gave the scalar form, Whittle (1963) who did not note the Burg identity (3.1g), and finally Wiggins and Robinson (1965) who, independently of Whittle, gave the complete solution.

Several of the special results described in Section 2 for scalar processes go over to vector or multichannel processes with obvious notational changes. However there are some important ones, such as the parameterization of covariance sequences by a set of reflection coefficients and adaptive methods (such as Burg's (1968)) for spectral estimation, that do not have obvious extensions. The main reason is that in the vector LWR algorithm we have two different candidates for the role of scalar reflection coefficient, viz., the quantities $-\Delta_{m+1}R_m^{-e}$ and $-\Delta^T_{m+1}R_m^{-e}$, which are equal in the scalar case.

It turns out that a way to almost completely restore the analogy with the scalar case is to use the same type of normalization as used in Section 2.6 to get a transmission line model for scalar processes; in addition to mathematical simplicity, we thereby also get a nice physical picture.

The normalized LWR algorithm will be presented in Section 3.1. Its similarity to the scalar problem will allow us to extend, with appropriate changes, almost all of the results of Section 2 to the multichannel case. In Section 3.2, we elaborate one of the special results on the Cholesky and Christoffel-Darboux formulas.

3.1 The Normalized LWR Algorithm

The LWR recursions can be cast in an even simpler form, both conceptually and computationally, by introducing certain normalizations.

Let

$$\bar{A}_m(z) = R_m^{-e/2} A_m \;\; ; \;\; \bar{B}_m(z) = R_m^{-r/2} B_m(z) \tag{3.2}$$

where $\{R_m^{e/2}, R_m^{r/2}\}$ are *square-root factors* of $\{R_m^e, R_m^r\}$, i.e., they are matrices such that

$$R_m^e = (R_m^{e/2})(R_m^{e/2})^T := R_m^{e/2} R_m^{eT/2} \tag{3.3a}$$

and similarly for R_m^r. If $R_m^{e/2}$ is symmetric, then it is in fact a (matrix) square root of R_N^e, but for computational reasons, we often prefer to use nonsymmetric (in fact, triangular) square-root factors. Note that

$$R_m^{-e} = R_m^{-eT/2} R_m^{-e/2} \tag{3.3b}$$

In any case, we can now rewrite the recursions (3.1a) as

$$\begin{bmatrix} R_{m+1}^{e/2} & 0 \\ 0 & R_{m+1}^{r/2} \end{bmatrix} \begin{bmatrix} \bar{A}_{m+1}(z) \\ \bar{B}_{m+1}(z) \end{bmatrix} = \begin{bmatrix} I & -\Delta_{m+1} R_m^{-r} \\ -\Delta_{m+1}^T R_m^{-e} & I \end{bmatrix} . \tag{3.4}$$

$$\cdot \begin{bmatrix} R_m^{e/2} & 0 \\ 0 & R_m^{r/2} \end{bmatrix} \begin{bmatrix} z\bar{A}_m(z) \\ \bar{B}_m(z) \end{bmatrix} .$$

We can rearrange this as

$$\begin{bmatrix} \bar{A}_{m+1}(z) \\ \bar{B}_{m+1}(z) \end{bmatrix} = \begin{bmatrix} P_{m+1}^{-1/2} & \\ & Q_{m+1}^{-1/2} \end{bmatrix} \begin{bmatrix} I & -k_{m+1} \\ -k_{m+1}^T & I \end{bmatrix} \begin{bmatrix} z\bar{A}_m(z) \\ \bar{B}_m(z) \end{bmatrix}$$

where

$$P_{m+1}^{-1/2} := R_{m+1}^{-e/2} R_m^{e/2} \; , \quad Q_{m+1}^{-1/2} := R_{m+1}^{-r/2} R_m^{r/2} \tag{3.5}$$

and

$$k_{m+1} := R_m^{-e/2} \Delta_{m+1} R_m^{-rT/2} \tag{3.6}$$

The relations (3.1e,f)

$$R_{m+1}^e = R_m^e - \Delta_{m+1} R_m^{-r} \Delta_{m+1}^T \; , \quad R_{m+1}^r = R_m^r - \Delta_{m+1}^T R_m^{-e} \Delta_{m+1}$$

then yield the expressions

$$P_m = I - k_m k_m^T \; , \quad Q_m = I - k_m^T k_m \; , \tag{3.7}$$

which allows us finally to rewrite the recursions (3.1a) in normalized form as

$$\begin{bmatrix} \bar{A}_{m+1}(z) \\ \bar{B}_{m+1}(z) \end{bmatrix} = \bar{\Theta}_{m+1} \begin{bmatrix} z\bar{A}_m(z) \\ \bar{B}_m(z) \end{bmatrix} \; , \quad \bar{A}_0(z) = R_0^{-1/2} = \bar{B}_0(z) \; , \tag{3.8}$$

where

$$\bar{\Theta}_{m+1} := \begin{bmatrix} (I - k_m k_m^T)^{-1/2} & 0 \\ 0 & (I - k_m^T k_m)^{-1/2} \end{bmatrix} \begin{bmatrix} I & -k_m \\ -k_m^T & I \end{bmatrix} . \tag{3.9}$$

The normalizing factors $\{R_m^{e/2}, R_m^{r/2}\}$ can also be recursively computed as

$$R_{m+1}^{e/2} = R_m^{e/2}[I - k_{m+1} k_{m+1}^T]^{1/2} \; , \quad R_{m+1}^{r/2} = R_m^{r/2}[I - k_{m+1}^T k_{m+1}]^{1/2} \tag{3.10}$$

with

$$R_0^{e/2} = R_0^{1/2} = R_0^{r/2}$$

These equations, (3.8)-(3.10), define the ***normalized LWR algorithm***, which has several interesting and useful features.

The first thing to note is that a single $p \times p$ matrix k_{m+1} suffices to define the recursion from $\{\bar{A}_m(z), \bar{B}_m(z)\}$ to $\{\bar{A}_{m+1}(z), \bar{B}_{m+1}(z)\}$, while two such matrices $\{-\Delta_{m+1}R_m^{-r}, -\Delta_{m+1}^T R_m^{-e}\}$ are needed in the unnormalized case. The matrices $\{k_m\}$ also allow a simpler and numerically better behaved updating of the error variances $\{R_m^e\}$ and $\{R_m^r\}$. Thus numerical errors may cause loss of positive-definiteness in using the updates (3.1a) of Section 1. In the normalized form, we propagate the square-root factors, which will help to ensure that the variances computed by "squaring" will be positive definite.

The fact that (by our nondegeneracy assumption) $\{R_m^e, R_m^r\}$ must be ***positive definite*** implies that (cf. (3.8))

$$k_m k_m^T < I \quad , \quad k_m^T k_m < I \tag{3.11}$$

The eigenvalues of $k_m k_m^T$, which are known to be equal to those of $k_m^T k_m$, are known as the ***singular values*** of the matrix k_m. The largest singular value of k_m is known as the ***spectral*** norm of k_m and is denoted $\|k_m\|$.

Therefore we can now assert that for each m, $\{R_0, R_1, \ldots, R_m\}$ being a nondegenerate covariance sequence implies that the matrices $\{k_i, i = 1, \ldots, m\}$ each have spectral norm strictly less than unity or equivalently have all their singular values strictly less than unity. As in the scalar case (Lemma 2.1), the converse is also true.

We see that the normalization has brought the LWR algorithm much closer in form to the scalar process case. Therefore we might expect that most of the special properties we discussed for scalar processes will also hold, with appropriate notational changes, for the multichannel case.

In fact, this is true of the concepts of orthogonal polynomials, the mapping between reflection coefficients and covariance sequences, the covariance extension problem, maximum entropy spectra, covariance matching properties, Schur algorithm, lattice filter implementations, etc.

For space and other reasons, we shall not elaborate on these here, except for a few remarks on orthogonal polynomials.

Matrix Orthogonal Polynomials. If

$$P(z) = \sum_{i=0}^{n} P_i z^i \ , \quad Q(z) = \sum_{1}^{n} Q_i z^i$$

are elements in the space of matrix polynomials of degree less than or equal to n, we shall define their (matrix) inner product as

$$\begin{aligned} <P,Q> &= \sum_{i,j=0}^{n} P_i R_{i-j} Q_j^T \\ &= \mathbf{P} \mathbf{R}_n \mathbf{Q}^T \end{aligned} \tag{3.12}$$

where

$$\mathbf{P} = [P_0 \ \cdots \ P_n] \ , \quad \mathbf{Q} = [Q_0 \ \cdots \ Q_n]$$

and $\mathbf{R}_n$ is a nonsingular positive definite $n \times n$ block Toeplitz covariance matrix. This inner product has the following characterizing properties:

a) $<P,Q> = <Q,P>^T$

b) $<z^l P,Q> = <P,z^{-l} Q>$ (3.13)

c) $<\Sigma \alpha_i P_i(z), \Sigma \beta_i Q_i(z)> = \sum_{i,j} \alpha_i <P_i, Q_j> \beta_j^T$
for all constant (independent of z) matrices $\{\alpha_i, \beta_j\}$.

We can now check that the polynomials $\{A_i(z)\}$ defined above form a sequence of orthogonal polynomials,

$$<A_n, A_m> = R_n^e \delta_{nm} \ , \quad <\bar{A}_n, \bar{A}_m> = \delta_{nm} \tag{3.14}$$

Let us define a new covariance sequence $\{R_l^b\}$ by the relation

$$R_l^b = R_{-l} \ .$$

Then it can be checked that the polynomials $\{z^n B_n(z^{-1})\}$ are orthogonal with respect to the covariance sequence $\{R_l^b\}$.

For more details on these orthogonal polynomials and especially their asymptotic properties, we refer to Vieira (1977), Delsarte, Genin, Kamp (1978), Youla and Kazanjian (1978).

In the next section, we pursue one particular (Christoffel-Darboux) property in more detail.

Problem 3.1. Inner Product Polynomial Spaces. The space, say P_m, of real (or complex) polynomials of degree less than or equal to m forms a linear vector space under addition and multiplication by real (or complex) numbers. We can define an inner product in this space by the formula

$$<p(\cdot),q(\cdot)> = \sum_{i,j=0}^{m} p_i q_i (R_m^{-1})_{ij}$$

where

$$p(z) = \sum_0^m p_k z^k \quad , \quad q(z) = \sum_0^m q_k z^k \ .$$

1. Verify that this is indeed an inner product, i.e., that it has the properties of linearity, symmetry, and nondegeneracy.
2. Show that

$$<A_k(\cdot),A_l(\cdot)> = R_k^e \delta_{kl} \ .$$

Problem 3.2. Reproducing Kernels. We have defined

$$K_m(z,\omega) = \sum_{i,j=0}^{m} z^i \omega^j (R_m^{-1})_{ij}$$
$$= \sum_{i=0}^{m} \bar{A}_i'(z)\bar{A}(\omega) \ .$$

Show that

1. $K_m(z,\cdot)\varepsilon\, P_m$
2. $<K_m(z,\cdot),p(\cdot)> = p(z)$, for every $p\ \varepsilon\ P_m$. This property leads to the name "reproducing kernel" for K_m.
3. Uniqueness. If $L(\cdot,\cdot)$ is another reproducing kernel for P_m prove that $L(z,\omega) = K_m(z,\omega)$.

3.2 The Cholesky and Christoffel-Darboux Formulas for R_m^{-1}; Computation of Quadratic Forms

We shall show in this section that the LWR algorithm and the connection with orthogonal polynomials will give us several useful representations of the block covariance matrix $\mathbf{R}_m^{-1}$, which arises as the coefficient matrix in computing quadratic and bilinear forms in several statistical applications.

We begin with the so-called Cholesky decompositions.

Theorem 3.2-A. Upper-Lower Cholesky Factorization of $\mathbf{R}_m^{-1}$. The unique UDL (upper-diagonal-lower) factorization of $\mathbf{R}_m^{-1}$ is

$$\mathbf{R}_m^{-1} = \mathbf{L}_A^T \mathbf{R}_m^{-e} \mathbf{L}_A \tag{3.15}$$

where

$\mathbf{L}_A$ = a block lower triangular matrix with $i-th$ row equal to $\mathbf{A}_i$

$\mathbf{R}_m^e$ = diagonal matrix with diagonal entries $\{R_0^e = R_0,\ldots,R_m^e\}$.

By defining

$$\mathbf{L}_{\bar{A}} = \mathbf{R}^{-e/2}\mathbf{L}_A$$

we obtain the unique upper-lower Cholesky decomposition

$$\mathbf{R}_m^{-1} = \mathbf{L}_{\bar{A}}^T \mathbf{L}_{\bar{A}} \quad . \tag{3.16}$$

Proof. Note that by the definition of the vector $\mathbf{A}_i$ we must have

$$\mathbf{A}_i \mathbf{R}_m = [0 \ \cdots \ R_i^e \ x \ \cdots \ x]$$

where the x's denote some generally nonzero entries. By "stacking up" such relations, we can see that

$$\mathbf{L}_A \mathbf{R}_m = \mathbf{R}_m^e \mathbf{U}$$

where $\mathbf{R}_m^e$ = diag $\{R_0^e,\ldots,R_m^e\}$ and $\mathbf{U}$ is an upper triangular matrix with ones on the diagonal. Therefore,

$$\mathbf{R}_m = \mathbf{L}_A^{-1} \mathbf{R}_m^e \mathbf{U} \quad .$$

Because $\mathbf{R}_m$ is symmetric, we also have

$$\mathbf{R}_m = \mathbf{U}^T\mathbf{R}_m^e\mathbf{L}_A^{-T}$$

Now these are two so-called LDU decompositions of a given positive definite matrix and by the uniqueness of such decompositions (see, e.g., Stewart (1973)), we must have

$$\mathbf{U}^T = \mathbf{L}_A^{-1}$$

Therefore

$$\mathbf{R}_m = \mathbf{L}_A^{-1}\mathbf{R}_m^e\mathbf{L}_A^{-T}$$

and

$$\mathbf{R}_m^{-1} = \mathbf{L}_A^T\mathbf{R}_m^{-e}\mathbf{L}_A$$

as claimed in the theorem. ▪

Theorem 3.2-B. Lower-Upper Cholesky Factorizations of $\mathbf{R}_m^{-1}$. The unique LDU factorization of $\mathbf{R}_m^{-1}$ is

$$\mathbf{R}_m^{-1} = \mathbf{U}_B^T\mathbf{R}_m^{-r}\mathbf{U}_B$$

where

$\mathbf{R}_m^r = \text{diag}\,\{R_0^r = R_0,\ldots,R_m^r\}$

$\mathbf{U}_B$ = a block upper triangular matrix with the $i-th$ row = $\mathbf{B}_i$

The unique lower-upper Cholesky factorization is

$$\mathbf{R}_m^{-1} = \mathbf{U}_{\overline{B}}^T\mathbf{U}_{\overline{B}} \tag{3.17}$$

where

$\mathbf{U}_{\overline{B}}$ = an upper triangular matrix with $i-th$ row = $\overline{\mathbf{B}}_i = R_i^{-r/2}\mathbf{B}_i$.

Proof. Similar to that of Theorem 3.2-A. ▪

Computation of Quadratic Forms. An important application of the Cholesky decomposition formulas is to the (recursive) computation of quadratic forms, such as

$$\Lambda_m = Y_m^T \mathbf{R}_m^{-1} Y_m \quad . \tag{3.18}$$

where Y_m is a vector, say of data samples,

$$Y_m^T = [y_0^T, \ldots, y_m^T] \quad .$$

Such quantities arise in statistical questions associated with Gaussian processes with covariance $\mathbf{R}_m$, e.g., in hypothesis testing. The decomposition (3.16) shows that we can rewrite Λ_m as

$$\Lambda_m = \|\mathbf{L}_{\bar{A}} Y_m\|^2 = \sum_{i=0}^{m} e_i^T R_i^{-e} e_i$$

where

$$e_i = \mathbf{A}_i Y_m, \text{ the } i-th \text{ row of } \mathbf{L}_A Y_m \quad .$$

We may recall that

e_i = the innovation at time i of the process
$\{y_0, \ldots, y_i\}$, with covariance matrix $\mathbf{R}_i$.

The value of this representation is that if we have one more observation, y_{m+1}, we need only to compute the innovation e_{m+1} and add in an extra term to obtain Λ_{m+1}. The reader should check, however, that such a statement is true for *any* process, once we have a Cholesky decomposition of the inverse of its covariance matrix. Our particular assumption of a *stationary* process, and hence a *Toeplitz* covariance matrix $\mathbf{R}_m$, only gives us the advantage that the innovations or equivalently the predictor coefficients $\{A_i\}$, can be more easily determined than in the general case, namely with $O(N^2)$ rather than $O(N^3)$ computations and, moreover, in a nice recursive fashion; also we can have lattice filter implementations.

One might wonder if the Toeplitzness of $\mathbf{R}_m$ cannot be reflected in a more direct way into the structure of $\mathbf{R}_m^{-1}$, and in fact the answer is yes. A particularly nice result would be to have Toeplitz Cholesky factors, say

$$\mathbf{R}_m^{-1} = T_l^T(\mathbf{a}) T_l(\mathbf{a})$$

where $T_l(\mathbf{a})$ is a lower-triangular Toeplitz matrix with first column $\mathbf{a}$. In this case,

$$\mathbf{L} = \|T_l(\mathbf{a})Y\|^2$$

where as the reader can verify, the Toeplitzness of $T_l(\mathbf{a})$ allows the product $T_l(\mathbf{a})Y_m$ to be evaluated as the convolution of the vectors $\mathbf{a}$ and Y_m. Since convolution can be done with $O(N \log N)$ operations, by FFT or fast convolution algorithms (see, e.g., Aho et al. (1978)), the Toeplitzness will allow a considerable reduction over the $O(N^2)$ operations needed to compute $\mathbf{L}_A Y_m$ for a general non-Toeplitz matrix $\mathbf{L}_A$.

In examining the formula (3.16) for $\mathbf{R}_m^{-1}$, one might wonder if it would be true that somehow $\mathbf{L}_{\bar{A}}$ could be replaced by a Toeplitz matrix, say one determined by the last row $\bar{A}_m$:

$$\begin{bmatrix} I & & & \\ \bar{A}_{m,1} & \ddots & & \\ \vdots & \ddots & \ddots & \\ \bar{A}_{m,m} & \cdots & \bar{A}_{m,1} & I \end{bmatrix} = T_l(\bar{A}_m), \text{ say} \tag{3.19}$$

Unfortunately, this is not true:

$$\mathbf{R}_m^{-1} \neq T_l^T(\bar{A}_m^T)\,T_l(\bar{A}_m) \ ,$$

But it is a remarkable fact that a "small" modification does yield a very similar correct result. This is the identity

$$\mathbf{R}_m^{-1} = T_l^T(\bar{A}_m)\,T_l(\bar{A}_m) - T_l^T(\bar{B}_m Z)\,T_l(\bar{B}_m Z) \tag{3.20}$$

where

Z = the lower shift matrix with ones on the main lower diagonal and zeros elsewhere . (3.21)

Note that

$$\bar{B}_m Z = [\bar{B}_{m,1} \ \cdots \ \bar{B}_{m,m} \ \ 0] \ . \tag{3.22}$$

The point of the formula (3.20) is that though we now have two factors,

$T_l(\bar{A}_m)$ and $T_l(\bar{B}_m Z)$, rather than just one, namely $\mathbf{L}_{\bar{A}}$, in the usual Cholesky decomposition, the two factors are *Toeplitz*. Thus for example, now

$$\Lambda_m = Y_m^T \mathbf{R}_m^{-1} Y_m$$

can be evaluated via two convolutions as

$$\Lambda_m = \| T_l(\bar{A}_m) Y_m \|^2 - T_l(\bar{B}_m Z) Y_m \|^2 \tag{3.23}$$

which for reasonably large m will still be simpler to compute than the single $\| L_{\bar{A}} Y_m \|^2$, $L_{\bar{A}}$ non-Toeplitz.

Formula (3.20) can be proved in several different ways, as can three other closely related variants. Formula (3.20) was perhaps first explicitly written down by Gohberg and Semencul (1962) for scalar Toeplitz matrices, and the extension to block Toeplitz matrices was obtained by Gohberg and Heinig (1974). Kailath, Vieira and Morf (1978) pointed out that such formulas had an interesting history, having been encountered before in several statistical problems. Moreover, they noted that when translated into polynomial form, these formulas could be immediately recognized as the classical Christoffel-Darboux formulas for orthogonal polynomials.

We shall pursue this latter direction here to obtain a proof of (3.20) and related formulas. Purely matrix-theoretic proofs can also be given.

The Reproducing Kernel. The first step is to introduce a function, called the reproducing kernel function (for reasons explained in Problem 3.2),

$$\begin{aligned} K_m(z,w) &:= V_m(z) \mathbf{R}_m^{-1} V_m^T(w) \\ &= [I \;\; zI \;\; \cdots \;\; z^m I] \mathbf{R}_m^{-1} [I \;\; wI \;\; \cdots \;\; w^m I]^T \\ &= \sum_{i,j=0}^{m} (zI)^i (\mathbf{R}_m^{-1})_{i,j} (wI)^j \;. \end{aligned} \tag{3.24}$$

Now if we use one of the Cholesky formulas for $\mathbf{R}_m^{-1}$, say (3.15), we can write

$$K_m(z,w) = \sum_0^m \bar{A}_k^T(z) \bar{A}_k(w) \;. \tag{3.25}$$

By using the LWR recursions for the $\{\bar{A}_k(z)\}$, and some algebraic rearrangement (see below), we can derive the alternative expressions

$$(1 - zw)K_m(z,w) = \overline{B}_m^T(z)\overline{B}_m(w) - zw\overline{A}_m^T(z)\overline{A}_m(w) \tag{3.26}$$

and

$$(1 - zw)K_m(z,w) = \overline{B}_{m+1}^T(z)\overline{B}_{m+1}(w) - \overline{A}_{m+1}^T(z)\overline{A}_{m+1}(w) \tag{3.27}$$

which are known as the Christoffel-Darboux formulas (see, e.g., Geronimus [1961, p. 8]. We shall see that from these two expressions we can derive four different formulas for $\mathbf{R}_m^{-1}$, including (3.20). But let us first establish (3.26) and (3.27).

Proof of the Christoffel-Darboux Formulas. Let us temporarily introduce a quantity

$$F_m(z,w) := [\overline{A}_m^T(z) \;\; \overline{B}_m^T(z)]J[\overline{A}_m^T(w) \;\; \overline{B}_m^T(w)]^T$$

where J is the block diagonal matrix

$$J = [I_p \oplus - I_p] \; .$$

We now recall the LWR recursion

$$\begin{bmatrix}\overline{A}_m(z)\\ \overline{B}_m(z)\end{bmatrix} = \overline{\Theta}_m \begin{bmatrix}z\overline{A}_{m-1}(z)\\ \overline{B}_{m-1}(z)\end{bmatrix}$$

where

$$\overline{\Theta}_m^T \; J \; \overline{\Theta}_m = J = \overline{\Theta}_m \; J \; \overline{\Theta}_m^T \; .$$

Therefore we can rewrite $F_m(z,w)$ as

$$\begin{aligned}F_m(z,w) &:= \overline{A}_m^T(z)\overline{A}_m(w) - \overline{B}_m^T(z)\overline{B}_m(w) &\quad (3.28)\\ &= zw\overline{A}_{m-1}^T(z)\overline{A}_{m-1}(w) - \overline{B}_{m-1}(z)\overline{B}_{m-1}(w) &\\ &= F_{m-1}(z,w) + (zw - 1)\overline{A}_m^T(z)\overline{A}_{m-1}(w) &\quad (3.29)\end{aligned}$$

Continuing this downwards recursion, we obtain the relation

$$F_m(z,w) = F_0(z,w) + (zw - 1)\sum_0^{m-1} \overline{A}_k^T(z)\overline{A}_k(w)$$

But

$$\bar{A}_0(z) = R_0^{-1/2} = \bar{B}_0(z)$$

and hence

$$F_0(z,w) = 0 \ .$$

Therefore, recalling (3.25), we note that

$$(1 - zw)K_{m-1}(z,w) = -F_m(z,w)$$

or equivalently that

$$(1 - zw)K_m(z,w) = -F_{m+1}(z,w) \ .$$

Using the two expressions (3.28) and (3.29) for $F_m(z,w)$ then gives us the Christoffel-Darboux formulas (3.26) and (3.27). ▪

We now return to the result we were seeking.

Proof of (3.14). To obtain (3.20) we first rewrite (3.28) as

$$\left[1 - \frac{1}{zw}\right]K_m(z,w) = \bar{A}_m^T(z)\bar{A}_m(w) - z^{-1}\bar{B}_m^T(z)w^{-1}\bar{B}_m(w) \tag{3.30}$$

Then the reader can check that comparing coefficients of $\{z^i w^j,\ i,j=0,1,\ldots,m-1\}$ on both sides of (3.30) will give us the matrix equation

$$\mathbf{R}_m^{-1} - Z^T\mathbf{R}_m^{-1}Z = \bar{A}_m^T\bar{A}_m - (\bar{B}_m Z)^T(B_m Z) \tag{3.31}$$

where Z is the lower-shift matrix defined earlier in (3.21). Equation (3.31) is a so-called Lyapunov equation for $\mathbf{R}_m^{-1}$ and solving it by successive substitution gives the formula (3.20). ▪

However, it should now be clear that there are some other ways in which $\mathbf{R}_m^{-1}$ could be expressed. For example, we could just compare coefficients on both sides of (3.26) to obtain the matrix equation

$$\mathbf{R}_m^{-1} - \bar{Z}\mathbf{R}_m^{-1}\bar{Z}^T = \bar{B}_m^T\bar{B}_m - (\bar{A}_m Z)^T(\bar{A}_m Z) \tag{3.32}$$

which has the unique solution

$$\mathbf{R}_m^{-1} = T_u^T(\overline{B}_m)T_u(\overline{B}_m) - T_u^T(A_m Z^T)T_u(A_m Z^T) \tag{3.33}$$

where

$T_u(\overline{B}_w)$ = an upper-triangular block Toeplitz matrix with first row $\overline{B}_m$.

This formula is the modified version of the Cholesky formula (3.17). We have not yet used the Christoffel-Darboux formula (3.27). Working with it, we can obtain the formulas

$$\mathbf{R}_m^{-1} = T_l^T(\overline{A}_{m+1}^o)T_l(\overline{A}_{m+1}^o) - T_l^T(\overline{B}_{m+1}^o)T_l(\overline{B}_{m+1}^o) \tag{3.34}$$

and

$$\mathbf{R}_m^{-1} = T_u^T(\overline{B}_{m+1}^{oo})T_u(\overline{B}_{m+1}^{oo}) - T_u^T(\overline{A}_{m+1}^{oo})T_u(\overline{A}_{m+1}^{oo}) \tag{3.35}$$

where the superscript 'o' denotes deletion of the first (block) row entry, viz.,

$$\overline{B}_{m+1}^o := [\overline{B}_{m+1,1} \cdots \overline{B}_{m+1,m+1}] \, , \; \overline{A}_{m+1}^o := [\overline{A}_{m+1,m} \cdots I] \tag{3.36}$$

and the superscript 'oo' denotes deletion of the last (block) entry, viz.,

$$\overline{B}_{m+1}^{oo} := [I \cdots B_{m+1,m}] \, , \; A_{m+1}^{oo} := [\overline{A}_{m+1,m+1} \cdots A_{m+1,1}] \tag{3.37}$$

4. ADAPTIVE FILTERING AND NONSTATIONARY PROCESSES

All the work described so far assumed that the covariance lags $\{R(0),R(1),\ldots\}$ of the stationary random process were available. But often we are only given a finite data sample, $\{y(0),\ldots,y(N)\}$, drawn from an underlying assumed stationary process, $\{y_0,\ldots,y_N\}$. Subscripts will be used to denote random variables and parentheses to denote sample values taken by these variables.

One method of proceeding in this case is to use the data sample to obtain estimates of the covariance lags as

$$\hat{R}(k) = \frac{1}{N}\sum_{t=0}^{N} y(t)y(t+k) \ ,$$

with the assumption that data outside the interval $[0,N]$ is zero, i.e.,

$$y(k) = 0 \ , \quad k < 0$$

and

$$y(N + l) = 0 \ , \quad l > 0 \ .$$

This assumption allows application of the stationary process theory, but can often be quite unrealistic, especially for short data samples. Therefore, in the speech community, the so-called "covariance" method was proposed in which no data outside the interval $(0,N)$ was used. In this case, the "estimated" covariance matrix of the data could no longer be taken as stationary, and there was no longer any guarantee that the optimum prediction filter could be represented in lattice form or even computed efficiently.

4.1 Least-Squares Fitting of AR Models to Data

To be more specific, consider the following prototypical adaptive filtering problem: give a set of observations $\{y_0,\ldots,y_T\}$, we wish to choose a set of coefficients $\{a_1,\ldots,a_n\}$, $n < T$, such that

$$\sum_t e_{n,t}^2 = \text{minimum}$$

where $e_{n,t}$ is the error residual in trying to predict y_t from n past values

$$e_{n,t} = y_t + a_1 y_{t-1} + \cdots + a_n y_{t-n} \ .$$

The problem arises when $0 \le t < n$, because then we do not have all n previous values. One choice is to only compute residuals for $t \ge n$, but for a number of reasons we often accommodate the case $t < n$, by assuming that the missing data $\{y_{-n}, y_{-n+1},\ldots,y_{-1}\}$ are zero. One reason is the hope that for $T \gg n$, the arbitrariness of this initial assumption will not affect the overall solution very much. This is not unreasonable, but sometimes a potentially more drastic assumption is made: namely residuals are computed for $T < t \le T + n$ by assuming that the "future" date $\{y_{T+1},\ldots,y_{T+n}\}$ is also zero. We shall see the reason for this presently, but first let us display the set of residuals we have defined in matrix form as

$$\mathbf{e} = \mathbf{Y}\mathbf{a}$$

and

$$\mathbf{e}^T = [e_{n,0} \quad e_{n,1} \cdots e_{n,T+n}]$$

where

$$\mathbf{a} = [1 \quad a_1 \ldots a_n]$$

$$\mathbf{Y} = \begin{bmatrix} y_0 & & & & \\ y_1 & y_0 & & & \\ \vdots & & \cdot & & \\ \vdots & & & \cdot & \\ - & - & - & - & - \\ y_n & \cdot & \cdot & \cdot & y_0 \\ y_t & \cdot & \cdot & \cdot & y_{t-n} \\ y_T & \cdot & \cdot & \cdot & y_{T-n} \\ - & - & - & - & - \\ & & \cdot & & \\ & & & \cdot & \\ & & & & y_T \end{bmatrix}$$

The least-squares solution, $\hat{a}$, that minimizes $\|e\|^2$ can now be computed as the solution of the so called "normal" equations

$$[\mathbf{Y}^T\mathbf{Y}]\mathbf{a} = [R_n^e \ 0 \ \cdots \ 0]^T \ ,$$

where $R_n^e = \min\|e\|^2$. We can now see one reason for extending the residuals $\{e_{n,t}\}$ to both sides of the given sample $\{y_0,\ldots,y_T\}$: only in this case, will the coefficient matrix $Y^T Y$ be Toeplitz, and in particular be the sample covariance matrix as described before; under this assumption, all the nice features (fast computational algorithm, lattice filter implementations) of the stationary stochastic theory become available.

However, in many problems it is quite unreasonable to assume that missing data is zero and therefore we may wish to only use the residuals $\{e_{n,n},\ldots,e_{n,T}\}$; or perhaps allow only assumptions on past missing data and thus use $\{e_{n,0},\ldots,e_{n,T}\}$. In either case, we shall no longer have $Y^T Y$ as a Toeplitz matrix, and therefore it would seem that the nice results of the stationary theory will no longer apply.

This would be disappointing except that closer examination shows that the situation can be saved in large measure by introducing the concept of ***displacement rank*** α (or ***α-stationarity***) (Kailath et al. (1979a,b)). The number α (and the displacement inertia) provide, roughly speaking, a measure of nonstationary of the process; for such processes, the reflection coefficients can be updated with time according to fairly simple formulas, which will have a complexity $O(N^2\alpha)$. It turns out that for the least-squares problem, we have $\alpha = 2$ for the fully windowed (Toeplitz) case, $\alpha = 3$ for the prewindowed case and $\alpha = 4$ for the nonwindowed (or covariance) case.

These exact results should be contrasted with the earlier work, where the nice properties of the lattice filter had led earlier investigators to ***arbitrarily impose the lattice structure*** and to compute the reflection coefficients under a variety of criteria involving the so-called "forward" and "backward" residuals obtained by passing the data samples $\{y(0),...,y(N)\}$ through the lattice filter. The first such calculations were given by Burg (1967), and a somewhat wider class of solutions were described by Griffiths (1977) and Makhoul (1978). The results obtained were all admittedly suboptimal: First, because of the a priori assumption of a lattice filter and secondly because, except in the work of Burg (1968), the methods proposed to compute the lattice coefficients did not minimize any particular optimization criterion (also see the comments in the Appendix of Morf Vieira, Lee and Kailath (1978)). Consequently, few results were available on the properties (e.g, bias, consistency, or accuracy) of the estimates used.

Therefore it came as a surprise when Morf, Vieira and Lee (1977) produced a lattice structure, with adaptively updated time-variant reflection coefficients, that at each time instant exactly minimized a particular least-squares criterion. This important result led to a host of refinements, generalizations and applications. Several authors, e.g., Friedlander (1982a,b), Samson (1982), Ljung and Soderstrom (1983), Lee (1980), Lee et al. (1981), Porat et al. (1982), Kailath (1982), Lev-Ari (1983), Lev-Ari and Kailath (1984), Lev-Ari, Cioffi and Kailath (1984), Cioffi (1984) have contributed to a simple unifying presentation of the several results now known on lattice filters and related fast adaptive algorithms.

We shall show first that an exact lattice filter structure can be obtained for any process, stationary or not; the price to be paid is that we will have time-variant gains. In the stationary case, the gains will essentially be constant (going from zero to a fixed value at a predetermined time); and moreover can be computed with effort $O(N^2)$ rather than $O(N^3)$ as in the general case. In the nonstationary case, the assumption of α-stationary will reduce the effort to $O(N^2\alpha)$.

It will be useful to introduce some notations and definitions before proceeding further with our discussion.

We study an indexed (by time) collection of "vectors" $\{y_i, i=0,1,\ldots,t,\ldots\}$ in some Hilbert space, with given inner products

$$<y_i, y_j> = R_{i,j} \ .$$

and such that the Gramian matrix

$$\mathbf{R} = [R_{i,j}]$$

is a symmetric positive definite matrix. The $\{y_i\}$ could be random variables in a probability space where the inner product is defined by expectation,

$$<y_i, y_j> = E\{y_i y_j^*\} \ .$$

The asterisk denotes complex conjugation (or Hermitian transpose, when applied to matrix quantities). However, this stochastic interpretation is not at all necessary, a fact that will be crucial in our later analysis of deterministic least-squares problems.

A single element in the Hilbert space will be called a ***vector***, and a collection of such elements -- a ***vector aggregate*** (VA). Thus, if $\{y_i, i \geq 0\}$ is a *multichannel* process, each y_i is itself a vector aggregate.

The inner product $<x,x>$ will be called the *Gramian* of the vector aggregate x. A vector aggregate whose Gramian is the identity matrix will be called *normalized* and denoted by an overbar. The elements of a normalized vector aggregate are mutually orthogonal and have unit norm.

4.2 Lattice Filters for General Nonstationary Processes

The structure of the collection of vectors $\{y_i\}$ will be explored by studying the family of finite-order *residuals*

$$e_{m,t} := y_t - \widehat{y}_{t\,|\,U} \ ,$$

where U is the vector aggregate

$$U = \{y_{t-m}, \ldots, y_{t-1}\}$$

and

$\hat{y}_{t|U}$ = the projection of y_t on the space spanned by the set U .

We shall call

$e_{m,t}$ = the $m-th$ order forward residual at time t .

The structure of the family of residuals, $\{e_{m,t}, m \leq t, t = 0,1,...\}$, can be exposed by first seeking, for each fixed t, how to determine order-updates of these residuals, i.e., knowing $e_{m,t}$ we shall try to determine

$$e_{m+1,t} = y_t - \hat{y}_{t|\{U,y_{t-m-1}\}}$$

in some convenient way. It is reasonable to seek to use our knowledge of $e_{m,t}$ by making the orthogonal decomposition

$$\{U,y_{t-m-1}\} = U \oplus \{r_{m,t-1}\}$$

where

$r_{m,t-1} := y_{t-m-1} - \hat{y}_{t-m-1|U}$, the $m-th$ order backward residual at $t - m - 1$.

Then we can write

$$\begin{aligned}\hat{y}_{t|\{U,y_{t-m-1}\}} &= \hat{y}_{t|U} + \hat{y}_{t|r_{m,t-1}} \\ &= \hat{y}_{t|U} + <y_t,\bar{r}_{m,t-1}>\bar{r}_{m,t-1} .\end{aligned}$$

It follows that we can write

$$\begin{aligned}e_{m+1,t} &= e_{m,t} - <y_t,\bar{r}_{m,t-1}>\bar{r}_{m,t-1} \\ &= e_{m,t} - <e_{m,t},\bar{r}_{m,t-1}>\bar{r}_{m,t-1} \\ &= \|e_{m,t}\|(\bar{e}_{m,t} - <\bar{e}_{m,t},\bar{r}_{m,t-1}>\bar{r}_{m,t-1}) \\ &= \|e_{m,t}\|(\bar{e}_{m,t} - k_{m+1,t}\bar{r}_{m,t-1})\end{aligned}$$

where we used the fact that $r_{m,t-1} \perp U$ to obtain the second equality, and where we defined the generalized "reflection coefficients"

$$k_{m+1,t} := \langle \bar{e}_{m,t}, \bar{r}_{m,t-1} \rangle \quad . \tag{4.1}$$

To also normalize $e_{m+1,t}$, we need to compute its norm, for which we note that the last equality yields

$$\|e_{m+1,t}\|^2 = \|e_{m,t}\|(I - k_{m+1,t}k^*_{m+1,t})\|e_{m,t}\|^*$$

so that

$$\|e_{m+1,t}\| = \|e_{m,t}\|(I - k_{m+1,t}k^*_{m+1,t})^{1/2} \tag{4.2}$$

Then we can rewrite the order-update formula for the forward residuals in normalized form as

$$\bar{e}_{m+1,t} = (I - k_{m+1,t}k^*_{m+1,t})^{-1/2}(\bar{e}_{m,t} - k_{m+1,t}\bar{r}_{m,t-1}) \tag{4.3a}$$

This, of course, leaves us with the problem of getting $\bar{r}_{m,t-1}$. But a similar recursion can be set up for it. Thus, and more briefly, we can write

$$\begin{aligned} r_{m+1,t} &= y_{t-m-1} - \hat{y}_{t-m-1|\{U,y_t\}} \\ &= r_{m,t-1} - \langle y_{t-m-1}, \bar{e}_{m,t} \rangle \bar{e}_{m,t} \\ &= \|r_{m,t-1}\|(\bar{r}_{m,t-1} - \langle \bar{r}_{m,t-1}, \bar{e}_{m,t} \rangle \bar{e}_{m,t}) \\ &= \|r_{m,t-1}\|(\bar{r}_{m,t-1} - k^*_{m+1,t}\bar{e}_{m,t}) \end{aligned}$$

This yields

$$\|r_{m+1,t}\| = \|r_{m,t-1}\|(I - k^*_{m+1,t}k_{m+1,t})^{1/2}$$

so that

$$\bar{r}_{m+1,t} = (I - k^*_{m+1,t}k_{m+1,t})^{-1/2}(\bar{r}_{m,t-1} - k_{m+1,t}\bar{e}_{m,t}) \tag{4.3b}$$

The recursions (4.3) can be written as

$$\begin{bmatrix} \bar{e}_{m,t} \\ \\ \bar{r}_{m,t} \end{bmatrix} = \Theta\{k_m(t)\} \begin{bmatrix} \bar{e}_{m-1,t} \\ \\ \bar{r}_{m-1,t-1} \end{bmatrix} \tag{4.4}$$

where

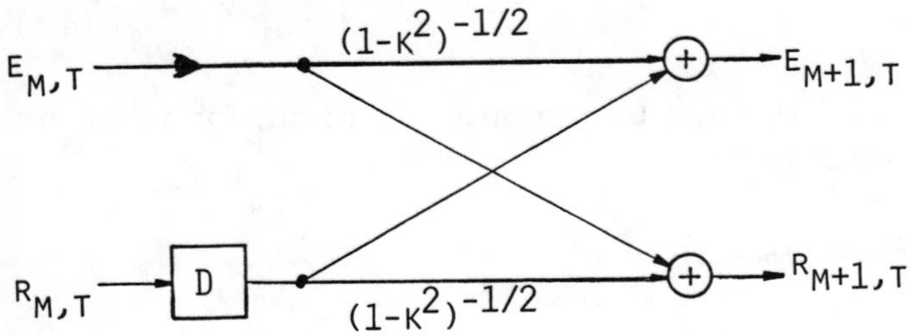

(a) Basic Lattice Section

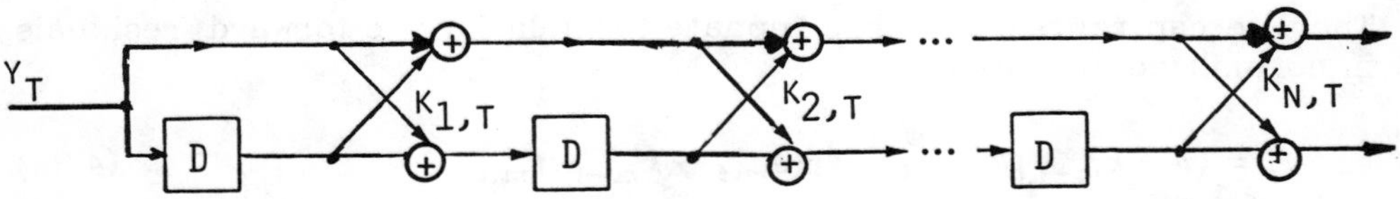

(b) Lattice filter with time-varying sections

FIGURE 6: General time-variant lattice filter for nonstationary processes.

$$\Theta\{k\} = \begin{bmatrix} (I - kk^*)^{1/2} & 0 \\ 0 & (I - k^*k)^{1/2} \end{bmatrix}^{-1} \begin{bmatrix} I & -k \\ -k & I \end{bmatrix} \tag{4.5}$$

As in the stationary case, we can combine the relations (4.4) to get a cascaded lattice structure, with each lattice section specified by a time-variant reflection coefficient, $k_m(t)$ (see Figure 6). The inputs to the first section will be

$$\bar{e}_{0,t} = \bar{y}_t = \|y_t\|^{-1} y_t = \bar{r}_{0,t}$$

and the first section will have reflection coefficient

$$k_1(t) = \langle \bar{e}_{0,t}, \bar{r}_{0,t-1} \rangle = \langle \bar{y}_t, \bar{y}_{t-1} \rangle = R_{t,t}^{-1/2} R_{t,t-1} R_{t-1,t-1}^{-1/2}$$

Also note that by Schwarz's inequality,

$$\|k_{m+1}(t)\| \leq \|\bar{e}_{m,t}\| \, \|\bar{r}_{m,t-1}\|^* = 1 \ . \tag{4.6}$$

The Stationary Case. We can recover our previous results by imposing the assumption of stationarity, which means that the inner products are invariant under shift in the indices, i.e.,

$$<y_i, y_j> = <y_{i-l}, y_{j-l}> , \quad l = 1,2,... \tag{4.7}$$

In this case, it follows that the reflection coefficients $\{k_{m+1}(t)\}$ are *independent* of t, so that we have a cascade lattice filter with essentially constant or *time-invariant sections*.

$$k_m(t) = \begin{cases} 0 , & t < m \\ k_m , & t \geq m \end{cases} \tag{4.8}$$

4.3 Simplifications in the Near Stationary Case

In the nonstationary case, when (4.7) does not hold, it seems unavoidable that the reflection coefficients must be time-variant. However, it is reasonable that the complexity of the time-variation should depend upon the degree of nonstationarity, measured in some sense. It turns out that the concept of displacement rank provides a meaningful way of classifying nonstationary processes, in that for a process with displacement rank α, each reflection coefficient can be updated with $O(\alpha)$ multiplications. For N observations and N reflection coefficients, this requires $O(N^2\alpha)$ multiplications as compared to the $O(N^3)$ that would be required if we just used the general formulas given above without attention to the displacement rank.

The time-update formulas can be compactly stated:

$$k_{m+1}(t) = (1 - \eta_m(t)\eta_m(t)^*)^{1/2} k_{m+1}(t)(1 - \mu_m(t-1)\mu_m(t-1)^*)^{1/2} + \\ + \eta_m(t)\mu_m^*(t-1) \tag{4.9a}$$

where $\{\eta, \mu\}$ are α-dimensional row vectors obeying the recursions

$$\eta_{m+1}(t) = F\{\eta_m(t), k_{m+1}(t), \mu_m(t-1)\} \tag{4.9b}$$

$$\mu_{m+1}(t) = F\{\mu_m(t-1), k_{m+1}^*(t), \eta_m(t)\} \tag{4.9c}$$

where the function $F(\ )$ is defined as

$$F\{A,B,C\} = (1 - BB^*)^{-1/2}(A - BC^*)(1 - CC^*)^{-*/2} \tag{4.9d}$$

In fact, we may note that the first equation is a rearrangement of the formula

$$k_{m+1,t-1} = F\{k_{m+1,t}, \eta_{m,t}, \mu_{m,t-1}\}$$
$$= (I - \eta\eta^*)^{-1/2}(k_{m+1,t} - \eta\mu^*)(I - \mu\mu^*)^{-*/2} \tag{4.10}$$

where the subscripts and arguments for μ and η have been omitted for simplicity.

The displacement rank itself is given as

$$\alpha = \text{rank of } \{\mathbf{R} - Z\mathbf{R}Z^*\} \tag{4.11}$$

where $\mathbf{R}$ is the covariance or Gramian matrix and Z is the "shift" matrix with ones on the first subdiagonal and zeros elsewhere.

The three recursions (4.9) can be pictorially represented as in Figure 7. The block named D denotes a "delay operator" and the numbers indicated denote implementation of the respective equations.

Fixed-Order Transversal Filters. The above lattice representations of the estimation procedure have the desirable feature that they generate estimated residuals of all orders up to a given maximum. For instance in the linear prediction example, the residuals $e_{m,t}$ for $m = 0,\ldots,M$ represent errors in the one step prediction from a growing set of past y values. However, in many adaptive-filtering applications (such as channel equalization, echo cancellation, system identification, etc.), the order of the desired estimate is known or upper-bounded a priori. In these cases, the lower order residuals are unnecessary, and significant computational and implementational simplifications result when the adaptive filter directly computes the maximum, fixed-order residuals recursively in time. It has been shown (see Cioffi and Kailath (1984)) that we also have lattice-like recursions for the impulse responses of fixed-order transversal filters.

$$\begin{bmatrix} A_t \\ D_t \end{bmatrix} = \begin{bmatrix} (I - \eta(t)\eta^*(t))^{1/2} & \eta^*(t) \\ +\lambda\eta^*(t) & (I - \eta^*(t)\eta(t))^{1/2} \end{bmatrix} \begin{bmatrix} A_{t-1} \\ [C_{t-1} | 0] \end{bmatrix}$$

$$\begin{bmatrix} B_t \\ [0 | C_t] \end{bmatrix} = \begin{bmatrix} (I - \mu(t)\mu^*(t))^{-1/2} & (I - \mu(t)\mu^*(t))^{-1/2}\mu(t) \\ (I - \mu(t)\mu^*(t))^{-1/2}\mu(t)^* & (I - \mu^*(t)\mu(t))^{-1/2} \end{bmatrix} \begin{bmatrix} B_{t-1} \\ D_t \end{bmatrix}$$

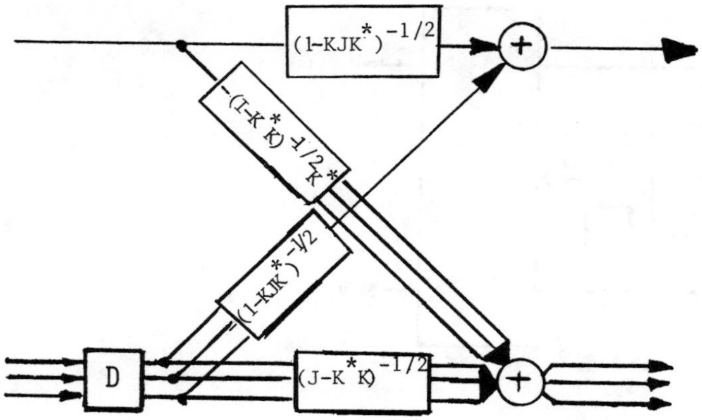

FIGURE 7: Representation of the time-update calculations.

Note that the quantities $\eta(t)$ and $\mu(t)$ act like reflection coefficients in these lattice-like recursions. $\{A_t, B_t\}$ are the impulse responses at time t of the filters producing the fixed order residuals and $\{C_t, D_t\}$ are certain auxiliary filters.

For reasons of space, we shall not elaborate here how these general results can be specialized to obtain several of the adaptive lattice- and transversal-filter algorithms known in the literature, e.g., the so-called prewindowed algorithms (Lee et al. (1981)), the prediction residuals form of Satorius and Pack (1979), the covariance algorithms (Porat et al. (1982)), the so-called fast Kalman (fixed order transversal filter) algorithms (Cioffi and Kailath (1984)), and so on. A detailed discussion can be found in the paper of Lev-Ari, Kailath and Cioffi (1984).

Time-Invariant Implementations. Perhaps surprisingly, the displacement rank can be used to reduce the complexity in a different way--by allowing completely time-invariant gains but of a higher dimension. That is, we shall still have a cascade of lattice sections, but each section will be defined by an $(\alpha-1)$-dimensional row vector rather than a scalar--see Figure 8. These row vectors will be called generalized Schur coefficients.

As shown, the input to the 'Delay' is a $(\alpha-1)\times 1$ signal; the 'Delay' itself is $z^{-1}I_{\alpha-1}$, where $I_{\alpha-1}$ denotes an identity matrix of dimension $(\alpha-1)\times(\alpha-1)$. It can be replaced by $\left(\frac{z-\alpha_i}{1-\alpha_i^* z}\right)^{-1} \times I_{\alpha-1}$, to yield an ARMA lattice structure. The generalized Schur coefficients $\{K_i\}$ are row vectors of dimension $\alpha-1$, satisfying

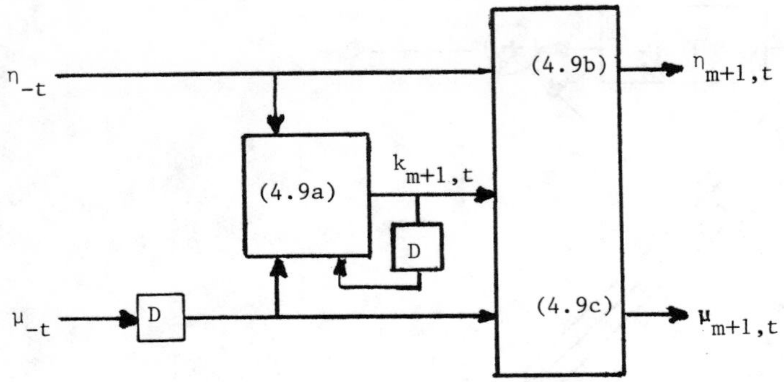

FIGURE 8: Generalized time-invariant lattice section.

$1 - K_i J K_i^* \geq 0$

where α is the displacement rank of the covariance matrix $\mathbf{R}$ of and J is defined through the signature of $\mathbf{R}$ as:

$\text{sgn}\{\mathbf{R} - Z\ \mathbf{R}\ Z^*\} = \{1, -J\}$

A detailed theory of such constant gain lattice filters can be found in Lev-Ari (1083) and Lev-Ari and Kailath (1984).

REFERENCES

1. Ahmed, H. M., Delosme, J. M., and Morf, M., Highly Concurrent Computing Structures for Digital Signal Processing and Matrix Arithmetic, *IEEE Computer Magazine*, February 1982.

2 Akhiezer, N. I., and Krein, M. G., Some Questions in the Theory of Moments, (in Russian, 1938). English Translation, *Amer. Math. Soc.*, Providence, R.I., 1962.

3. Akhiezer, N. I., *The Classical Moment Problem and Some Related Questions in Analysis*, (in Russian, 1961), Hafner, N.Y., 1965.

4. Anderson, T. W., *Multivariate Analysis*, J. Wiley & Sons, N.Y., 1958.

5. Bruckstein, A., and Kailath, T., Spatio-Temporal Scattering and Inverse Problems, Tech. Rept., Inform. Systems Lab., Stanford University, 1983.

6a. Burg, J. P., Maximum Entropy Spectral Analysis, paper presented at the 37th Ann. Int. Meeting, Soc. of Explor. Geophys., Oklahoma City, OK, October 31, 1967.

6b. Burg, J. P., A New Analysis Technique for Time Series Data, NATO Adv. Study Inst. on Signal Processing, 1968. Both are reprinted in *Modern Spectrum Analysis*, D. G. Childers, ed., IEEE Press, 1978.

7: Burg, J. P., Maximum Entropy Spectral Analysis, Ph.D. Dissertation, Stanford University, Stanford, CA, 1975.

8. Cioffi, J., and Kailath, T., Fast, Recursive-Least-Squares Transversal Filters for Adaptive Filtering, *IEEE Trans. ASSP*, 1984.

9. Claerbout, J. F., *Fundamentals of Geophysical Data Processing with Applications to Petroleum Prospecting*, McGraw-Hill, New York, 1976.

10. Delsarte, P., Genin Y., and Kamp, Y., Orthogonal Polynomial Matrices on the Unit Circle, *IEEE Trans. on Circuits and Systems*, vol. CAS-25, no. 3, pp. 149-160, 1978.

11. Delsarte, P., Genin, Y., and Kamp, Y., A Method of Matrix Inverse Triangular Decomposition, Based on Contiguous Principal Submatrices, *J. Linear Algebra and its Applications*, vol. 31, pp. 199-212, 1980.

12. Despain, A. M., Fourier Transform Computers Using CORDIC Iterations, *IEEE Trans. Comput.*, vol. C-23, pp. 993-1001, Oct 1974.

13. Dewilde, P., Vieira, A. C., and Kailath, T., On a Generalized Szegö-Levinson Realization Algorithm for Optimal Linear Predictors Based on a Network Synthesis Approach, *IEEE Trans. Circuits Systems*, vol. CAS-25, no. 9, pp. 663-675, 1978.

14. Doob, J. L., *Stochastic Processes*, Wiley, New York, 1953.

15. Durbin, J., The Fitting of Time-Series Models, *Rev. Int. Inst. Stat.*, 28, pp. 233-244, 1960.

16. Flanagan, J. L., *Speech Analysis Synthesis and Perception*, Springer-Verlag, New York, 1972.

17. Friedlander, B., Lattice Filters for Adaptive Processing, *Proc. IEEE*, vol. 70, no. 8, pp. 829-867, August 1982a.

18. Friedlander, B., Lattice Methods for Spectral Estimation, *Proc. IEEE*, vol. 70, no. 9, pp. 990-1017, September 1982b.

19. Geronimus, L. Y., Polynomials Orthogonal on a Circle and Their Applications, (in Russian, 1948), *Amer. Math. Soc. Translations*, Series One, vol. 3, pp. 1-78, 1954.

20. Geronimus, L. Y., On Asymptotic Properties of Polynomials Orthogonal on the Unit Circle, (in Russian, 1950), *Amer. Math. Soc. Translations*, Series One, vol. 3, pp. 79-106, 1954.

21. Geronimus, L. Y., *Orthogonal Polynomials*, (in Russian, 1958), Consultant's Bureau, New York, 1961.

22. Gohberg, I. C., and Semencul, A., Toeplitz Matrices Composed of the Fourier Coefficients of Functions with Discontinuities of Almost Periodic Type, *Mat. Issled.*, 5,4 (in Russian), 1970.

23. Gohberg, I. C., and Heinig, G., Inversion of Finite Toeplitz Matrices with Entries From Noncommutative Algebra, *Revue Romaine de Math. Pures et Appliquees*, 19, pp. 623-665, 1974.

24. Goupillaud, P. L., An Approach to Inverse Filtering of Near-Surface Layer Effects from Seismic Record, *Geophysics*, vol. 26, pp. 754-760, 1961.

25. Gray, A. H., Jr., and Markel, J. D., A Normalized Digital Filter Structure, *IEEE Trans. Acoustics, Speech, Signal Proc.*, vol. ASSP-23, no. 3, pp. 268-277, 1975.

26. Grenander, U., and Szegö, G., *Toeplitz Forms and Their Applications*, University of California Press, Berkeley, CA, 1958.

27. Griffiths, L. J., A Continuously Adaptive Filter Implemented As A Lattice Structure, *Proc. IEEE Int. Conf. Acoust., Speech, Signal Proc.*, Hartford, CN, pp. 683-686, May 1977.

28. Haviland, G. L., and Tuszynski, A. A., A CORDIC Arithmetic Processor Chip, *IEEE Trans. Computers*, vol. C-29, no. 2, pp. 68-79, Feb. 1980.

29. Itakura F., and Saito, S., Digital Filtering Techniques for Speech Analysis and Synthesis, in *Proc. 7th Int. Cong. Acoust.*, Budapest, Paper 25-C-1, pp. 261-264, 1971.

30. Jury, E. I., *Inners and Stability of Dynamic Systems*, J. Wiley, N.Y., 1974.

31. Kailath, T., *Linear Systems*, Prentice-Hall, Inc., N.J., xxi + 682 pages, 1980.

32. T. Kailath, Time-Variant and Time-Invariant Lattice Filters for Nonstationary Processes, *Proc. Fast Algorithms for Linear Dynamical Systems*, pp. 417-464, Aussois, France, Sept. 21-25, 1981. Reprinted *Outils et Modeles Mathematiques Pour L'Automatique, L'Analyse De Systems Et Le Traitement Du Signal*, vol. 2, ed. I. D. Landau, 417-464, CNRS, France, 1982.

33. Kailath, T., *Lectures on Wiener and Kalman Filtering*, CISM Monographs, no. 140, Springer-Verlag, iv + 165 pages, 1981.

34. Kailath, T., Kung, S. Y., and Morf, M., Displacement Ranks of Matrices and Linear Equations, *Journal of Mathematical Analysis and Applications*, vol. 68, no. 2, pp. 395-407, April 1979.

35. Kailath, T., Kung, S. Y., and Morf, M., Displacement Ranks of a Matrix, *Bull. Amer. Math. Soc.*, vol. 1, no. 5, pp. 769-773, Sept. 1979.

36. Kailath, T., Vieira, A., and Morf, M., Inverses of Toeplitz Operators, Innovations, and Orthogonal Polynomials, *SIAM Review*, vol. 20, no. 1, pp. 1006-1019, 1978.

37. Kelly, J. L., Jr., and Lochbaum, C. C., Speech Synthesis, *Proc. 4th International Congress Acoust.*, Paper G42, pp. 1-4, 1962.

38. Kung, S. Y., and Hu, H., A Highly Concurrent Algorithm and Pipelined Architecture for Solving Toeplitz Systems, *IEEE Trans. on Acoust., Speech and Signal Processing*, vol. ASSP-31, no. 1, pp. 66-76, 1983.

39. Lee, D. T. L., Canonical Ladder Form Realizations and Fast Estimation Algorithms, Ph.D. Dissertation, Department of Electrical Engineering, Stanford University, Stanford, CA., August 1980.

40. Lee, D. T. L., Morf, M., and Friedlander, B., Recursive Square-Root Ladder Estimation Algorithms, *IEEE Joint Special Issue on Adaptive Signal Processing*, vol. CAS-28, no. 6, pp. 627-641, June 1981.

41. Le Roux, J., and Gueguen, C., A Fixed Point Computation of Partial Correlation Coefficients, *IEEE Trans. Acoust., Speech and Signal Proc.*, vol. ASSP-25, pp. 257-259, 1977.

42. Lev-Ari, H., Parameterization and Modeling of Nonstationary Processes, Ph.D. Dissertation, Stanford University, Stanford, CA, 1983.

43. Lev-Ari, H., Kailath, T., and Cioffi, J., Least-Squares Adaptive-Lattice and Transversal Filters: A Unified Geometric Theory, *IEEE Trans. on Inform. Thy.*, (special issue), March 1984.

44. Lev-Ari, H., and Kailath, T., Lattice Filter Parametrization and Modeling of Nonstationary Processes, *IEEE Trans. Inform. Thy.*, vol. IT-30, no. 1, pp. 2-16, January 1984.

45. Levinson, N., The Wiener RMS (Root-Mean-Square) Error Criterion in Filter Design and Prediction, *J. Math. Phys.*, vol. 25, no. 4, pp. 261-278, January 1947.

46. Ljung, L., and Soderstrom, T., *Analysis of Recursive Identification Algorithms*, MIT Press, 1983.

47. Makhoul, J., Stable and Efficient Lattice Methods for Linear Prediction, *IEEE Trans. Acoust., Speech and Signal Proc.*, vol. ASSP-25, pp. 423-428, 1977.

48. Makhoul, J., A Class of All-Zero Lattice Digital Filters: Properties and Applications, *IEEE Trans. Acoust., Speech, Signal Proc.*, vol. ASSP-26, no. 4, pp. 304-314, Aug. 1978.

49. Markel, J. D., and Gray, A. H., Jr., *Linear Prediction of Speech*, Springer-Verlag, New York, 1976.

50. Marple, S. L., Jr., A New Autoregressive Spectrum Analysis Algorithm, *IEEE Trans. Acoust., Speech, Signal Proc.*, vol. ASSP 28, no. 4, pp. 441-454, August 1980.

51. Mason, S.H., and Zimmerman, H. A., *Electronic Circuits, Signals and Systems*, J. Wiley, N.Y., 1960.

52. Morf, M., Lee, D. T., Nickolls, J. R., and Vieira, A., A Classification of Algorithms for ARMA Models and Ladder Realizations, *Proc. 1977 IEEE Conf. on Acoust. Speech, Signal Processing*, Hartford, CT, pp. 13-19, April 1977. Reprinted in *Modern Spectrum Analysis*, ed. D. G. Childers, IEEE Press Series, Modern Spectrum Analysis, N.Y., pp. 262-268, 1978.

53. Morf, M., Vieira, A., and Kailath, T., Covariance Characterization by Partial Autocorrelation Matrices, *Annals Stat.* vol. 6, pp. 643-648, May 1978.

54. Morf, M., Vieira, A., Lee, D. T., and Kailath, T., Recursive Multichannel Maximum Entropy Spectral Estimation, *IEEE Trans. Geoscience Electr.*, vol. GE-16, no. 2, pp. 85-94, April 1978. Reprinted in *Modern Spectrum Analysis*, ed. D. G. Childers, IEEE Press, N.Y., 1978.

55. Nash, J. G., Nudd, G. R., and Hansen, S., Concurrent VLSI Architectures for Toeplitz Linear Systems Solvers, *Proc. 1982 GOMAC Conf.*, Orlando, FL, November 2-4, 1982.

56. Parzen, E., Multiple Time Series Modeling, in *Multivariate Analysis II*, ed. P. R. Krishnaian, Academic Press, N.Y., pp. 389-409, 1969.

57. Porat, B., Friedlander, B., and Morf, M., Square Root Covariance Ladder Algorithms, *IEEE Trans. on Automatic Control*, vol. AC-27, no. 4, pp. 813-829, August 1982

58. Rao, S. K., and Kailath, T., Orthogonal Digital Filters for VLSI Implementation, *IEEE Trans. Circuits and Systems*, October 1984.

59. Robinson, E. A., *Multichannel Time-Series Analysis with Digital Computer Programs*, Holden-Day, San Francisco, CA, 1967.

60. Robinson, E. A., in *Applications of Digital Signal Processing*, ed. A. V. Oppenheim, Prentice-Hall, 1978.

61. Samson, C., A General Method for the Derivation of Lattice Algorithms, *Outils et Modeles Mathematiques Pour L'Automatique, L'Analyse De Systems et Le Traitement Du Signal*, vol. 2, ed. I. D. Landau, 530-545, CNRS, France, 1982.

62. Satorius, E. H., and Pack, J., Application of Least Squares Lattice Algorithm to Adaptive Equalization, *IEEE Trans. on Communications*, pp. 136-142, May 1981.

63. Schur, I., Uber Potenzreihen, die in Innern des Einheitskreises Beschrankt Sind, *Journal fur die Reine und Angewandte Mathematik*, vol. 147, pp. 205-232, Berlin, 1917.

64. Stewart, G. W., *Introduction to Matrix Computations*, Academic Press, 1973.

65. Szegö, G., *Orthogonal Polynomials*, Colloqium Publications, no. 23, American Math. Soc., Providence, RI, 2nd ed., 1958, 3rd ed., 1967.

66. Ulrych, T. J., and Clayton, R. W., Time Series Modeling and Maximum Entropy, *Phys. Earth and Plan. Inter.*, August 1976.

67. Vieira, A. C. G., Matrix Orthogonal Polynomials, with Applications to Autoregressive Modeling and Ladder Forms, Ph.D. Dissertation, Department of Electrical Engineering, Stanford University, Stanford, CA., December 1977.

68. Vieira, A., and Kailath, T., On Another Approach to the Schur-Cohn Criterion, *IEEE Trans. Circuits Systems*, vol. CAS-24, pp. 218-220, 1977.

69. Volder, J. E., The CORDIC Trigonometric Computing Technique, *IRE Trans. on Electronic Computers*, vol. EC-8, no. 3, pp. 330-334, Sept. 1959.

70. Walker, G., On Periodicity in Series of Related Terms, *Proc. Royal Soc.*, A131, p. 518, 1931.

71. Whittle, P., On the Fitting of Multivariable Autoregressions, and the Approximate Canonical Factorization of a Spectral Density Matrix, *Biometrika*, vol. 50, pp. 129-134, 1963.

72. Wiggins, R., and Robinson, E. A., Recursive Solution to the Multichannel Filtering Problem, *J. Geophys. Res.*, vol. 70, no. 8, pp. 1885-1891, April 1965.

73. Willsky, A. S., *Digital Signal Processing and Control and Estimation Theory: Points of Tangency, Areas of Intersection, and Parallel Directions*, MIT Press, Cambridge, MA, 1979.

74. Youla, D., and Kazanjian, N., Bauer-Type Factorization of Positive Matrices and the Theory of Matrices Orthogonal on the Unit Circle, *IEEE Trans. Circuits & Syst.*, vol. CAS-25, no. 1, pp. 57-69, 1978.

75. Yule, G. U., On a Method of Investigating Periodicities in Disturbed Series, With Special Reference to Wölfer's Sunspot Numbers, *Phil. Trans.*, vol. A226, p. 267, 1927.

3

Spectral Estimation with Applications

S. LAWRENCE MARPLE, JR.

A summary of several modern spectral estimation methods is presented in this tutorial. Most of the methods may be explained in the context of time series modeling. A few methods involve nonparametric treatment. Techniques discussed include classical spectral estimation, autoregressive (maximum entropy), ARMA, Prony, maximum likelihood, Pisarenko and MUSIC methods. Several applications from the areas of radar, sonar, well-logging, radio astronomy, speech, and beamforming are presented.

TABLE OF CONTENTS

1. INTRODUCTION

The unifying approach employed in this tutorial is to view each spectral estimation technique as based on the fitting of measured data to an assumed model. The variations in performance among the various spectral estimates may often be attributed to how well the assumed model matches the process under analysis. Different models may yield similar results, but one may require fewer model parameters and be therefore more efficient in its representation of the process.

Estimation of the power spectral density (PSD), or simply the spectrum, of discretely sampled deterministic and stochastic processes is usually based on procedures employing the Fast Fourier transform (FFT). This approach to spectral analysis is computationally efficient and produces reasonable results for a large class of signal processes. In spite of these advantages, there are several inherent performance limitations of the FFT approach. The most prominent limitation is that of frequency resolution, i.e., the ability to distinguish the spectral responses of two or more signals. The frequency resolution in Hertz is roughly the reciprocal of the time interval in seconds over which sampled data is available. A second limitation is due to the implicit windowing of the data that occurs when processing with the FFT. Windowing manifests itself as "leakage" in the spectral domain, i.e., energy in the main lobe of a spectral response "leaks" into the sidelobes, obscuring and distorting other spectral responses that are present.

These two performance limitations of the FFT approach are particulary troublesome when analyzing short data records. Short data records occur frequently in practice because many measured processes are brief in duration or have slowly time-varying spectra that may be considered constant only for short record lengths. In radar, for example, only a few data samples are available from each received radar pulse. In sonar, the motion of targets results in a slowly time-varying spectral response due to Doppler effects.

In an attempt to alleviate the inherent limitations of the FFT approach, many alternative spectral estimation procedures have been proposed within the last decade. The apparent improvement in resolution provided by these techniques have fostered their popularity, even though classical FFT-based spectral estimation has been shown to often provide better performance at very low signal-to-noise ratios. Even in those cases where improved spectral fidelity is achieved by use of an alternative spectral estimation procedure, the computational requirements of that alternative method may be significantly higher than FFT processing. This may make some modern spectral estimators unattractive for real-time implementation.

The nonparametric classical one-dimensional time (or space) series spectral estimation methods are presented in Section 3. Parametric spectral estimation techniques are covered in Section 4, which covers many of the so-called modern spectral estimation methods. The maximum likelihood spectral estimate, Pisarenko's method, and the MUSIC method are covered in Section 5. A short discussion of multidimensional maximum entropy spectral estimation may be found in Section 6. Various applications of modern spectral estimates are illustrated in Section 8, demonstrating some of the performance improvements achievable with these methods. Additional spectral estimation methods are presented in greater detail in reference [1]. The time series models developed here for spectral estimation find use in other digital signal processing fields such as communications and image processing; see the chapters by Kailath, Proakis, Jain, and Kung for more details.

2. SPECTRAL DENSITY BASICS

Traditional spectrum estimation, as currently implemented using the FFT, is characterized by many tradeoffs in an effort to produce statistically reliable spectral estimates. There are tradeoffs in windowing, time-domain averaging, and frequency-domain averaging of sampled data obtained from random processes in order to balance the needs to reduce sidelobes, to perform effective ensemble averaging, and to ensure adequate spectral resolution. The spectrum analysis of a random process is, in concept, not obtained directly from the process $x(t)$ itself, but is based on knowledge of the autocovariance function of a zero mean process

$$R_{xx}(\tau) = E[x(t+\tau)x^*(t)] \tag{2.1}$$

The Wiener-Khinchin theorem relates $R_{xx}(\tau)$ via the Fourier transform to $P(f)$, the power spectral density (PSD),

$$P(f) = \int_{-\infty}^{\infty} R_{xx}(\tau) \exp(-j2\pi f\tau)\,d\tau \tag{2.2}$$

As a practical matter, one does not usually know the statistical autocovariance function. Thus an additional assumption often made is that the random process is ergodic in the first and second moments. This property permits the substitution of time averages for ensemble averages. For an ergodic process, then, the statistical autocovariance function may be equated to

$$R_{xx}(\tau) = \lim_{T\to\infty} \frac{1}{2T}\int_{-T}^{T} x(t+\tau)x^*(t)\,dt \tag{2.3}$$

It is possible to show with the use of (2.3), that (2.2) may be equivalently expressed as

$$P(f) = \lim_{T\to\infty} E\left\{\frac{1}{2T}\left|\int_{-T}^{T} x(t)\exp(-j2\pi ft)\,dt\right|^2\right\} \tag{2.4}$$

The expectation operator is required since the ergodic property of $R_{xx}(\tau)$ does not necessarily imply that the Fourier transform of the process $x(t)$ is also ergodic; this means that the limit in (2.4) without the expected value will not converge in any statistical sense.

Attempting to estimate $P(f)$ with finite data sets using (2.4) without taking into consideration the expectation operation and the limit operation can lead to meaningless spectral estimates if no statistical averaging is performed; i.e., the variance of the PSD estimate will not tend to zero as T increases without bound.

3. CLASSICAL METHODS

Two spectral estimation techniques based on Fourier transform operations have evolved [2]. The PSD estimate based on the indirect approach via an autocorrelation estimate was popularized by Blackman and Tukey. The other PSD estimate, based on the direct approach via an FFT operation on the data, is the one typically referred to as the periodogram.

With a finite data sequence, only a finite number of discrete autocorrelation function values, or lags, may be estimated. Blackman and Tukey proposed the spectral estimate

$$\hat{P}_{BT}(f) = \Delta t \sum_{m=-M}^{M} \hat{R}_{xx}(m)\exp(-j2\pi fm\Delta t) \tag{3.1}$$

based on the available biased autocovariance lag estimates $\hat{R}_{xx}(m)$,

$$\hat{R}_{xx}(m) = \frac{1}{N}\sum_{k=0}^{N-m} x_{k+m}x_k^* \tag{3.2}$$

where $-1/(2\Delta t) \leq f \leq 1/(2\Delta t)$ and ^ denotes an estimate. This spectral estimate is the discrete-time version of the Wiener- Khinchin expression (2.2).

The direct method of spectrum analysis is the modern version of Schuster's periodogram. A sampled data version of expression (2.4), for which measured data is available only for samples $x_0, \ldots, x_{N-1}$, is

$$\hat{P}_{PER}(f) = \frac{1}{N\Delta t}\left|\Delta t \sum_{n=0}^{N-1} x_n \exp(-j2\pi fn\Delta t)\right|^2 \tag{3.3}$$

also defined for the frequency interval $-1/(2\Delta t) \leq f \leq 1/(2\Delta t)$. Note that the expectation operation in (2.4) has been ignored for the moment. Use of the FFT will permit evaluation of (6) at the discrete set of N equally spaced frequencies $f_m = m\Delta f$ Hz, for $m = 0, 1, \ldots, N-1$ and $\Delta f = 1/N\Delta t$,

$$\hat{P}_m = \hat{P}_{PER}(f_m) = \frac{1}{N\Delta t}\left|X_m\right|^2 \tag{3.4}$$

where X_m is the discrete Fourier transform (DFT). $\hat{P}_m$ is identical to the *energy* spectral density [squared modulus of the Fourier transform of a time function] except for the division by the time interval of $N\Delta t$ seconds required to make $\hat{P}_m$ a *power* spectral density.

In order to reduce the bias of the Blackman-Tukey spectral estimate due to the implied rectangular window over a finite number of lag estimates, various windows other than rectangular are often applied to the lag estimates in order to suppress the sidelobes, and therefore reduce the bias. In order to emulate ensemble averaging needed to make the periodogram a statisitically consistent spectral estimate, two basic approaches have evolved. The averaging of modified periodograms is one technique that breaks the original time sequence into segments, windows each segment (in order to reduce the bias), calculates a periodogram of each modified segment, and then averages the segment periodograms. A particular version of this averaging method using the FFT algorithm was proposed by Welch (see his paper in [3]). A second method is to compute the periodogram as in (3.4), and then average over adjacent frequency bins (i.e., low pass filter the periodogram). More details on windowing and classic spectral estimation computation may be found in reference [4]. The two methods of periodogram averaging may not be possible to perform in situations where only short data sequences are available.

In situations in which the time series involve a process with time-varying charactersitics that can only be considered relatively constant for only short time intervals (a short time frame), the short-time Fourier transform (STFT) is often used as a time-frequency representation for such signals. Speech and sonar data are two signal types that involve time-varying characteristics. The STFT is defined as

$$X(nL,f) = \sum_{m=-\infty}^{\infty} x(m\Delta t)\, w[(nL-m)\Delta t] \exp(-j2\pi fm\Delta t) \tag{3.5}$$

where $w(n\Delta t)$ is a frame analysis window and L is an integer which denotes the separation in time between adjacent short-time frames. Thus, for a fixed n, $X(nL,f)$ represents the Fourier transform at time nL of windowed data samples. Taking the modulus of $X(nL,f)$, as in (3.3), at each L produces a short-time spectral estimate [see 5].

The performance of classical spectral estimates at a given frequency f may be characterized by the stability-time-bandwidth product inequality

$$\Delta S\, \Delta T\, \Delta f > 1 \tag{3.6}$$

where ΔT is the time interval over which data has been measured, Δf is the resolution in Hertz, and ΔS is the stability factor, defined as the ratio of the spectral estimate variance over the spectral estimate mean. In order to have a stable spectral estimate for a fixed data set of ΔT seconds duration, ΔS must be made small. However, expression (3.6) indicates this can only be achieved by giving up resolution (accept a larger value for Δf). Thus, spectral estimation involves a tradeoff between statisitical stability and resolution.

The conventional Blackman-Tukey and periodogram approaches to spectral estimation have the following advantages: 1) computationally efficient if only a few lags are needed (BT) or if the FFT is used (periodogram), 2) PSD estimate directly proportional to the power for sinusoid processes, and 3) a good model for some applications (the model is a sum of harmonically-related sinusoids). The disadvantages of these techniques are: 1) suppression of weak signal main-lobe responses by strong signal sidelobes, 2) frequency resolution limited by the available data record duration, independent of the characteristics of the data or its SNR, 3) introduction of distortion in the spectrum due to sidelobe leakage, 4) need for some sort of pseudo ensemble averaging to obtain statistically consistent periodogram spectra, and 5) the appearance of negative PSD values with the BT approach when some autocovariance sequence estimates are used.

4. PARAMETRIC METHODS

Often one has more knowledge about the process from which the data samples are taken, or at least is able to make a more reasonable assumption other than to assume the data is zero outside the window. Use of *a priori* information (or assumptions) may permit selection of an exact model for the process that generated the data samples, or at least a model that is a good approximation to the actual underlying process. It is then usually possible to obtain a better spectral estimate based on the model by determining the parameters of the model from the observations. The lack of a true model of the process under measurement does not mean one cannot try the parametric methods; use of these models may still yield reasonable results. One major motivation for the current interest in the modeling, or parametric, approach to spectral estimation is the higher frequency resolution achievable with these modern techniques over that achievable with the traditional techniques previously discussed. The degree of improvement in resolution and spectral "fidelity", if any, will be determined by the ability to fit an assumed model with a few parameters to the measured data. The selection of a model for the spectral estimate is intimately tied to estimation and identification techniques employed in linear system theory (see, for examples, the chapter by Kailath in this book).

Many deterministic and stochastic discrete-time processes encountered in practice are well approximated by a rational transfer function model. In this model, an input driving sequence $\{n_n\}$ and the output sequence $\{x_n\}$ that is to model the data are related by the linear difference equation,

$$x_n = \sum_{l=0}^{q} b_l n_{n-l} - \sum_{k=1}^{p} a_k x_{n-k} \tag{4.1}$$

This most general linear model is termed an autoregressive-moving average (ARMA) model. The interest in these models stems from their relationship to linear filters with rational transfer functions.

The system function $H(z)$ between the input n_n and output x_n for the ARMA process is the rational function

$$H(z) = \frac{B(z)}{A(z)} \tag{4.2}$$

where

$$A(z) = z - \textit{transform of AR branch} = \sum_{m=0}^{p} a_m z^{-m} \tag{4.3}$$

$$B(z) = z - \textit{transform of MA branch} = \sum_{m=0}^{q} b_m z^{-m}$$

The PSD of the ARMA output process driven by a white noise process is then

$$P_{ARMA}(f) = P_x(f) = \sigma^2 |B(f)/A(f)|^2 \tag{4.4}$$

where $A(f) = A(\exp[j2\pi f \Delta t])$ and $B(f) = B(\exp[j2\pi f \Delta t])$. Specification of the parameters $\{a_k\}$ (termed the autoregressive coefficients), the parameters $\{b_k\}$ (termed the moving-average coefficients), and σ^2 is equivalent to specifying the spectrum of the process $\{x_n\}$. Without loss of generality, one can assume $a_0 = 1$ and $b_0 = 1$ since any filter gain can be incorporated into σ^2.

4.1 Autoregressive Spectral Estimation

If all the $\{b_i\}$, except $b_0 = 1$, are zero, then

$$x_n = - \sum_{k=1}^{p} a_k x_{n-k} + n_n \tag{4.5}$$

and the process is strictly an autoregression of order p driven by white noise process n_n. The process is termed AR in that the sequence x_n is a linear regression on itself with n_n representing the error. With this model, the present value of the process is expressed as a weighted sum of past values plus a noise term of variance σ^2. The PSD is

$$P_{AR}(f) = \frac{\sigma^2}{|A(f)|^2} = \frac{\sigma^2}{\left| 1 + \sum_{k=1}^{p} a_k \exp(-j2\pi f k \Delta t) \right|^2} \tag{4.6}$$

This model is sometimes termed an all-pole model due to fact that the only frequency dependence of the spectrum is in the denominator. To estimate the PSD one need only estimate $\{a_1, a_2, \ldots, a_p, \sigma^2\}$.

Known autocovariance case

A relationship between the AR parameters and the autocovariance function can be developed (see the chapter by Kailath for details of this development). This relationship is known as the Yule-Walker normal equations. In matrix form, they are compactly expressed as

$$\begin{bmatrix} R_{xx}(0) & R_{xx}(-1) & \ldots & R_{xx}(-p) \\ R_{xx}(1) & R_{xx}(0) & \ldots & R_{xx}(-p+1) \\ \cdot & \cdot & & \cdot \\ \cdot & \cdot & & \cdot \\ \cdot & \cdot & & \cdot \\ R_{xx}(p) & R_{xx}(p-1) & \cdots & R_{xx}(0) \end{bmatrix} \begin{bmatrix} 1 \\ a_1 \\ \cdot \\ \cdot \\ \cdot \\ a_p \end{bmatrix} = \begin{bmatrix} \sigma^2 \\ 0 \\ \cdot \\ \cdot \\ \cdot \\ 0 \end{bmatrix} \tag{4.7}$$

To determine the AR parameters and σ^2, one must then solve (4.7) with the $p+1$ estimated autocovariance lags $R_{xx}(0), \ldots, R_{xx}(p)$ and use the fact that $R_{xx}(-m) = R_{xx}^*(m)$. A computationally efficient algorithm known as the Levinson recursion can solve (4.7) with order p^2 operations (Zohar, [6]).

An alternative representation of (4.6) is

$$P_{AR}(f) = \sum_{n=-\infty}^{\infty} r_{xx}(n) \exp(-j2\pi fn\Delta t) \tag{4.8}$$

where

$$r_{xx}(n) = \begin{cases} R_{xx}(n), & \text{for } |n| \leq p \\ -\sum_{k=1}^{p} a_{pk}\, r_{xx}(n-k), & \text{for } |n| > p \end{cases} \tag{4.9}$$

Thus, the autoregressive model produces one of many possible extensions of the autocovariance sequence that one might devise from the available autocovariance lag values. From this, it is easy to see that the AR PSD preserves the known lags and recursively extends the lags beyond the window of known lags. The AR PSD function (4.8) summation is identical to the BT PSD function (3.1) up to lag p, but continues with an infinite extrapolation of the autocovariance function rather than windowing it to zero. Thus AR spectra do not exhibit the traditional sidelobes due to windowing. Also, it is this implied extrapolation through (4.9) that is responsible for the high resolution property of the AR spectral estimator.

An alternative interpretation for the autocovariance extension given by (4.9) was provided by Burg (see his papers in [3]). Burg argued that the autocovariance extrapolation should be selected to yield a positive definite autocovariance function with maximum entropy. Thus, the process with such an autocovariance sequence would be the "most random" one possible based on knowledge of only the autocovariance lag values from 0 to p. In particular, if one assumes a Gaussian random process, then the entropy per sample is proportional to

$$\int_{-\frac{1}{2}\Delta t}^{\frac{1}{2}\Delta t} \ln P_x(f)\, df \tag{4.10}$$

where $P_x(f)$ is the PSD of x_n. $P_x(f)$ is found by maximizing (4.10) subject to the constraints that the $(p+1)$ known lags satisfy the Wiener-Khintchin theorem,

$$\int_{-\frac{1}{2}\Delta t}^{\frac{1}{2}\Delta t} P_x(f) \exp(-j2\pi fn\Delta t)\, df = R_{xx}(n) \qquad \text{for } n = 0, 1, \ldots, p\ . \tag{4.11}$$

The solution is found by the Lagrange multiplier techniques; the result is a PSD that is equivalent in form to (4.6), the AR PSD.

Given data case

In most practical situations, one has data samples rather than known autocovariance lags available for the spectral estimation procedure. By operating directly on the data without making an autocovariance estimate, it is possible to obtain better AR parameter estimates, and therefore better AR spectral estimates. Least squares linear prediction techniques are used in this case. Most of the many algorithms for AR coefficient estimation fall into two broad categories. They either use forward only linear predictions for the estimates, or they employ a combination of forward and backward linear prediction. In the later case, it is often assumed that the backward linear prediction parameters are identical to the forward linear prediction coefficients, as it must be for known autocovariances of statistically stationary processes, only reversed in time and conjugated (for complex stochastic processes). In this later case, the Burg algorithm is probably the most well-known, followed by the "Least Squares" algorithm of Ulrych-Clayton and Nuttall. Of the forward-only algorithms, the "covariance" method is probably the most well-known. All of the above methods have fast computational algorithms that have been developed for them with order p^2 operations to solve the associated normal equations. So-called doubling algorithms can bring this computational burden down to order $p \log^2 p$.

Consider first the forward only linear prediction least squares techniques. Assume the data sequence $x_0, \ldots, x_{N-1}$ is used to find the p-th order AR parameter estimates. The forward linear prediction will have the form

$$\hat{x}_n = -\sum_{k=1}^{p} a_{pk} x_{n-k}\ . \tag{4.12}$$

The prediction is forward in the sense that the prediction for the current sample is a weighted sum of p previous samples. The forward linear prediction error is

$$e_{pn} = x_n - \hat{x}_n = \sum_{k=0}^{p} a_{pk} x_{n-k} \tag{4.13}$$

where $a_{p0} = 1$ by definition. We may compute e_{pn} for $n = 0$ to $n = N + p - 1$ if one assumes the terms outside the measurement interval are zero, i.e., $x_n = 0$ for $n < 0$ and $n > N-1$. There is an implied windowing of the data sequence in order to extend the index range for e_{pn} from 0 to $N + p - 1$. Using a matrix formulation for (4.13),

$$\begin{bmatrix} e_0 \\ \vdots \\ e_p \\ \vdots \\ e_{N-1} \\ \vdots \\ e_{N+p-1} \end{bmatrix} = \mathbf{X}_2 \; \mathbf{X}_1 \begin{bmatrix} x_0 & & \bigcirc \\ \vdots & \ddots & \\ x_p & & x_0 \\ \vdots & & \vdots \\ x_{N-1} & & x_{N-p-1} \\ & \ddots & \vdots \\ \bigcirc & & x_{N-1} \end{bmatrix} \mathbf{X}_3 \; \mathbf{X}_4 \begin{bmatrix} 1 \\ a_{p1} \\ \vdots \\ a_{pp} \end{bmatrix} \tag{4.14}$$

or

$$\mathbf{E} = \mathbf{X}\,\mathbf{A} \tag{4.15}$$

The prediction error energy is simply

$$E_p = \sum_n |e_{pn}|^2 = \sum_n \left| \sum_{k=0}^{p} a_{pk} x_{n-k} \right|^2 . \tag{4.16}$$

The summation range for E_p is purposely not specified for the moment. To minimize E_p, the derivatives of E_p with respect to the $\{a_{pk}\}$ are set to zero and the resultant equations solved for the AR (linear prediction) parameters. The result is

$$\sum_{k=0}^{p} a_{pk} \left[\sum_n x_{n-k} x_{n-i}^{*} \right] = 0, \quad \text{for } 1 \leq i \leq p \tag{4.17}$$

with minimum error energy,

$$E_p = \sum_{k=0}^{p} a_{pk} \left[\sum_n x_{n-k} x_n^{*} \right] . \tag{4.18}$$

Expressions (4.17) and (4.18) can be reformulated in normal equation form

$$(\mathbf{X}_k^H \mathbf{X}_k)\mathbf{A} = (E_p \; 0 \; 0 \; \cdots \; 0)^T \tag{4.19}$$

for which four special indexing ranges for k=1,2,3,4 can be selected,as indicated in (4.14). Note that (4.19) has a similar structure as (4.7); however, the data matrix product $\mathbf{X}_k^H \mathbf{X}_k$ is not necessarily Toeplitz as it was in the Yule-Walker equations.

If data matrix $\mathbf{X}_1$ is selected, the normal equations (4.19) are termed the "covariance" equations, often used in speech processing. If the data matrix $\mathbf{X}_2$ is selected, the resulting normal equations are called the "autocorrelation" equations since the product matrix $\frac{1}{N}\mathbf{X}_2^H X_2$ reduces exactly to the Yule-Walker equations (4.7) in which the biased autocovariance estimates have been substituted. Note that a data window has been assumed for this case, as the zeros in the $\mathbf{X}_2$ matrix indicate. This data window reduces the resolution of AR spectra estimated with data matrix $\mathbf{X}_2$. If the data matrix $\mathbf{X}_3$ is selected, the normal equations are termed the "prewindowed" normal equations due to the zero value assumptions made for the missing data prior to data sample x_0. If the data matrix $\mathbf{X}_4$ is selected, the normal equations are termed the "postwindowed" normal equations since a zero data assumption is made for the data beyond x_{N-1}.

If a process is stationary, the coefficients of the backward prediction error filter are identical to the forward prediction error filter in the known autocovariance case, but conjugated and reversed in time. The use of backward prediction errors in the given data case was introduced by Burg in 1967. The Burg algorithm, separate and distinct from the maximum entropy viewpoint discussed earlier, may be viewed as a constrained least squares minimization. Assuming a stationary process, the forward linear prediction error will be given by (4.13) and the backward linear prediction error will be defined by

$$b_{pn} = \sum_{k=0}^{p} a_{pk}^{*} x_{n-p+k} \tag{4.20}$$

where both errors are defined over only available data, $p \leq n \leq N-1$. To obtain estimates of the AR (linear prediction) parameters, Burg minimized the sum of the forward and backward prediction error energies,

$$E_p = \sum_{n=p}^{N-1} |e_{pn}|^2 + \sum_{n=p}^{N-1} |b_{pn}|^2 \tag{4.21}$$

subject to the constraint that the AR parameters satisfy the Levinson recursion

$$a_{pk} = a_{p-1,k} + a_{pp} a_{p-1,p-k}^{*} \tag{4.22}$$

for all orders from 1 to p. This constraint was motivated by Burg's desire to ensure a stable AR filter (poles within the unit circle). The Levinson constraint forces E_p to become a function of only the unknown coefficient a_{pp}. Setting the derivative of E_p with respect to a_{pp} to zero then yields

$$a_{ii} = \frac{-2 \sum_{k=i}^{N-1} b_{i-1,k-1}^{*} e_{i-1,k}}{\sum_{k=i}^{N-1} (|b_{i-1,k-1}|^2 + |e_{i-1,k}|^2)} \tag{4.23}$$

It is simple to show from (4.23) that $[a_{ii}] \leq 1$ for all i, which is sufficient to guarantee a stable filter.

The Burg algorithm has several problems associated with it, including spectral line splitting and biases of the frequency estimate. These were corrected by the so-called "least squares" algorithm, independently suggested by Ulrych-Clayton and Nuttall. By simply minimizing E_p of (4.21) by differentiating with respect to all the a_{pk}, not just a_{pp}, a set of normal equations similar to (4.19) can be obtained. A fast computational algorithm has been developed to solve this normal equation. For more details on fast least squares algorithms, see the Kailath chapter in the sections on the prewindowed case and lattice filters.

In addition to the batch data processing just presented there are analogous algorithms for time sequential on-line adaptation of the AR parameters on a sample-by-sample basis. These algorithms fall under one of the four categories: recursive least squares (RLS) method, fast Kalman method, gradient lattice method, and exact least squares lattice method. More details on the lattice methods are provided in the chapter on adaptive equalizers by Proakis. VLSI implementation of some of the fast algorithms for solving normal equations may be found in the chapter by Kung.

Since the best choice of AR order p is not generally known *a priori*, it is usually necessary in practice to postulate several possible model orders. Based on these, one then computes some error criterion that indicates which model order to select. Too low a guess for order results in a highly smoothed spectral estimate. Too high an order introduces spurious detail into the estimate. One intuitive approach would be to construct AR models of increasing order until the computed prediction error power reaches a minimum. However, all the least squares estimation procedures highlighted previously have the property that the prediction error energies decrease monotonically with increasing order p. As a result, several criteria have been developed that modify the prediction squared error total with a penalty function that prevents too high an order selection. The FPE, AIC, and CAT functions are three of the more popular order selection criteria [1,3].

4.2 ARMA Spectral Estimation

Recall that the ARMA model assumes that a time series x_n can be modeled as the output of a p pole and q zero filter excited by white noise, i.e.,

$$x_n = - \sum_{k=1}^{p} a_k x_{n-k} + \sum_{k=0}^{q} b_k n_{n-k} \tag{4.24}$$

where $R_{nn}(k) = \sigma^2$ for k=0, else it is zero, and $b_0 = 1$. The poles of the filter are assumed to be within the unit circle of the z-plane. The zeros of the filter may lie anywhere in the z-plane.

Once the parameters of the ARMA(p,q) model are identified, the spectral estimate is obtained as

$$P_x(f) = |H(\exp[j2\pi f]|^2 \, P_n(f) \tag{4.25}$$

$$= \frac{\sigma^2 \Delta t \left| 1 + \sum_{k=1}^{q} b_k \exp(-j2\pi fk\Delta t) \right|^2}{\left| 1 + \sum_{k=1}^{p} a_k \exp(-j2\pi fk\Delta t) \right|^2}$$

Many ARMA parameter estimation techniques have been formulated theoretically, which usually involve many matrix computations and/or iterative optimization techniques. These approaches are normally not practical for real-time processing. Suboptimum techniques have therefore been developed to make the computational load more manageable. These techniques are usually based on a least squares criterion and require solutions of linear equations. These methods generally estimate the AR and MA parameters separately rather then jointly as required for optimal parameter estimation. The AR parameters can first be estimated independently of the MA parameters if one uses the so-called high-order (or modified) Yule-Walker equations. A final point in favor of the suboptimal linear approaches is that iterative optimization techniques are not guaranteed to converge or may converge to the wrong solution.

Assuming the autocovariance lags of a process known to be an ARMA(p,q) process are available, then the extended Yule-Walker equations that yield the AR parameters only is given by

$$\begin{bmatrix} R_{xx}(q) & R_{xx}(q-1) & \cdots & R_{xx}(q-p+1) \\ R_{xx}(q+1) & R_{xx}(q) & \cdots & R_{xx}(q-p+2) \\ \cdot & \cdot & & \cdot \\ \cdot & \cdot & & \cdot \\ \cdot & \cdot & & \cdot \\ R_{xx}(q+p-1) & R_{xx}(q+p-2) & \cdots & R_{xx}(q) \end{bmatrix} \begin{bmatrix} a_1 \\ a_2 \\ \cdot \\ \cdot \\ \cdot \\ a_p \end{bmatrix} = - \begin{bmatrix} R_{xx}(q+1) \\ R_{xx}(q+2) \\ \cdot \\ \cdot \\ \cdot \\ R_{xx}(q+p) \end{bmatrix} \tag{4.26}$$

A fast algorithm for solving (4.26) has been developed by Zohar [6]. Once the AR parameter esimates have been found, the MA parameters may by found by filtering the data with the all-zero filter $A(z)$, where

$$A(z) = 1 + \sum_{k=1}^{p} a_k z^{-k}$$

to yield a purely MA process. The usual techniques to find the MA parameters may then be used (see reference [1] for details).

A second technique for estimating the ARMA parameters utilizes the identity

$$\frac{B(z)}{A(z)} = \frac{1}{C(z)}$$

where

$$C(z) = 1 + \sum_{k=1}^{\infty} c_k z^{-k}$$

to equate an ARMA model to an infinite order AR model. The $\{c_k\}$ may thus be estimated using an high order AR fit with AR estimation algorithms only, and then relating them as above to the ARMA parameters [7].

A third technique based upon least squares input-output identification has been proposed by many researchers. The normal equations exhibit a nonlinear character due to the unknown cross covariance between the input and output. If n_n is unobservable, the $R_{nx}(k)$ cannot be estimated. If, however, n_n were known or could be roughly estimated, so that $R_{nx}(k)$ could be estimated, then the ARMA parameters could be found as the solution of a set of linear equations. In practice, n_n is estimated from x_n in a boot-strap approach, for example, with a lattice filter configuration [8].

4.3 Prony's Method

Prony's method, a technique for modeling data of equally spaced samples by a linear combination of exponentials, is not a spectral estimation technique in the usual sense, but a spectral interpretation is provided in this section. The original procedure by Baron de Prony exactly fitted an exponential curve having p exponential terms (each term with two parameters -- an amplitude A_i and an exponent α_i where $A_i \exp[\alpha_i t]$) to $2p$ data measurements. For the case where only an *approximate* fit with p exponentials to a data set of N samples is desired, such that $N > 2p$, a least squares estimation procedure is used. This procedure is called the extended Prony method.

The model assumed in the extended Prony method is a set of p exponentials of arbitrary amplitude, phase, frequency, and damping factor. The discrete-time function

$$\hat{x}_n = \sum_{m=1}^{p} b_m z_m^n, \quad \text{for } n=0, \ldots, N-1 \tag{4.27}$$

is the model to be used for approximating the measured data $x_0, \ldots, x_{N-1}$. For generality, b_m and z_m are assumed complex and

$$b_m = A_m \exp(j\theta_m) \tag{4.28}$$

$$z_m = \exp[(\alpha_m + j2\pi f_m)\Delta t]$$

where A_m is the amplitude, θ_m is the phase in radians, α_m is a damping factor, f_m is the oscillation frequency in Hertz, and Δt represents the sample interval in seconds. Finding $\{A_m, \theta_m, \alpha_m, f_m\}$ and p that minimize the squared error

$$E = \sum_{n=0}^{N-1} \left| x_n - \hat{x}_n \right|^2 \tag{4.29}$$

is a difficult nonlinear least squares problem. An alternative suboptimum solution that does not minimize (4.29) but yet still provides satisfactory results, is based on Prony's technique. Prony's method solves two sequential sets of linear equations with an intermediate polynomial rooting step that concentrates the nonlinearity of the problem in the polynomial rooting procedure.

The key to the Prony technique is to recognize that (4.27) is the homogeneous solution to a constant coefficient linear difference equation, the form of which is found as follows. Define the polynomial $\Psi(z)$ as

$$\Psi(z) = \prod_{k=1}^{p} (z - z_k) = \sum_{i=0}^{p} a_i z^{p-i}, \quad a_0 = 1 \tag{4.30}$$

Thus $\Psi(z)$ has the complex exponentials z_k of (4.28) as its roots and complex coefficients a_i when multiplied out. Based on (4.27), one way of expressing $\hat{x}_{n-m}$ is

$$\hat{x}_{n-m} = \sum_{l=1}^{p} b_l z_l^{n-m} \tag{4.31}$$

for $0 \leq n - m \leq N-1$. Multiplying (4.31) by a_m and summing over the past $p + 1$ products yields

$$\sum_{m=0}^{p} a_m \hat{x}_{n-m} = \sum_{l=1}^{p} b_l \sum_{m=0}^{p} a_m z_l^{n-m} \tag{4.32}$$

defined for $p \leq n \leq N-1$. If in (4.32) the substitution $z_l^{n-m} = z_l^{n-p} z_l^{p-m}$ is made, then

$$\sum_{m=0}^{p} a_m \hat{x}_{n-m} = \sum_{l=1}^{p} b_l z_l^{n-p} \sum_{m=0}^{p} a_m z_l^{p-m} = 0 \tag{4.33}$$

The zero result in (4.33) follows by recognizing that the final summation above is just the polynomial $\Psi(z_l)$ of (4.30), evaluated at one of its roots. Expression (4.33) then yields the recursive difference equation

$$\hat{x}_n = - \sum_{m=1}^{p} a_m \hat{x}_{n-m} \tag{4.34}$$

defined for $p \leq n \leq N-1$. The approximation error $e_n = x_n - \hat{x}_n$ can be substituted into (4.34), and the moving average error ϵ_n defined as

$$\epsilon_n = \sum_{m=0}^{p} a_m e_{n-m}, \quad \text{for } n=p, \ldots, N-1 \tag{4.35}$$

so that

$$x_n = - \sum_{m=1}^{p} a_m x_{n-m} + \epsilon_n \tag{4.36}$$

The so-called extended Prony method then seen to supoptimally minimize $\sum_{n=p}^{N-1} \left|\epsilon_n\right|^2$, rather than the true optimum obtained by minimizing $\sum_{n=p}^{N-1} \left|e_n\right|^2$. Thus the extended Prony parameter estimation procedure reduces to that of an AR/linear prediction parameter estimation for which all the least squares approaches discussed in section 4.1 can be applied.

Once the z_i have been determined from the polynoomial rooting, expression (4.27) reduces to a set of linear equations in the unknown b_m parameters, expressible in matrix form as

$$\mathbf{\Phi} \, \mathbf{B} = \hat{\mathbf{X}} \tag{4.37}$$

where

$$\Phi = \begin{bmatrix} 1 & 1 & \cdots & 1 \\ z_1 & z_2 & \cdots & z_p \\ \cdot & \cdot & & \cdot \\ \cdot & \cdot & & \cdot \\ \cdot & \cdot & & \cdot \\ z_1^{N-1} & z_2^{N-1} & \cdots & z_p^{N-1} \end{bmatrix}$$

$$\mathbf{B} = [\ b_1 \ \cdots \ b_p\ ]^T$$

$$\hat{\mathbf{X}} = [\ \hat{x}_0\ , \ldots,\ \hat{x}_{N-1}\ ]^T$$

Note that Φ is a Van der Monde matrix form. A least squares minimization of $\sum (x - \hat{x})^2$ yields the well-known solution

$$\mathbf{B} = [\Phi^H \Phi]^{-1} \Phi^H \mathbf{X} \tag{4.38}$$

Here, the H denotes the complex conjugate transpose operation. Determining the a_i parameters by a least squares estimation, rooting the polynomial, and solving for the b_j parameters (or residues) constitutes the entire extended Prony procedure.

It is possible to "define" a Prony spectrum by simply taking the modulus of the Fourier transform of (4.27) to yield

$$\hat{P}_{PRONY}(f) = \left| \hat{X}(f) \right|^2 \tag{4.39}$$

where

$$\hat{X}(f) = \sum_{m=1}^{p} A_m \exp(j\theta_m) \frac{2\alpha_m}{[\alpha_m^2 + (2\pi[f - f_m])^2]} \tag{4.40}$$

Other kinds of Prony "spectrums" could have been defined, but this seems to make the most sense as we shall see in the applications section.

5. NONPARAMETRIC METHODS

5.1 Maximum Likelihood Spectral Estimation

The maximum likelihood spectral estimate (MLSE) falls into the category of a nonparametric technique in the sense that no model parameters are explicitly computed. The original concept was developed by Capon for frequency-wavenumber analysis [9]. A filter model analogy will be used to describe this method. The MLSE was originally developed for seismic array frequency-wavenumber analysis. In this method, one estimates the PSD by effectively measuring the power out of a set of narrow-band filters. MLSE is actually a misnomer in that the spectral estimate is not necessarily a true maximum likelihood estimate of the PSD; it may more appropriately be termed the Capon spectral estimate after its inventor. The name MLSE will be retained here only for historic reasons. The difference between MLSE and conventional BT/periodogram spectral estimation is that the shape of the narrow-band filters in MLSE are, in general, different for each frequency, whereas they are fixed with the BT/periodogram procedures. The filters adapt to the process second order statistics for which a PSD estimate is sought. In particular, the filters are finite impulse response (FIR) types with p weights (taps),

$$\mathbf{A} = [\, a_0 \; a_1 \; \cdots \; a_{p-1} \,]^T \tag{5.1}$$

The coefficients are chosen so that at the frequency under consideration, f_0, the frequency response of the filter is unity (i.e., an input sinusoid at that frequency would be undistorted at the filter output) and the variance of the output process is minimized. Thus the filter should adjust itself to reject components of the spectrum not near f_0 so that the output power is due mainly to frequency components close to f_0. To obtain the filter, one minimizes the output variance σ^2, given by

$$\sigma^2 = \mathbf{A}^H \mathbf{R}_{xx} \mathbf{A} \tag{5.2}$$

subject to the unity frequency response constraint (so that the sinusoid of frequency f_0 is filtered without distortion),

$$\mathbf{E}^H \mathbf{A} = 1 \tag{5.3}$$

where $\mathbf{R}_{xx}$ is the autocovariance matrix of x_n, and $\mathbf{E}$ is the vector

$$\mathbf{E} = [\, 1 \; \exp(j2\pi f_0 \Delta t) \; \cdots \; \exp(j2\pi [p-1] f_0 \Delta t) \,]^T$$

The solution for the filter weights is easily shown to be

$$\mathbf{A}_{OPT} = \frac{\mathbf{R}_{xx}^{-1} \mathbf{E}}{\mathbf{E}^H \mathbf{R}_{xx}^{-1} \mathbf{E}} \tag{5.4}$$

and the minimum output variance is then

$$\sigma_{MIN}^2 = \frac{1}{\mathbf{E}^H \mathbf{R}_{xx}^{-1} \mathbf{E}} \tag{5.5}$$

It is seen that the frequency response of the optimum filter is unity at $f = f_0$ and that the filter characteristics change as a function of the underlying autocovariance function. Since the minimum output variance is due to frequency components near f_0, then $\sigma_{MIN}^2 \Delta t$ can be interpreted as a PSD estimate. Thus, the MLSE PSD is defined as

$$\hat{P}_{ML}(f_0) = \frac{1}{\mathbf{E}^H \mathbf{R}_{xx}^{-1} \mathbf{E}} \tag{5.6}$$

To compute the spectral estimate, one only needs an estimate of the autocovariance matrix.

The MLSE and AR PSD have been related analytically as follows (see Burg paper in reference [3])

$$\frac{1}{\hat{P}_{ML}(f)} = \frac{1}{p} \sum_{m=1}^{p} \frac{1}{\hat{P}_{AR}^{(m)}(f)} \tag{5.7}$$

where $\hat{P}_{AR}^{(m)}(f)$ is the AR PSD for an mth order model and $\hat{P}_{ML}(f)$ is the MLSE PSD, both based upon a known autocovariance matrix of order p.

5.2 Pisarenko Harmonic Decomposition

In general, a $2p$-th order difference equation of real coefficients of the form

$$x_n = - \sum_{m=1}^{2p} a_m x_{n-m} \tag{5.8}$$

can represent a deterministic process consisting of p real sinusoids of the form $\sin(2\pi f_i \Delta t)$. In this case, the $\{a_m\}$ are coefficients of the symmetric polynomial

$$z^{2p} + a_1 z^{2p-1} + \ldots + a_1 z + 1 = \tag{5.9}$$

$$\sum_{i=1}^{p} (z - z_i)(z - z_i^*) = 0$$

with unit modulus roots that occur in complex conjugate pairs of the form $z_i = \exp(j2\pi f_i \Delta t)$, where the f_i are arbitrary frequencies such that $[f_i] \leq \frac{1}{2}\Delta t$, and $i = 1, \ldots, p$. For sinusoids in additive white noise w_n, the observed process is

$$y_n = x_n + w_n = - \sum_{m=1}^{2p} a_m x_{n-m} + w_n \tag{5.10}$$

where $E(w_n w_{n+k}) = \sigma^2 \delta_k$, $E(w_n) = 0$, and $E(x_n w_n) = 0$ since the noise is assumed to be uncorrelated with the sinusoids. Substituting $x_{n-m} = y_{n-m} - w_{n-m}$ into (5.10), it is possible to express (5.10) as

$$\sum_{m=0}^{2p} a_m y_{n-m} = \sum_{m=0}^{2p} a_m w_{n-m} \tag{5.11}$$

where $a_0 = 1$ by definition. Expression (5.11) has the structure of an ARMA(p,p) model. However, this ARMA has a special symmetry in which the AR parameters are identical to the MA parameters in the model.

If the autocovariance function of y_n is known, the ARMA parameters can be found as the solution to an eigenequation, as is now shown. An equivalent matrix expression for (5.11) is

$$\mathbf{Y}^T \mathbf{A} = \mathbf{W}^T \mathbf{A} \tag{5.12}$$

where

$$\mathbf{Y}^T = [\, y_n \; y_{n-1} \; \cdots \; y_{n-2p} \,]$$

$$\mathbf{A}^T = [\, 1 \; a_1 \; \cdots \; a_{2p-1} \; a_{2p} \,] \qquad a_{2p-i} = a_i$$

$\mathbf{W}^T = [\ w_n\ w_{n-1}\ \cdots\ w_{n-2p}\ ]$.

Premultiplying both sides of (5.12) by the vector **Y** and taking the expectation yields

$$E[\mathbf{Y}\,\mathbf{Y}^T]\,\mathbf{A} = E[\mathbf{Y}\,\mathbf{W}^T]\,\mathbf{A} \quad . \tag{5.13}$$

Defining

$\mathbf{X}^T = [\ x_n\ \cdots\ x_{n-2p}\ ]$

then

$$E[\mathbf{Y}\,\mathbf{Y}^T] = \mathbf{R}_{yy} = \begin{bmatrix} R_{yy}(0) & \cdots & R_{yy}(-2p) \\ \vdots & \ddots & \vdots \\ R_{yy}(2p) & \cdots & R_{yy}(0) \end{bmatrix} \tag{5.14}$$

$$E[\mathbf{Y}\mathbf{W}^T] = E[(\mathbf{X} + \mathbf{W})\mathbf{W}^T] = E[\mathbf{W}\,\mathbf{W}^T] = \sigma^2\,\mathbf{I}$$

Here $\mathbf{R}_{yy}$ is the Toeplitz autocovariance matrix for the observed process and **I** is the identity matrix. The fact that $E[\mathbf{X}\,\mathbf{W}^T] = 0$ follows from the assumption that the sinusoids are uncorrelated with the noise. Expression (5.13) is then rewritten as

$$\mathbf{R}_{yy}\,\mathbf{A} = \sigma^2\,\mathbf{A} \tag{5.15}$$

which is an eigenequation where the noise variance (σ^2) is an eigenvalue of the autocovariance matrix $\mathbf{R}_{yy}$. The ARMA parameter vector **A** is the eigenvector associated with this eigenvalue that has been scaled so that the first and last elements are unity (**A** is symmetric). Equation (5.16) will yield the ARMA parameters of the Pisarenko technique. It turns out [1] that the noise variance eigenvalue is the smallest eigenvalue when the process is sinusoids in white noise and the autocovariance is exactly known.

Once the eigenvector **A** has been determined from the solution to (5.16), it remains to find the frequencies and powers to complete the harmonic decomposition (actually, nonharmonic). The frequencies are obtained by rooting the polynomial (5.9). Once the frequencies are known, the sinuoidal powers are obtained from the autocovariance lags, since

$$R_{yy}(k) = \sum_{i=1}^{p} P_i \cos\,(2\pi f_i k \Delta t) \tag{5.16}$$

for $k \neq 0$. A set of linear equations using p of the lags can then be used to find the sinusoid powers $\{P_i\}$.

5.3 MUSIC Technique

A closely related eigenvector-eigenvalue decomposition of the autocovariance matrix to produce a spectral estimate is the Multiple Signal Classification (MUSIC) method [10]. If m represents the number of narrowband signal components, form a matrix $\mathbf{S}_p$ of the eigenvectors formed from the minimum $(p-m)$ eigenvalues of the $p \times p$ autocovariance matrix $\mathbf{R}_{xx}$. The MUSIC spectral estimate is then simply

$$\hat{P}_{MUSIC}(f) = \frac{1}{\mathbf{E}^H \mathbf{S}_p \mathbf{S}_p^H \mathbf{E}} \tag{5.17}$$

where **E** was defined previously for the MLM spectral estimate.

This technique has also been used for high resolution beamforming. Using a uniform linear array of sensors, time samples are replaced by space samples. If the substitution

$$2\pi f = 2\pi (d/\lambda) \sin \theta$$

is made for frequency f, where d is the sensor spacing and λ is the signal wavelength, then (5.17) may be used for high resolution beamforming. This is a time-space duality between uniform time samples and uniform linear array space samples. Kung, in another chapter, discusses methods of solving eigenequations involving the Toeplitz matrix.

6. MULTIDIMENSIONAL MEM [11]

The maximum entropy method (MEM) has generated an enormous amount of acitivity in the field of multidimensional high resolution spectral estimation. Unlike the one-dimensional case where MEM and AR were equivalent, in the M-D case the true maximum entropy estimate is distinctly different from the spectrum derived by AR modeling. In fact, the computation of the MEM spectral estimate appears to require the solution of a non-linear optimization problem. Recent research has been directed at the problem of computing the true MEM estimate. The existing approaches either attack the non-linear problem with a general optimization algorithm or assume an approximation to simplify the calculations. While the optimum MEM spectrum was shown to be the inverse of a positive multivariate polynomial, this polynomial may not be factorable as the magnitude-squared of a finite order polynomial. Since a spectrum based on a multidimensional AR model will always be of the form $[A(\mathbf{k})]^{-2}$, the MEM is, in the M-D case, more general than the AR spectral estimate. Note, however, the M-D AR estimator will usually not satisfy the correlation matching property (expression (4.9) in one dimension), whereas the true M-D MEM spectrum will. On the other hand, experiments with the M-D AR estimate show that it does have potential as a high-resolution M-D spectral estimator.

7. SUMMARY OF METHODS

Table I provides a summary of all the techniques discussed in this paper, plus a few additional ones not covered. A brief overview of key properties, equation numbers from reference [1] for details of computation of each spectral estimate, and a list of key literature references, also from [1], to assist the reader to readily implement any of the techniques may be found in the Table. If the method only finds sinusoidal frequencies, the method is listed in the Table as a discrete (D) line type of spectrum; otherwise it is continuous (C). The time series model associated with each method, if appropriate, is listed in the Table. This table serves as a guideline for choosing a spectral method appropriate for the user's data.

Figure 1 illustrates typical spectra of the spectrum estimation methods listed in Table I. Each spectral estimate is based on the same 64-point real sample sequence from a process consisting of three sinusoids and a colored noise process, obtained by filtering a white Gaussian noise process. Two of the sinusoids are very close together at .2 and .21; these are frequency values expressed as fractions of the sampling frequency (i.e., fractional frequencies range between 0. and .5). A weak sinusoid may be seen at fractional frequency .1 . The colored noise process has a true spectrum of a raised cosine, as illustrated in Figure 1a. The spectral plots in Figure 1 are not intended to demonstrate relative performance of the various methods, but only to point out features of each technique.

8. APPLICATIONS

Several applications of some of the modern spectral estimation methods are described in this section. The range of applications is quite diverse, covering the fields of radar, sonar, geomagnetism, radio astronomy, and well logging.

8.1 Radar

The traditional purpose of radar is to measure range to a target. Modern spectral estimation has now provided the tools to improve the range resolution of radars that use so-called "chirp" (linear frequency modulation) emissions for ranging. The conventional range resolution of a radar is given by

$$\Delta R = \textit{Range Resolution} = \frac{c}{2}\,(\textit{Time Delay Resolution}) \approx \frac{c}{2\,BW}$$

where c is the speed of light and BW is the radar pulse bandwidth. The selection of chirp modulation of radar pulses turns range time processing into frequency analysis processing for range. Consider a chirp pulse return from a point target delayed Δt seconds from the transmitted chirp pulse, as illustrated in Figure 2. If these signals are mixed (product) at the receiver, the result is a constant beat frequency over the window of the time overlap (a pulsed sinusoid). The frequency for a point target at range R is

$$f_R = \gamma \Delta t = \left[\frac{2\gamma}{c}\right] R$$

where γ is the chirp rate in Hz/second [note that $BW = \gamma T$, where T is the pulse width]. Thus, the range of each point target becomes coded in terms of the beat frequency associated with the time delay of the returning radar pulse.

This principle of encoding range in terms of frequency is used in synthetic aperture radar (SAR) imagery [12]. An example of processing one line of a set of actual SAR radar data is shown in Figure 3. It was known that five spatially small (point) targets of varying radar reflectivity were in this line. A Prony spectrum and an FFT periodogram spectral estimate were applied to the 128 in-phase and quadrature (complex) data

TABLE I. SUMMARY OF MODERN SPECTRAL ESTIMATORS. Note that references and equation numbers listed in the table may be found in reference [1].

TECHNIQUE	KEY REFS.	DISCRETE LINE OR CONTINUOUS SPECTRUM	ALGORITHM PROCEDURE	ROUGH COMPUTATIONAL COMPLEXITY (ADDS / MULTS)	MODEL(S)	ADVANTAGES AND DISADVANTAGES	REMARKS
Periodogram** FFT Version	1–4 262	D	3.3, 3.4	$N \log_2 N$	Sum of harmonically related sinusoids	Output directly proportional to power Most computationally efficient Resolution roughly the reciprocal of the observation interval Performance poor for short data records Leakage distorts spectrum and masks weak signals	Harmonic least squares fit Requires some type of frequency domain statistical averaging to stabilize spectrum (e.g., Welch method) Windowing can reduce sidelobes at expense of resolution
Blackman-** Tukey (BT)	1–4	C	3.1, 3.2	Lag Ests.: NM PSD Est.*: MS	Identical to MA with windowing of the lags	Most computationally efficient if M<<N Resolution roughly the reciprocal of the observation interval Leakage distorts spectrum and masks weak signals	Negative PSD values in spectrum may result with some window weightings and autocorrelation estimates (e.g., unbiased)
Autoregressive (AR) Yule-Walker Version	1, 13 15, 16	C	4.5, 4.7	Lag. Ests.: NM AR Coeffs.: M^2 PSD Est.*: MS	Autoregressive (all-pole) process	Model order must be selected Better resolution than FFT or BT, but not as good as other AR methods Spectral line splitting occurs Implied windowing distorts spectrum No sidelobes	Model applicable to seismic, speech, radar clutter data Minimum-phase (stable) linear prediction filter guaranteed if biased lag estimates computed AR related to linear prediction analysis and adaptive filtering Models peaks in spectrum better than valleys
Autoregressive (AR) Burg-Algorithm Version	1, 3	C	4.21–4.23	AR Coeffs.: $NM + M^2$ PSD Est.*: MS	Autoregressive (all-pole) process	High resolution for low noise levels Good spectral fidelity for short data records Spectral line splitting can occur Bias in the frequency estimates of peaks No implied windowing No sidelobes Must determine order	Stable linear prediction filter guaranteed Adaptive filtering applicable Uses constrained recursive least squares approach
Autoregressive (AR) Least squares or forward-backward linear prediction version	1, 3	C	4.19	AR Coeffs.: $MN + M^2$ PSD Est.*: MS	Autoregressive (all-pole) process	Sharper response for narrowband processes than other AR estimates No spectral line splitting observed Bias reduced in the frequency estimates Must determine order No sidelobes	Stable linear prediction filter not guaranteed, though stable filter results in most instances Based on exact recursive least squares solution with no constraint
Moving Average (MA)	1, 3 7, 8	C	———	MA Coeffs.: Nonlinear Simult.Eqn.Set Lag Ests.: NM PSD Est.*: MS	Moving average (all-zero) process	Broad spectral responses (low resolution) Must determine order Has sidelobes	Generalized form of BT technique
ARMA (Yule Walker Version)	1, 7 8	C	4.26	Lag Ests.: NM Coeff. Computation: M^3 PSD Est.*: MS	ARMA process (Rational Transfer Function) (MA order ≠ AR order)	Must determine AR & MA orders	Models all rational transfer function processes Requires accurate lag estimates to obtain good results

TABLE I. SUMMARY OF MODERN SPECTRAL ESTIMATORS. Note that references and equation numbers listed in the table may be found in reference [1]. (Continued)

Pisarenko Harmonic Decompo-sition (PHD)	10, 12	D	5.12–5.16	Lag Ests.: NM Eigen eqn.: M^2 to M^3 Poly. Rooting: Dependent on Root Algorithm Powers: M^3	Special ARMA with equal MA and AR coefficients Sum of nonharmonically undamped sinusoids in additive white noise	Must determine order Does not work well in high noise levels Eigenequation and rooting are computationally inefficient	Requires accurate lag estimates to obtain good results Spurious spectral lines if order selected too high
Prony's Method (Extended)	1,3	C	4.27–4.37	AR Coeffs.: M^2 + NM Poly. Rooting: Dependent on Root Algorithm Amp. Coeffs.: M^3 PSD Est.*: MS	Sum of nonharmonically related damped exponentials ARMA with equal MA and AR coefficients and equal orders (p=q)	Must determine order Output linearly proportional to power Requires a polynominal rooting Resolution as good as AR techniques, sometimes better No sidelobes	Uses least squares estimates to obtain exponential parameters First step same as AR least sqaures estimation
Prony Spectral Line Decom-position	1,3	D	4.37–4.40	Coeffs.: M^3 Rooting: Function of root algorithm used Amp. Coeffs.: M^3	Sum of nonharmonically related sinusoids	Must determine order Output linearly proportional to power Requires a polynominal rooting Resolution as good as AR techqnives, sometimes better No sidelobes	Uses least squares estimation
Capon Maximum Likelihooc (MLSE)	9	C	5.1–5.7	Lag Ests.: NM Matrix Inversion M^3 PSD Ests.*: MS	Forms an optimal bandpass filter for each spectral component	Resolution better than BT; not as good as AR Statistically less variability in MLSE spectra than AR spectra	MLSE is related to AR spectra (see Eqn. (2.166))

** Computer programs may be found in Programs for Digital Signal Processing, edited by Digital Signal Processing Committee of the IEEE ASSP Society, IEEE Press, 1979.
* FFT could be used to generate $S = 2^{\nu}$ values of the PSD.
N=Number of data samples
S=Number of Spectral Samples Computed (usually S>>M)
M=Order of Model (or number of Autocorrelation Lags)

samples from this line. Figure 3 clearly indicates that the Prony method suppresses the sidelobes that obscure the five targets partially in the periodogram estimate, thereby making the number of targets and their ranges difficult to determine.

A similar application for radar involves measuring the spin rate of a spinning target [13]. The interest of this application is to determine the spin rate. Since the effective return pulse frequency is constantly changing due to the doppler effect, only a short time sequence can be processed in which the radar frequency can be considered effectively time stationary. Figure 4 illustrates radar velocity spectra versus time when processed by the periodogram (FFT) and autoregressive (using Burg algorithm) spectral estimation methods. The main lobe from each PSD line in the plot locates the range rate (velocity) of the object during a given processing interval. The object in this example also had a precession about its axis as well as a spin rate, thereby complicating and smearing the resulting spectra, which made accurate velocity estimation impossible with the periodogram approach (broad spectral smearing). However, individual reflection points on the target that move linearly in velocity as a function of time are visible on the MEM (AR) spectral plot, but are not obvious on the periodogram plot. By measuring the linear rate of change, one can then infer the spin rate. The AR PSD estimates using the Burg algorithm were thus able to detect the correct spin velocity and give an indication of the precessional envelope, as Figure 4 shows.

One last radar application is the use of AR PSD estimation to model continuous wave (CW) radar clutter [14] in order to classify the different forms of clutter as encountered in an air traffic control radar environment. Clutter here means unwanted echoes on a radar display due to reflection of the transmitted wave from such objects as ground, weather disturbances, or migrating flocks of birds. The spectral density of a CW radar clutter process has been shown to have a Gaussian shape, which can be closely represented by an AR process of relatively low order (perhaps as low as third order). The results shown in Figure 5 were generated from clutter data taken from an airport surveillance radar ASR-8 in Canada. Again, the Burg algorithm was used. The spectral spread, the sidelobe levels, and peak shift away from 0 are all a function of the type of clutter (ground, weather, or birds). This then permits the possibility of on-line classification of various types of clutter. Note that the modified (Welch) periodogram method does not, unlike the AR estimate, indicate all the distinguishing features that are observable in the AR spectra (such as peak lobe shift) This makes clutter classification more difficult with periodogram processing.

8.2 Speech

The use of AR spectral estimation for speech processing is one of the earliest uses of modern spectral estimation. Figure 6 is an example of overlapped periodogram and AR PSD estimates from a segment of speech. Note the smoother appearance of the AR PSD and its ability to track the formant frequencies of speech signals.

8.3 Sonar

Modern spectral estimation has found application as a spectral line tracker for passive sonar data systems. Figure 7 is a block diagram of the sonar signal processing system. Data from a deployed sonar beamformer system that has been complex demodulated down to baseband and then sampled was put through the two processing chains shown in Figure 7. A long-term periodogram shown in Figure 8 clearly brings out the single sonar target in the ocean noise. The long-term average used to produce the periodogram improves the detection, but smoothes out any small variations in target movement. If the data is processed in shorter time segments and a target line tracker implemented as illustrated in Figure 7, the resulting line tracking performance is shown in Figure 8, which compares the periodogram and AR line tracking methods. The data had a Rayleigh fading sinusoidal target signal at about -15 dB signal-to-noise ratio in the ocean noise. Note that the $\alpha-\beta$ trackers shown in Figure 7 are second order recursive filters that resemble the predictor-corrector form of a Kalman filter. The AR tracker actually is a combination FFT pre-processor followed by AR analysis; it is described in more detail in reference [16]. The box symbols in Figure 9 denote the raw frequency estimates (no $\alpha-\beta$ tracker smoothing), while the solid curves represent the smoothed estimate out of the $\alpha-\beta$ tracker. Figure 9 shows that the AR line tracker has picked up some frequency variations not seen by the periodogram tracker, which may also be local colored ocean noise effects as well as target movement.

8.4 Radio Astronomy

One particular method of two-dimensional maximum entropy spectral estimation has been applied by Wernecke [17] to the production of radio brightness maps of stellar RF sources. The data for the radio brightness maps is obtained by earth aperture systhesis of noisy interferometer measurements. The brightness maps shown in Figure 10 are based on data from the Stanford University five-element radio telescope [17] that operates at a 2.8 cm wavelength. The maps are of the galactic radio source Cygnus A (3C405). Sidelobe ringing present in the direct Fourier transformed (periodogram) map is very high, but is negligent in the maximum entropy plot. The objects seem to have steep outer edges, with a more gradual contouring in the center. This structure was confirmed by maps made with other techniques at other observatories.

8.5 Microwave Antenna Arrays

The use of AR (maximum entropy) spectral estimates has increased the angular resolvability of microwave linear antenna arrays. Consider the application for inphase-quadrature data from a 12-element vertical antenna array at Stanford University, used for reception of transhorizon microwave signals, as shown in Figure 11. This array has been used to study lower atmospheric influences on microwave signal propagation beyond the horizon. It was found that an aircraft landing at San Francisco airport reflected some of the microwave signal energy, the result of which was clearly resolvable by the angular array response obtained by AR spectral processing (see Figure 12). It was not clearly resolved by usual FFT beamforming processing.

8.6 Geomagnetism

An interesting, and very early, application of AR PSD estimation was the determination of the periodicity(ies) of polarity reversals in the earth's magnetic field [19]. Figure 13a is a plot of the percentage of normal polarity in world-wide palaeomagnetic investigations, which is obviously the type of short data record where modern spectral estimation should shine. Figure 13b and c are spectra processed by both the periodogram and Burg algorithm AR methods. The geomagnetic field exhibits plainly two inherent periodicities, the weaker of which had not been reported before in the literature. It turned out that this weaker component was correlated with a 280 m.y. rotation in the Milky Way.

8.7 Well Logging

High resolution spatial spectrum analysis using the Prony method has been applied to sonic well logging. In Figure 14a, an array of receivers in a sonic device receives sonic waves propagating in a well borehole, having been generated by a transmitter several feet away. The velocities of the various components of the received waveforms at each receiver can be used to determine properties of the rock through which the sonic wave travels. Wave components which propagate at lower velocities arrive later in time and often overlap earlier wave arrivals. The procefure outlined here makes estimates of the various wave component velocities (or slowness $p = 1/v$) as a function of well depth in order to make a log. Since the spatial extent of the array aperture is only 8 receivers, high resolution spectral estimation is needed for the spatial dimension of signal processing.

The processing procedure is to estimate the spatial frequencies present in the complex-valued sequence $S(f,k)$ using a variation of Prony's method, from which wave velocities may be inferred. $S(f,k)$ is obtained by computing, at a single temporal frequency f, the discrete Fourier transform of the kth received waveform, $s(t,x)$, after a windowing procedure has been performed to attenuate interfering signals [see Figure 14b]. A variation of Prony's method makes a spatial estimate of frequency, from which velocity (the inverse of slowness) may be calculated [see Figure 14c]. The estimates of temporal and spatial frequencies for various times are used in a clustering procedure to identify individual wave components [see Figure 14d]. For the illustrated example, this high-resolution method was able to separate the two closest wave components indicated by the arrows. The conventional analysis was not able to distinguish them.

9. CURRENT PROBLEMS IN SPECTRAL ESTIMATION

Problems of current interest to spectral estimation researchers include fast algorithms and efficient algorithmic structures for real-time hardware implementation of high resolution spectral estimation techniques, improved ARMA modeling techniques, and better eigenanalysis approaches for spectral estimation and linear array beamforming. High resolution multidimensional spectral estimation continues to hold much research appeal, although it is primarily of academic rather than practical interest due to the complex formulations and solutions involved.

REFERENCES

[1] Kay, S.M. and Marple, S.L. Jr.,"Spectrum Analysis -- A Modern Perspective",*Proceedings of the IEEE*,vol. 69,pp. 1380-1419, November 1981

[2] Jenkins, G.M. and Watts, D.G.,*Spectral Analysis* and *Its Applications*, Holden-Day, Inc. 1968.

[3] Childers, D.G.,editor,*Modern Spectrum Analysis*,IEEE Press, 1978.

[4] Geckinli, N.C. and Yavuz, D., *Discrete Fourier Transformation and Is Applications to Power Spectral Estimation* ,Elsevier Scientific Publishing Company, Amsterdam, 1983.

[5] Rabiner, L.R. and Gold, B.,*Theory* and *Applications of Digital Signal Processing*, Prentice-Hall,Inc., 1975.

[6] Zohar, S.,"FORTRAN Subroutines for the Solution of Toeplitz Sets of Linear Equations," *IEEE Trans. on Acoustics, Speech,* and *Signal Processing*, vol. ASSP-27, pp.656-658, December 1979.

[7] Cadzow, J.A.,"ARMA Modeling of Time Series," *IEEE Trans. on Pattern Analysis* and *Machine Intelligence*, vol.PAMI-4, pp.124-128, March 1982.

[8] Friedlander, B.,"Efficient Algorithms for ARMA Spectral Estimation," *IEE Proc.*, vol.130, Pt.F, pp.195-201, April 1983.

[9] Capon, J.,"High Resolution Frequency-Wavenumber Spectrum Analysis," *Proc. IEEE*, vol.57, pp.1408-1418, August 1969.

[10] Schmidt, R.,"Multiple Emitter Location and Signal Parameter Estimation," RADC Spectrum Estimation Workshop Record, pp.243-258, October 1979.

[11] McClellan, J.H.,"Multidimensional Spectral Estimation," *Proc. IEEE*, vol.70, pp.1029-1039, September 1982.

[12] Marple, S.L. Jr.,"Exponential Energy Spectral Density Estimation", in *Proceedings of the* 1980 *IEEE ICASSP*, 9-11 April 1980, Denver, CO., pp. 588-591

[13] Bowling, S.B.,"Linear Prediction and Maximum Entropy Spectral Analysis for Radar Applications",MIT Lincoln Laboratory Project Report RMP-122, 24 May 1977

[14] Haykin, S.S., Kesler, S., and Currie, B.,"An Experimental Classification of Radar Clutter",*Proceedings of the IEEE*,vol. 67, pp. 332-333, February 1979.

[15] Makhoul, J.,"Linear Prediction: A Tutorial Review", *Proceedings of the IEEE*, vol. 63, pp. 561-580, April 1975

[16] Kay, S.,"Final Report: Linear Spectral Prediction for Passive Sonar", Raytheon Company, Submarine Signal Division, Internal Research Report, April 1980

[17] Wernecke, S.J.,"Two-Dimensional Maximum Entropy Reconstruction of Radio Brightness",*Radio Sciences*, vol. 12, pp. 831-844, September- October 1977

[18] Thorvaldsen, T., Waterman, A.T., and Lee, R.W.,"Maximum Entropy Angular Response Patterns of Microwave Transhorizon Signals", *IEEE Transactions on Antennas and Propagation,* vol. AP-28, pp. 722-724,September 1980

[19] Ulrych, T.,"Maximum Entropy Power Spectrum of Long Period Geomagnetic Reversals", *Nature*, vol. 235, pp. 218-219, 28 January 1972

[20] Parks, T.W., McClellan, J.H., and Morris, C.F., "Algorithms for Full-Waveform Sonic Logging," ASSP Spectrum Estimation Workshop II, 10-11 November 1983, Tampa, Florida, pp.186-191.

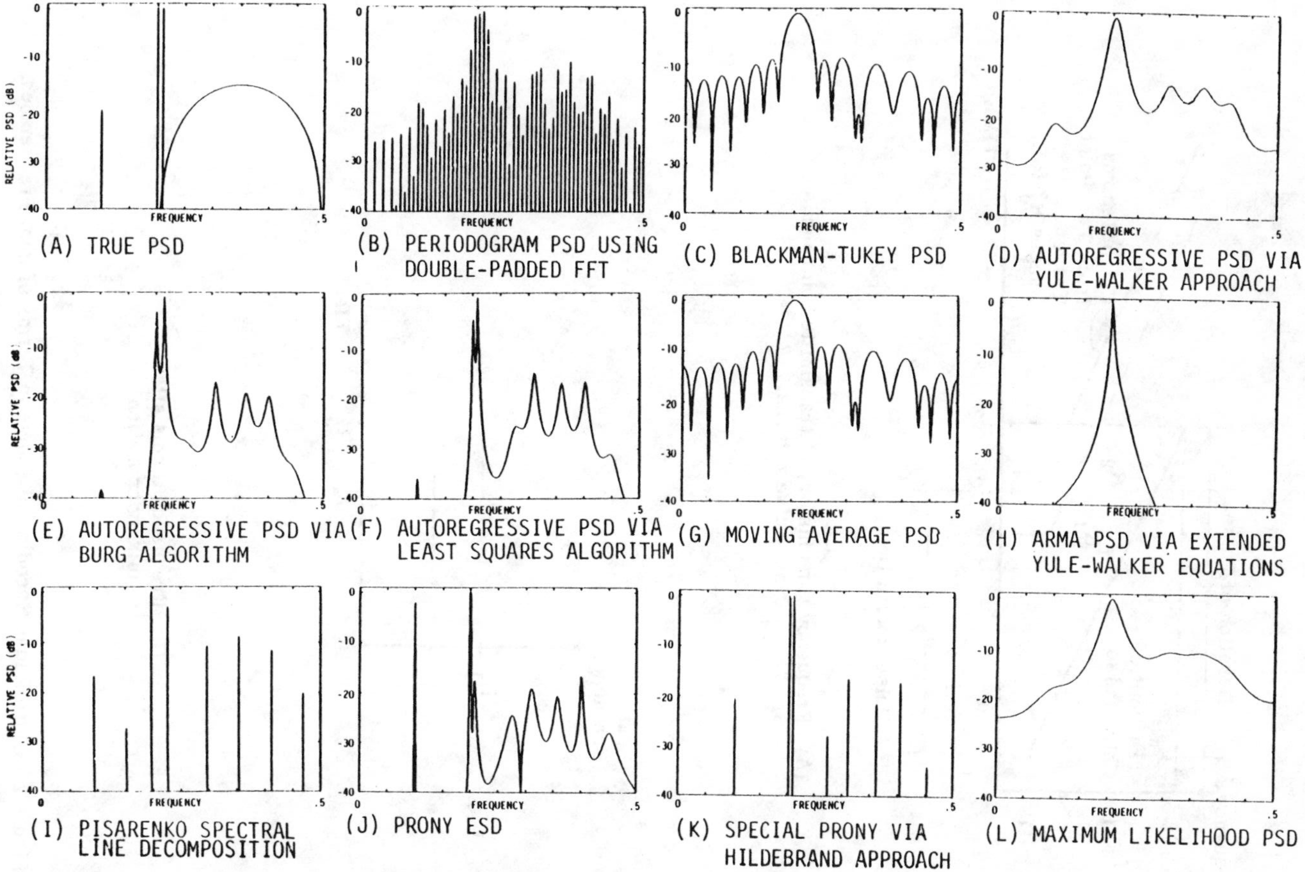

Figure 1. Examples of Various Spectral Estimates for the Same 64-Point Sample Sequence (from reference [1]).

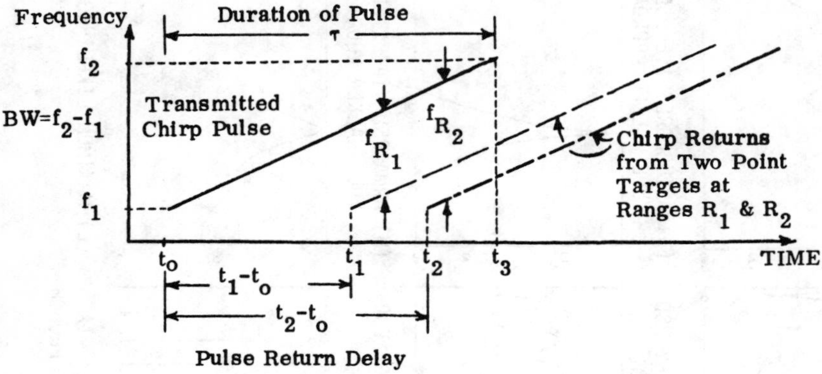

(A) Frequency time history of transmitted chirp pulse and pulse returns.

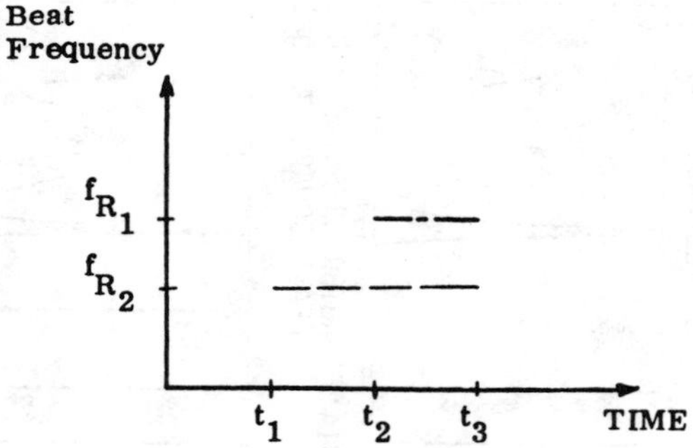

(B) Frequency terms after mixing operation

Figure 2. Chirp Radar Encoding of Range in Terms of Beat Frequencies.

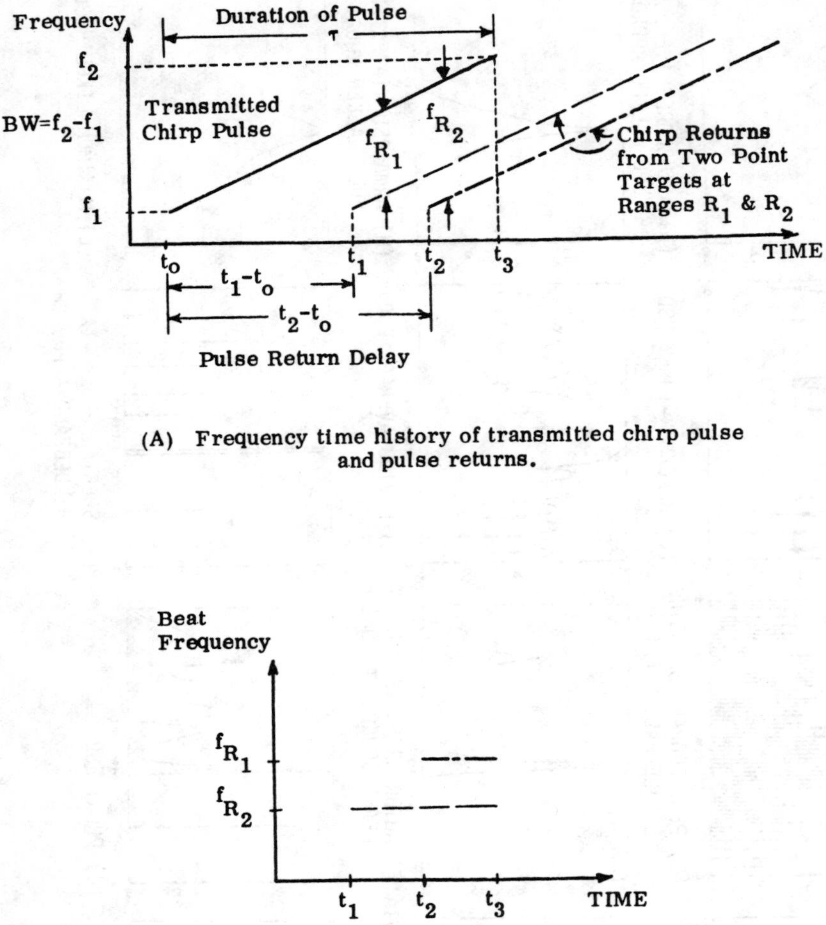

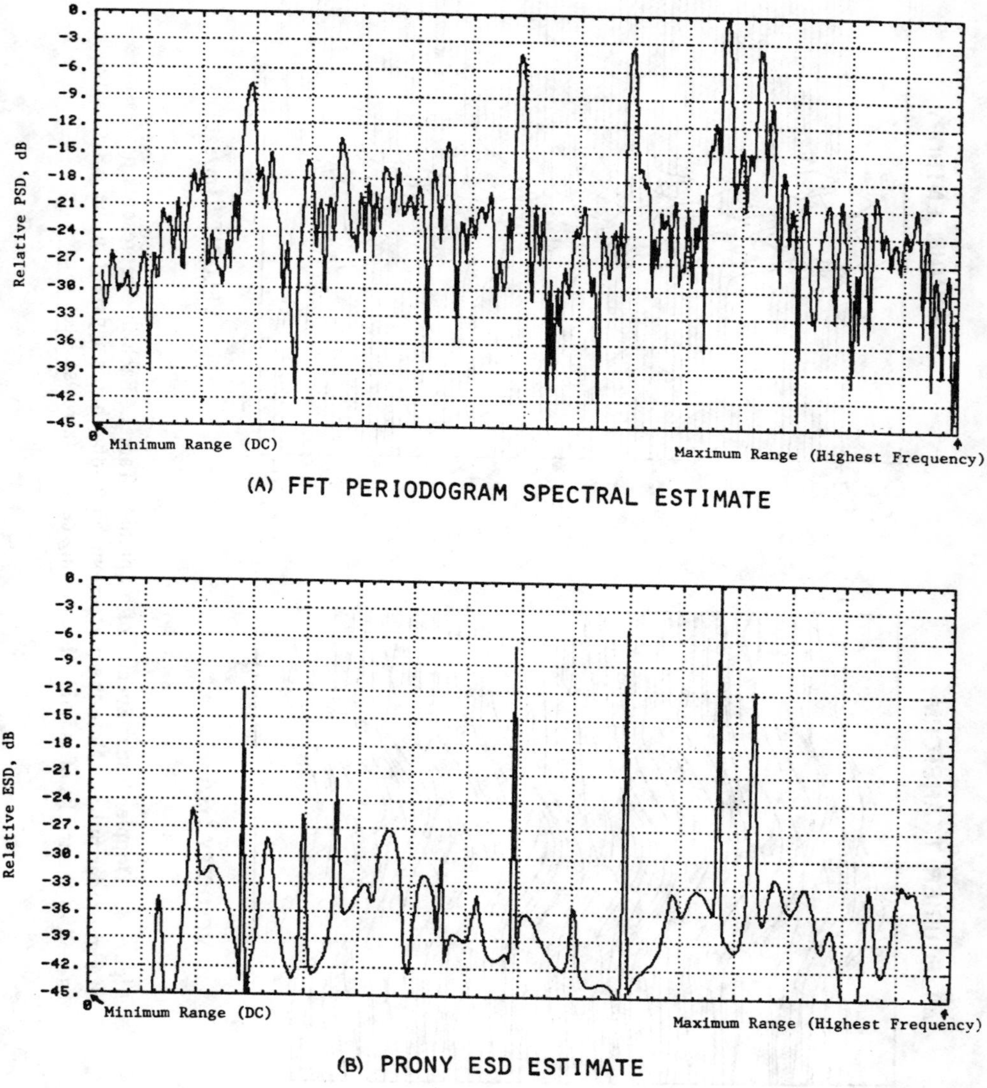

Figure 3. Doppler Radar Spectra of Five Spatially Small (Point) Targets [Ref. 12].

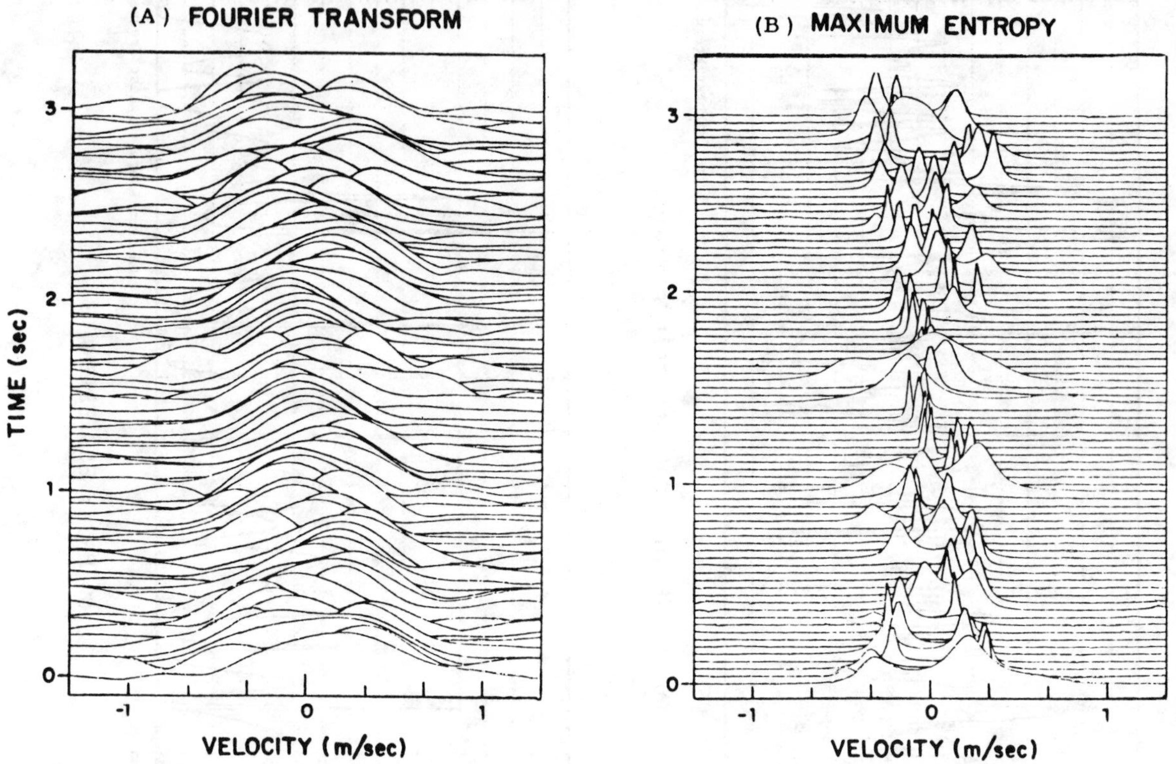

Figure 4. Doppler Radar Spectra of a Spinning Target. Each line represents processing of eight radar pulses. Note the precessional envelope that is apparent in the AR spectra, as well as the tracks of individual scattering centers (arrows) [Ref 13].

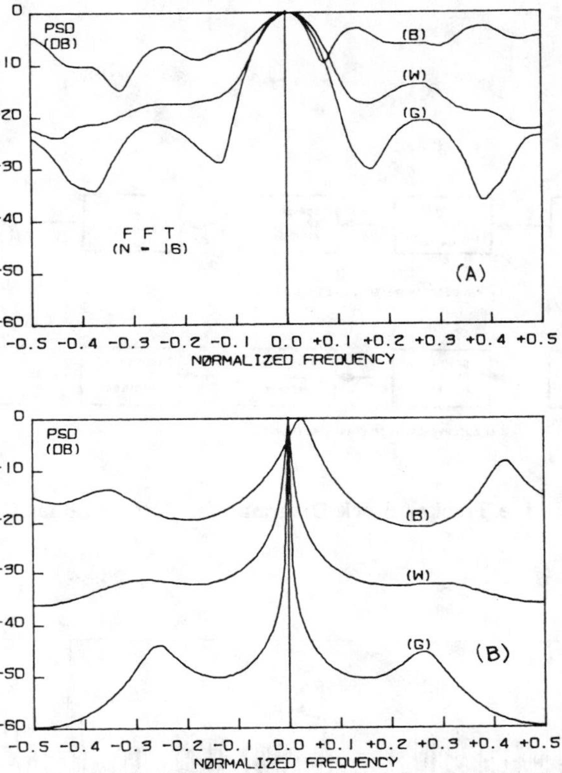

Figure 5. Spectral Estimates of Actual Ground (G), Weather (W), and Bird (B) Radar Clutter Signals . (a) Welch Periodogram Spectral Estimate. (b) AR Spectral Estimate [Ref. 14].

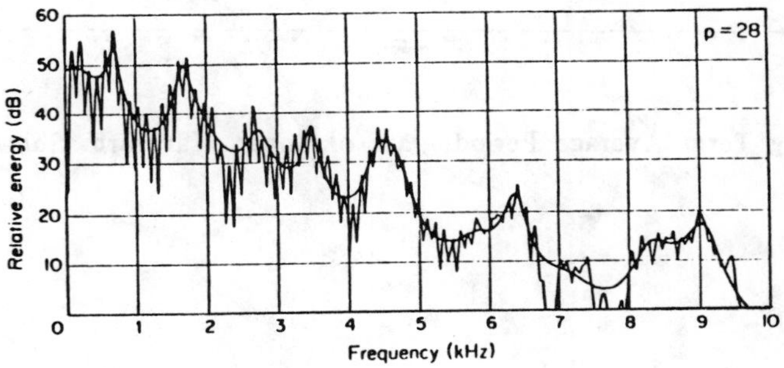

Figure 6. Comparison of Periodogram and AR(28) PSD Estimates on a Section of Speech [Ref. 15].

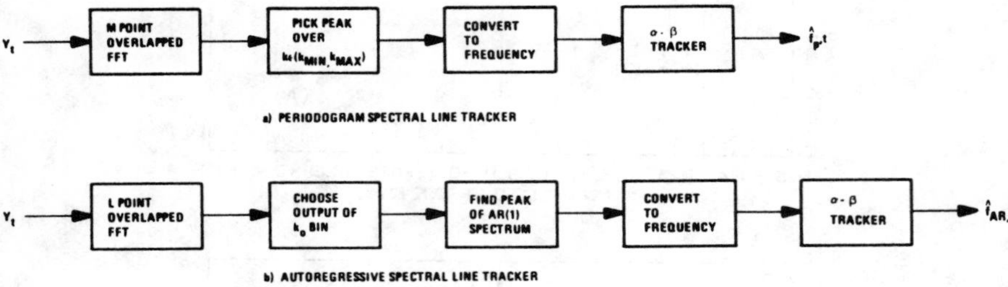

Figure 7. Spectral Line Tracker Block Diagrams for Passive Sonar Data Processing [Ref. 16].

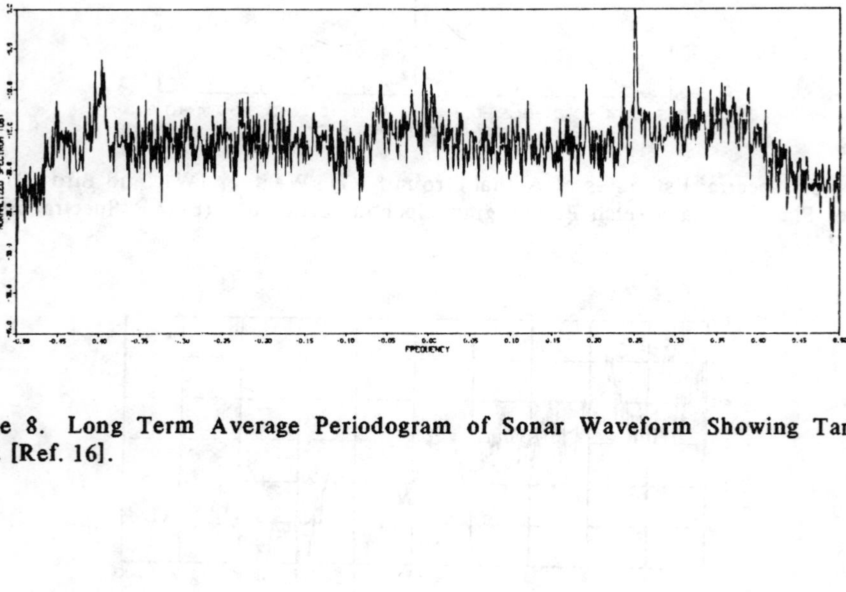

Figure 8. Long Term Average Periodogram of Sonar Waveform Showing Target at Right [Ref. 16].

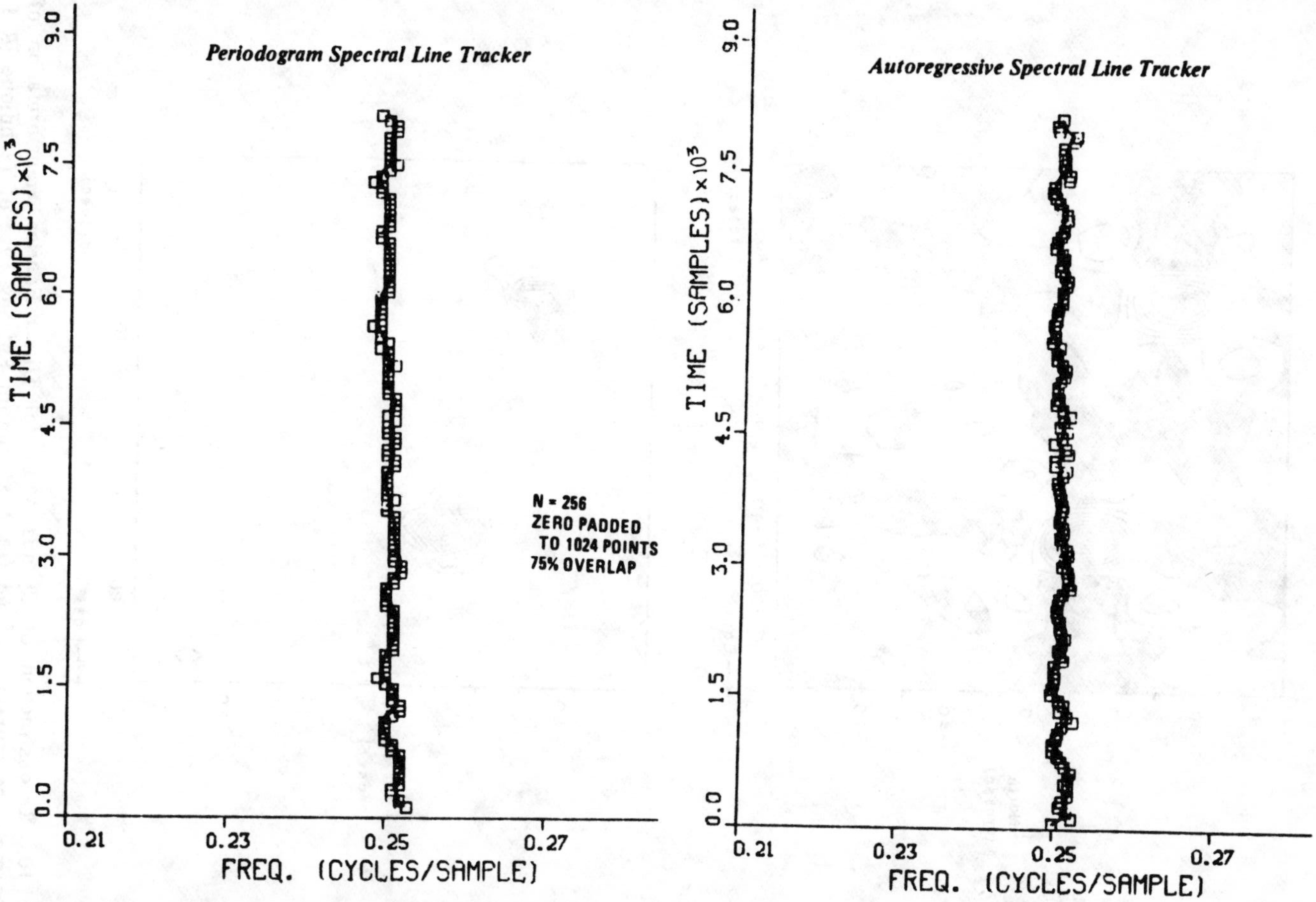

Figure 9. Performance of Spectral Line Trackers Using Actual Sonar Data with a Single Target Present [Ref. 16].

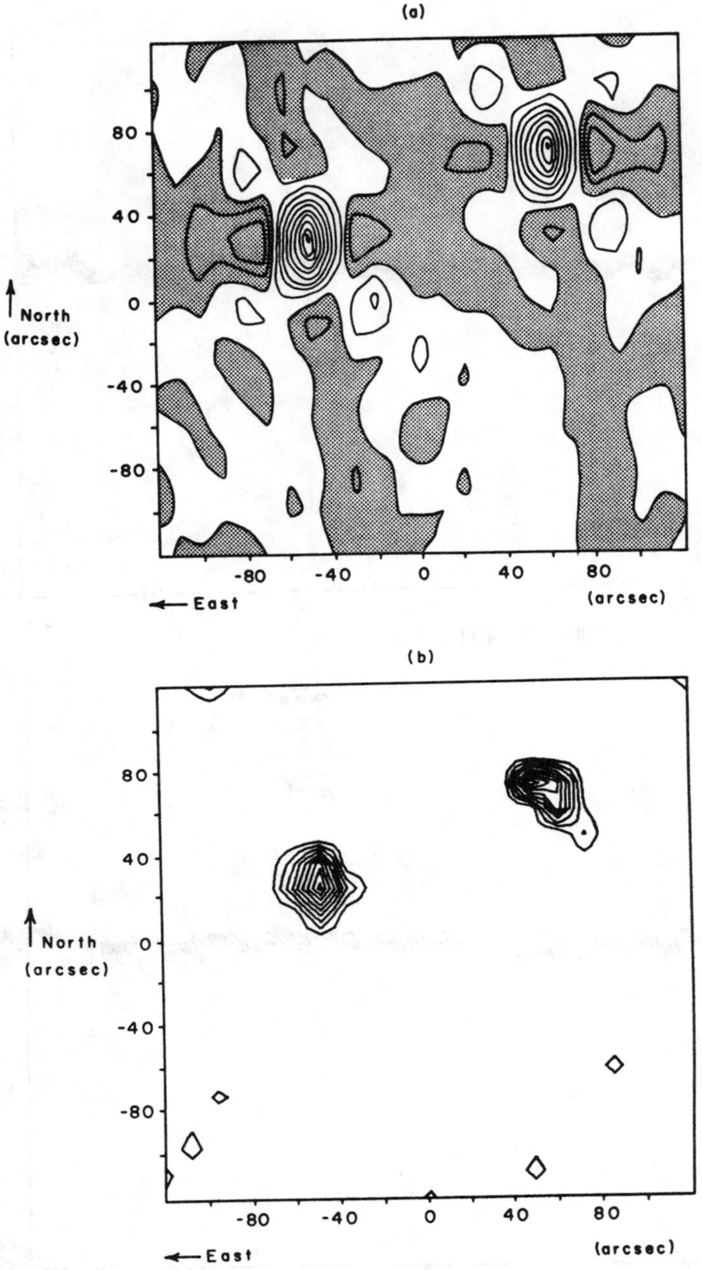

Figure 10. Reconstruction of Cygnus A by (a) Direct Fourier Transformation of Interferometer Measurements, and (b) A Maximum Entropy Type of Technique [Ref. 17].

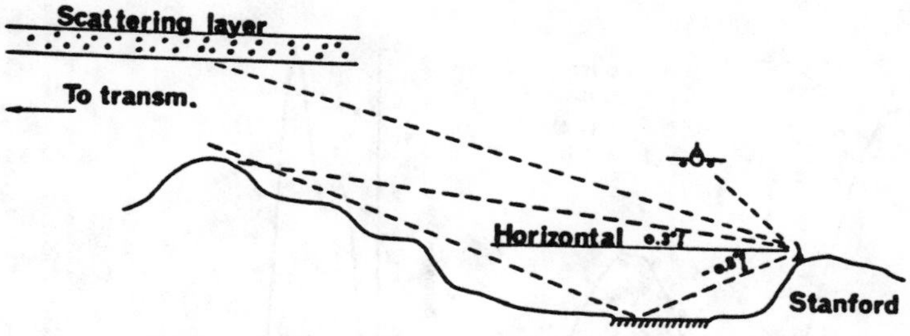

Profile of terrain near receiving end of transhorizon propagation path.

Figure 11. Profile of Terrain at Receiving Antenna Array for Transhorizon Propagation Path Measurement System [Ref. 18].

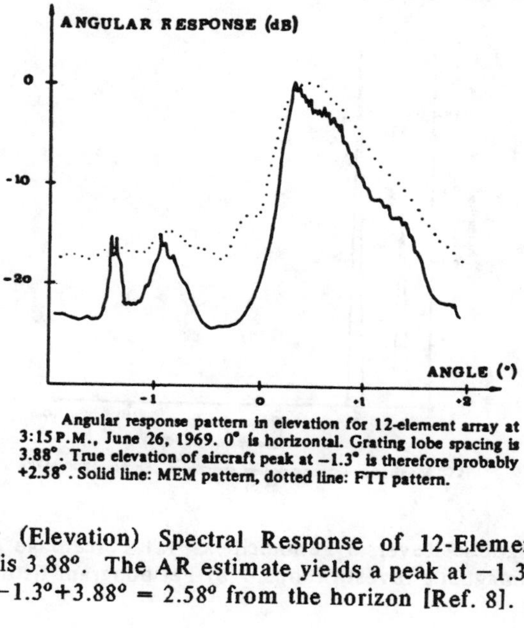

Angular response pattern in elevation for 12-element array at 3:15 P.M., June 26, 1969. 0° is horizontal. Grating lobe spacing is 3.88°. True elevation of aircraft peak at −1.3° is therefore probably +2.58°. Solid line: MEM pattern, dotted line: FTT pattern.

Figure 12. Angular (Elevation) Spectral Response of 12-Element Antenna Array. Grating lobe spacing is 3.88°. The AR estimate yields a peak at $-1.3°$ due to an aircraft, which is probably at $-1.3°+3.88° = 2.58°$ from the horizon [Ref. 8].

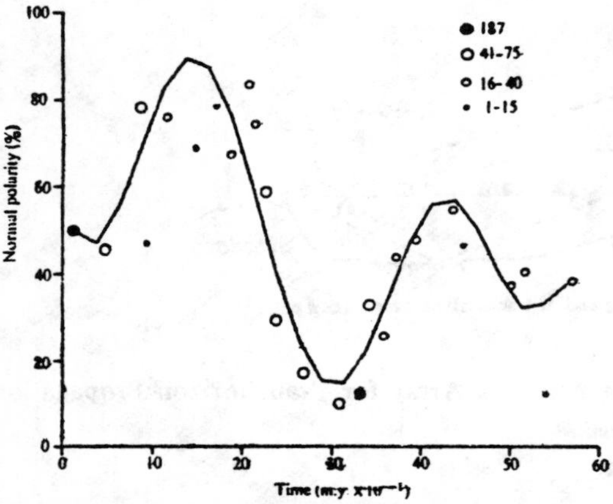

(a) Percentage of normal polarity in world-wide palaeomagnetic investigations for the Phanerozoic.

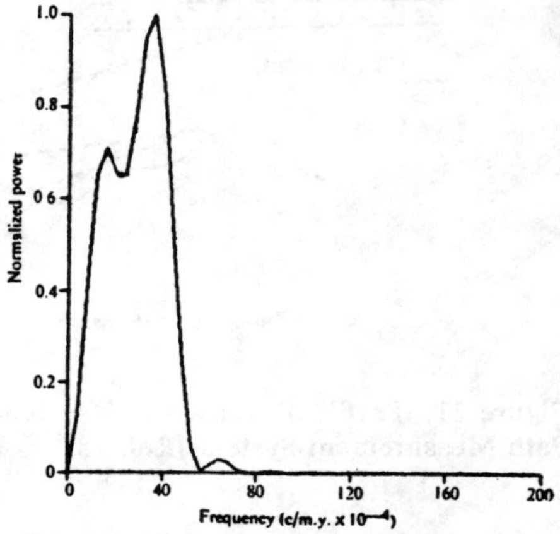

(b) Power spectrum computed as the square of the modules of the Fourier transform. Sampling interval, 25 m.y.

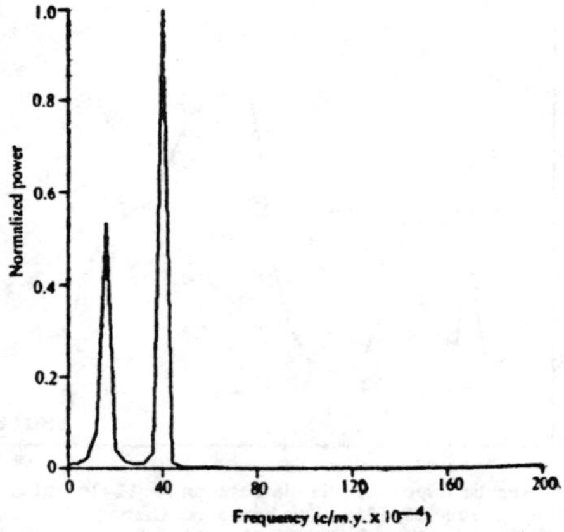

(c) Maximum entropy power spectrum. Sampling interval, 25 m.y.

Figure 13. Geomagnetic Reversal Periodicities. (a) Percentage of Normal Polarity in World-Wide Palaeomagnetic Investigations. (b) Periodogram Estimate. (c) AR Spectral Estimate [Ref. 19].

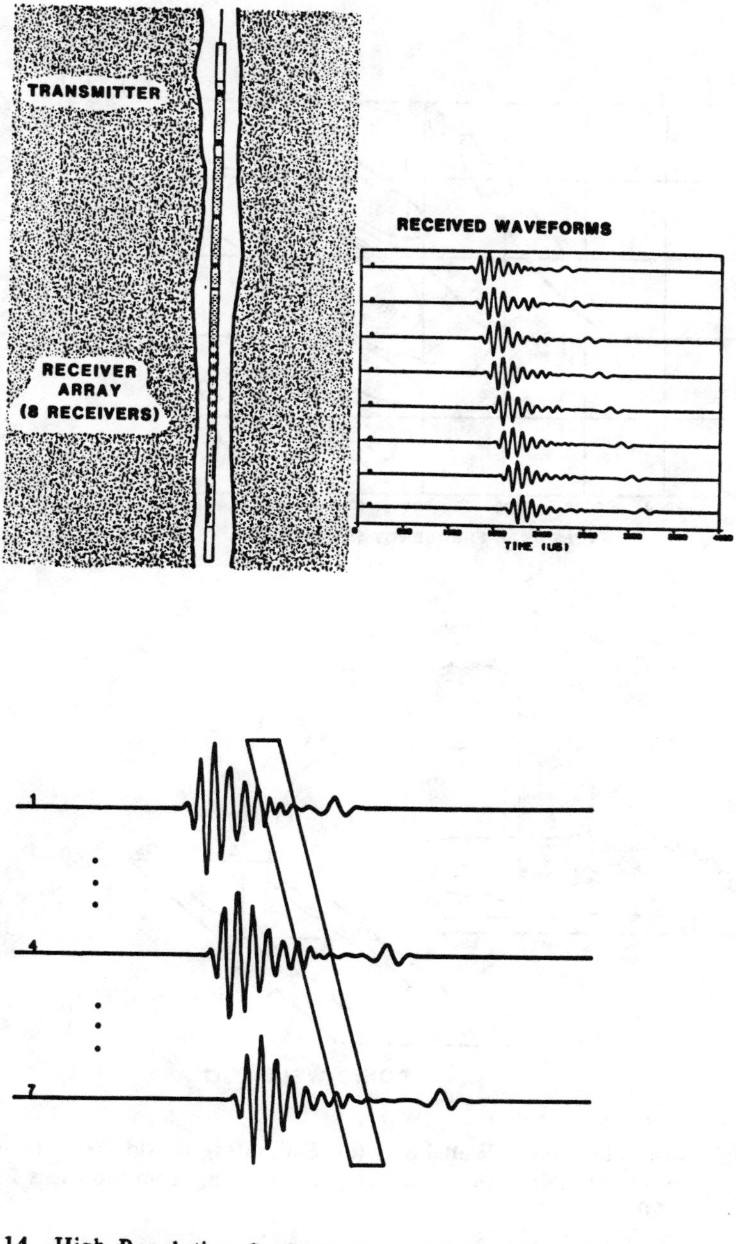

Figure 14. High Resolution Sonic Well Log. (a) Logging Device (Tool) with Typical Receiver Array Waveforms. (b) Frequency-Wavenumber Spectrum for Waveforms from *a*.

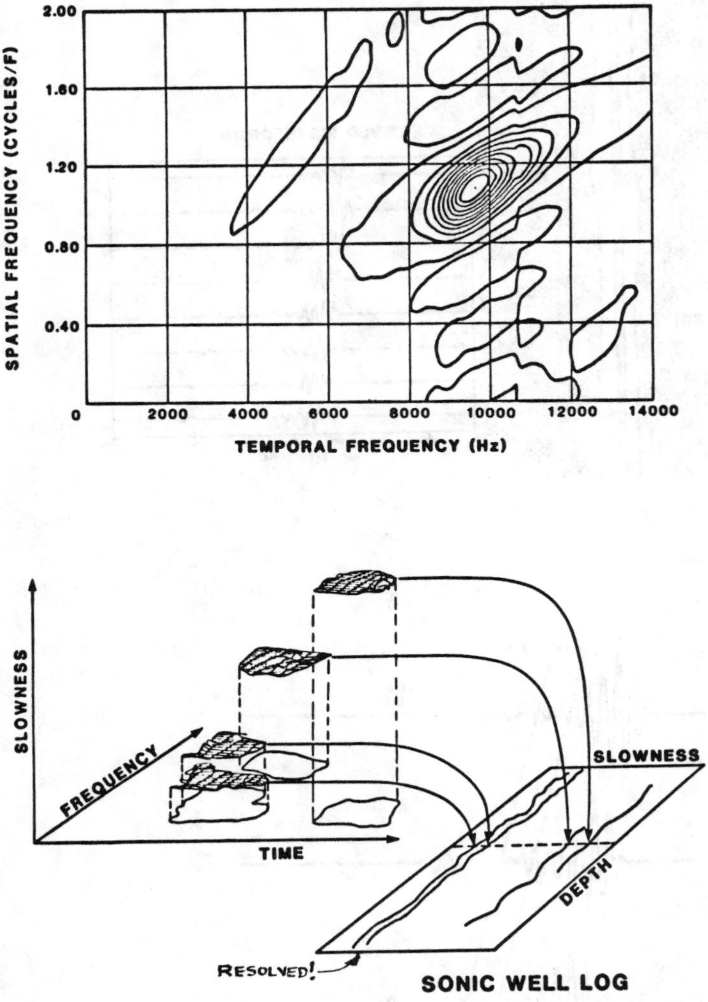

Figure 14. High Resolution Sonic Well Log. (c) Estimates of Slowness (Inverse Velocity) Vs. Time at 10 kHz. (d) Producing a Sonic Log from Slowness Estimates by Clustering [Ref. 20].

4

Advanced Digital Filters

P. DEWILDE

We present some techniques to realize advanced high quality filters. Most filters discussed will be of IIR type and will be suited for high quality selective filtering or for linear prediction. The synthesis techniques are numerical and algebraic leading to orthogonal filters and wave digital filters. In the case of orthogonal filters, the system map (state, input) to (next state, output) is an orthogonal map while in the case of wave digital filter, the filter realization is algebraically minimal with a transfer contrained in magnitude. We give an introductory discussion to both types of filters, explore their properties, present a synthesis method and explore various connections, e.g. with estimation theory. It turns out that the now famous Schur method plays a central role and brings a strong unity to the theory.

TABLE OF CONTENTS

I. INTRODUCTION

Digital Filters. The term "digital filtering" is used in many circumstances, some of which have great technical significance. By way of motivation, let me give a few instances where filtering is of prime import for today's electrical products.

Example 1: the CODEC. A CODEC transforms an analog voice signal into a digitally coded form (a Pulse Code Modulated-PCM-signal), and vice-versa. The following sequence of operations is necessary for the voice signal to PCM conversion fig. (1.1):

- a low-pass filtering of the signal down to a signal whose frequency range is restricted essentially to less than half the sampling frequency. This filter is called an anti-aliasing filter and it is intended to restrict distortion during A/D conversion.

- a A/D conversion and a coding of the result - after use of a companding law needed to increase the dynamic range.

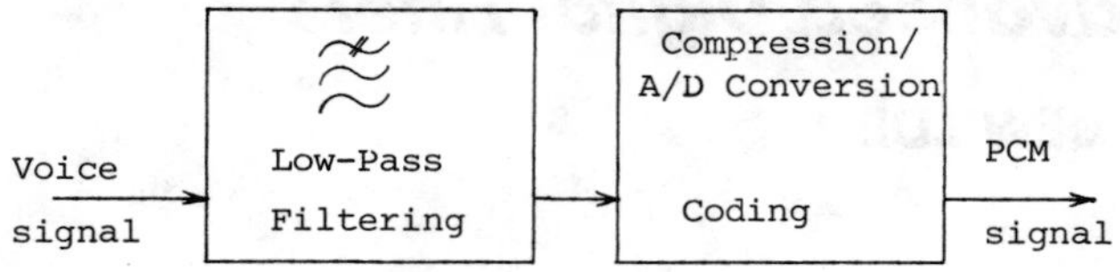

Fig. 1.1 Basic Scheme of CODEC operation.

The low-pass filtering may be done directly on the analog voice signal using a classical analog filter. The analog filter may be of medium quality and selectivity (say a 6th degree filter) and needs a cut-off frequency of ca. 3 kHz. It can be either an LC circuit with large L's and C's due to the low frequency range or an active filter with opamps and again bulky capacitors. None of these solutions is really attractive because the circuits proposed are not suited for integration.

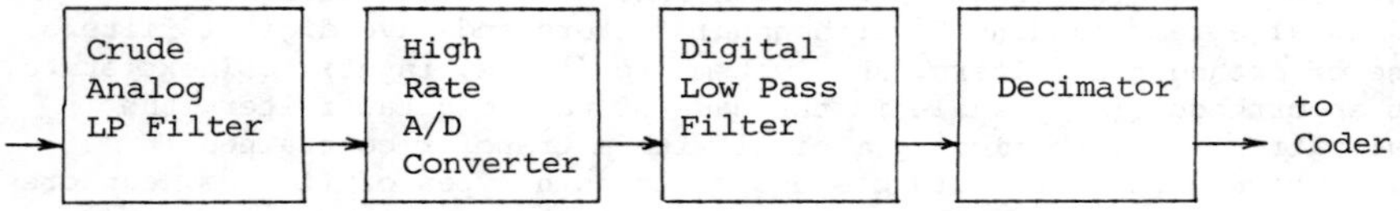

Fig. 1.2 A digital (VLSI-suited) realization of the low-pass filtering and A/D conversion.

A solution which is better suited for VLSI signal processing is shown in fig. 1.2.

Instead of performing an accurate low-pass filtering down to a frequency of 3 kHz to prevent aliasing of the signal during A/D conversion, we use a much higher sampling rate together with a crude analog LP filter (say 30 kHz cutoff frequency with 68 kHz sampling rate). Because the original voice signal is limited in frequency to 20 kHz, a very simple LP filter - say an RC circuit - will do. Moreover, it will not be bulky because of the higher frequency. After the high rate D/A converter follows an accurate digital low pass filter followed by a decimator, which brings down the sampling rate to the desired one by picking one sample every so many. Both the high rate D/A converter and the digital low pass filter are easy to integrate. Because the CODEC circuitry is to be produced in very large quantities one will opt for the most efficient filter circuit. In the present day technology that would be a wave-digital filter (WDF) consisting of an optimized sequence of multiplier coefficients with very few bits and some adder circuits. The efficiency of the WDF is due mainly to its low sensitivity with the consequence that multiplier coefficients may crudely be approximated by a few shifts and adds/substracts (typically two or three).

<u>Example 2</u>: Analysis and Reproduction of Speech. Another potential large-volume application of digital filtering is the analysis and reproduction of speech. Many companies are busy developing systems for voice mail to complement present day telephone circuits or computer networks. A voice mail system will normally use a speech estimation and a speech modelling filter. The orthogonal filters which are discussed here are well suited for that purpose: their structure is highly modular and adaptive and they originate in a natural way from the speech analysis algorithm.

ARMA (IIR) filters. In the applications discussed above and in many other applications it will prove necessary to use the most general type of filter possible: the ARMA (Autoregressive-moving response) filter. This type of filter has a rational transfer function with non-trivial numerator and nontrivial denominator. There are a number of canonical methods to realize such filters. These methods have many defects and are mostly unsuited for filter design. The more advanced techniques which we shall discuss in the course of the paper will take care of the defects. By way of illustration, we treat an example of a canonical design and discuss its problems.

A canonical IIR design. Suppose that we wish to design a 3^d order low-pass filter with cut-off frequency $\theta_c = \pi/10$ and suppose that we have choosen a Butterworth characteristic. A standard s-plane Butterworth filter with cut-off frequency at j=1 along the ω-axis in the complex s-plane has the transfer function:

$$\frac{1}{s^3+2s^2+2s+1} \tag{1.1}$$

1. The order of the filter is 3.
2. The transfer is maximally flat at s=0.
3. There is a transmission zero of degree 3 at $s=\infty$.

These properties may be transferred to the unit circle by means of a degree one (bilinear) conformal transformation which maps the right half s-plane onto the interior of the unit circle and keeps the degree and the analytical properties of the frequency response intact:

$$s = K\,\frac{1-z}{1+z} \tag{1.2}$$

where K is a constant which must be chosen so that the cut-off frequency is at the position $e^{\pm i\theta_c}$:

$$i = K\,\frac{1-e^{-i\theta_c}}{1+e^{-i\theta_c}} \rightarrow K = 1/\tan(\theta_c/2) \tag{1.3}$$

This leads to the transfer function (see chapter 2 of this paper for the conventions used):

$$H(z) = \frac{1+3z+3z^2+\ z^3}{345.1 - 819.3z + 665.8z^2 - 183.6z^3} \tag{1.4}$$

which is analytic inside the unit disc. We interpret now the variable z as a unit delay and rewrite (1.4) in the form of a difference equation which will lead us to a canonical realization:

$$Y(z) = \frac{1}{345.1}\ [U(z)+z[3U(z)+819.3Y(z)+z[3U(z)-665.8Y(z)+z[U(z)+183.6Y(z)]]]\ .$$

For the Butterworth amplitude response of fig. 13., we have now obtained the canonical realisation of fig. 1.4.

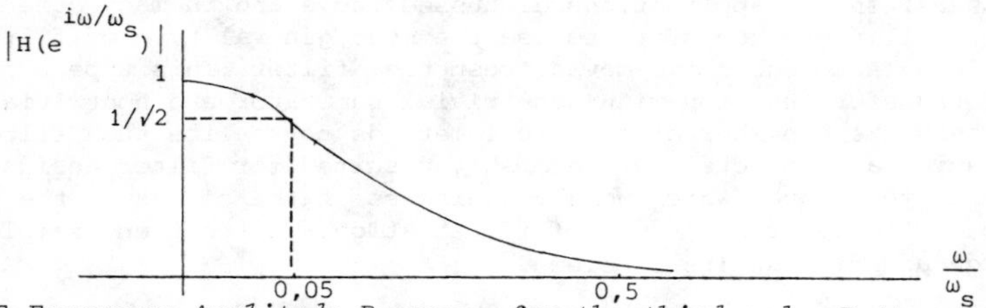

Fig.1.3 Frequency Amplitude Response for the third order Butterworth Filter with cut-off frequency at 00.5 of the sample frequency ω_s.

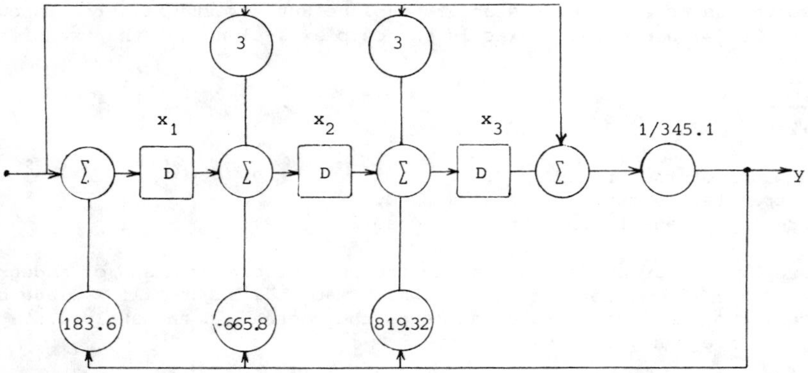

Fig. 1.4 Realization of (1.4) in a canonical form.

The state equations for fig. 1.4 are (with the state assignment shown):

$$\begin{bmatrix} x_1(t+1) \\ x_2(t+1) \\ x_3(t+1) \end{bmatrix} = \begin{bmatrix} 0 & 0 & 0.532 \\ 1 & 0 & -1.929 \\ 0 & 1 & 2.374 \end{bmatrix} \begin{bmatrix} x_1(t) \\ x_2(t) \\ x_3(t) \end{bmatrix} + \begin{bmatrix} 1.532 \\ 1.071 \\ 3.374 \end{bmatrix} u(t)$$

$$y(t) = \begin{bmatrix} 0 & 0 & \frac{1}{345.1} \end{bmatrix} \begin{bmatrix} x_1(t) \\ x_2(t) \\ x_3(t) \end{bmatrix} + \frac{1}{345.1} u(t)$$

The realization so obtained is said to be in the "observer canonical form" because a. the A matrix is in companion form and b. the C matrix is essentially a unit vector. A dual form, with A,B,C changed to A^T, C^T, B^T would yield the alternative "controller canonical form" with the same transfer function.

In the passband of the filter, in the vicinity of z=1, this filter realization will exhibit a high sensitivity to variations of the multiplier coefficients. A variation of 0.1% in one of the lower multiplier coefficients will result in a variation of ca. 10% in the amplitude. This can be seen from the fact that the value H(1)=1 is obtained by cancellation between large quantities:

$$345.1 - 819.3 + 665.8 - 183.6 = 8$$

a mistake of 1 in 819.3 (i.e. ca. 0.1%) results in a mistake of 1 in 8 in the frequency response (i.e. ca. 10%).

The consequence, however, is that the precision in which the filter is implemented must be at least 100 * better that the precision of the data in order to obtain a comparable precision in the output sequence. Because of the

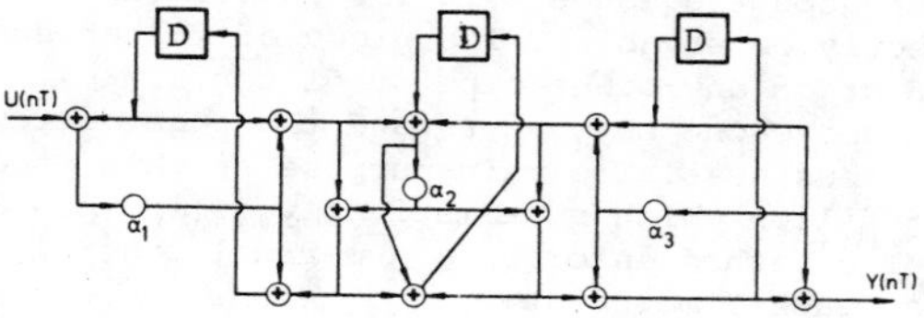

$\alpha_1 = \alpha_3 = 0.137$

$\alpha_2 = 0.0212$

Fig. 1.5 Wave Digital (Low Sensitivity) realization of 1.4 [*1*].

poor quality of canonical realizations we shall henceforth consider only more advanced types of IIR filters.

Using the theory of Wave Digital Filter or of Orthogonal Filters, it is possible to produce a filter realization for H(z) for which the sensitivity is exactly one. Such a realization is given in fig. 1.5. A low sensitivity here is due to Orchard's theorem which will briefly be discussed later in this paper.

Although the flow diagram in fig. 15. looks more elaborate than in fig. 1.4, the coefficients are all in the same range and the sensitivity is optimal. Moreover, there are only three multipliers as compared to 7 in the canonical realization. Conclusion: the WDF is highly superior to the canonical filter.

Limit Cycli. Another problem with filters is the occurence of limit cycli. Those are stray signals which circulate in the filter without damping. They are a consequence of the inevitable rounding - a nonlinear operation. Take for example the circuit of fig. 1.6, taken from [1].

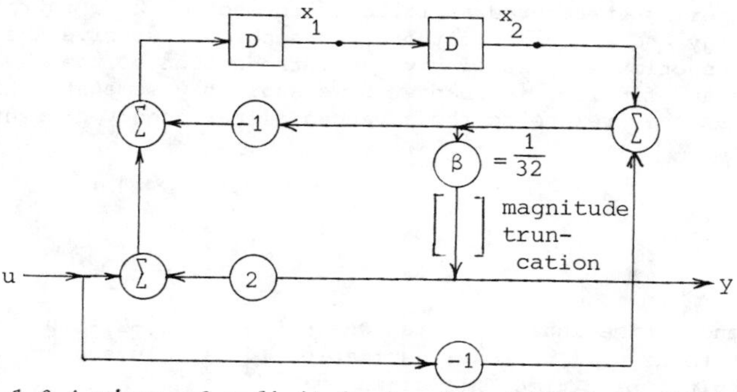

Fig. 1.6 A piece of a digital circuit exhibiting a limit cycle [*1*].

Working with integer arithmetic and magnitude truncation, a u and y sequence for fig. 1.4 is given as follows:

time	0	1	2	3	4	5	6	7	8	9	10	11	12	13	14	15
u	16	0	0	0	0	0	0	0	0	0	0	0	0	0	0	0
x_1	0	32	0	-30	0	30	0	-30	0	30	0	-30	0	30	0	-30
x_2	0	0	32	0	-30	0	30	0	-30	0	30	0	30	0	30	0
y	0	0	1	0	0	0	0	0	0	0	0	0	0	0	0	0

At time t=0 a fatal rounding occurs after the multiplier β and a limit cycle is generated in the internal loop of the two delay elements. The circulating signal does not appear in the output sequence - it is 'unobservable' (although the system itself is perfectly observable). The piece of filter shown in fig. 1.4 may be a part of a WDF or of any other type of filter.

From the examples above one may understand that the design of a digital filter is a non-trivial business. The main purpose of this paper is to develop a method for designing filter which systematically avoids problems with sensitivity and limit-cycli. The method enforces a numerical control on the detailed algebraic operations at each time point.

Aside from sensitivity and limit-cycli problems, a few more arise which we shall discuss to some extend. In a filter-realization one wishes to utilize an arithmetic system which will keep the hardware as simple as possible, e.g. fixed-point - say a transition between pass-band and stop-band amplitude of 80 db to be achieved in a frequency-region of a few percent of the total frequency range - one may expect a fairly large range of numerical values making the use of fixed-point arithmetic difficult. The classical remedy is the use of scaling at strategic points in the filter structure combined with the use of a word size which is larger than the input/output wordsize (say 24 bits for 16 bits). The inclusion of scaling in the algorithm is a major nuisance. Any filter structure which can avoid it, is highly desirable.

We have nowadays an attractive array of LSI components. Array multipliers are nice and fast but still quite bulky. To obtain fast throughput they must be used in pipeline. Another new building block which has not yet been used extensively is the CORDIC module. It executes a rotation of a two-dimensional vector over a given angle. Similarly, it is capable to compute the rotation as well. Although the CORDIC module is in principle even more complex than a multiplier (especially with respect to time complexity), it is an attractive module because of its versatility (the many functions it can compute), its programmability and the fact that it may be executed as a pipeline.

Structure of the paper

The remainder of the paper is structured as follows: in section 2 we introduce basic notions, terminology and algebraic methods; in section 3 we give the orthogonal filter theory; in section 4 we shall give an introduction to the theory of wave-digital filters and finally, in chapter 5 we shall pay some attention to VLSI design aspects as they relate to the advanced filter theory presented in these notes.

2. BASIC NOTIONS AND USAGE

Time series

$u(t), y(t), x(t)$ will denote time-indexed series where $t = .. -3,-2,-1,0,1,..$. Most series of interest to us will be square summable. If $u(t)$ is scalar we shall say that the energy in $u(t)$ up to time T is given by:

$$E(t) = \sum_{t=-\infty}^{T} u^2(t);$$

if $u(t)$ is vector-valued of dimension m then the energy in $u(t)$ is simply

$$\|u(t)\|^2 = \sum_{k=1}^{m} |u_k(t)|^2$$

and

$$E(t) = \sum_{t=-\infty}^{T} \|u(t)\|^2$$

Sometimes we write

$$\|u\|^2 = E(\infty)$$

for the grand total energy in u. The $-\infty$ in the series is seldom of importance - most time series start at some finite time and the infinite summation is actually finite.

We shall write $X(e^{i\theta})$ for the Fourier Transform of x(t) defined as

$$X(e^{i\theta}) = \sum_{k=-\infty}^{\infty} x(k)e^{ik\theta} \tag{2.1}$$

Of course $X(e^{i\theta})$ has a meaning only when (2.1) converges. It is known [2] that for a signal with finite energy (2.1) will converge (in the mean). x(k) can be recovered from (2.1) by the rule:

$$x(k) = \frac{1}{2\pi}\int_{-\pi}^{\pi} X(e^{i\theta})e^{-ik\theta}d\theta \tag{2.2}$$

Moreover, there is an "Energy-isomorfism" between a time-series and its Fourier transform given by the Parseval formula:

$$\sum_{k=-\infty}^{\infty} |x(k)|^2 = \frac{1}{2\pi}\int_{-\pi}^{\pi} |X(e^{i\theta})|^2 d\theta \tag{2.3}$$

Z-transforms

A 'one-sided' time series u where u(t) = 0 for t < 0 has a 'formal' z-transform defined by:

$$U(z) = \sum_{k=0}^{\infty} u(k)z^k \tag{2.4}$$

(2.4) need not converge. However, suppose that the u(k) have some bound $|u(k)| \leq M$ for all k, then (2.4) will surely converge for z in the open unit disc of the complex plane. We use:

$$D = \{z \mid |z| \leqslant 1\}$$
$$T = \{z \mid |z| = 1\}$$
$$E = \{z \mid |z| \geqslant 1\}$$

and say that U(z) is analytic in D. Clearly, by filling in $z = e^{i\theta}$ (2.4) reduces to a Fourier transform. However, the unit circle T often will not belong to the domain of convergence of the z-transform - it is only its boundary.

Digital Filters

Fig. 2.1 respresents the global structure of a digital filter.
The Arithmetic Unit (AU) defines a map :

$$\begin{bmatrix} x(t) \\ y(t) \end{bmatrix} \to \begin{bmatrix} x(t+1) \\ y(t) \end{bmatrix} = \mathcal{A}\begin{bmatrix} x(t) \\ u(t) \end{bmatrix} \tag{2.5}$$

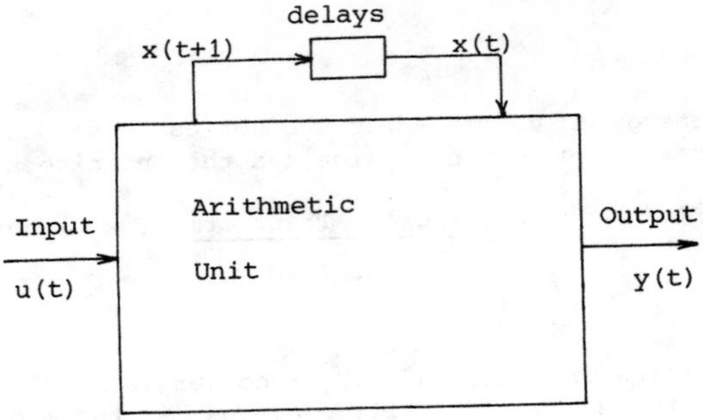

Fig. 2.1 The general structure of a digital filter

consisting of the actual filter operations. Suppose that the input is an m-dimensional time series, the output an n-dimensional series and the state δ dimensional, then h maps a $\delta + m$ vector on a $\delta + n$ vector.
If the operations of h are stationary and linear, then h is represented by a constant

$(\delta+n) * (\delta+m)$ matrix:

$$h = \begin{array}{c} \begin{array}{cc} \delta & m \end{array} \\ \left[\begin{array}{c|c} A & B \\ \hline C & D \end{array}\right] \end{array} \begin{array}{c} \delta \\ n \end{array}$$

We shall call h the system transition matrix. The <u>impulse-response</u> of the filter is the time series:

$D, CB, CAB, CA^2B, \ldots$

and its z-transform:

$$\begin{aligned} T(z) &= D + Cz(1 + zA + z^2A^2 + \ldots)B \\ &= D + Cz(1-zA)^{-1}B \end{aligned}$$

$(1-zA)^{-1}$ is formally the inverse of the series $1 + zA^2 + z^2A^2 + \ldots$ for an algebra of time series with support bounded in the past (checked by computing $(1-zA) * [1+zA+z^2A^2+\ldots]$). The filter is said to be "stable" if all eigenvalues of A are inside the unit disc A. In that case $(1-Az)^{-1}$ will be analytic inside D. The arithmetic unit must realize the operation described by equation (2.5). If $T(z)$ is a scalar transfer function of degree δ, then it has only $2\delta + 1$ algebraic parameters. The matrix representation (2.5) allows for $(\delta+1)^2$ algebraic parameters, δ^2 to many. We shall say that a realization in <u>algebraically canonical</u> if it uses only the minimal number of algebraic parameters. One also says that a realization is <u>canonical</u> or <u>system-canonical</u> if its state vector x is minimal. In the case of digital filter theory it is not all that essential that the state be minimal. Minimization of the number of operations and keeping them numerically under control is much more important.

3. ORTHOGONAL FILTERS AND EMBEDDINGS

We shall say that a filter is orthogonal if its transition map $\mathcal{T}$ is an orthogonal matrix. For an orthogonal filter the transition function T(z) is very special as is indicated by the following theorem:

Theorem 3.1

If a transfer function T(z) *has an orthogonal realization* $\mathcal{R}$, *then* T(z) *has the following properties:*

1. T(z) is rational
2. for $z \in \mathbb{D}$ $\quad \tilde{T}(z)T(z) \le 1$
3. for $z \in \mathbb{T}$ $\quad \tilde{T}(z)T(z) = 1$

Conversely, if T(z) *satisfies the properties 1) - 3) then it has a realization* $\mathcal{R}$ *which is orthogonal.*

Theorem 3.1 is relatively easy to proof - see the litterature [3],[4]. From it, it follows that the transfer function of an orthogonal filter is an "all-pass function". As a consequence one may understand that a given, stable transfer function T(z) cannot directly be realized as an orthogonal filter. The method we shall follow to achieve our goal anyway goes via a detour called 'embedding'. It means that we shall realize T(z) as a partial transfer function in a system of higher I/O dimension which itself is orthogonal. T(z) is then obtained by imposing appropriate boundary conditions. Before embarking on our embedding theory, let me emphasize the importance of implementing a filter function as an orthogonal computation. For one thing, an ε-error in the input vector $\begin{bmatrix} x \\ u \end{bmatrix}$ will result in an ε-error in the output vector $\begin{bmatrix} x+ \\ u \end{bmatrix}$ (x+ indicates the next state). The same is true moreover for the global map from input to output. One may state that the transition is implemented with optimal numerical conditioning. Another advantage consists in the fact that with an implementation in which the rounding is done systematically so that the norm of the output vector is slightly less than the norm of the input vector, one will have that no limit cycli can possibly occur (they are squelched by the contractivity). Putting both facts together one achieves optimal conditioning (in the sense given to it in numerical algebra) and absence of limit cycli. Since optimal conditioning is closely related to optimal sensitivity one may expect good behaviour of the filters in that sense as well. Sensitivity is a delicate subject which we shall discuss in more detail at the end of this section.

Embedding

We treat only the scalar case, i.e. where input and output dimensions are 1 (the former n=m=1). Let

$$T(z) = \frac{p(z)}{q(z)} \tag{3.1}$$

be a transfer function of degree n where q(z) is a stable polynomial (i.e. its zeros are in $\mathbb{E}$) and T(z) is of arbitrary degree n. For future convenience we shall represent p(z) as:

$$p(z) = \prod_{i=1}^{n} (1-\bar{\omega}_i z) \tag{3.2}$$

The zeros p(z) are the points $1/\bar{\omega}_i$. The representation (2.6) allows us to treat eventual zeros at infinity in an explicit manner.

Method I (direct embedding with minimal dimension)

Suppose that $|T(z)| \le 1$ for $z \in \mathbb{D}$. Then one may attempt to produce a 2 * 2 transfer matrix for which T(z) is the 1,1 entry:

$$\Sigma(z) = \begin{bmatrix} T(z) & \Sigma_{12}(z) \\ \Sigma_{21}(z) & \Sigma_{22}(z) \end{bmatrix} \tag{3.3}$$

The complementary transfer functions Σ_{12}, Σ_{21} and Σ_{22} may be obtained from T by using theorem 3.2. Let f(z) be any polynomial of degree n and let

$$f(z) = f_n z^n + f_{n-1} z^{n-1} + \dots + f_0.$$

We define $f_*(z)$ and $f^*(z)$ by the formulas:

$$f_*(z) = \bar{f}_n z^{-n} + \bar{f}_{n-1} z^{-n+1} + \dots + \bar{f}_0 = [f(\bar{z}^{-1})]^-$$

$$f^*(z) = z^n f_*(z) = \bar{f}_n + \bar{f}_{n-1} z + \dots + \bar{f}_0 z^n$$

$f^*(z)$ is called the 'reverse' polynomial of $f_*(z)$. Notice that the definition of "*" is dependent on n.

Theorem 3.2

*Every 2 * 2 orthogonal transfer function has the general form*

$$\Sigma(z) = \begin{bmatrix} d_1(z) & 0 \\ 0 & d_2(z) \end{bmatrix} \frac{1}{f} \begin{bmatrix} h & g \\ -g^* & h^* \end{bmatrix} \tag{3.4}$$

where $d_1(z)$ *and* $d_2(z)$ *are scalar all-pass functions, f,g,h are polynomial of degree* n *and* f *has its zeros in* $\mathbb{E}$, *and they satisfy:*

$-ff_* = gg_* + hh_*$
-g and h have no common zero.

The proof is elementary and omitted.

The second matrix in (3.4) is the actual 2 * 2 orthogonal transfer function while d_1 and d_2 are but scalar all-passes on the first and second port.

It follows that in most cases one may restrict discussions to the form:

$$\frac{1}{f} \begin{bmatrix} h & g \\ -g^* & h^* \end{bmatrix}$$

If we wish to realize T(z) as h/f we identify h=p and f=q. g is next computed from:

$$gg^* = ff^* - hh^* \tag{3.5}$$

by spectral factorization. Notice that (3.5) can only be solved when $|T(z)| \le 1$ in $\mathbb{D}$.

Suppose that we have succeeded to realize $\Sigma(z)$ by means of an orthogonal filter - the existence of such a filter is acertained by theorem 3.1 - with the

I/O relation defined in fig. 3.1, then T(z) is obtained by imposing the boundary condition $b_2=0$,

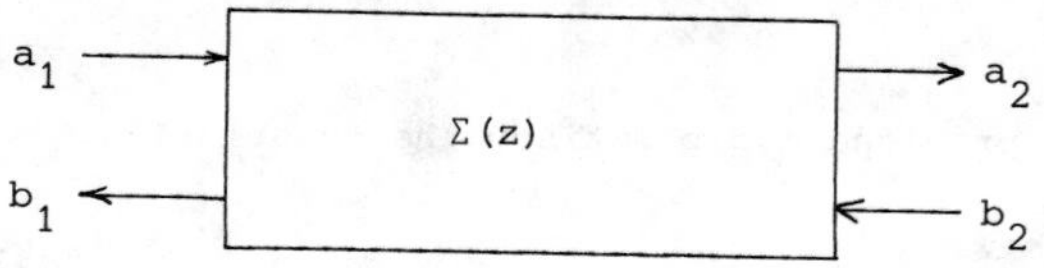

$$\begin{bmatrix} A_2 \\ B_1 \end{bmatrix} = \Sigma(z) \begin{bmatrix} A_1 \\ B_2 \end{bmatrix}$$

Fig. 3.1 I/O relation for Σ(z).

and neglecting the output pin b_1.

<u>Method II</u> [Embedding with lossless boundary condition]

Another way to realize T(z) is by imposing the boundary condition $a_2 = b_1$, i.e. by realizing fig. 3.2

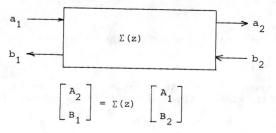

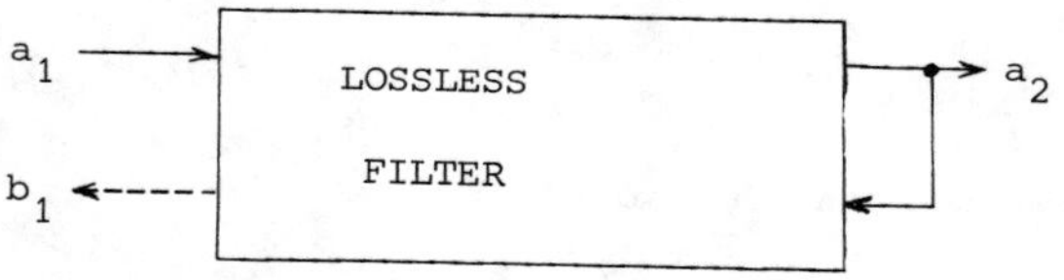

Fig. 3.2 Embedding with a lossless boundary condition.

and looking at the transfer $a_1 \to a_2$. With Σ(z) as in formula (3.4) - say with $d_1 = d_2 = 1$, we shall have:

$$\begin{bmatrix} A_2 \\ B_1 \end{bmatrix} = \frac{1}{f} \begin{bmatrix} h & g \\ -g_* & h_* \end{bmatrix} \begin{bmatrix} A_1 \\ B_2 \end{bmatrix}$$

and the transfer $A_1 \to A_2 = T(z)A_1$ becomes:

$$T(z) = \frac{h}{f-g} \tag{3.6}$$

Identify (for $T(z) = \frac{p}{q}$):

$$h = p$$
$$f - g = q$$

and define a new polynomial d with

$$f + g = d$$

we find easily that

$$\Sigma(z) = \frac{1}{d+q}\begin{bmatrix} 2p & d-q \\ q^*-d^* & 2p^* \end{bmatrix} \tag{3.7}$$

is again an orthogonal filter provided d,q,p satisfy the equation:

$$dq^* + qd^* = 2pp^* \tag{3.8}$$

(3.8) is called a "Lyapunov Equation" for reasons explained in appendix A. The meaning of d may become a little clearer if we divide (3.8) by qq^* to obtain:

$$\frac{1}{2}\left[\frac{d}{q} + \frac{d^*}{q^*}\right] = \frac{p}{q} \cdot \frac{p^*}{q^*} \tag{3.9}$$

Recall that q is a stable polynomial. Hence d/q is analytic in $\mathbb{D}$. (2.13) now says that on $\mathbb{T}$ we have

$$\text{Re}\ \frac{d}{q} = \left|\frac{p}{q}\right|^2 \geq 0 \text{ for } z = e^{i\theta}$$

But since d/q is analytic in a NBH of the unit disc we shall have

$$\text{Re}\ \frac{d}{q} \geq 0$$

everywhere in $\mathbb{D} \cup \mathbb{T}$. It follows that the function

$$C(z) = \frac{d(z)}{q(z)}$$

has positive real part in $\mathbb{D}$ - it is called a p.r. function in $\mathbb{D}$ or a Caratheodory function. Another consequence of (3.9) is that d/q is uniquely determined by p/q except for an arbitrary imaginary constant which may be taken zero if d/q is to be real. Since given q and arbitrary p, (3.9) can always be solved for d, we are able to determine an embedding of type (3.7) for an arbitrary stable T(z) which then may be realized by the overall realization shown in fig. 3.2.

Although both methods of embedding are used extensively, we shall concentrate on method II because it is generally applicable and ties in nicely with stochastic filtering theory.

Cascade filters of minimal dimension

An additional constraint on the realization of orthogonal filters is that they form a cascade of elementary sections. This cascade condition is added in order to increase the overall accuracy of the filter in the stopband. As we shall see, the orthogonal structure takes care of the passband, while the stopband (which is often very critical because way below the passband in amplitude) must be taken care off by another mechanism. Our filters then get the structure shown in fig. 3.3, where each section realises a specific transmission zero of the system. Typically the elementary sections will be of degree one or two and will have to be computed from the original Σ(z) or T(z). We look first at the concrete form of these 'elementary sections'.

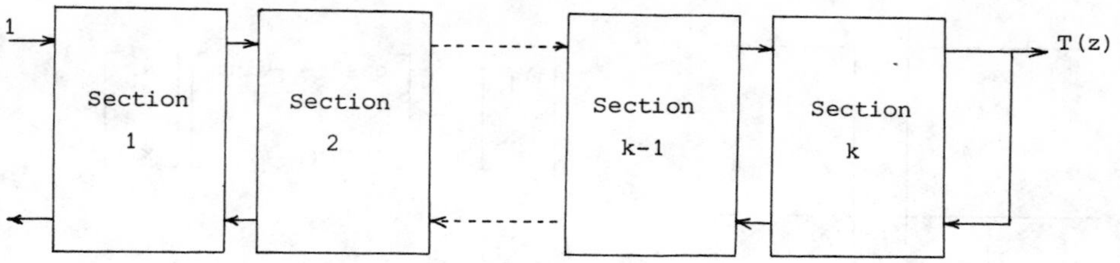

Fig. 3.3 Cascade realization of an orthogonal filter.

Elementary sections

A problem which may occur with the architecture shown in fig. 3.3 is the occurence of delay-free loops between the sections. In order to avoid those we shall make sure that sections are designed in such a way that there is no direct time dependence (no 'D-term') between b_2 and a_2, i.e. $\Sigma_{12}(0) = 0$, $d(0) = q(0)$, $g(0) = 0$. Such sections are called 'normalized' and are depicted by the symbol shown in fig. 3.4.

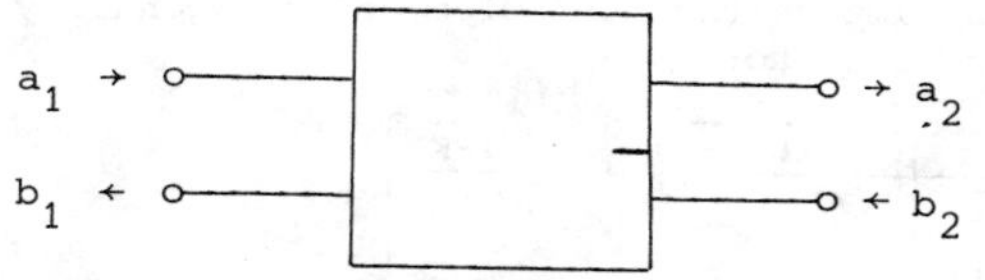

Fig. 3.4 Normalized orthogonal section.
(there is no instantaneous dependence of a_2 on b_2).

Degree one normalized section

Due to the previous theory, a degree one normalized section must have the form:

$$\Sigma(z) = \frac{1}{1-\bar{a}z} \begin{bmatrix} (1-\bar{\alpha}z)\sqrt{K} & gz \\ -\bar{g} & (z-\alpha)\sqrt{K} \end{bmatrix} \tag{3.10}$$

where some relations between the coefficients shown must exist in accordance with (2.9):

$$a \in \mathbf{D}\ ,\ K < 1$$

$$a = \alpha . K$$

$$|g|^2 = (1-K)(1-|\alpha|^2 K)$$

α is called the "transmission zero" of the section. Clearly, our section satisfies the condition of normalization mentioned earlier.

To realize Σ we must agree on the building blocks which may be used. We avoid multipliers and adders because these elements do not result in orthogonal transformations. In numerical analysis one would have recourse to orthogonal transformations like Givens rotations or Householder reflections. It is quite possible to realize a Givens rotation in hardware. The result is called a CORDIC-module and it executes an elementary rotation over a given angle θ on a two dimensional vector. In fig. 3.5 we depict the symbol for the basic Givens rotor and the corresponding algebraic action.

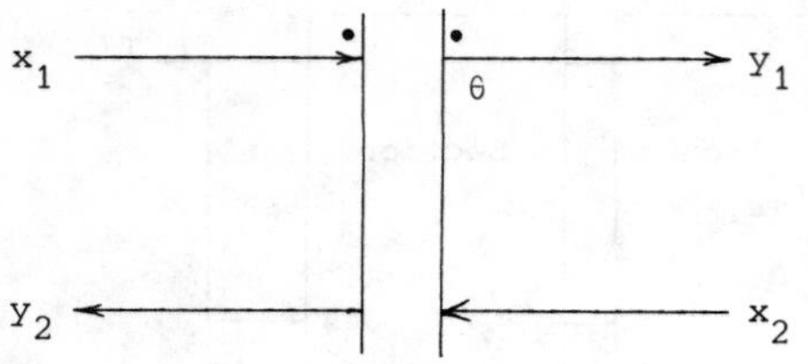

$$\begin{bmatrix} y_1 \\ y_2 \end{bmatrix} = \begin{bmatrix} \cos\theta & -\sin\theta \\ \cos\theta & \cos\theta \end{bmatrix} \begin{bmatrix} x_1 \\ x_2 \end{bmatrix}$$

Fig. 3.5 An elementary rotation and its symbol representing a CORDIC module.

If one now attempts to realize (3.10) one will have to deduce a transition matrix for it. There are general methods for doing that. In this case things are pretty easy. Since A is scalar one surely must have $A = \bar{a}$. D is 2 * 2 and clearly equal to $\Sigma(0)$. Hence:

$$D = \begin{bmatrix} \sqrt{K} & 0 \\ -\bar{g} & -\sqrt{K} \end{bmatrix}$$

Finally, we have CB from $\Sigma(z)$ and may impose the condition of orthogonality to determine C and B individually. This leads to:

$$\mathcal{T} = \left[\begin{array}{c|c} \overbrace{A}^{1} & \overbrace{B}^{2} \\ \hline C & D \end{array}\right]\begin{matrix} \}\ 1 \\ \}\ 2 \end{matrix} = \left[\begin{array}{c|cc} \bar{a} & -\bar{\alpha}\sqrt{K(1-K)} & g/\sqrt{1-K} \\ \hline \sqrt{1-K} & \sqrt{K} & 0 \\ \bar{g}\sqrt{\dfrac{K}{1-K}} & -g & -\alpha\sqrt{K} \end{array}\right]$$

$\mathcal{T}$ is orthogonal as announced and may now further be decomposed using elementary Given rotations as follows:

$$= \begin{bmatrix} \bar{\alpha}\sqrt{K} & 0 & -\dfrac{g}{\sqrt{1-K}} \\ 0 & 1 & 0 \\ \dfrac{\bar{g}}{\sqrt{1-K}} & 0 & \alpha\sqrt{K} \end{bmatrix} \begin{bmatrix} 1 & & \\ & 1 & \\ & & 1 \end{bmatrix} \begin{bmatrix} \sqrt{K} & -\sqrt{1-K} & 0 \\ \sqrt{1-K} & \sqrt{K} & 0 \\ 0 & 0 & 1 \end{bmatrix}$$

This leads to the realization shown in fig. 3.6.

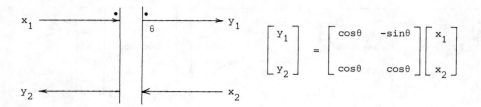

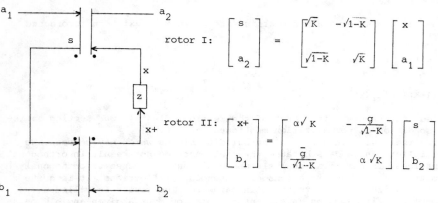

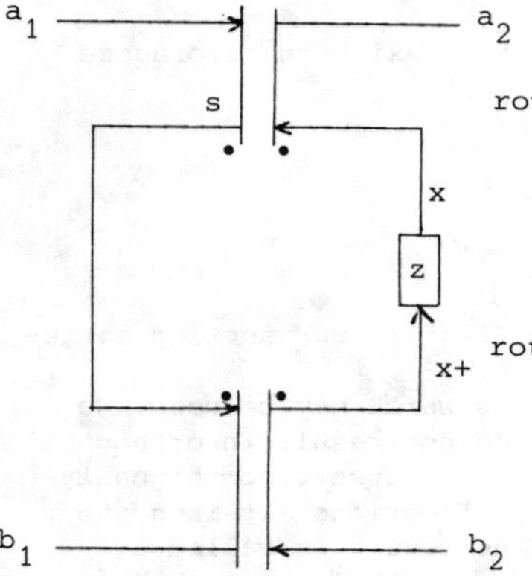

rotor I: $\begin{bmatrix} s \\ a_2 \end{bmatrix} = \begin{bmatrix} \sqrt{K} & -\sqrt{1-K} \\ \sqrt{1-K} & \sqrt{K} \end{bmatrix} \begin{bmatrix} x \\ a_1 \end{bmatrix}$

rotor II: $\begin{bmatrix} x+ \\ b_1 \end{bmatrix} = \begin{bmatrix} \alpha\sqrt{K} & -\dfrac{g}{\sqrt{1-K}} \\ \dfrac{\bar{g}}{\sqrt{1-K}} & \alpha\sqrt{K} \end{bmatrix} \begin{bmatrix} s \\ b_2 \end{bmatrix}$

Fig. 3.6 Realization of the first degree elementary section using only CORDIC modules.

The elementary section of fig. 3.6 is complex (see the equation for rotor II). If one wishes real realizations then a degree one section with possible complex poles and transmission zeros will not be enough as building block. A similar theory as discussed for the degree one section can be developed and will lead to an elementary real section of degree two. Its realization is shown in fig. 3.7. For brevity's sake we skip its construction. The degree one and degree two elementary sections suffice as basic building blocks to construct any orthogonal filter by means of a cascade [5].

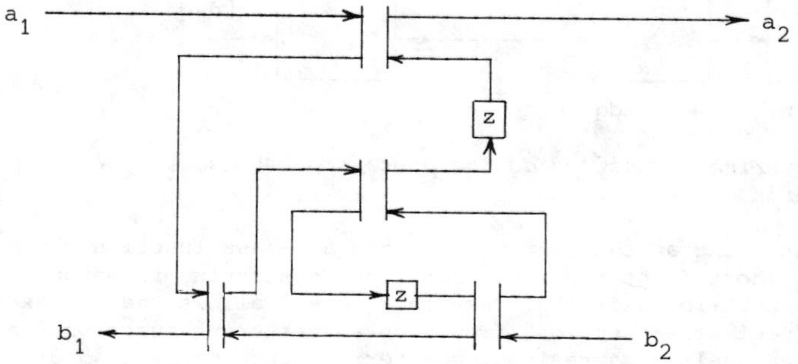

Fig. 3.7 Realization of a second degree real elementary section using CORDIC modules.

The case where $\alpha = 0$ is special and known as a "Levinson ladder section". It produces the diagram of fig. 3.8 which is a simplified version of fig. 3.6.

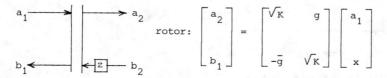

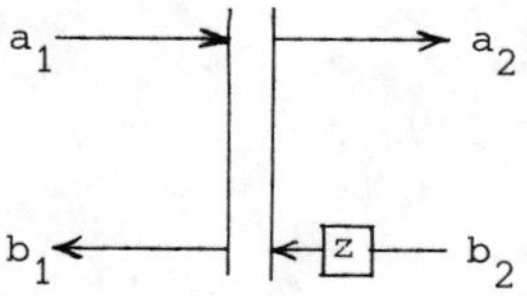

rotor: $$\begin{bmatrix} a_2 \\ b_1 \end{bmatrix} = \begin{bmatrix} \sqrt{K} & g \\ -\bar{g} & \sqrt{K} \end{bmatrix} \begin{bmatrix} a_1 \\ x \end{bmatrix}$$

Fig. 3.8 Realization of the Levinson ladder section as a special case of fig. 3.6.

<u>Recursive Synthesis</u>

In the first part of this section, the global approach to synthesis has been described. From a given transfer function $T(z) = p(z)/q(z)$ we have constructed a lossless $\Sigma(z)$. While it is possible to produce a global orthogonal synthesis for $\Sigma(z)$ by computing a state space and related operators, this method does not seem too attractive as compared to a more simple recursive approach which peels off section by section and is known as "Schur's method". This method starts out from the knowledge of one of the port scattering functions deduced from $T(z)$. Since $T(z)$ is a port to port transfer (see e.g. fig. 3.3), there will not be a direct relation between the scattering at one of the ports and the transfer between ports. Analyzing the situation further and using (2.11)-(2.12) we obtain the situations depicted in fig. 3.7.

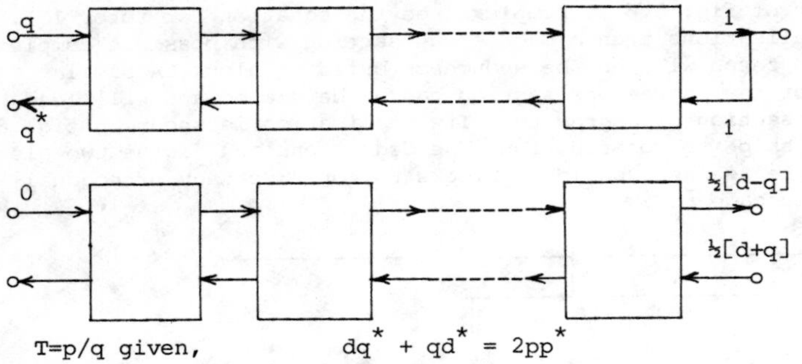

Fig. 3.7 Scattering situations at the ports of the cascade filters used in Schur's algorithm.

The input scattering at the left port is the all-pass function $V = q^*/q$ when the right port is shorted. At the right port, the scattering function is $S=(d-q)/(d+q)$ when the left port is absorbing. The Schur method allows one to "extract a section" or "peel off a section" from such a scattering function, leaving a remainder which itself is a residual scattering function (fig. 3.8).

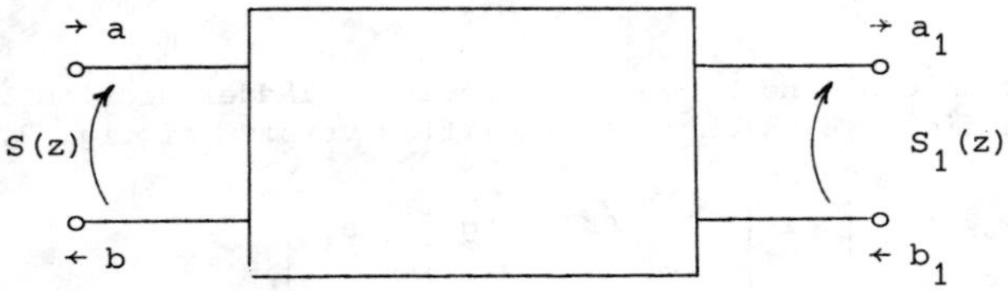

Fig. 3.8 The Schur Peeling procedure.

The elementary section is characterized by a transmission zero α (see 2.14), which in principle may be arbitrary in $\mathbb{C}$.

Before embarking on the details of the Schur procedure, let us investigate the effects of the individual transmission zeros on the overall structure. This is most easily done by using a new transfer matrix which describes the relations between the quantities at one port in function of the quantities at another. For instance, with reference to fig. 3.8, we define the "Chain Scattering Matrix" $\Theta(z)$ by the relation:

$$\begin{bmatrix} A(z) \\ B(z) \end{bmatrix} = \Theta(z) \begin{bmatrix} A_1(z) \\ B_1(z) \end{bmatrix} \tag{3.11}$$

It is easy to compute $\Theta(z)$ in function of the quantities in $\Sigma(z)$:

$$\Theta(z) = \frac{1}{p} \begin{bmatrix} \frac{q+d}{2} & \frac{q-d}{2} \\ \frac{q^* - d^*}{2} & \frac{q^* + d^*}{2} \end{bmatrix} \tag{3.12}$$

For the cascade case shown in fig. 3.9 we see that it is easy to express the overall CSM in function of the CSM's of the individual sections, yielding

$$\Theta(z) = \theta_0\theta_1(z)\ldots\theta_n(z) = \theta_0 \frac{1}{\prod_{i=1}^{n} p_i(z)} \overrightarrow{\prod_{i=1}^{n}} \begin{bmatrix} \frac{q_i+d_i}{2} & \frac{q_i-d_i}{2} \\ \frac{q_i^*-d_i^*}{2} & \frac{q_i^*+d_i^*}{2} \end{bmatrix} \quad (3.13)$$

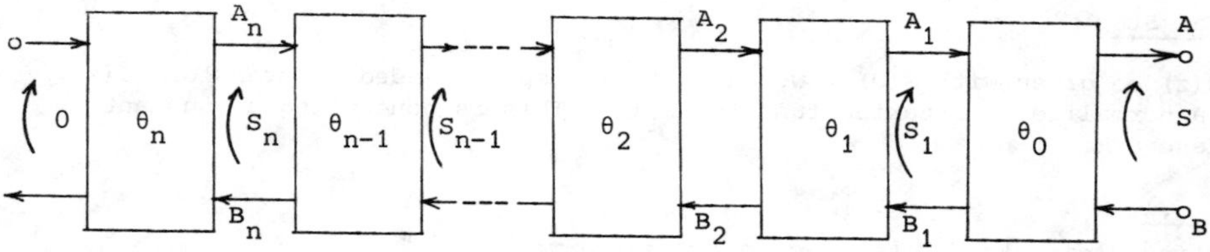

Fig. 3.9 Cascade of Elementary sections.

θ_0 is included as a constant CSM to allow an initialization of the right-port scattering function which would not be general enough if one had started with the normalized S_1 ($S_1(0)=0$). We see that the overall $p(z)=\prod_{i=1}^{n} p_i(z)$ is the product of the individual $p_i(z)$'s. Hence: the transmission zeros of the overal scattering matrix are the transmissionzeros of the individual transmission zeros ω_i in the sections. Conversely, in order to realize a given $p(z) = \prod_{i=1}^{n} (1-\bar{\omega}_i z)$ by means of cascaded orthogonal filter, it will be necessary to realize the individual transmission zeros ω_i in the sections. By means of a Schur extraction, we shall be able to peel of a section leaving a passive residu. In principle, the transmission zero of the section is arbitrary. If it is choosen to be one of the points ω_i, then it turns out that also the degree of the original scattering function is reduced. The overall method now becomes:

1. Start out with $S(z) = (d-q)/(q+d)$
2. Since $S(0)$ may not be zero, it may be necessary to include a constant CSM θ_0 which transforms $S(z)$ into $S_1(z)$ with $S_1(0) = 0$ (a normalization).
3. Choose ω_1 among the $\{\omega_i\}$. If ω_1 is real, extract a degree-one real elementary section with transmission zero ω_1 using Schur's procedure. If ω_1 is complex, same procedure but now with a degree-two real section. The residu will be $S_2(z)$ with $S_2(0) = 0$ and degree $S_2(z)$ lower than degree $S_1(z)$.
4. Repeat 3 until $S_{n+1} = 0$.

The result is as follows: we have obtained a cascaded orthogonal filter with input scattering function $S(z) = (d-q)/(d+q)$ and with transmission zeros $p(z) = \prod_{i=1}^{n} (1-\bar{\omega}_i z)$. It is now easy to see that $T(z) = p(z)/q(z)$ has been realized automatically. Indeed: $p(z)$ surely will be the numerator by construction. As for the denominator because $C(z) = d(z)/q(z)$, because we have

$$T(z)T_*(z) = \frac{1}{2}\{C(z) + C_*(z)\}$$

and because degree $T(z)$ = degree $C(z)$ we must find $k.q(z)$ with k some constant of modulus 1. Since the procedures are real $k = \pm 1$. The ultimate sign is dependent on the final implementation.

The previous discussion reduces the synthesis to the peeling off of one layer given input scattering function $S_i(z)$ and the desired transmission zero ω_i. We make a distinction between $\omega_i \in \mathbb{D}^i \cup \mathbb{E}$ and $\omega_i \in \mathbb{T}$. In the first case we call the procedure a 'Schur extraction' in the second case a 'Brune extraction'. We discuss both in sequence.

Schur Method

Initial Step

Let $S(z)$ be given with $S(0) \neq 0$. An initial step is needed to transform $S(z)$ into a normalized scattering function $S_1(z)$. This is done using a constant lossless CSM.

Lemma 3.1

A constant lossless CSM has the following form:

$$\theta_0 = \begin{bmatrix} d_1 & \\ & d_2 \end{bmatrix} \frac{1}{\sqrt{1-|\rho|^2}} \begin{bmatrix} 1 & \rho \\ \bar{\rho} & 1 \end{bmatrix} \tag{3.13}$$

where $|\rho| < 1$ *(the reflection coefficient) and* $d_1 = d_2 = 1$.

The proof is elementary (express $\theta J \theta_* = J$). d_1 and d_2 are usually choosen 1 but it is convenient to have them around.

When θ_0 is applied on $S(z)$, see fig. 3.9, we have (with $A = S$ and $B = 1$):

$$\frac{1}{\sqrt{1-|\rho|^2}} \begin{bmatrix} 1 & \rho \\ \bar{\rho} & 1 \end{bmatrix} \begin{bmatrix} S \\ 1 \end{bmatrix} = \frac{1}{\sqrt{1-|\rho|^2}} \begin{bmatrix} \rho + S \\ \bar{\rho}S + 1 \end{bmatrix} \tag{3.14}$$

so that

$$S_1 = \frac{S + \rho}{1 + \bar{\rho}S} \tag{3.15}$$

and we shall have $S_1(0) = 0$ when ρ is

$$\rho = -S(0) \tag{3.16}$$

To satisfy $|\rho| < 1$ it is necessary that $|S(0)| < 1$. Because $S(z)$ is contractive in $\mathbb{D}$ two cases are possible:

1. $|S(0)| = 1$. Then, because of the maximum modulus theorem [2] $S(z) = c$ where c is constant of modulus 1. The normalization is not possible and the synthesis terminates with a pure reflection ± 1.

2. $|S(0)| < 1$ and the normalization may be executed leaving $S_1(z)$ with $S_1(0)=0$.

Degree 1 - Schur Reduction

Let be given $S_i(z)$ and a transmission zero ω_i, with - say - $\omega_i \in \mathbb{D}$ (or in $\mathbb{E}$). S_i is normalized: $S_i(0) = 0$. First we execute a constant transformation such that $S_i(\omega_i) = 0$

$$S_i' = \frac{S_i + \rho}{1+\bar{\rho}S_i}$$

using a constant CSM as in (3.13).

We now have $S_i' = \left(\frac{z-\omega_i}{1-\bar{\omega}_i z}\right) S''$ with S_i'' contractive together with S_i' because S_i'' is analytic in a neighbourhood of $\mathbb{D}$ and it has the same magnitude as S_i' on $\mathbb{T}$. Moreover, the degree of S_i'' will be lower than the degree of S_i' if ω_i is a transmission zero. The latter fact is somewhat tricky to prove but has hopefully been made acceptable in the previous discussion. Finally we renormalize S_i'' by executing again a constant transformation:

$$S_{i+1} = \frac{S_i'' + \beta}{1 + \bar{\beta}S_i''}$$

with $\beta = -S_i''(0) = \frac{S_i''(0)}{\omega_i} = \frac{\rho}{\omega_i}$

The endresult of the subsequent CSM transformations is as follows (where all indiced i have been dropped):

$$\theta(z) = \frac{1}{\sqrt{1 - \left|\frac{\rho}{\omega}\right|^2}} \begin{bmatrix} 1 & \frac{\rho}{\omega} \\ \frac{\bar{\rho}}{\bar{\omega}} & 1 \end{bmatrix} \cdot \begin{bmatrix} 1 & \\ & \frac{z-\omega}{1-\bar{\omega} z} \end{bmatrix} \cdot$$

$$\cdot \frac{1}{\sqrt{1-|\rho|^2}} \begin{bmatrix} 1 & \rho \\ \bar{\rho} & 1 \end{bmatrix}$$

$$= \sqrt{\frac{1-|\rho|^2}{1-\left|\frac{\ell}{\omega}\right|^2}} \cdot \frac{1}{1-\bar{\omega} z} \circ$$

$$\circ \begin{bmatrix} 1 - z\bar{\omega} \cdot \frac{1-\left|\frac{\rho}{\omega}\right|^2}{1-|\rho|^2} & z.\bar{\omega} \cdot \frac{1-|\omega|^2}{1-|\rho|^2} \\ \frac{\bar{\rho}}{\bar{\omega}} \cdot \frac{1-|\omega|^2}{1-|\rho|^2} & z-\omega \cdot \frac{1-\left|\frac{\rho}{\omega}\right|^2}{1-|\rho|^2} \end{bmatrix}$$

With $a = \omega . \dfrac{1-\left|\frac{\rho}{\omega}\right|^2}{1-|\rho|^2}$, $\alpha=\omega$, $K = \dfrac{1-\left|\frac{\rho}{\omega}\right|^2}{1-|\rho|^2}$, $g = \frac{\rho}{\omega}\dfrac{1-|\omega|^2}{1-|\rho|^2}$ and converting back to a $\Sigma_i(z)$ matrix, we obtain the form (3.10) which has been realized as fig. 3.6.

There is only one case when the Schur extraction breaks down and that is when the intermediate scattering function S''_i is of modulus one. Then the second CSM coordinate rotation cannot be executed, because the corresponding reflection coefficient is not contractive. In that case we shall have $S_i(z) = c.z.$ with $|c| = 1$ and the synthesis terminates trivially with a pure reflection. In the case where we use real sections we terminate with $S_i(z) = \pm z$ - a pure delay with a reflector. Clearly that case will never occur unless the original $S(z)$ is itself all-pass or lossless.

Degree 2 - Schur Reduction

The degree 2 Schur reduction with transmission zeros ω and $\bar{\omega}$ can be executed much along the same line as the degree 1 reduction. A few observations greatly simplify the procedures. In the degree 1 case one may have remarked that all the parameters in the section are function of ω and $\rho=-S(\omega)$ - the initial reflection coefficient. Other values of $S(z)$ do not participate. Because of Schwarz' lemma (a derivative of the maximum modulus theorem) we also have

$$\left|\frac{\rho}{\omega}\right| < 1$$

(with = 1 possibly when $S(z)$ is lossless itself). In the case of the degree 2 extraction the same observation holds. In fact, it is an instance of a more general principle - the interpolation theorem which we briefly formulate. Let $\Sigma(z)$ be the scattering matrix of a lossless filter with $\omega_i \in \mathrm{D}$ and which is loaded - say at the left port, into a (further unknown) passive scattering function $S_L(z)$. The input scattering function (at the right port) is given by:

$$S = \frac{d-q}{d+q} + \frac{2p}{d+q} \cdot S_L \cdot \left[1 - \frac{q^* - d^*}{q+d} S_L\right]^{-1} \frac{2p}{d+q} \qquad (3.17)$$

The second term in (3.23) is analytic in D because the zeros of d+q must be in E and because

$$1 - \frac{q^* - d^*}{q+d} S_L$$

cannot be zero in D when S_L is contractive. Hence for all the zeros ω_i in p we have that

$$S(\omega_i) = \frac{d(\omega_i) - q(\omega_i)}{d(\omega_i) + q(\omega_i)}$$

independently of the load S_L. It follows that the function $(d-q)/(d+q)$ interpolates S in the points ω_i. For the case of the degree 2 Schur reduction, we are able to determine the degree 2 real section just from the values of ω and $\rho = -S(\omega)$.

Brune sections

Let p have a zero ω on T. Then p^* has the same zero because $(1-\bar{\omega}z)^* = (z-\omega) = -\omega(1-\bar{\omega}z)$. Such a zero is a "PLL" on T. We have $\mathrm{Re}Z(\omega) = 0$, and the corresponding scattering function has $S(\omega) = 1$. Let $c=S(\omega)$ and consider $S_1 = \bar{c}S$, then $S_1(\omega) = 1$ and $Z_1 = (1+S_1)(1-S_1)^{-1}$ has a pole at ω. Because Z_1 is p.r. the pole has some special properties:

Lemma

Suppose that Z is rational and p.r. in D *and that it has a pole at* $\omega \in$ T *then it has the form*

$$Z = Z_1 + \rho \frac{\omega+z}{\omega-z} \tag{3.18}$$

where $\rho > 0$, *degree* Z_1 = *degree* $Z-1$ *and* Z_1 *is p.r.*

The proof is classical and can be found e.g. in [6].

The Brune extraction now proceeds as follows:

1. Suppose that S is given and normalized (i.e. $S(0) = 0$). Let $c = S(\omega)$, define $S_1 = \bar{c}S$. Then $S_1(\omega) = 1, S_1(0) = 0$ and $Z_1 = (1+S_1)(1-S_1)^{-1}$ has a pole at ω.

2. Let $Z_1 = \dfrac{1+S_1}{1-S_1} = \rho\,\dfrac{\omega+z}{\omega-z} + Z_2$

   then, by the lemma, Z_2 is p.r. with degree lower than Z_1. Notice that $\rho \le 1$ with $\rho = 1$ iff $S(z) = \frac{c}{\omega} \cdot z$ because of the normalization on Z_1.

3. Let $Z_2 = Z_1 - \dfrac{\omega+z}{\omega-z}$. This corresponds to the following CSM operation:

$$\begin{bmatrix} Z_2-1 \\ Z_2+1 \end{bmatrix} = \begin{bmatrix} 1+\frac{\rho}{2}\frac{\omega+z}{\omega-z} & -\frac{\rho}{2}\frac{\omega+z}{\omega-z} \\ \frac{\rho}{2}\frac{\omega+z}{\omega-z} & 1-\frac{\rho}{2}\frac{\omega+z}{\omega-z} \end{bmatrix} \begin{bmatrix} Z_1-1 \\ Z_1+1 \end{bmatrix} \tag{3.19}$$

4. Finally, we must renormalize:

$$\frac{1}{\sqrt{1-|\beta|^2}} \begin{bmatrix} 1 & \beta \\ \bar{\beta} & 1 \end{bmatrix} \left\{ 1 + \frac{\rho}{2}\frac{\omega+z}{\omega-z} \begin{bmatrix} 1 & -1 \\ 1 & -1 \end{bmatrix} \right\} \begin{bmatrix} \bar{c} & 0 \\ 0 & 1 \end{bmatrix} \tag{3.20}$$

   and choose

$$\beta = \frac{\rho/2}{1-\rho/2}$$

Notice that $|\beta| < 1$ iff $\rho < 1$ as expected.
The total section now becomes:

$$\frac{1}{\sqrt{1-\rho}} \begin{bmatrix} 1-\frac{\rho}{2} & \frac{\rho}{2} \\ \frac{\rho}{2} & 1-\frac{\rho}{2} \end{bmatrix} \left\{ 1 + \frac{\rho}{2}\,\frac{\omega+z}{\omega-z} \begin{bmatrix} 1 & -1 \\ 1 & -1 \end{bmatrix} \right\} \begin{bmatrix} 1 & \\ & \bar{c} \end{bmatrix}$$

$$= \frac{1/\sqrt{1-\rho}}{1-\bar{\omega}z} \begin{bmatrix} \bar{c}\bar{\omega}(\rho-1)z+\bar{c} & -\rho\bar{\omega}z \\ \rho\bar{c} & -(\bar{\omega}z)+(1-\rho) \end{bmatrix}$$

This can be brought to the form $\Theta(z)$ corresponding to (2.14) if one premultiplies with the J-unitary matrix

$$\begin{bmatrix} c & 0 \\ 0 & -\omega \end{bmatrix}$$

to obtain the final Brune section. The realization is again as indicated before, with the definitions:

$K = (1-\rho)$

$\alpha = K\omega = (1-\rho)\omega$

$\gamma = \rho\bar{\omega}c$

This brings us to a complete theory of cascade synthesis of a transfer function $T(z) = p(z)/q(z)$ by Schur/Brune extractions. The global algorithm becomes:

1. Compute either $S(z) = (d-q)(d+q)^{-1}$ or $V(z) = q^{*}(z)/q(z)$ (the latter being the easiest). Initialize.

2. Select a zero ω_i in $p^{*}(z)$ and perform either a Schur/Darlington reduction (if ω_i is in $\mathbb{D} \in \mathbb{E}$) or a Brune reduction (if $\omega_i \in \mathbb{T}$).

3. Compute the residual scattering function

4. Repeat 2) and 3) until all zeros of $p(z)$ have been exhausted.

5. Realize each section with the modules shown in fig. 3.6-3.8.

Method III [Non cascade realizations]

Some very promising novel structures for orthogonal filter design using VLSI arithmetic units have recently been pointed out in [19]. Here use is made of a Schur algorithm on a embedding of larger than minimal size. In contrast to the previous case, one does not have the full scattering specification at the input port, only a partial set of waves is available. The inverse scattering algorithm now becomes what we have called a "partial lossless inverse scattering" of PLIS problem. The resulting structure is not cascade in the sense defined above: transmission zeros are not localized in specific parts of the algorithm. Hence lower sensitivity in the stopband is to be expected. This is, however, compensated by a very attractive overall architecture which lends itself to pipelining in a structured way. Sensitivity and accuracy of these filters have, to my knowledge , not yet been studied. They are, however, in some sense similar to alternative types of WDF's as discussed in [1]. The synthesis of the basic PLIS structure is simple and given in the following paragraphs.

Suppose that $S(z) = h(z)/f(z)$ is given and that $|S(z)| \leq 1$ in $\mathbb{D}$, and let $g(z)$ be determined by

$$f * f = h * h + g * g \tag{3.22}$$

(there is some ambiguity in g as to the phase but this is unimportant: take any minimal g). We shall realize the $2 * 1$ matrix Σ

$$\begin{bmatrix} b \\ \hat{b} \end{bmatrix} = \Sigma a \quad , \Sigma = \frac{1}{f} \begin{bmatrix} h \\ g \end{bmatrix} \tag{3.23}$$

Clearly, the transfer $a \to [b,\hat{b}]$ is lossless and the structure of fig. 3.1 should be able to handle the situation:

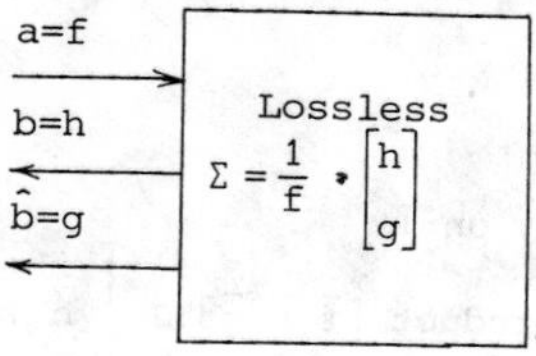

Fig. 3.10 The lossless two port realizing (3.23)

We have here the specification of a lossless network with one incident wave and two reflected waves, in fact of a two-port network. We shall attempt to realize is as a four-port network with a lossless load at the output port.

Because Σ in (3.23) is a lossless scattering matrix, we have in $\mathbb{D}$:

$$|f(e^{i\theta})|^2 \geqslant |h(e^{i\theta})|^2 + |g(e^{i\theta})|^2 \tag{3.24}$$

with equality on $\mathbb{T}$ and the property that if there is equality anywhere in $\mathbb{D}$, then f,h and g are constant (maximum modulus theorem). f,h and g are polynomial, and may be represented as:

$$\begin{bmatrix} f \\ h \\ g \end{bmatrix} = \begin{bmatrix} f_n z^n + \dots + f_0 \\ h_n z^n + \dots + h_0 \\ g_n z^n + \dots + g_0 \end{bmatrix} \tag{3.25}$$

with $|f_0|^2 > |h_0|^2 + |g_0|^2$. Let

$$J = \begin{bmatrix} 1 & & \\ & -1 & \\ & & -1 \end{bmatrix} \tag{3.26}$$

and construct a J-unitary matrix θ_0 such that

$$\theta_0 \begin{bmatrix} f_0 \\ h_0 \\ g_0 \end{bmatrix} = \begin{bmatrix} \sqrt{|f_0|^2 - |h_0|^2 - |g_0|^2} \\ 0 \\ 0 \end{bmatrix} \tag{3.27}$$

This is done by choosing:

$$\theta_0 = \frac{1}{\sqrt{1-|\rho_2|^2}} \begin{bmatrix} 1 & 0 & \rho_2 \\ 0 & 1 & 0 \\ \rho_2 & 0 & 1 \end{bmatrix} \frac{1}{\sqrt{1-|\rho_1|^2}} \begin{bmatrix} 1 & \rho_1 & 0 \\ \rho_1 & 1 & 0 \\ 0 & 0 & 1 \end{bmatrix} \tag{3.28}$$

with $\rho_1 = -h_0/f_0$, $\rho_2 = -g_0/\sqrt{|f_0|^2 - |h_0|^2}$

We have:

$$\begin{bmatrix} f'(z) \\ h'(z) \\ g'(z) \end{bmatrix} = \theta_0 \begin{bmatrix} f(z) \\ h(z) \\ g(z) \end{bmatrix}$$

with $|f'(z)|^2 \geq |h'(z)|^2 + |g'(z)|^2$ in $\mathbb{D}$ with equality on $\mathbb{T}$

(because θ_0 will not change the norm in the inner product $[\bar{f}\ \bar{h}\ \bar{g}]\, J \begin{bmatrix} f \\ h \\ g \end{bmatrix}$).

I claim that also $f'_n = 0$. This follows from the equality on $\mathbb{T}$ and the remark that the coefficient of z^n in the inner product must hence be zero. Thus:

$$\begin{bmatrix} f' \\ h' \\ g' \end{bmatrix} = \begin{bmatrix} f'_{n-1}z^{n-1} + \dots + 1 \\ h'_n z^n + h'_{n-1}z^{n-1} + \dots + 0 \\ g'_n z^n + h'_n z^{n-1} + \dots + 0 \end{bmatrix} \tag{3.29}$$

The final Schur step is now as before in the 2 * 2 case: a simple shift on the top row yielding

$$\begin{bmatrix} z & & \\ & 1 & \\ & & 1 \end{bmatrix} \begin{bmatrix} f' \\ h' \\ g' \end{bmatrix} = z \begin{bmatrix} f'_{n-1}z^{n-1} + \dots + 1 \\ h'_n z^{n-1} + \dots + h'_1 \\ g'_n z^{n-1} + \dots + z'_1 \end{bmatrix} \triangleq z \begin{bmatrix} f'' \\ h'' \\ g'' \end{bmatrix} \tag{3.30}$$

with as before

$$|f''(z)|^2 \geq |h''(z)|^2 + |g''(z)|^2 \tag{3.31}$$

with equality on $\mathbb{T}$, and the degree reduced by 1. This may now continue for n steps until one obtains a constant $[f_0\ \ h_0\ \ g_0]$ with, of course, $|f_0|^2 = |h_0|^2 + |g_0|^2$ i.e. a lossless termination. Collecting the data we obtain the structure of Fig. 3.2.

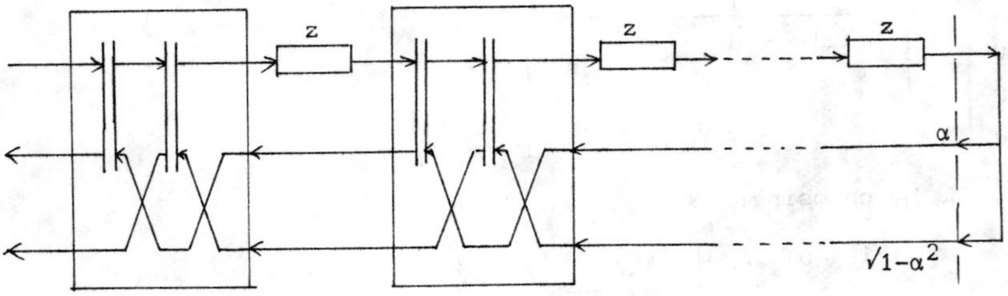

Fig. 3.11 Overall architecture of the PLIS filter realizing S(z) = h(z)/f(z).

The structure of fig. 3.11 is a generalization of the classical Schur ladder filter, it will have the same realization properties as the Schur filter (see the discussion in [13]).

4. WAVE DIGITAL FILTERS: AN INTRODUCTION

Another approach to cascade, low sensitivity filtering is due to Fettweis [7], [8]. His "Wave Digital Filters" are sibblings of the orthogonal filters presented above - in fact they are an "unnormalized" form of them with a minimal number of multipliers. They are well-suited for special purpose selective filter synthesis because they can be realized with a very small number of very small multipliers. As disadvantages we mention that they are wasteful in the number of adders, may exhibit limit-cycli (which can be suppressed), are hard to design, are not easily pipelined (see chapter 5) and are not well suited for adaptive filtering.

The original design method proposed by Fettweiss (a method derived from section 3 is an alternative) has some aspects in common with the classical design of microwave filters, which serves as the basic ingredient to guide one's intuition. One is lead to introduce a new frequency variable $w = \tanh \frac{p}{\omega_n}$ where ω_n is some normalizing frequency. Transmission-line transfer functions are then easily characterized by simple functions of w while open and shorted lines of quarter wavelength exhibit w-plane impedances wL and 1/wC respectively. An informal analogy is obtained by mapping a "unit-length" transmission line to a unit return-delay. In the same way, delays can be used to represent w-plane inductors and capacitors.

Essential in the working of microwave filters are the reflections occuring between transmission line segments. We shall show - following Fettweis - how one can translate w-plane filters into z-plane filters with a similar structure. Since w-plane (equivalently p-plane) filter theory is well-understood and documented [9],[10],[11], one may use these results directly in digital filter synthesis. The transformation $w = \tanh \frac{p}{\omega_n}$ maps the open right half p-plane into the open right half w-plane preserving losslessness.

The Bilinear Transformation

A standard transformation which maps the open right half w-plane onto $\mathcal{D}$ is given by

$$z = \frac{1-w}{1+w}\ , \quad w = \frac{1-z}{1+z} \tag{4.1}$$

We shall use this relation between w and z throughout.

The Wave Formalism

Let C be the value of a w-plane capacitor, i.e. an element with a v-i relation given by fig. 4.1.

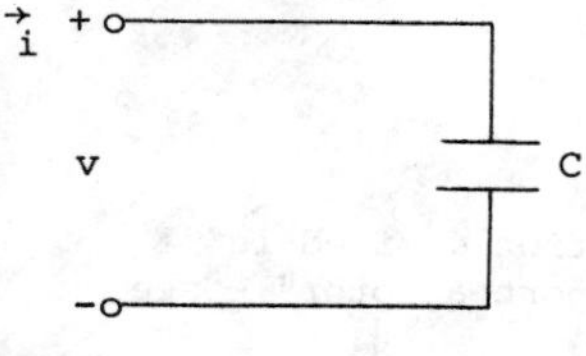

$$v = \frac{1}{w\,C}\,i$$

Fig. 4.1 A w-plane capacitor.

(v and i might be imagined as the actual voltage and current in a microwave circuit). Suppose that v and i are quantities existing at the end of a transmission line with given characteristic impedance R then one defines an incident wave:

$$a = v + Ri = \left[\frac{1}{wC} + R\right] i \tag{4.2}$$

and reflected wave

$$b = v - Ri = \left[\frac{1}{wC} - R\right] i \tag{4.3}$$

The scattering functions - normalized to R - relates a to b and is given by:

$$S = \frac{1-w}{1+w}$$

Hence for this special choice of wave-normalization the scattering function becomes a pure delay and we have the z-plane relations of fig. 4.2.

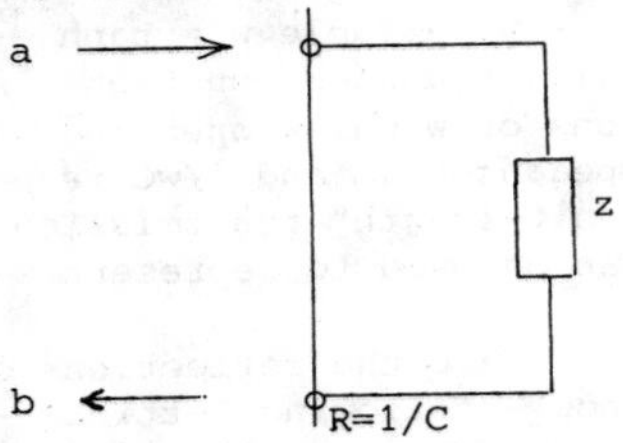

Fig. 4.2 Wave Digital Representation of the Capacitor (wave normalization is shown).

Similarly, for an inductor we find:

$$S = \frac{wL - R}{wL + R}$$

and with L = R, S = -z we obtain fig. 3.

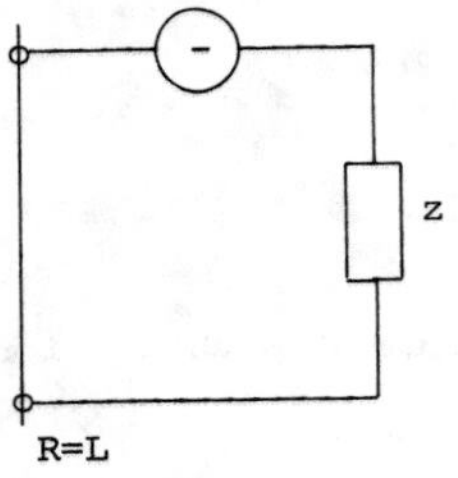

Fig. 4.3 Wave Digital Representation of the inductor.

The Two-Port Adaptor

Suppose now that we have a case where the port normalization desired is different from the one given. Then we need a so-called "port adaptor" - see fig. 4.4 for reference.

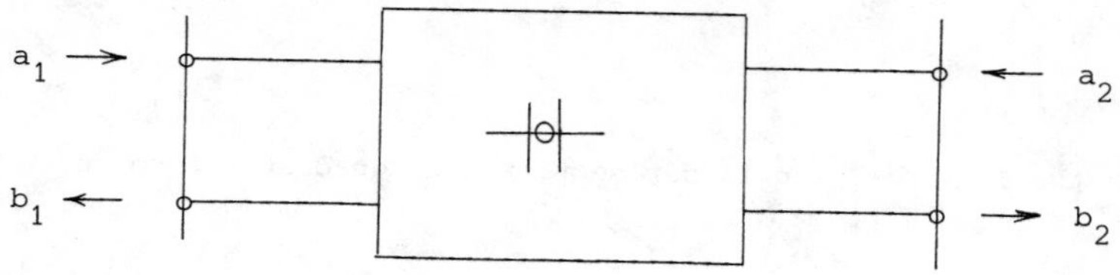

Fig. 4.4 The Two-port Adaptor transforms a R_1-wave normalization into an R_2-normalized wave system.

invariant are the values of v and i. We have:

$$\begin{cases} a_1 = v + R_1 i \\ b_1 = v - R_1 i \end{cases} \qquad \begin{cases} a_2 = v - R_2 i \\ b_2 = v + R_2 i \end{cases}$$

(Notice wave-directions: we do it this way in order to be consistent with later usage). The wave-scattering is represented by a matrix which transfers $\begin{bmatrix} a_1 \\ a_2 \end{bmatrix}$ to $\begin{bmatrix} b_1 \\ b_2 \end{bmatrix}$ and is given by:

$$S = \begin{bmatrix} \rho & 1-\rho \\ 1+\rho & -\rho \end{bmatrix} \qquad \rho = \frac{R_2 - R_1}{R_2 + R_1}$$

and a WDF realization is given in fig. 4.5.

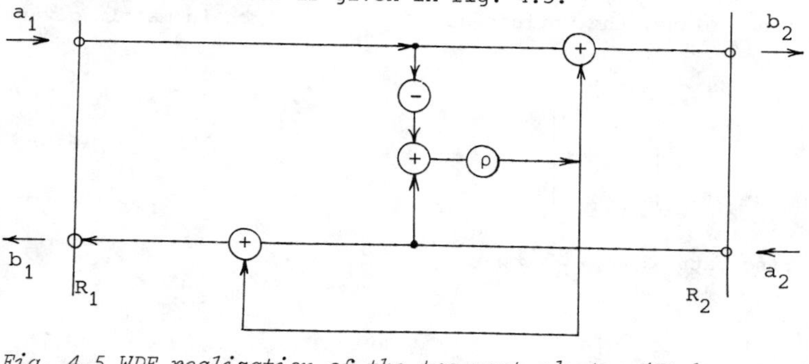

Fig. 4.5 WDF realization of the two-port adaptor (Itakura section).

Notice that fig. 4.5 is nothing else but the Itakura-Saito denormalized version of the Levinson ladder section with reflection coefficient ρ. Notice also that S is not unitary (it does not correspond to an orthogonal or normalized section). The corresponding CSM $\Theta(z)$ is defined by:

$$\begin{bmatrix} a_1 \\ b_1 \end{bmatrix} = \Theta \begin{bmatrix} b_2 \\ a_2 \end{bmatrix}$$

and is given by:

$$\Theta = \frac{1}{1+\rho}\begin{bmatrix} 1 & \rho \\ \rho & 1 \end{bmatrix} \tag{4.5}$$

as expected for the Itakura-Saito ladder section. The quadratic form for Θ gives:

$$\Theta^T J \Theta = \frac{1+\rho}{1-\rho} J$$

and the energy is off by a factor $(1+\rho)/(1-\rho)$.

Unit elements

In contrast to analog signals, digital signals can easily be 'delayed'. The basic delay element is the "unilateral unit element" (UE) shown in fig. 4.6.

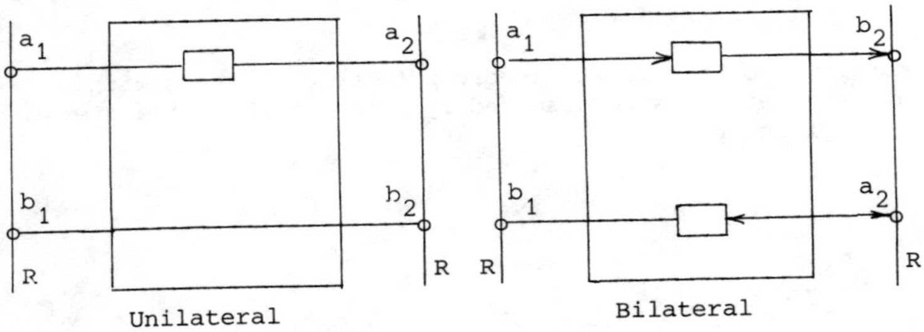

Fig. 4.6 Unit Delay Elements

Translated back to the w-plane, the unilateral UE produces a chain matrix given by:

$$\begin{bmatrix} v_1 \\ i_1 \end{bmatrix} = \frac{1}{1-w}\begin{bmatrix} 1 & wR \\ \frac{w}{R} & 1 \end{bmatrix}\begin{bmatrix} v_2 \\ -i_2 \end{bmatrix} \tag{4.6}$$

with w-plane scattering matrix:

$$S = \begin{bmatrix} 0 & 1 \\ \frac{1-w}{1+w} & 0 \end{bmatrix} \tag{4.7}$$

The chain-matrix of the bilateral UE is

$$\begin{bmatrix} v_1 \\ i_1 \end{bmatrix} = \frac{1}{\sqrt{1-w}}\begin{bmatrix} 1 & wR \\ \frac{w}{R} & 1 \end{bmatrix}\begin{bmatrix} v_2 \\ -i_2 \end{bmatrix} \tag{4.8}$$

and corresponds to the w-plane representation of a lossless transmission line of characteristic impedance R and unit electrical length. When the unilateral or bilateral UE is loaded in a w-plane inductor we obtain as input impedance:

$$Z_1 = \frac{v_1}{i_1} = \frac{(L+R)w}{1+\frac{L}{R}w^2}$$

which is equivalent to the impedance of a parallel resonant circuit. An easy realization for such a circuit is shown in fig. 4.7.

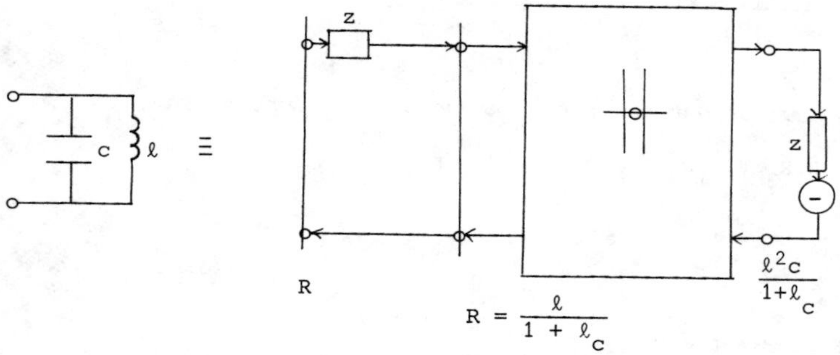

Fig. 4 A parallel resonant circuit realized as a UE followed by an adaptor-delay combination.

Similarly for a series resonant circuit we obtain the representation of fig. 4.8.

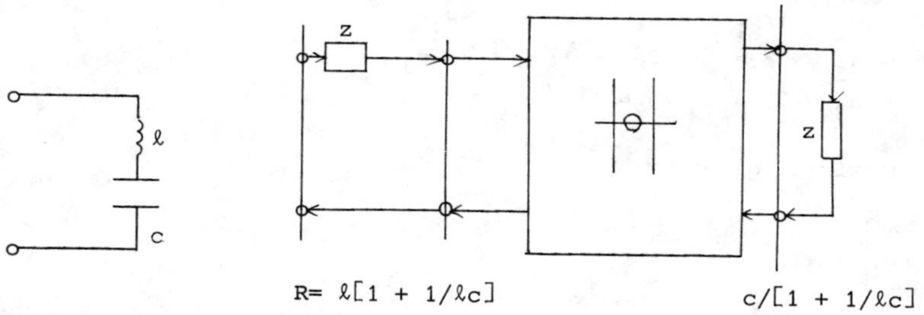

Fig. 4.8 A series resonant circuit realized as a UE followed by an adaptor-delay combination.

Series/Parallel Adaptors

In order to accomodate more general circuits, it is necessary to have more general adaptor circuits available. For our purposes 3-port series and parallel adaptors will suffice. The parallel adaptor translates the non-dynamical circuit of fig. 4.9 into a wave digital circuit. The equations are:

$$\begin{cases} a_1 = v + R_1 i_1 \\ b_1 = v - R_1 i_1 \end{cases} \quad \begin{cases} a_2 = v + R_2 i_2 \\ b_2 = v - R_2 i_2 \end{cases} \quad \begin{cases} a_3 = v + R_3 i_3 \\ b_3 = v - R_e i_3 \end{cases}$$

$$i_1 + i_2 + i_3 = 0.$$

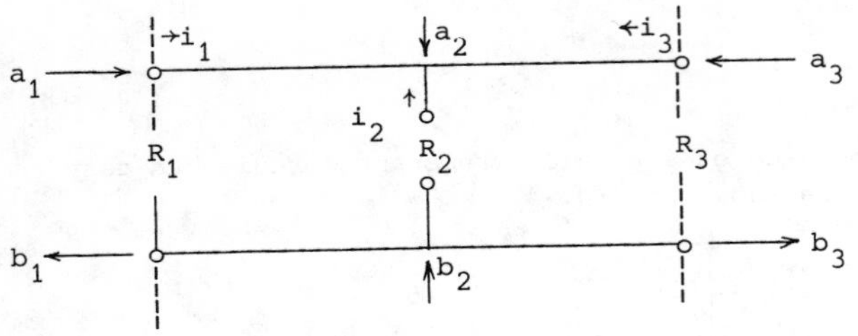

Fig. 4.9 A parallel adaptor between three characteristic impedances.

Define
$$\begin{cases} \Sigma = R_1R_2 + R_2R_3 + R_1R_3 \\ \rho_1 = R_2R_3/\Sigma \\ \rho_2 = R_1R_3/\Sigma \\ \rho_3 = R_1R_2/\Sigma \end{cases}$$

Then
$$v = \rho_1 a_1 + \rho_2 a_2 + \rho_3 a_3$$
$$R_1 i_1 = a_1 - v$$
$$R_2 i_2 = a_2 - v$$
$$R_3 i_3 = a_3 - v$$

Hence
$$\begin{cases} b_1 = 2v - a_1 = (2\rho_1 - 1)a_1 + 2\rho_2 a_2 + 2\rho_3 a_3 \\ b_2 = 2v - a_2 = 2\rho_1 a_1 + (2\rho_2 - 1)a_2 + 2\rho_3 a_3 \\ b_3 = 2v - a_3 = 2\rho_1 a_1 + 2\rho_2 a_2 + (2\rho_3 - 1)a_3 \end{cases}$$

The scattering matrix is given by:

$$\begin{bmatrix} b_1 \\ b_2 \\ b_3 \end{bmatrix} = S \begin{bmatrix} a_1 \\ a_2 \\ a_3 \end{bmatrix} \quad \text{with}$$

$$S = \begin{bmatrix} 2\rho_1-1 & 2\rho_2 & 2\rho_3 \\ 2\rho_1 & 2\rho_2-1 & 2\rho_3 \\ 2\rho_1 & 2\rho_2 & 2\rho_3-1 \end{bmatrix} \tag{4.9}$$

where $\rho_1 + \rho_2 + \rho_3 = 1$.

We shall always need S in a form where one of the ports - say port 3 - is "adapted" i.e. such that b_3 is not a direct function of a_3. Then

$$2\rho_3 = 1 \quad \text{or} \quad R_3 = R_1R_2/(R_1+R_2)$$

and let $\rho \triangleq 2\rho_1$, so that

$$S = \begin{bmatrix} \rho-1 & 1-\rho & 1 \\ \rho & -\rho & 1 \\ \rho & 1-\rho & 0 \end{bmatrix} \tag{4.10}$$

and the parallel adaptor is realized as in fig. 10.

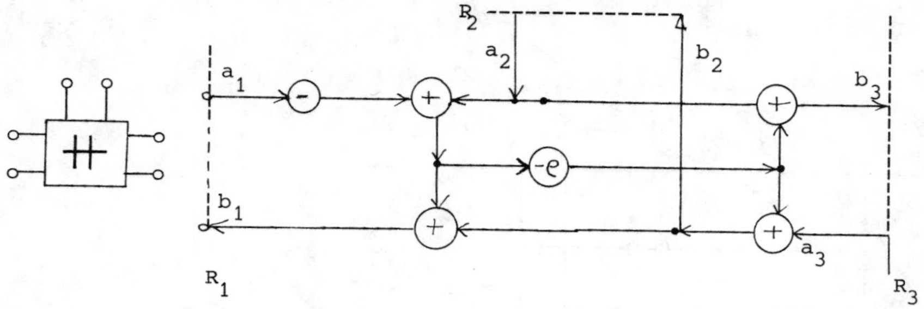

Fig. 4.10 The parallel adaptor with matched port 3 realized as a WDF.

Similarly, for the series adaptor we obtain the sequence of representations and formulas now given:

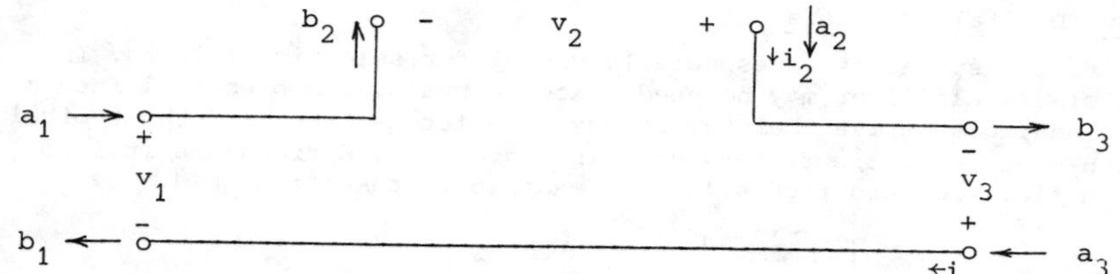

Fig. 4.11 The series adaptor between three characteristic impedances.

$$\begin{cases} a_1 = v_1 + R_1 i \\ b_1 = v_1 - R_1 i \end{cases} \qquad \begin{cases} a_2 = v_2 + R_2 i \\ b_2 = v_2 - R_2 i \end{cases} \qquad \begin{cases} a_3 = v_3 + R_3 i \\ b_3 = v_3 - R_3 i \end{cases}$$

Define:

$$\begin{cases} \Sigma = R_1 + R_2 + R_3 \\ \rho_1 = R_1/\Sigma \\ \rho_2 = R_2/\Sigma \qquad \rho_1 + \rho_2 + \rho_3 = 1. \\ \rho_3 = R_3/\Sigma \end{cases}$$

We have:

$$\begin{bmatrix} b_1 \\ b_2 \\ b_3 \end{bmatrix} = \begin{bmatrix} 1-2\rho_1 & -2\rho_1 & -2\rho_1 \\ -2\rho_2 & 1-2\rho_2 & -2\rho_2 \\ -2\rho_3 & -2\rho_3 & 1-2\rho_3 \end{bmatrix} \begin{bmatrix} a_1 \\ a_2 \\ a_3 \end{bmatrix} \tag{4.11}$$

With adapted port 3: $2\rho_3 = 1$ or

$R_3 = R_1 + R_2$

and $\rho \triangleq 2\rho_1$:

$$S = \begin{bmatrix} 1-\rho & -\rho & -\rho \\ \rho-1 & \rho & \rho-1 \\ -1 & -1 & 0 \end{bmatrix} \tag{4.12}$$

realized by:

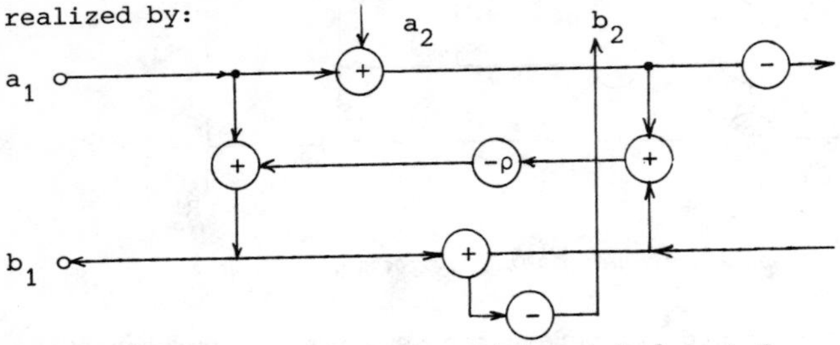

Fig. 4.12 The series adaptor with matched port 3.

Classical Wave Digital Synthesis

Wave digital filter synthesis is especially useful for selective filters. In that case, a classical filter may be used which is translated piecemeal into a WDF using the adaptors derived before. For example let us translate the typical lowpass fourth-order elliptic filter shown in fig. 4.13 and translate it into a WDF shown in fig. 4.14 and fig. 4.15. The example is taken from [1].

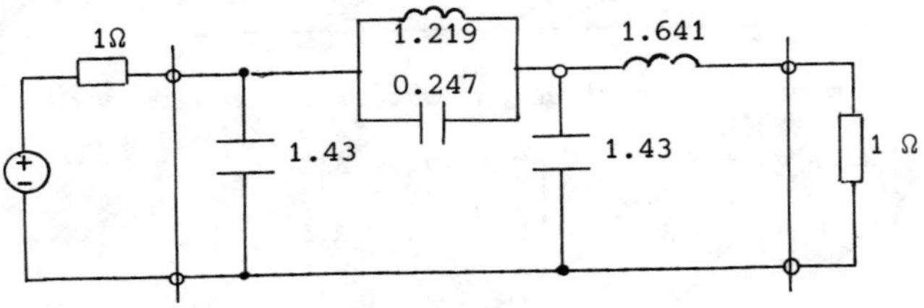

Fig. 4.13 A Fourth-order elliptic function lowpass filter (notice that the realization is non-minimal).

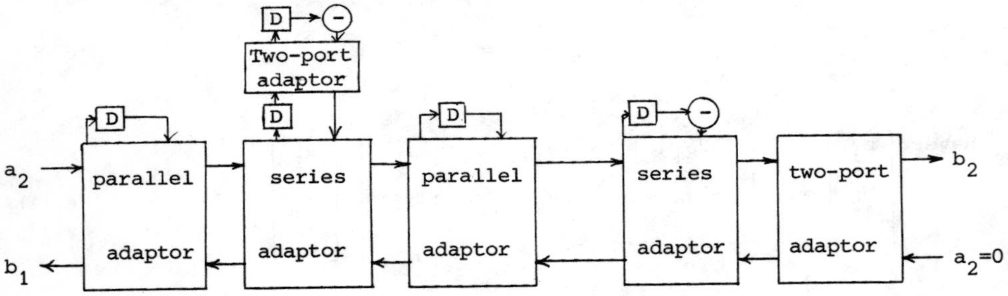

Fig. 4.14 Translation of fig. 4.13 into a series-parallel combination with matched adaptors.

In fig. 4.14 the adapted port is indicated with a bar. One sees that no delay-free loop occur.

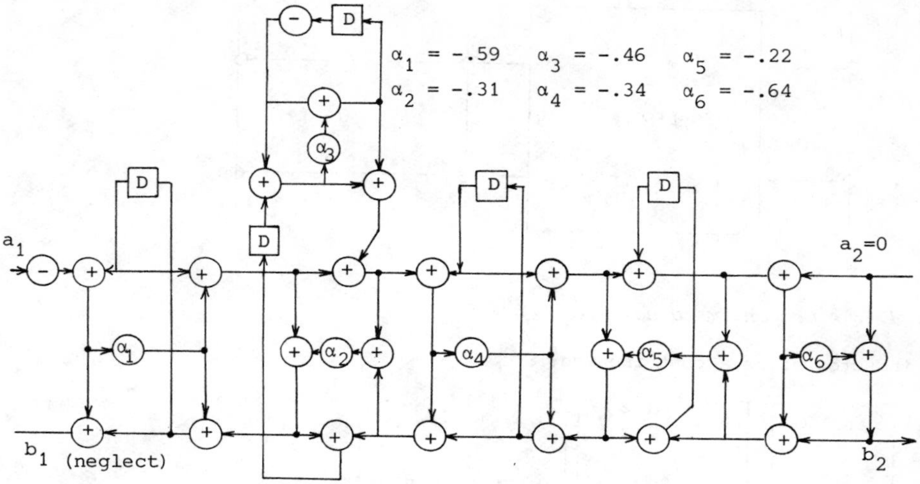

Fig. 4.15 Detailed WDF flow diagram corresponding to the filters in fig. 4.13 and 4.14 [1].

Wave Digital Filters to Orthogonal Filters

Up to now we have followed the classical WDF theory. Now we wish to establish the connection with the theory of chapter 3. As a result of our analysis we shall obtain a clear insight in the algebraic relations between WDF and orthogonal filter.

An example will illustrate the situation. Take the case of a series adaptor connected to an inductor and with the port normalizations shown in fig. 4.16. We shall show that this circuit is equivalent to a Brune section with TZ at z=-1.

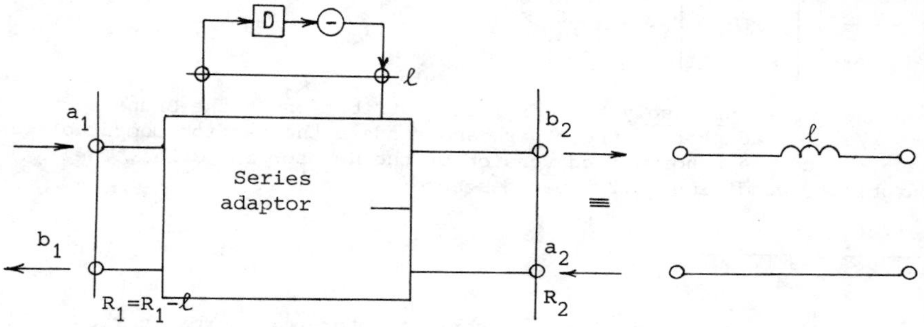

Fig. 4.16 A series WDF inductor.

The normalized CSM matrix for this case produces the transfer:

$$\begin{bmatrix} a_1 \\ b_1 \end{bmatrix} = \Theta_d \begin{bmatrix} b_2 \\ a_2 \end{bmatrix}$$

We shall compute it on the analog circuit following the mechanism depicted in fig. 4.7.

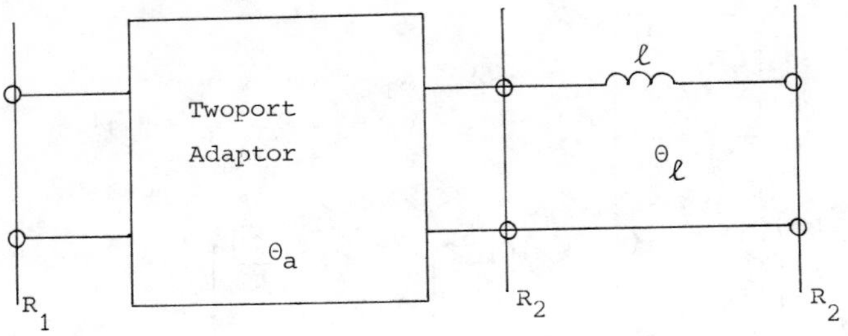

Fig. 4.17 A mechanism to compute $\Theta_d = \Theta_a\Theta_l$.

The inductor gives a contribution:

$$\Theta_\ell = \begin{bmatrix} 1 & -R_2 \\ 1 & -R_2 \end{bmatrix} \begin{bmatrix} 1 & \frac{1-z}{1+z}\,\ell \\ 0 & 1 \end{bmatrix} \begin{bmatrix} 1/2 & 1/2 \\ \frac{1}{2}R_2 & -\frac{1}{2}R_2 \end{bmatrix}$$

$$= 1 + \frac{-1+z}{-1-z} \cdot \frac{\ell}{2R_2} \begin{bmatrix} 1 & -1 \\ 1 & -1 \end{bmatrix}$$

while the contribution of the adaptor is simply (compare with (4.5)):

$$\Theta_a = \frac{1}{1+\rho} \begin{bmatrix} 1 & \rho \\ \rho & 1 \end{bmatrix}, \ \rho = \frac{R_2-R_1}{R_2+R_1}$$

We see that Θ_ℓ is <u>precisely</u> the Herglotz contribution in the Brune section with weight ℓ/R. The adaptor on the other hand renders the section computable and corresponds to the unnormalized version of the β-rotor in (3.26). The computability condition (3.20) becomes here:

$$\rho = \frac{\ell/2R_2}{1-\ell/2R_2} = \frac{R_2 - R_1}{R_2 + R_1}$$

hence $R_2 = R_1 + \ell$ as requested by the three adaptor-port normalizations.

We have shown, for the case of the series inductor, that there is a one-one correspondence between the Brune method discussed in chapter 2 and the WDF realization method presented here. This was of course to be expected (both being instances of the same algebraic procedure), and may be generalized which we presently do. Section 2 handles J-unitary Θ matrices while in WDF's the energy is not conserved. However, a simple "port-normalization" restores the energy level. Indeed and refering back to the w-plane - we see that the quantities $a_k/2\sqrt{R_k}$ and $b_k/2\sqrt{R_k}$ produce wave-energies independent of the normalization:

$$\left[\frac{a_k}{2\sqrt{R_k}}\right]^2 - \left[\frac{b_k}{2\sqrt{R_k}}\right]^2 = \text{Re } v_k \bar{i}_1$$

It follows that the relation between the normalized Θ_n and the denormalized Θ_d in a given section is given by (see fig. 4.18)

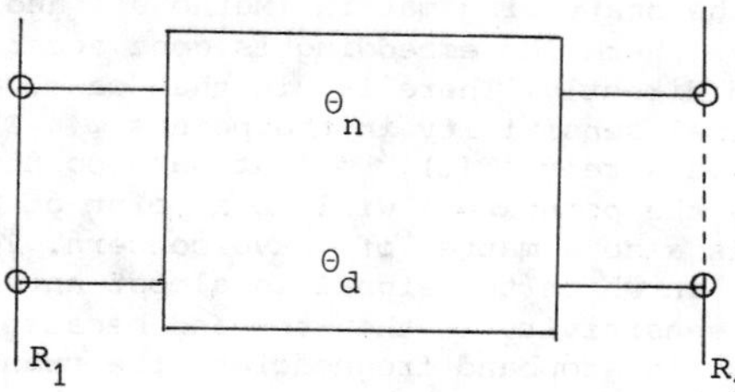

Fig. 4.18 Normalized and Denormalized CSM w.r.t. to port normalization.

$$\Theta_n = \begin{bmatrix} \sqrt{\frac{R_2}{R_1}} & 0 \\ 0 & \sqrt{\frac{R_2}{R_1}} \end{bmatrix} \Theta_d$$

As a result the basic Schur/Darlington/Brune synthesis procedure of the previous chapter can be used also to design WDF's recursively, and conversely behind each cascade WDF synthesis there is a Schur reduction. Based on this insight, it is not hard to see that cascade WDF synthesis will not produce algebraically minimal sections in all instances. For example, the Brune sections consisting of a series or parallel resonant circuit are but a subclass of all possible Brune sections. The more general Brune section cannot be realized by a section containing only one multiplier.

5. IMPLEMENTATION AND DISCUSSION

Both orthogonal and wave digital cascaded filters have good sensitivity properties both in the passband and in the stopband. The passband properties are due to a modification of Orchards theorem [9] which runs as follows. With reference to fig. 4.1.

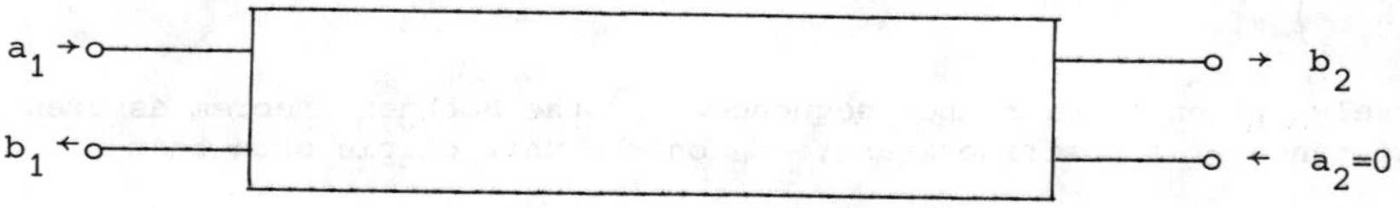

Fig. 5.1 Discussing Orchard's theorem.

a reflection zero is characterized by a point $\omega \in T$ for which there is full transmission, i.e., $b_1=0$ and $\|b_2\| = \|a_1\|$. If fig. 4.1 represents an algebraically minimal WDF or an orthogonal filter, then any variation in a multiplier or in an angle will not change the passive characters of the device. It follows that, when $b_1=0$, the sensitivity of the transmission to a variation of any parameter must be of second order, because a non-zero first order sensitivity would result in a crossing of the boundary $|\Sigma_{21}|= 1$ which is impossible when the device is lossless.

Crucial in the argument, however, is the condition $b_1=0$. Due to the variation in the coefficients it may be that the point ω in not a reflection zero. If, however, it is true that there is a point ω' in the neighbourhood of ω for which $b_1=0$, then one will have very low sensitivity of the pass-band response because points in the neighbourhood have second-order sensitivity for their transfer. Orchards argument is strictly speaking only valid for the case where the transfer function $T(z)$ is embedded in the scattering matrix (Methods I and III of chapter 3: $T(z) = \Sigma_{11}(z)$). For the case where the embedding is done according to method II, the argument does not hold directly. There is, in that case, a derived transfer function 2p/d+q, with optimal sensitivity in the points $\omega \in T$ where q=d, i.e. points for which $C(\omega) = 1$. While surely $|T(\omega)|^2 = 1$ at such points, the converse is not true and typically only the point $\omega = 1$ will be a point of low sensitivity. In the stopband, sensitivity is also a matter of grave concern. For many filters, the stopband is a large region in which the signal is almost entirely cancelled. A cascade filter achieves low sensitivity in the stopband because it physically realizes zero transfer at specific stopband frequencies (the transmission zeros) using exact cancellation in specific sections.

Because of their low sensitivity to variations in coefficient values, both WDF's and orthogonal filters may be realized with elements of low precision. For the case of a WDF one will try to find for each multiplier coefficient, a representation consisting of a few operations of the type shift then add or subtract. In practice one may find that three operations are sufficient per multiplier for most filtering applications. Filters which are optimized in this way cannot be made adaptive - of course - but they are very compact and of high quality. For the CODEC application mentioned in the introduction, they may be the best filters available in terms of surface and performance.

Similarly, in the case of orthogonal filters one may also exploit the low sensitivity of the frequency response w.r. to angle variations. However, orthogonal filters derive their attractiveness much more from their modular structure and the fact that they offer filters with standard numerical computing techniques. The key to an orthogonal filter realization is the basic Givens rotation implemented in a device called the "CORDIC"-chip which executes precisely an orthogonal rotation on a two-dimensional vector. For a description of the actual hardware implemetation of the CORDIC module in filtering applications, I refer to [13]. The most compact CORDIC implementation I know is due to Udo and Deprettere [14].

Finally I wish towrap up our discussion with the connection between orthogonal filtering and estimation theory.
To a given transfer function $T(z) = \frac{p(z)}{q(z)}$ corresponds a spectrum $W(e^{i\theta}) = |T(e^{i\theta})|^2$ and a "covariance sequence":

$$c_k = \int_{-\pi}^{\pi} W(e^{i\theta}) e^{-ik\theta} \frac{d\theta}{2\pi} \tag{5.1}$$

Conversely, given a covariance sequence $\{c_k\}$ the Bochner theorem assures us of the existence of a positive measure $d\mu$ on the unit circle such that

$$c_k = \int_{-\pi}^{\pi} e^{-ik\theta} \frac{d\theta}{2\pi} \quad d\mu = W(e^{i\theta})d\theta + d\mu_s \tag{5.2}$$

(with μ_2 representating a 'point-spectrum').
Hidden behind the covariance sequence there is a stochastic $\{x_t\}$ such that

$$c_k = Ex_t\bar{x}_{t-k} \qquad 5.3)$$

The Kolmogoroff isomorfism maps the stochastic process $\{x_t\}$ to $L^2(d\mu)$ according to the rule:

$$x_t \leftrightarrow e^{-it\theta} \quad E\, x_t\bar{x}_{t-k} = (e^{-it\theta}, e^{-i(t-k)\theta})_{L^2(d\mu)}$$

$$= e^{-ik\theta} d\mu = c_k \tag{5.4}$$

Consider now the space

$$\mathcal{M} = \left\{ \frac{\text{polynomials of degree} \le n}{q(z)} \right\} \tag{5.5}$$

in $L^2(d\mu)$. Because $q(z)$ is strictly stable one has that

$$\mathcal{M} \subset \overline{\text{span}}\ \{1, z, z^2 \ldots\}$$

in $L^2(d\mu)$. We decompose $L^2(d\mu)$ as:

$$\text{Past} = \overline{\text{span}}\ \{z, z^2, \ldots\}$$

$$\text{Present} = \text{span}\ \{1\}$$

$$\text{Future} = \overline{\text{span}}\ \{z^{-1}, z^{-2}, z^{-3}, \ldots\}$$

and consider

$$\mathcal{M}' = \{f \in \mathcal{M} \mid f(0) = 0\}$$

the subspace of $\mathcal{M}$ laying strictly in the past. $\mathcal{M}'$ plays the role of an "information subspace" in the past of the process $\{x_t\}$. If one wishes to perform a linear least squares estimation of the present $X_0 \approx 1$ using information in the past $\mathcal{M}'$, then one will have to project the present on $\mathcal{M}$. I claim:

Projection 1 on past $\mathcal{M}' = 1 - T(0)T^{-1}(z)$

Estimation error $= T(0)T^{-1}(z)$

Normalized estimation error $= T^{-1}(z)$

The proof of these assertions is based on the fact that $T^{-1}(z) \in \mathcal{M}$ and $T^{-1}(z) \perp \mathcal{M}'$. Indeed, let $f \in \mathcal{M}'$, you have:

$$(f, T^{-1}) = \int_{-\pi}^{\pi} f(e^{i\theta}) T^{-1}(e^{i\theta}) |T(e^{i\theta})|^2 \frac{d\theta}{2\pi}$$

$$= \int_{-\pi}^{\pi} f(e^{i\theta}) T(e^{i\theta}) \frac{d\theta}{2\pi}$$

$$= f(0)T(0) = 0$$

We may conclude that $T^{-1}(z)$ can be interpreted as an innovation for x_0 based on information contained in $\mathcal{M}'$. The principles which we have been touching on in this discussion may be generalized to α-stationary processes and even to more general situations see [14] and [15].

APPENDIX

LYAPUNOV EQUATIONS AND HAMILTONIANS

In this appendix it is shown how a minimal realization for a strictly stable transfer function T(z) may be embedded in a Hamiltonian which will yield the impedance function C(z) as well as the maximum phase counterpart R(z) of T(z). We use here a state-space method similar to what is presented in [17]. A direct method is also possible - see [18].

Let $\begin{bmatrix} F & G \\ H & J \end{bmatrix}$ be a minimal realization for T.

Since T is strictly stable, all the eigenvalues of F will be strictly in $\mathcal{D}$, the transfer relation being:

$$T(z) = J + H(z^{-1} - F)^{-1}G.$$

Let us look at the embedding of T expressed in the following way (we shall deduce the values of the elements presently)

$$\begin{bmatrix} x+ \\ b_1 \\ v_2 \end{bmatrix} = \begin{bmatrix} A & B_1 & B_2 \\ C_1 & D_{11} & D_{12} \\ C_2 & D_{21} & D_{22} \end{bmatrix} \begin{bmatrix} x \\ a_1 \\ i_2 \end{bmatrix} = \mathcal{T} \begin{bmatrix} x \\ a_1 \\ i_2 \end{bmatrix} \qquad (A.1)$$

T(z) is obtained through the boundary condition $i_2 = 0$ or equivalently $a_2 = b_2$, in which case we also have $v_2 = a_2$.

If A.1 is to be a lossless embedding, then it must satisfy a conservation of energy relation which we presently deduce from the actual form of the energy quadratic form:

$$||x||^2 - ||x+||^2 + 2(\tilde{i}_1 v_1 + \tilde{v}_1 i_1) + ||b_2||^2 - ||a_2||^2 = 0 \qquad (A.2)$$

Let $\mathcal{T}$ be the transition matrix in (A.1) and let moreover:

$$\Pi = \begin{bmatrix} 1 & & \\ & 1 & \\ & & 0 \end{bmatrix}, \quad \Pi' = \begin{bmatrix} 0 & & \\ & 0 & \\ & & 1 \end{bmatrix}$$

then (A.2) may be rewritten as:

$$[\tilde{x} \ \tilde{b}_2 \ \tilde{i}_2] \{\Pi - \tilde{\mathcal{T}}\Pi\mathcal{T} + 2\Pi'\mathcal{T} + 2\tilde{\mathcal{T}}\Pi'\} \begin{bmatrix} x \\ b_2 \\ i_2 \end{bmatrix} = 0$$

will preserve energy iff:

$$\Pi - \tilde{\mathcal{T}}\Pi\mathcal{T} + 2[\Pi'\mathcal{T} + \tilde{\mathcal{T}}\Pi'] = 0$$

and with a repartioning of $\mathcal{T}$ as:

$$= \begin{bmatrix} A & B_1 & B_2 \\ C_1 & D_{11} & D_{12} \\ C_2 & D_{21} & D_{22} \end{bmatrix} = \begin{bmatrix} \mathcal{A} & \mathcal{B} \\ \mathcal{C} & \mathcal{D} \end{bmatrix}$$

this becomes:

$$\left\{ \begin{array}{l} 1 - \mathcal{A}\tilde{\mathcal{A}} = 0 \\ 2\tilde{\mathcal{C}} + \mathcal{A}\mathcal{B} = 0 \\ 2(\tilde{\mathcal{D}}+\mathcal{D}) - \tilde{\mathcal{B}}\mathcal{B} = 0 \end{array} \right\} \tag{A.3}$$

We now identify the values in the transition matrix. Puting the boundary condition $i_2 = 0$ and neglecting b_1 we have as candidate realization for T(z):

$$\begin{bmatrix} x+ \\ a_2 \end{bmatrix} = \begin{bmatrix} A & B_1 \\ C_2 & D_{21} \end{bmatrix} \begin{bmatrix} x \\ a_1 \end{bmatrix} \tag{A.4}$$

which we wish to identify with the original F,G,H,J. From (A.3) it should be clear that the only restriction on (A.4) is that $[A \quad B_1]$ be isometric:

$$A\tilde{A} + B_1\tilde{B}_1 = 1 \tag{A.5}$$

Indeed, (A.5) must be satisfied as part of the unitary condition on $\mathcal{A}$. Conversely, once (A.5) is satisfied we shall be able to satisfy all the conditions in (A.3) by first augmenting $[A \quad B_1]$ to a unitary matrix, and next taking

$$\mathcal{B} = 2\tilde{\mathcal{A}}\tilde{\mathcal{E}}$$

and decomposing

$$\tilde{\mathcal{D}} + \mathcal{D} = \frac{1}{2}\tilde{\mathcal{B}}\mathcal{B}.$$

Hence we are able to embed T iff $[A \quad B_1]$ is isometric. Starting from F,G,H,J we must transform

$$[F \quad G]$$

into an isometric operator using a similarity transformation $S^{-1} \ldots S$ on the state:

$$\begin{bmatrix} A & B_1 \\ C_2 & D_{21} \end{bmatrix} \sim \begin{bmatrix} S^{-1}F\,S & S^{-1}G \\ HS & J \end{bmatrix}$$

This now leads to a Lyapunov equation:
Let $P = S\tilde{S} > 0$, then P must be the solution of

$$FP\tilde{F} + G\tilde{G} = p \tag{A.6}$$

Since F has all its eigenvalues in $\mathbb{D}$, the unique solution of (A.6) is given (theoretically) by:

$$P = \sum_{k=0}^{\infty} F^k (G\tilde{G}) \tilde{F}^k \tag{A.7}$$

where $F^n G\tilde{G}\tilde{F}^n \to 0$ as fast as $|\lambda|^{2n}$ where λ is the biggest eigenvalue of F. P is nonsingular because the controllability matrix $[G, FG, F^2G \ldots]$ has full rank.

Starting from a realization

$$\begin{bmatrix} F & G \\ H & J \end{bmatrix}$$

for T(z) where $[F \quad G]$ is isometric, $F\tilde{F} + G\tilde{G} = 1$, we may now explicitely compute the embedding as follows:

$$= \begin{bmatrix} F & G & 2\tilde{F}[\tilde{H} - (1-G\tilde{G})^{-1/2} G\tilde{J}] \\ -\tilde{G}(1-G\tilde{G})^{-1/2}F & (1-\tilde{G}G)^{1/2} & 2\{\tilde{G}\tilde{H} + (1-\tilde{G}G)^{1/2}\tilde{J}\} \\ H & J & H\tilde{H} + J\tilde{J} \end{bmatrix} \tag{A.8}$$

The realization for C(z) satisfying

$$\frac{1}{2}(C(z) + C_*(z)) = T(z)T_*(z)$$

is deduced from (A.8):

$$\begin{bmatrix} F & 2\tilde{F}\{\tilde{H} - (1-G\tilde{G})^{-1/2} G\tilde{J}\} \\ H & H\tilde{H} + J\tilde{J} \end{bmatrix} \tag{A.9}$$

BIBLIOGRAPHY

1. Nouta, R., "Studies in Wave Digital Filter Theory and Design", Ph.D. Thesis, Delft Univ. of Techn., Oct. 1978.

2. Hoffman, K., "Banach Spaces of Analytic Functions", Englewood Cliffs, N.J., Prentice Hall, 1962.

3. Dewilde, P., "Stochastic Modelling with Orthogonal Filters", in Outils et Modèles Mathématique pour l'Automatique, l'Analyse des Systèmes et le Traitement du Signal", Vol. 2, Ed. du CNRS, 1982, pp. 331-398.

4. Dewilde, P. and H. Dym, "Lossless Inverse Scattering with rational digital filters: theory and applications", Delft Univ. of Technology, Dec. 1982, to be published.

5. Deprettere, E. and P. Dewilde, "Orthogonal Cascade Realization of Real Multiport Digital Filters", Circuit Theory and Applications, Vol. 8, pp. 245-272, (1980).

6. Guillemin, E.A., "Theory of Linear Physical Systems", J. Wiley & Sons, 1963.

7. Fettweis, A., "Digital Filter Structures related to Classical Filter Networks", A.E.U. band 25, pp. 79-89, 1981.

8. Fettweis, A., "Pseudopassivity sensitivity and stability of wave digital filters", IEEE Trans. on CT, Vol. 19, pp. 668-672, 1972.

9. Temes, G.C. and J.W. La Patra, "Circuit Synthesis and Design", Mc Graw Hill, 1977.

10. Belevitch, V., "Classical Network Theory", Holden Day, San Francisco, 1968.

11. Newcomb, R.W., "Linear Multiport Synthesis", Mc Graw Hill Book Co., 1966

12. Boite, R. and H. Leich, "Les Filtres Numeriques", Masson, Paris, 1980.

13. Dewilde, P., E. Deprettere and R. Nouta, "Parallel and Pipelined VLSI Implementation of Signal Processing Algorithms", in VLSI and Signal Processing, S.Y. Kung editor (to appear).

14. Deprettere, E. and R. Udo, "The pipeline CORDIC", Dept. of Electrical Eng., Delft, 1983.

15. Lev Ari, H. and T. Kailath, "Parametrization and Modelling of Non-Stationary Process", Stanford Univ., Dept. of EE, 1982.

16. Dewilde, P., J.T. Fokkema and I. Widya, "Inverse Scattering and Linear Prediction, the Time continuous case", in Stochastic Systems, Nato Advanced Study Inst., Series, Reidel, Dordrecht, 1980.

17. Anderson, B.D.O. and S. Youngpanitherd, "Network Analysis and Synthesis", Prentice Hall, 1973.

18. Dewilde, P. and H. Dym, "Lossless Chain Scattering Matrices and Optimum Linear Prediction: the Vector Case", Circuit Theory and Applications, Vol. 9, 135-175 (1981).

19. Rao, S.K. and T. Kailath, "Orthogonal Digital Filters for VLSI Implementation" preprint, Dept. of Electrical Engineering, Stanford Univ., May 1983,

5

Speech Coding and Processing

JOHN MAKHOUL

A review is undertaken of various methods and techniques for the coding of speech signals. The methods described code the speech signal in the range from 64000 bits/s with pulse code modulation, down to only a few hundred bits per second with vector quantization. The principles underlying the various redundancy removal techniques are explained and illustrated. The chapter ends with a brief review of speech synthesis, speech recognition, and speech hardware.

TABLE OF CONTENTS

1. INTRODUCTION

With recent developments in the design of very-large-scale-integration (VLSI) circuits and the appearance of signal-processing computers on a chip, it has become practical to implement many digital processing algorithms for speech processing in working real-time systems. In this chapter we review some of the important speech processing principles and techniques that are in use today with heavy emphasis on speech coding.

The chapter begins by a description of how speech is produced and how it is prepared for digital processing. The major part of the chapter is then devoted to methods of speech coding. The first set of methods under the heading of waveform coding treat speech like any other waveform, and the coding methods described can be applied to other non-speech signals as well. By making use of certain properties of speech perception one then is able to reduce the data rate further with a relatively small decrease in speech quality. We then describe the application of speech processing techniques in speech synthesis and speech recognition. The chapter ends with a very brief description of recent developments in speech hardware. The subject of speech enhancement in the presence of acoustic background noise is important in speech coding and recognition applications. This topic is not covered here; the reader is referred to a book edited by Lim [41] for descriptions of various speech enhancement algorithms.

2. THE SPEECH SIGNAL

2.1 Speech Production

Speech sounds can be classified into voiced and unvoiced or fricated sounds [1,2]. Voiced sounds are produced as a result of quasi-periodic vibration of the vocal cords. Unvoiced sounds are produced by a turbulent flow of air created at a constriction somewhere along the vocal tract, thereby generating an incoherent acoustic noise which excites the vocal tract. Certain sounds, such as voiced fricatives [z] and [v], are produced using both voiced and unvoiced excitations.

Beyond the type of excitation, a speech sound is determined by the specific shape of the vocal tract. Changes in that shape are effected by constrictions formed by the tongue, velum, pharynx or larynx, and produce sounds with different spectra.

2.2 Digital Processing

To perform any type of digital processing on the speech signal, one must first sample and digitize the analog waveform [see Lim, this volume]. A low-pass filter with cut-off frequency W is used to filter out all energy components above W Hz, followed by a digital-to-analog (D/A) converter which samples the speech, usually at the Nyquist rate of 2W Hz, and digitizes each sample to a number of bits. The frequency W depends on the application and is usually between 3-5 kHz for most applications. With the speech signal sampled and digitized, we are ready to perform various digital signal processing operations for different applications. One major application is speech coding, which is discussed next.

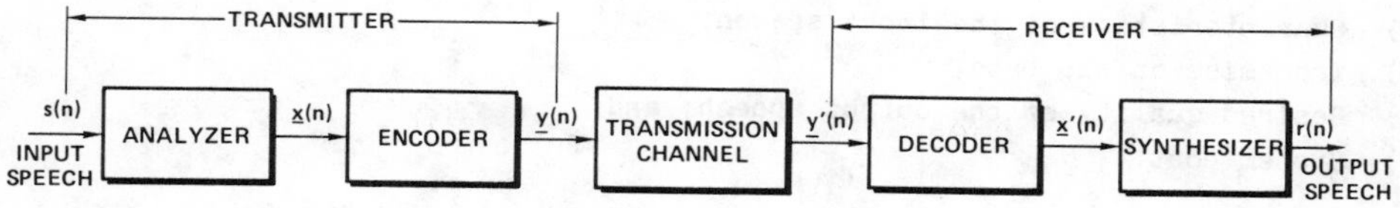

FIGURE 1. Components of a speech compression system.

3. GENERAL CODING SYSTEMS

3.1 Basic Coding System

Fig. 1 shows the basic components of a speech compression (or voice coding) system. The first component analyzes the speech signal s(n) that has been low-pass filtered and time-sampled, and extracts a vector of unquantized parameters $\underline{x}(n)$. These parameters are then quantized and encoded in the encoder as $\underline{y}(n)$ and are transmitted through the transmission channel. In a noiseless channel $\underline{y}'(n)=\underline{y}(n)$. The parameters $\underline{y}'(n)$ are decoded in the decoder to produce an estimate $\underline{x}'(n)$ of the analysis parameters $\underline{x}(n)$. The last component in Fig. 1 uses the parameters $\underline{x}'(n)$ to synthesize the reconstructed signal r(n) which is an approximation to the original signal s(n). A compression system attempts to minimize the number of bits/second in $\underline{y}(n)$ while maintaining good quality in the synthesized speech.

The nature of the synthesizer determines the type of voice coder and dictates the type of analysis to be performed. Fig. 2 depicts the two major components of the synthesizer: excitation (or source) and transfer function. Once a synthesizer model is chosen, any reduction in transmission rate is accomplished by the encoder. The encoder in Fig. 1 performs the two functions, quantization and encoding. The quantization process converts the extracted parameters into a set of numbers using specified quantization schemes. The encoding process encodes these numbers into a sequence of binary digits for transmission. The encoding can be as simple as direct binary encoding, or as complicated as desired for the minimization of the average transmission rate. In order to maximize the cost effectiveness of a speech compression system, its design must take into consideration four important parameters:

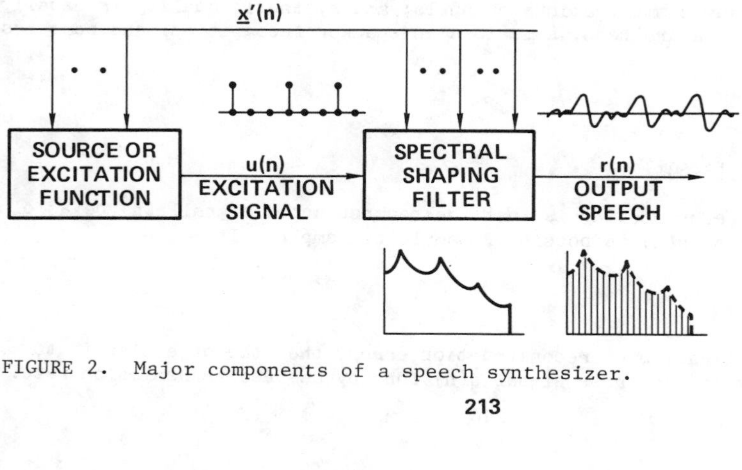

FIGURE 2. Major components of a speech synthesizer.

(a) Characteristics of the input speech,
(b) Transmission bit rate,
(c) Desired quality of the output speech, and
(d) System cost.

Given (a), one attempts to achieve (b) and (c) while minimizing (d).

3.2 Redundancy Removal

Speech compression is essentially a process of redundancy removal, and bit rate reduction is achieved by identifying and removing the different forms of redundancy. In speech transmission, the main types of redundancy are:

(a) Lack of flatness of short-term spectrum,
(b) Quasi-periodicity during voiced segments,
(c) Limitations on the shape and rate of movement of the vocal tract, and
(d) Non-uniform probability distributions of the values of transmission parameters.

The first three are due to physical properties of the speech production mechanism, and the last is a function of the particular speech coding method used.

The redundancy due to adjacent speech samples being correlated is exhibited in a nonflat spectrum. This redundancy is removed by appropriate spectral filtering. Most waveform coding techniques attempt to take advantage of this type of redundancy by spectral flattening methods. The removal of redundancy due to quasi-periodicity (pitch) of voiced segments is used in a number of speech coding systems. The third type of redundancy (c) allows the transmission of spectral filter parameters at a slow rate of once every 10-30 ms. The fourth type of redundancy (d) is removed by proper encoding. At a given transmission data rate, the speech quality of a coder is optimized by judiciously exploiting these four forms of redundancy.

Table 1 gives a summary of some of the popular systems for speech coding at various data rates. The systems will be discussed in the sections below, and a glossary of symbols is given at the end of the chapter for easy reference. In Table 1, toll quality is in reference to a high-quality telephone system; communications quality implies the existence of some noise and distortion as would be experienced in many communications channels; and synthetic quality is usually exhibited by lack of naturalness, some loss in speaker recognition, and some loss in intelligibility.

4. WAVEFORM CODING [3-6]

The objective in waveform coding is that the reconstructed signal r(n) be as close to the original signal s(n) as possible, sample by sample. If

$$q(n) = r(n)-s(n) \tag{1}$$

represents the quantization or reconstruction error, then the objective is to minimize the energy of the error signal q(n) for any desired transmission rate.

TABLE 1. Examples of speech coding systems at different data rates. (See glossary of symbols at end of chapter.)

DATA RATE (bits/s)	SYSTEM	SPEECH QUALITY
64000	PCM	Toll
32000	CVSD	Toll
16000	APC, ATC	Toll
9600	BBC	Communications
2400	LPC	Synthetic
<800	VQ	Synthetic

Waveform coding is used in a number of fields and the principles employed are common. Although the discussion below focuses on speech coding, the same techniques can be used to code other signals as well.

4.1 Pulse Code Modulation (PCM)

This is the simplest waveform coding scheme known. In this scheme, each signal sample is quantized into one of $L=2^R$ levels, or R bits/sample. The total bit rate is then 2WR bits/s (bits per second) if we assume a signal bandwidth of W Hz.

Uniform Quantization. The simplest PCM coding scheme employs a uniform quantizer. The range of signal values of s(n) is divided into L equal bins, each of width d, which is known as the quantization step size. All signal values falling within a particular bin are decoded to one value in that bin. If L is large, one can make the reasonable assumption that the quantization error q(n) will be uniformly distributed over an interval d in width. In that case, one can show that the signal-to-quantization-noise ratio is given approximately by:

$$E_{sq} = \frac{E_s}{E_q} = \frac{E_s}{d^2/12} \tag{2}$$

where E_s and E_q are the variances (or average energies) of the input signal and the quantization noise, respectively. In decibels, we shall write

$$S/Q = 10 \log_{10} E_{sq} = 10.8 - 20 \log_{10} d \text{ dB.} \tag{3}$$

If the range of signal values for the quantizer is taken to be $-4\sqrt{E_s}$, to $+4\sqrt{E_s}$, (i.e., ± 4 times the standard deviation), then

$$d = 8\sqrt{E_s}/L \tag{4}$$

and

$$S/Q = 6.02R - 7.2 \text{ dB.} \tag{5}$$

For R=8 bits, S/Q=41 dB which may be acceptable for many applications. Smaller values of R will lead to a quantization noise that is quite perceptible. For R<7, the noise becomes also correlated with the signal, which has been found to be perceptually objectionable.

Dithering. Dithering is employed to alleviate the signal-dependent character of the quantization noise. The process of dithering is as follows: a pseudorandom sequence is added to the signal before it is quantized, then at the receiver the same sequence is subtracted from the decoder output. The result is a quantization error that is almost white, preserves the same S/Q, and is less objectionable. Dithering has been helpful in the range $4 \leq R \leq 6$.

Nonuniform Quantization. Uniform quantization is most useful under two conditions: (1) The range of signal values is known, and (2) the distribution of signal values over that range is uniform. However, because speech is a nonstationary signal, with a ratio of more than 30 dB between strong and weak sounds, and because people speak with different sound levels, both conditions above are not met in practice. If one designs the quantizer for too small a range, then the overload or saturation noise can be substantial for loud sounds. If one designs the quantizer for too large a range, then the ordinary granular noise will be large and may mask weak sounds.

One method to improve performance under these circumstances is to design a nonuniform quantizer whose step size is small for small signal values and large for large signal values. If one could measure the probability density function (pdf) of speech, then a nonuniform quantizer could be designed to minimize the quantization noise for a given number of bits R. The design of such quantizers can be done interactively as originally proposed by Max [7], and the result is what is known as a Max quantizer. However, a Max quantizer can be sensitive to a mismatch between actual pdf and design pdf.

Two types of quasi-logarithmic quantizers have been proposed which are very insensitive to a quantizer mismatch relative to signal variance and signal pdf. These are the μ-law and A-law quantizers. Since nonuniform quantization is equivalent to uniform quantization over a nonlinear companding (mapping) of signal values, one can specify a nonuniform quantizer by an appropriate companding law. The companding laws for μ-law and A-law are given below. If x is an original signal value and c is its companded value, then:

μ-Law

$$c(x) = x_{max} \frac{\log_e(1+\mu|x|/x_{max})}{\log_e(1+\mu)} \operatorname{sgn} x \tag{6}$$

A-Law

$$c(x) = \begin{cases} \dfrac{A|x|}{1+\log_e A} \operatorname{sgn} x; & 0 \leq \dfrac{|x|}{x_{max}} \leq \dfrac{1}{A} \\ x_{max} \dfrac{1+\log_e(A|x|/x_{max})}{1+\log_e A} \operatorname{sgn} x; & \dfrac{1}{A} \leq \dfrac{|x|}{x_{max}} \leq 1 \end{cases} \tag{7}$$

x_{max} is the maximum amplitude of the signal s(n). Both companding laws are linear for small values of x and logarithmic for large values of x. The North American PCM standard uses μ=255 and the European PCM standard uses A=87.56. Both companders have a S/Q that is approximately equal to

$$S/Q = 6.02R - 10 \text{ dB}. \tag{8}$$

Even though this value of S/Q appears smaller than that given in (5), the value in (8) is applicable to a much wider range of signal variance than (5) is. In fact, logarithmic quantizers have been shown to provide approximately 24 dB improvement in S/Q over uniform quantizers (or a difference of 4 bits for the same S/Q value).

Adaptive PCM (APCM). With any fixed quantizer, the short-term S/Q can vary by as much as 30 dB over time because of the nonstationarity of the speech signal. An adaptive quantizer adapts the quantizer step size to the short-term amplitude of the signal. Fig. 3 shows a schematic diagram of an APCM system. The quantizer is assumed fixed and G is varied to be inversely proportional to the short-term root-mean-square (rms) value of the signal. For this type of quantizer one could use a simple uniform quantizer and S/Q would follow (5) on a short-term basis.

APCM systems have two major properties: For a given quantizer, the short-term S/Q is constant with time, and for moderate quantization (R>2), the spectral envelope of the quantization noise is expected to be flat.

One important consideration in APCM is how the receiver will know which value of G is being used at any particular time. There are two general methods: forward estimation and backward estimation. In forward estimation, the energy of the signal s(n) is usually computed over a block of about 10-30 ms, and G is set equal to

$$G_F = \frac{1}{\sqrt{E_s}} \tag{9}$$

The value of G is transmitted to the receiver for every block as side information, in addition to the quantized sample values. G is usually quantized logarithmically. In backward estimation, the energy of the reconstructed signal r(n) is measured into the past and G is set equal to

$$G_B = \frac{1}{\sqrt{E_r}} \tag{10}$$

where E_r is measured on a short-term basis. Usually, the value of G_B is updated every sample. Because r(n) is available at the transmitter as well as the receiver, the value of G_B need not be transmitted.

In general, forward adaptation is more optimal (has higher S/Q) than backward adaptation. However, forward adaptation contains a delay equal to the block length while backward adaptation is instantaneous. The propagation of channel errors is more serious with backward adaptation.

FIGURE 3. Adaptive PCM (APCM) system.

4.2 Adaptive Predictive Coding (APC)

One can improve the performance of a signal quantizer by taking advantage of the sample-to-sample signal correlation to predict a signal value from past samples of the signal. We shall begin the discussion by a brief review of linear prediction.

<u>Linear Prediction (LP)</u> [8-10]. In linear prediction, it is assumed that a speech sample s(n) can be approximately predicted as a linear combination of a number of immediately preceding samples:

$$s'(n) = - \sum_{k=1}^{p} a(k)s(n-k) \tag{11}$$

where $\{a(k),\ 1 \le k \le p\}$ is a set of real constants known as <u>predictor coefficients</u>, and p is the order of the predictor. Let e(n) denote the error between the actual value and the predicted value

$$e(n) = s(n)-s'(n) = s(n) + \sum_{k=1}^{p} a(k)s(n-k). \tag{12}$$

e(n) is also known as the LP residual. By taking the z transform in (12) and denoting z transforms of signals by the corresponding capital letters, we obtain

$$\begin{aligned} E(z) &= [1 + \sum_{k=1}^{p} a(k)z^{-k}]S(z) \\ &= A(z)S(z) \end{aligned} \tag{13}$$

where

$$A(z) = 1 + \sum_{k=1}^{p} a(k)z^{-k}. \tag{14}$$

Therefore, e(n) can be obtained by passing the signal s(n) through the all-zero filter A(z).

In LP analysis, the predictor coefficients are obtained as a result of minimizing the energy or variance of the residual e(n). By squaring (12), averaging over time, and taking the partial derivatives with respect to all a(k), one obtains the <u>normal equations</u>

$$\sum_{k=1}^{p} a(k)\, R(i-k) = -R(i), \quad 1 \le i \le p. \tag{15}$$

This is a set of p linear equations in p unknowns which can be solved easily for the predictor coefficients. The minimum residual energy is then given by:

$$E_e = R(0) + \sum_{k=1}^{p} a(k)R(k). \tag{16}$$

The filter A(z) is often known as a "whitening" filter since the residual e(n) will have a relatively flat spectral envelope. The larger p is, the whiter is the residual. From (13) we see that if E(z) has a flat spectrum, then the signal spectrum can be modelled by the spectrum of the all-pole filter

$$H(z) = 1/A(z). \tag{17}$$

A(z) is also known as the "inverse filter" since it is the inverse of the all-pole model H(z) of the signal spectrum.

A particularly efficient method for solving (15) is an iterative method, often known as the Levinson-Durbin recursion, which solves (15) for p=1, then p=2, etc., until the desired value of p is reached [see Kailath, this volume]. The solution is given below:

$$E(0) = R(0)$$

For $m = 1,2,\ldots,p$

$$K_m = -[R(m) + \sum_{k=1}^{m-1} a_{m-1}(k)R(m-k)]/E(m-1) \quad (18a)$$

$$a_m(m) = K_m$$

$$a_m(k) = a_{m-1}(k) + K_m a_{m-1}(k), \qquad 1 \le k \le m-1 \quad (18b)$$

$$E(m) = (1-K_m^2)E(m-1) \quad (18c)$$

The coefficients $\{a_m(k),\ 1 \le k \le m\}$ are the predictor coefficients for the optimal mth order predictor. The final solution is given by the coefficients $\{a_p(k),\ 1 \le k \le p\}$.

From (18c) one can write

$$E(m) = R(0) \prod_{i=1}^{m} (1-K_i^2). \quad (19)$$

The minimum error E(m) has to decrease as the predictor order increases, otherwise one would not increase the order. From (19) this means that we must have

$$|K_m|<1, \qquad 1 \le m \le p. \quad (20)$$

The intermediate quantities K_m are known as the <u>reflection</u> <u>coefficients</u>. In the statistical literature, in autoregressive modeling [see Marple, this volume], the negative of K_m are known as <u>partial</u> <u>correlation</u> (PARCOR) <u>coefficients</u> [11]. Equation (20) can be shown to be a necessary and sufficient condition for the all-pole synthesis filter H(z) to be stable, i.e., all poles are inside the unit circle. Filter stability is very important in speech synthesis, because an unstable filter can lead to "pops" and "clicks" in the synthetic signal. If $|K_p|=1$, then all the poles will be <u>on</u> the unit circle, which is an unstable condition.

Dividing E(p) in (19) by the energy $R(0)=E_s$ of the signal, we obtain the <u>normalized</u> <u>minimum</u> <u>error</u>

$$V(p) = \frac{E_e}{E_s} = \prod_{m=1}^{p} (1-K_m^2). \quad (21)$$

From (20) and (21), we have

$$0<V(p)\le 1 \quad (22)$$

which says that the residual variance is always less than or equal to the signal variance. The actual value of V(p) will depend on the shape of the all-pole filter H(z) which models the signal spectrum. In particular, one can show that

$$V(p) = \frac{\text{geometric mean of LP spectrum}}{\text{arithmetic mean of LP spectrum}} = \frac{\exp\left[\int_{-\frac{1}{2}}^{\frac{1}{2}} \log|H(f)|^2 \, df\right]}{\int_{-\frac{1}{2}}^{\frac{1}{2}} |H(f)|^2 \, df} \tag{23}$$

For relatively flat spectra, V(p) is close to 1, while for spectra with a large dynamic range, V(p) can be very small. In general, V(p) is small for voiced sounds but closer to one for unvoiced sounds.

<u>Basic APC System</u>. Fig. 4 shows the basic form of an APC system, where all signals are shown in the z domain. A(z) is the filter obtained as a solution to (15). W(z) is the APC residual, which is quantized as $\hat{W}(z)$ and transmitted. The adaptive quantizer is the same as the APCM system shown in Fig. 3. The residual quantization noise Q(z) is, therefore,

$$Q(z) = \hat{W}(z) - W(z) \tag{24}$$

and

$$R(z) = \hat{W}(z)/A(z) \tag{25}$$

is the reconstructed signal, which is available at the transmitter as well as the receiver (see Fig. 4). From Fig. 4 we can write

$$W(z) = S(z) + [A(z)-1]\hat{W}(z)/A(z). \tag{26}$$

Multiplying both sides of (26) by A(z), substituting for $\hat{W}(z)$ from (24), and rearranging terms, we have

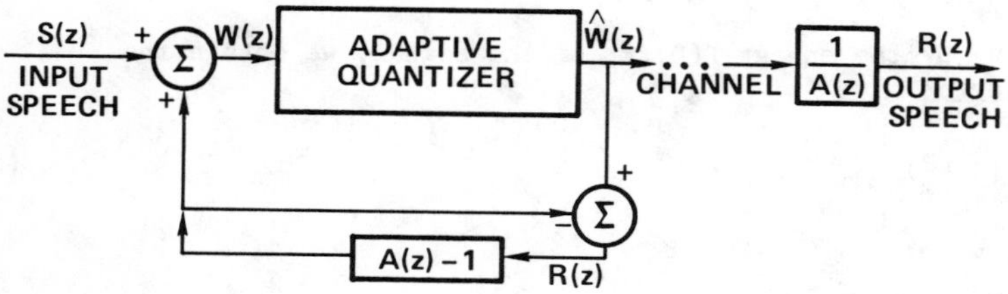

FIGURE 4. Basic adaptive predictive coding (APC) system.

$$W(z) = A(z)S(z) + [A(z)-1]Q(z). \tag{27}$$

But, from (13), we see that $A(z)S(z)$ is nothing but the LP residual $E(z)$. Therefore, the APC residual in (27) is equal to the LP residual augmented by a filtered version of the quantization noise. Equation (27) can be written in the time domain as:

$$w(n) = s(n) + \sum_{k=1}^{p} a(k)s(n-k) + \sum_{k=1}^{p} a(k)q(n-k). \tag{28}$$

Therefore, from (28), one can see that only <u>past</u> values of the noise are used in the computation of the APC residual.

From (27) and (24), the quantized residual $\hat{W}(z)$ is given by:

$$\hat{W}(z) = A(z)S(z) + A(z)Q(z). \tag{29}$$

Dividing (29) by $A(z)$ and substituting (25), we have

$$R(z) = S(z) + Q(z). \tag{30}$$

Therefore, the error $R(z)-S(z)$ in quantizing the signal $S(z)$ is equal to the residual quantization error $Q(z)$. Equation (30) is a hallmark of predictive and differential coding systems in general. This relationship is a direct result of including the quantizer in the prediction loop.

<u>Signal-to-Noise Analysis</u>. The major improvement of APC over APCM is an increase in the S/Q ratio. For APC, we have

$$E_{sq} = \frac{E_s}{E_q} = \frac{E_s}{E_w} \cdot \frac{E_w}{E_q} = E_{sw} \cdot E_{wq} \tag{31}$$

where E_s, E_w, and E_q are the average energies of the signal, residual, and quantization noise, respectively. E_{sw} is the signal-to-residual ratio and E_{wq} is the residual-to-noise ratio. Taking $10 \log_{10}$ of (31) one obtains

$$10 \log_{10} E_{sq} = 10 \log_{10} E_{sw} + 10 \log_{10} E_{wq}$$

or

$$S/Q = S/R + R/Q \text{ in decibels,} \tag{32}$$

where S/R and R/Q are the signal-to-residual and residual-to-noise ratios in decibels, respectively. One can show that E_{wq} is equal to the signal-to-noise ratio for APCM. Therefore,

$$(S/Q)_{APC} = S/R + (S/Q)_{APCM} \tag{33}$$

thus emphasizing the fact that S/Q in APC is larger than S/Q in APCM by a quantity equal to S/R. If in (27) we assume that the quantization noise is small relative to the LP residual $A(z)S(z)$, then $W(z)$ is approximately equal to $E(z)$ and, from (21),

$$E_{sw} = \frac{E_s}{E_w} \cong \frac{E_s}{E_e} = \frac{1}{V(p)} \tag{34}$$

From (22), E_{sw} is always greater than or equal to one. E_{sw} is then the increase in signal-to-noise ratio due to APC, which we shall denote by I_{APC}

$$I_{APC} = \frac{1}{V(p)} = \frac{\text{arithmetic mean of LPC spectrum}}{\text{geometric mean of LP spectrum}} \geq 1 \tag{35}$$

and

$$S/R = 10 \log_{10} I_{APC} = -10 \log_{10} V(p). \tag{36}$$

From the discussion following (23), we expect I_{APC} to be large for voiced sounds and not so large for unvoiced sounds. Fig. 5 shows I_{APC} plotted in decibels for a speech utterance, along with plots of the pitch and signal variance. In Fig. 5, A(z) was recomputed every 20 ms. Note the large variations in I_{APC} for different sounds.

We conclude that APC has three major properties:

1. For a given quantization step size, S/Q is always greater for APC than for APCM.
2. The increase in signal-to-noise ratio, I_{APC}, changes with time as a function of the signal spectrum. Spectra with larger dynamic ranges generally have large values of I_{APC}.
3. The output noise spectrum is flat.

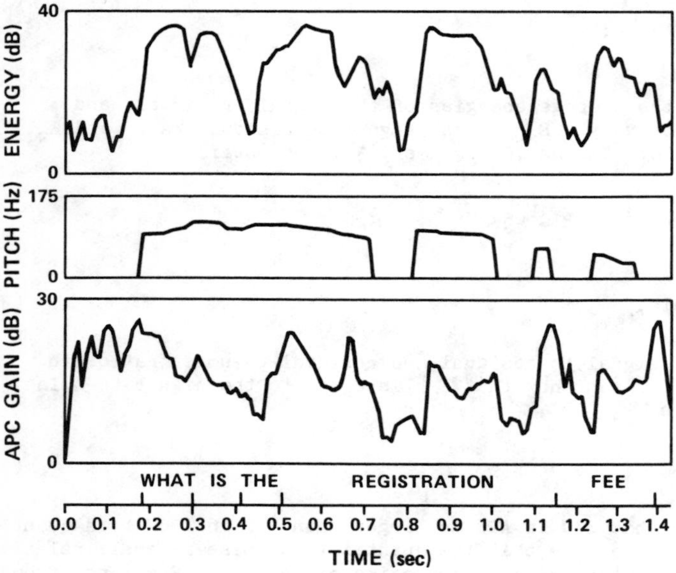

FIGURE 5. Plots of speech energy, pitch frequency, and APC gain for one utterance spoken by an adult male.

We note that the increase in S/Q, I_{APC}, is a direct result of including the quantizer as part of the predictor loop in Fig. 4. That increase would not be possible if the quantizer were not included in the loop.

<u>Differential Coding</u>. Differential PCM (DPCM) is a special case of APC where the predictor has the simple form $A(z)=1+a_1z^{-1}$ where a_1 is a fixed negative number. a_1 is computed by taking the long-term average of the signal spectrum and computing the optimal A(z) for p=1. Of course, higher-order (p>1) fixed predictors could be used as well, although little can be gained beyond p=4.

Adaptive DPCM (ADPCM), where the predictor is adapted to the signal, is essentially the same as the APC system shown in Fig. 4.

Delta modulation (DM) is basically a DPCM scheme with two main differences: (1) The waveform is sampled at a rate much larger than the Nyquist rate; and (2) the residual or the difference signal is quantized using a 2-level quantizer (1 bit). The over-sampling of the waveform increases the adjacent sample correlation, which in turn permits the simple use of a 2-level quantization strategy of transmitting only the sign of the residual. The bit rate of a DM system is therefore equal to the sampling rate of the waveform.

Adaptive delta modulation (ADM) uses a variable step size, which is determined based on the quantizer outputs. Since, with a 2-level quantizer, the observation of a single sample of quantizer output does not provide any indication of slope overload or granularity, it is necessary to employ a sequence of quantizer outputs for meaningful step-size adaptation. When the step size is adapted very smoothly in time, with a time constant on the order of 5-10 ms, we obtain a continuously-variable-slope-delta modulation (CVSD) coder. The slow adaptation makes the CVSD coder more tolerant to channel errors than ADM coders. Also, it has the effect of decreasing the granular noise in the output speech, at the cost of a significant increase of slope-overload distortion relative to ADM coders.

The CVSD coder produces a more "clean-sounding" speech than other ADM coders at bit rates less than 24 kbits/s; at these rates, ADM coders tend to have excessive granular noise. However, the quality of the CVSD coder at 16 kbits/s is lower than what one can obtain from an APC coder at the same bit rate.

<u>Pitch Prediction</u> [12]. The residual W(z) in the APC system shown in Fig. 4 contains fairly large excursions known as "pitch pulses" which are due to large prediction errors at the beginning of each pitch period in voiced sounds. These pulses cause the quantizer to be less efficient because of the need to balance overload noise and granular noise. The problem can be diminished if somehow one could reduce the amplitude of the pitch pulses. One method that has been quite successful is to use a pitch predictor C(z) as shown in Fig. 6. The most general form for C(z) that has been used to date is the 3-tap pitch predictor

$$C(z) = 1+c(M-1)z^{-(M-1)} + c(M)z^{-M} + c(M+1)z^{-(M+1)} \tag{37}$$

where M is the pitch period in samples, and c(M-1), c(M), c(M+1), are the three unknown coefficients. The first problem is to find the pitch period M. One simple method is to find the peak of the autocorrelation function of the signal beyond a small lag; the value of the pitch period is taken to be the lag corresponding to that peak. (Another method employs a "pseudo-autocorrelation" function which uses in its definition the magnitude of the difference instead of the product; the pitch period is then taken as the lag corresponding to the minimum value.) Once a value of M is estimated, the pitch predictor coefficients c(k) are computed from the signal S(z) as in ordinary LP. The equations to be solved are

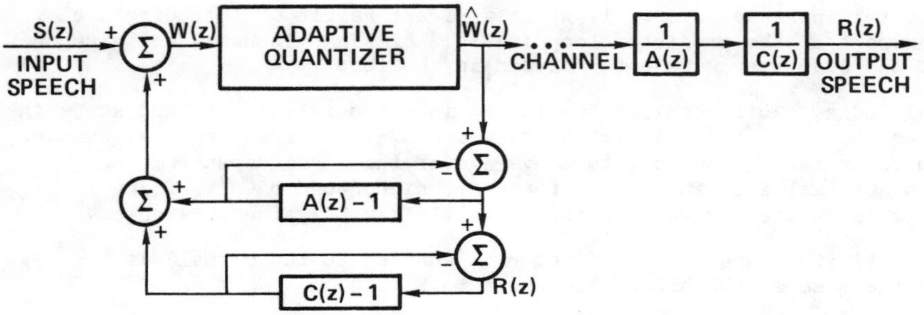

FIGURE 6. APC system with a pitch predictor C(z).

$$\sum_{k=M-1}^{M+1} c(k)R(i-k) = -R(i), \qquad M-1 \leq i \leq M+1 \tag{38}$$

where R(i) is the autocorrelation of the signal S(z). S(z) is then filtered by C(z) to obtain a residual E'(z) = C(z)S(z), which has approximately the same spectral envelope as the signal S(z) but has lost much of the harmonic structure. The spectral predictor coefficients of A(z) are then computed from the residual E'(z) using LP analysis.

The solution to (38) does not guarantee a stable synthesis pitch filter 1/C(z). In the small number of cases where that happens, one could use a 1-tap predictor, i.e., c(M-1)=c(M+1)=0, which is always guaranteed to be stable.

Using a pitch predictor has three majors effects:

1. It reduces the amplitudes of large "pitch pulses".

2. The output S/Q is increased due to an increase in signal-to-residual ratio.

3. Since the synthesis pitch filter has a relatively long impulse response, it causes a sustained propagation of any channel error in the decoded residual, which produces speech with a reverberant quality.

Noise Spectral Shaping [13-15]. The quantization noise in APC generally has a flat spectral envelope. If the short-term spectrum has a large dynamic range, as in voiced speech, it is possible that the noise level may be above the signal level in some frequency ranges. The result is a perceived hissing noise, which is undesirable. In noise spectral shaping, one shapes the quantization noise spectrum such that it remains below the signal spectrum as much as possible.

Fig. 7 shows a complete APC system with spectral noise shaping, which we shall call an APC-NS system. B(z) is the spectral noise shaping filter and it is in the noise feedback path since the quantization noise Q(z) is fed back through it. Systems that employ a noise feedback path are often given the name noise feedback coding systems. From Fig. 7 one can show that the output signal at the receiver is given by

$$R(z) = S(z) + B(z)\,Q(z). \tag{39}$$

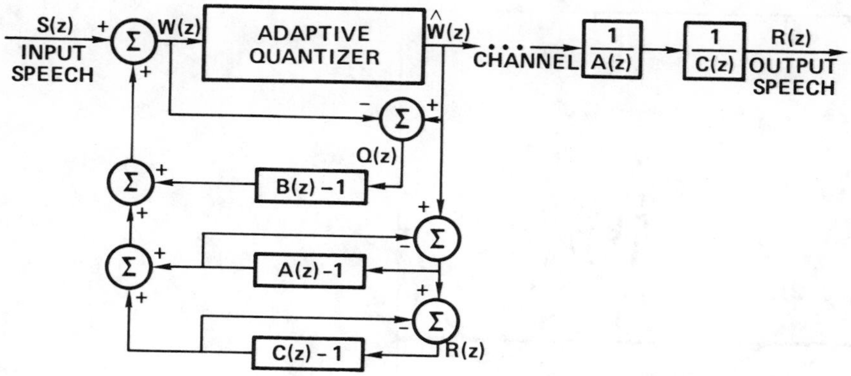

FIGURE 7. Complete APC-NS system: APC with spectral noise shaping.

Since Q(z) has a flat spectrum, the quantization noise B(z)Q(z) at the output has the same spectral shape as the filter B(z). Note that B(z) is not used at the receiver and so its parameters need not be transmitted.

Since B(z)-1 should be predictor operating on past values only, the first coefficient of B(z) must be 1. So,

$$B(z) = 1 + \sum_n b(n)z^{-n}$$

where the summation over n may be infinite, as in the case of a recursive filter. In fact, one of the better noise shaping filters that have been proposed is a pole-zero filter of the form [14]

$$B(z) = \frac{A(z/\alpha)}{A(z)}, \quad 0 \leq \alpha \leq 1 \tag{40}$$

where α is a constant.

The S/Q for the APC-NS system is affected by the filter B(z). One can show from (39) that the new S/Q is given by

$$(S/Q)_{NS} = S/Q - 10 \log_{10} \sum_n b^2(n) \tag{41}$$

where NS refers to the noise shaping case. It is clear from (41) that, since b(0)=1, $(S/Q)_{NS}$ is always less than S/Q. However, even if S/Q is reduced, with proper design the perceptual effect can be a reduced noise level and higher speech quality.

An example showing the noise spectrum in APC and in APC-NS is shown in Fig. 8. The filter B(z) used was the pole-zero filter given in (40).

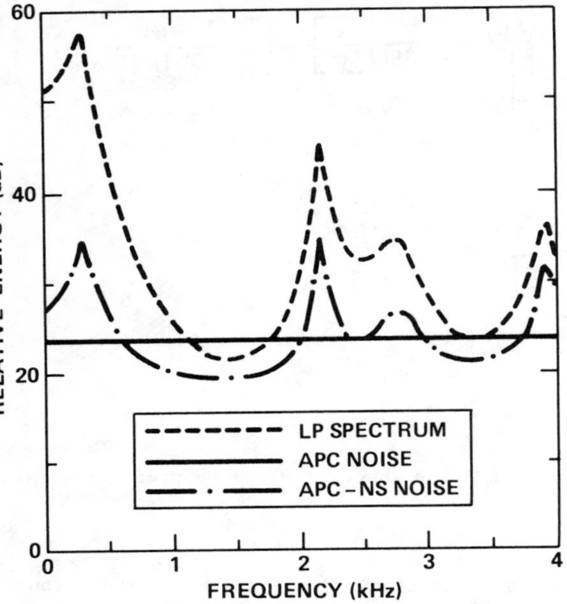

FIGURE 8. Comparison between the noise spectrum in APC and in APC-NS.

Quantization of Parameters. In APC, as in APCM, one can have forward prediction or backward prediction. In the discussion above we have largely assumed that one is performing forward prediction, where the various parameters are computed from a block of the speech signal. In this case, the parameters of the filters A(z), C(z) and the adaptive quantizer gain G need to be quantized and transmitted as side information in addition to the residual. In backward prediction, the filter parameters are computed from past values of the output signal r(n). Here, only the quantized residual needs to be transmitted. Backward prediction is generally less optimal than forward prediction and is affected more by channel errors. But forward prediction entails a certain amount of delay.

Quantization of the residual is performed in a manner similar to that described for APCM. More complicated schemes have been used to advantage in some systems.

Quantization of the predictor coefficients a(k) is best performed on the corresponding reflection coefficients K_m since one can quantize them and maintain the stability condition (20) easily. To achieve uniform spectral sensitivity to quantization, one should quantize K_m in a nonuniform fashion [16]. One method is to transform the K_m to the so-called log-area-ratios (LARs) using the transformation:

$$L_m = \log \frac{1+K_m}{1-K_m}, \quad 1 \leq m \leq p \tag{42}$$

and then use a uniform quantizer to quantize the L_m. The step sizes for lower-indexed coefficients are generally smaller than those with a higher index.

The pitch predictor parameters are quantized directly, and the pitch value M is usually quantized logarithmically.

Encoding [17]. After a parameter is quantized into one of a set of L levels, each value must be encoded into a digital bit stream. There are generally two types of encoding: fixed-length coding and variable-length coding. In fixed-length coding, each one of the L levels is coded with the same fixed number of bits. In variable-length coding, the code word for each level may have a different number of bits.

The most commonly used code in fixed-length coding is the natural binary code which starts with an all-zero word for level 1 and adds one for each level. Another important code that is used for coding signed values is the folded binary code. In this code, the most significant bit gives polarity information and the remaining bits represent the parameter magnitude in natural binary code. For parameters having a nonuniform pdf, folded binary code performs better than natural binary code across noisy channels.

The most important type of variable-length coding is what is known as entropy coding. If the set $\{y_i, i=1,2,\ldots L\}$ represents quantized levels of a parameter or a waveform y, and $P(y_i)$ is the probability associated with level y_i, then the minimum number of bits needed to transmit the L levels is given by the entropy of y_i:

$$H(y) = - \sum_{i=1}^{L} P(y_i)\log_2 P(y_i). \quad (43)$$

(In practice, $P(y_i)$ is estimated by collecting a histogram of how often each level y_i is encountered for a large sample of parameter values y; the histogram is then normalized so that the summation of $P(y_i)$ for $i=1,\ldots,L$ is equal to one.) This minimum H(y) can be approximately achieved by the use of Huffman coding [18], which attempts to code level y_i using $-\log_2 P(y_i)$ bits. The main idea of entropy coding is to use only a few bits to code high probability levels and a larger number of bits to code lower probability levels, so that the average number of bits is minimized. Note that fixed-length codes are only optimal for a parameter with a uniform distribution. It is worth emphasizing that the use of entropy coding minimizes the average data rate without losing any information. Also, entropy codes are prefix codes [17], so there is no ambiguity in decoding strings of these variable-length codes.

Another advantage of entropy coding is that it permits the use of a large number of levels for each parameter without necessarily increasing the bit rate substantially. Large-amplitude residual values, which would normally cause overload distortion in a quantizer with a small number of levels, can be coded using a variable-length code with little effect on the transmission rate because those samples generally have very low probability.

In spite of its advantages, variable-length coding causes problems when transmission is over fixed-rate, synchronous channels or over noisy channels in general. Transmission of variable-length codes across a synchronous channel requires the use of appropriate buffers at the send and receive ends of the channels, which introduces a delay into the system [19]. The buffers must be of sufficient size so that neither buffer exhausts its contents at any time. Another more serious problem is that a single bit error in the channel could cause misinterpretation of all subsequent parameters since the code boundaries would be lost. The result may be a total loss of synchronization between transmitter and receiver. Therefore, variable-length coding is largely useful over relatively error-free channels.

4.3 Adaptive Transform Coding (ATC) [20-22]

While APC is viewed chiefly as a time-domain coding technique, adaptive transform coding (ATC) may be viewed as an analogous coding technique in the frequency domain. In ATC, the time-domain signal is transformed into another domain, and the transform coefficients are quantized and transmitted. At the receiver, an inverse transformation gives the reconstructed signal. The signal is analyzed a block at a time, each block comprising about 25 ms of speech. The algorithm adapts the system parameters for each block. The discussion below deals with a single block in which the speech signal may be assumed to be stationary.

Let the signal s(n), $0 \leq n \leq N-1$, in a block of length N be placed in a vector $\underline{x}$:

$$\underline{x} = [s(0)\ s(1)\ \ldots\ s(N-1)]^T \tag{44}$$

where superscript T denotes transpose. The vector $\underline{x}$ is transformed using a linear unitary matrix $\underline{U}$ into the transform vector $\underline{y}$:

$$\underline{y} = \underline{U}\ \underline{x} \tag{45}$$

with

$$\underline{U}^{-1} = \underline{U}^T. \tag{46}$$

The elements of $\underline{y}$ are the transform coefficients that are to be quantized to form a vector $\hat{\underline{y}}$, which is to be transmitted. At the receiver, an inverse transform is applied to $\hat{\underline{y}}$ to produce the reconstructed signal vector $\hat{\underline{x}}$:

$$\hat{\underline{x}} = \underline{U}^{-1}\hat{\underline{y}} = \underline{U}^T\hat{\underline{y}}. \tag{47}$$

The problem of ATC is to design the quantizers of the elements of $\underline{y}$ such that the overall distortion is minimized, or equivalently, such that S/Q is maximized.

For unitary matrices, one can show that the expected value of the overall distortion at the output is equal to the total quantization noise in the transform coefficients:

$$D = E[(\underline{x}-\hat{\underline{x}})^T\ (\underline{x}-\hat{\underline{x}})] = E[(\underline{y}-\hat{\underline{y}})^T(\underline{y}-\hat{\underline{y}})]. \tag{48}$$

D is the total distortion and E(x) is the expected value of x. Therefore in computing the output noise or distortion, one can compute the quantization noise in the transform coefficients instead.

We shall assume that the transform coefficients y_i, $0 \leq i \leq N-1$, are quantized independently. The total distortion is then equal to the sum of the individual distortions D_i:

$$D = \sum_{i=0}^{N-1} D_i \tag{49}$$

Now, assuming also that the coefficients have the same probability distribution but different variances v_i, one needs b_i bits to code coefficient y_i if the distortion is not to exceed D_i, where

$$b_i = \delta + \frac{1}{2} \log_2 \frac{v_i}{D_i} , \qquad 0 \le i \le N-1 \tag{50}$$

where δ depends on the probability distribution and the type of quantizer chosen. (δ is zero for a Gaussian distribution.)

The problem then is to quantize $\underline{y}$ such that the total distortion D in (48) is minimized subject to the constraint of a fixed overall bit rate

$$b = \sum_{i=0}^{N-1} b_i = \text{constant} \tag{51}$$

The solution to this problem is that all transform coefficients should have the same distortion $D_i=D/N$, for all i. The optimum bit assignment is then given by

$$b_i = \frac{b}{N} + \frac{1}{2} \log_2 \frac{v_i}{\left(\prod_{i=0}^{N-1} v_i\right)^{1/N}} \tag{52}$$

Therefore, all transform coefficients are quantized using the same step size, but different numbers of bits, determined by b_i in (52). The bit-assignment calculation given by (52) typically involves an iterative procedure, for the following two reasons. First, the number of bits for those transform coefficients with very small variances may be computed to be negative according to (52). Such coefficients are not transmitted. The allocation of b bits among the remaining coefficients is accomplished by repeating the above optimization procedure. Second, the number of bits assigned for each coefficient should be an integer, but still satisfy the constraint (51). A practical method of solving (52) and (51) which requires only three steps is available [21]. Notice that in an ATC system, one computes the above bit allocation both at the transmitter and at the receiver, for every block of speech.

One can also show that with ATC the S/Q increase over APCM is given by

$$I_{ATC} = \frac{\frac{1}{N} \sum_{i=0}^{N-1} v_i}{\left(\prod_{i=0}^{N-1} v_i\right)^{1/N}} \tag{53a}$$

$$= \frac{\text{arithmetic mean of transform "spectrum"}}{\text{geometric mean of transform "spectrum"}} \tag{53b}$$

where "transform spectrum" is the power of the transform coefficients. The value of I_{ATC} depends on the particular transform chosen. The maximum value for I_{ATC} is obtained using the Karhunen-Loeve transformation (KLT) [20], in which case $v_i=\lambda_i$, the eigenvalues of the covariance matrix of the signal. However, it has been found in practice that the discrete cosine transform (DCT) gives results very close to KLT. Henceforth, we shall assume that the DCT is used. One definition of the DCT is given by:

$$y_0 = \sum_{k=0}^{N-1} x_k \quad ; \quad y_i = 2 \sum_{k=0}^{N-1} x_k \cos \frac{\pi(2k+1)i}{2N} , \qquad 1 \le i \le N-1 \tag{54}$$

The inverse DCT (IDCT) is then given by

$$x_k = \frac{1}{N} \sum_{i=0}^{N-1} y_i \cos \frac{\pi(2K+1)i}{2N}, \qquad 0 \le k \le N-1 \tag{55}$$

Efficient methods to compute (54) and (55) exist [23]. A significant computational advantage of DCT over KLT as regards their use in block-adaptive systems, such as ATC, is that DCT does not require computing correlations, eigenvalues and eigenvectors.

The only remaining issue is how to determine adaptively the bit assignment or the number of bits to be used in quantizing the individual transform coefficients. This can be done by obtaining a representation of the envelope of the transform spectrum once every frame or block, and using the spectral values in place of variances v_i for computing the bit assignment given by (52) and (51). The parameters characterizing this transform spectral envelope are transmitted at the block rate as side information. The receiver uses the same procedure as the transmitter for computing the bit assignment before it decodes the transform coefficients. Zelinski and Noll used a 16-channel step-wise linear approximation for the spectral envelope [20]. One can alternately obtain an LP representation, in which case the LP coefficients are transmitted to the receiver [22].

The reader has probably already noticed the similarity between (53b) and (35). In both ATC and APC, the increase in S/Q over APCM is given by the ratio of the arithmetic mean to the geometric mean of some spectrum related to the signal. Therefore, one expects I_{ATC} to be of the same order as I_{APC}. Also, it is interesting to note that having the distortions for the different coefficients be equal maximizes S/Q, and implies that the noise spectrum is flat. This is also true for APC.

Just as we were able to shape the noise spectrum in APC-NS we can also shape the noise spectrum in ATC. This is accomplished by appropriately modifying the assignment of bits to different coefficients. This may be done by employing a frequency-weighted mean-squared error criterion instead of (48).

ATC and APC are similar in many respects but have one major difference. This difference is in the coefficients that are quantized: APC quantizes in time, while ATC quantizes in frequency. Even though the total S/Q may be the same in both types of systems, the perception of the noise may be quite different. In APC the noise is uniformly distributed in the time domain, while in ATC the noise is uniformly distributed in the frequency domain.

Finally, the performance of ATC in the presence of channel errors should be considered. In this context, it should be emphasized that the received side information has the important role of determining the bit assignment or the number of bits with which individual transform coefficients were coded at the transmitter. The computed bit assignment is in turn used by the decoder to decode the received bit stream into transform coefficients. Thus, an error in the transmitted side information due to channel noise implies a bit assignment different from what was used at the transmitter, which leads to an erroneous interpretation of the received bit stream. This may have a drastic effect on speech quality. Clearly, protection of the side information bits against channel errors should mitigate the extent of the problem. In contrast, an error in the side information of a coder like APC may change the spectral envelope and cause perceivable distortion, but the degradation may be much more graceful than might happen in ATC.

4.4 Sub-Band Coding [21]

Sub-band coding is a hybrid between APC and ATC. In sub-band coding, the speech signal is divided into a number of frequency bands. Each band is heterodyned, decimated (down sampled), quantized, coded, and transmitted. At the receiver, the different bands are interpolated, heterodyned, and added to produce the reconstructed speech signal.

The S/Q ratio is controlled separately in each frequency band by specifying the quantization step size and the number of bits for that band. Thus, one can use more bits at low frequencies to increase S/Q there and remove the roughness or rumble in the coded speech. Therefore, one has some control over the shape of the noise spectrum, but the S/Q distribution with frequency is usually kept fixed in time.

Another possible mode of operation would be to change the number of bits for different bands dynamically, and thus have more control over the noise spectrum and over S/Q at different frequencies. In particular, it is possible to distribute the number of available bits such that the noise is the same at all frequencies. One can show that such a design maximizes the overall S/Q and results in a flat noise spectrum. In fact, one can also show that as the number of bands becomes large, the method just proposed gives results that are similar to an ATC coder. With appropriate distribution of bits over a large number of bands, one can also introduce noise spectral shaping. Therefore, assuming that the different bands cover the whole speech band, one can make the performance of sub-band coding arbitrarily close to that of ATC over the same band. A sub-band coder with a few bands can be considered to be a gross approximation to ATC. Therefore, unless the implementation is far cheaper than ATC, sub-band coding does not provide a useful alternative to ATC. In fact, since the bit allocation in sub-band coding is usually kept fixed, the results are not as good as ATC or APC at the same bit rate. Sub-band coding can produce toll-quality speech at or above 24 kbits/s, but not at 16 kbits/s.

4.5 Delayed-Decision Coding [6,24,25]

In our discussion of APC above we mentioned that, in forward prediction, the spectral parameters are computed for a block or frame of speech before the residual is computed and quantized. Compared to backward prediction, forward prediction constitutes a delayed decision in that the decision to transmit parameters is delayed an amount at least equal to the frame length. In delaying the decision, one is able to achieve more optimal performance. In quantizing the residual, however, we perform the quantization on an instantaneous basis. By doing so, we do not take full advantage of the delay already caused by performing forward prediction. While in computing the spectral parameters we attempt to minimize some global error criterion over the frame (albeit without considering the effects of quantization), we do not do so for the residual.

Clearly, optimal performance at the frame level is achieved by answering the question: Given a desired bit rate, what set of spectral parameters <u>and</u> residual values minimize a given error criterion over a frame of speech? Even if a simple mean-square error (mse) criterion is used, the problem is highly nonlinear in a high-dimensional space. For example, assuming a 16 kbits/s bit rate, a frame of 25 ms is coded into 400 bits. Therefore, one has to choose which of 2^{400} possible set of values produces the optimal solution, which is a highly impractical task! Some compromises are necessary.

The first compromise is to compute the side information (spectral and pitch information in APC) before the residual is coded. Such computation is effected as described earlier. Now, one is left with finding the set of quantized residual values that give the minimum output error. This problem is still formidable. For example, assuming an 8 kHz sampling rate and R=1 bit/sample, in a 25-ms frame one has 200 bits, or 2^{200} choices. In general, with N samples in a frame and R bits/sample, the problem is to choose the best of

$$L = 2^{NR} = 2^{b} \tag{56}$$

sets of residual vectors of length N, where

$$b = NR \tag{57}$$

is the number of bits in a frame.

It is important to stress here that our goal is to minimize the error between the reconstructed signal r(n) and the input signal s(n) over a frame. Therefore, in choosing which of a set of L residual vectors is optimal, we must compute r(n) for each of the residual vectors and then pick the residual vector that achieves the minimum error in the output r(n). Note that this problem is different from choosing the residual vector to minimize the error between that vector and the prediction residual e(n) as computed from (12). The solution to the latter problem will have the same S/Q as APCM. To achieve the increased S/R due to the APC gain, one must minimize the error in the reconstructed signal. In effect, one must include the quantization in the APC loop.

In performing delayed-decision residual coding, one must first generate the set of L residual vectors, or what is known as the codebook. One method would be to choose the L vectors from larger amounts of actual residual signals using clustering methods. Unfortunately, the computations involved may be excessive. However, since the APC residual can be approximated well by a Gaussian process, one can "populate" the codebook from a Gaussian random number generator. Storing L vectors in memory is prohibitive beyond 2^{20}, even for large computers. One solution to the storage problem is to generate the codebook vectors "on-line" at the transmitter and receiver simultaneously.

After the codebook vectors are determined, one must find the vector that minimizes the reconstruction error. When L is large, the problem becomes impractical very quickly. If L were reduced to a manageable size, such as $L=2^{20}$, the computations become on the order of a million operations, where each operation consists in computing r(n) over the block and then computing the reconstruction error. If R=1, then the block length would be 20 samples; or 2.5 ms at a sampling rate of 8 kHz. There is some evidence that the block length must extend to 10 ms or longer before the result in terms of improved speech quality can be perceived.

Instead of having an optimal solution over a short block, an alternate but superior compromise is to find a suboptimal solution over a longer block. The two popular methods used here are known as tree and trellis coding. Below we describe the two methods briefly.

Tree Coding. Fig. 9 shows a tree code with R=1 and N=3. For a general tree, the number of outgoing branches at each mode of the tree is 2^R branches. The letters on the tree nodes in Fig. 9 represent specific residual values in our problem. The code 010, for example, represents the residual vector (x_1, x_4, x_9), and the code 101 represents the vector (x_2, x_5, x_{12}). The total number of residual vectors in Fig. 9 is L=8. Because most of the elements on the tree are common to many vectors, the total number of residual values in the codebook is reduced. However, the total number of elements is still exponential with N. The element values can still be computed from a random number generator as suggested before.

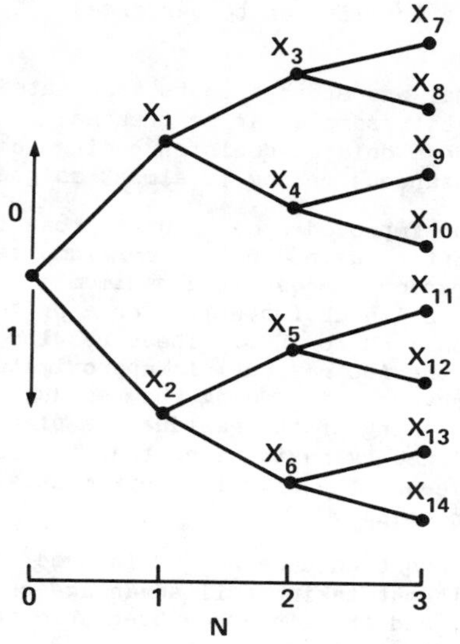

FIGURE 9. Tree coding for N=3 and R=1.

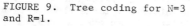

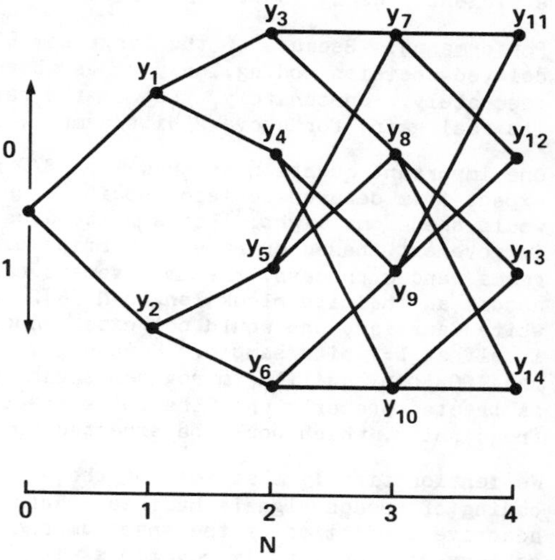

FIGURE 10. A trellis code for N=4, R=1, and intensity K=2.

Fig. 9 shows the structure of the codebook that results from a code tree, but it says nothing about how one searches for the optimal vector that minimizes the reconstruction error. An exhaustive search would be out of the question for very large L. What is usually done is to compute the different cumulative reconstruction errors by following the tree along all branches from left to right until the number of branches reaches some prespecified number M. Thereafter, RM branches are followed, and the M branches with the minimum errors are continued. In this manner, the computations increase only linearly with N. The result will be a suboptimal solution, of course, but for moderate values of M the solution may be close to optimal. This tree search algorithm is often known as the M-algorithm.

One could use the algorithm above to code the residual values in an iterative manner. Thus, after N stages of the tree are reached, the code of the first bit of the optimal vector up to that point is transmitted. The base of the tree now moves one level depending on whether a 0 or 1 was transmitted and half of the tree may be discarded. Only incremental computations to keep the tree N levels deep are needed. This iterative algorithm would be known as the (M,N)-algorithm.

For R<1 bit/sample, one increases the number of elements at each node. For example, in Fig. 9, if one were to place two elements at each node, the resulting R would be 0.5 bits/sample.

Trellis Coding. In Trellis coding, the codebook is structured further such that the number of distinct elements grows only linearly with N. Fig. 10 shows an example of a trellis code for N=4 and R=1. We note that after the first two stages, the number of levels remains at 4, i.e., there are 4 new elements at each following stage. In general, a trellis with 2^K levels is said to have an intensity of K. Thus, the intensity in Fig. 10 is K=2. The total number of code words in Fig. 10 is equal to L=16. For example, the code 0110 represents the sequence (y_1, y_4, y_9, y_{11}).

An exhaustive search of the optimal vector in a trellis can be performed efficiently using the Viterbi algorithm [31].

Performance. Because of the large computational and storage costs associated with delayed-decision coding, only cases where R<2 bits/sample can be treated adequately. Fortunately, that is the range where delayed-decision coding achieves its real gain, for above 2 bits/sample, instantaneous coding is almost as good.

One important question is: How much S/Q gain or improved speech quality can one expect from delayed-decision coding over instantaneous coding? Some examples would shed some light. For a white Gaussian random process, the maximum improvement, based on rate-distortion bounds, is 1.5 dB for R=1. For a white gamma random process, the improvement can reach 7 dB for R=1. These results assume an infinite block length N [6]. Since the APC residual is approximately white Gaussian, one would not expect much improvement in coding the residual itself at R=1 bits/sample. However, when the coding of the residual is placed in the APC loop, quality improvements can be attained by making sure that the signal is treated properly when the noise spectrum exceeds the signal spectrum at any frequencies, which would be expected to happen at R=1.

We mention finally that most of the work that has been done so far in tree/trellis coding of speech signals has been performed without taking full advantage of adaptive prediction of the spectrum and pitch, and therefore has been able to achieve some quality improvements with relatively small block lengths. However, greater improvements have been possible with a complete APC system and without tree coding. When tree coding is used with APC, a relatively large block length (on the order of 10 ms) is needed to realize a substantial improvement [25].

4.6 Multi-Pulse Coding

In APC, one generally quantizes each residual value using a fixed number of bits. In multi-pulse coding, only a relatively small number of residual samples (pulses) are transmitted, with the remaining samples being set to zero. The problem then becomes: Given that we want to represent the residual or excitation by M pulses in a frame N samples long (M<<N), determine the time positions and amplitudes of the M samples such that some error criterion is minimized (hopefully to maintain high speech quality). This problem is highly nonlinear and an optimal solution generally requires a number of iterations. However, certain suboptimal solutions appear to work quite well [42]. In [42], Atal and Remde report that only minimal improvements in speech quality are achieved for M/N > 0.1.

Multi-pulse coding is expected to have significant impact on data rates below 10 kbits/s. In the coming few years, a number of studies and algorithms should appear which explore the various dimensions of the method at different data rates.

5. BASEBAND CODING (BBC) [26]

In Section 4, we discussed different methods of waveform coding where the objective is to design a coder that attempts to match the reconstructed signal to the input signal on a sample-by-sample basis by minimizing a squared-error criterion. The speech quality of waveform coders begins to degrade rapidly as the bit rate goes much below 12 kbits/s. To achieve higher speech quality at lower bit rates, it is necessary to take advantage of certain properties of human speech perception. As we shall see, it will also be necessary to forego our desire to have the reconstructed signal match the input signal on a sample-by-sample basis, but rather require that the output speech have a quality similar to the input speech on a subjective basis.

Fig. 11 shows a block diagram of a baseband coding (BBC) system based on the linear prediction method of spectral analysis. The input signal with bandwidth W Hz is inverse filtered using the LP filter A(z), resulting in the usual LP residual. Then, instead of transmitting the full-bandwidth residual as done in APC, we only transmit a low-frequency portion of it B Hz wide (B<W), known as the baseband. The number of samples transmitted is 2B samples/s compared to 2W samples/s for APC. (Typically, a ratio of W/B=3 is used.) The result, of course, is a lower transmission rate.

At the receiver, we need a method to regenerate the high-frequency components of the residual from the baseband. High-frequency regeneration (HFR) methods make use of the perceptual fact that the voicing, pitch, and frication information is largely contained in the relatively "white" baseband. The HFR process generally involves some nonlinear operation on the baseband to produce a signal with a fullband width W and with a flat spectral envelope. One HFR method that has been popular in the past involves rectification and spectral flattening of the interpolated baseband. Another more recent HFR method, which is simpler

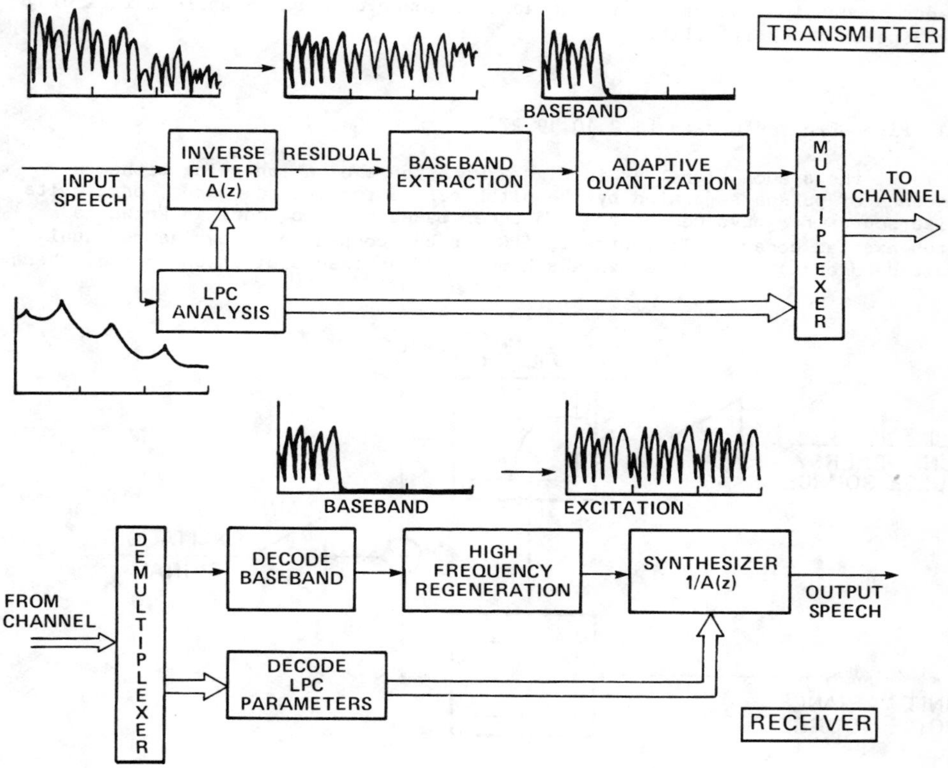

FIGURE 11. Baseband coding (BBC) system using linear predictive analysis and coding of a baseband of the residual.

computationally and produces generally higher speech quality, is that of spectral folding. In its simplest form, spectral folding can be performed by introducing (W/B)-1 zeros between consecutive baseband samples. This upsampling aliases the baseband into the higher bands resulting in a fullband spectrum that has a flat envelope. This simple method introduces certain "tonal noises" which can be masked by further appropriate processing [26].

The baseband can be quantized and coded in a number of ways. In particular, one can use APC with only a pitch loop. Another method is to use ATC [22]. In fact, by taking the transform of a block of the residual one could extract a baseband simply by discarding the high-frequency components. HFR could then be achieved simply by duplicating the baseband transform components at high frequencies.

Baseband coders are generally used to transmit speech at rates of 8-10 kbits/s with communications quality, where the distortion is audible but acceptable.

6. NARROWBAND CODERS

As the desired bit rate goes much below 8 kbits/s, the performance of baseband coders begins to degrade rapidly. Here, we make further use of our knowledge of speech production and perception to achieve lower data rates while maintaining good speech intelligibility.

6.1 Pitch-Excited Coders [1,2,10,19,27]

We model the speech signal as in Fig. 2 with the excitation being either a sequence of pulses separated by the pitch period for voiced sounds, or a white noise source for unvoiced sounds. A coder based on this model is known as a pitch-excited coder. (In contrast, APC and BBC coders are known as residual-excited coders.) Fig. 12 shows a schematic of a mixed-excitation source, where

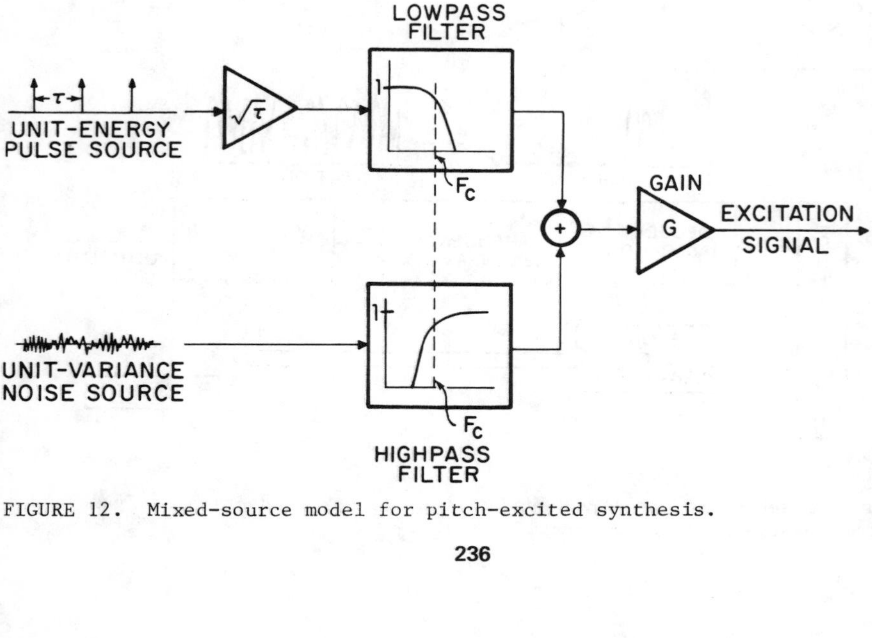

FIGURE 12. Mixed-source model for pitch-excited synthesis.

the pulse source excites low frequencies below F_c and the noise source excites high frequencies [28]. Most pitch-excited coders use a binary decision as to whether the speech is voiced or unvoiced. The model of Fig. 12 can also synthesize mixed-excitation sounds such as [z] and [v]. The gain G determines the energy level of the signal at each frame.

Many algorithms have been developed over the years to help distinguish voiced speech frames from unvoiced frames. For voiced sounds, then, a pitch value is determined by one of many pitch extraction methods. However, the voiced/unvoiced decision remains by far the most difficult in a pitch-excited coder. A small percentage of errors in that decision is readily perceived by the ear. Determining the cut-off frequency F_c for mixed excitation in Fig. 12 makes use of the relative height of the autocorrelation peak at the pitch period lag.

The transfer function in Fig. 2 is determined by a number of methods. The most popular are the channel and LPC methods. In a channel coder, the spectral envelope is represented by the output of a number of overlapping bandpass filters. In an LPC coder, the spectral envelope is determined by the all-pole filter H(z) in (17).

The analysis process in an LPC coder is to perform the LP analysis and the pitch extraction every 20 ms or so. The LP parameters, pitch, and gain are quantized and transmitted. At the receiver, the pitch and gain information are used to generate an excitation signal which excites the all-pole filter. The number of bits needed to quantize the parameters independently is the same as the side information transmitted in an APC coder. Typically, pitch is quantized to 6 bits, the gain to 5 bits, and the LP coefficients are quantized as log-area-ratios (LARs) in (42) to about 37 bits for a 10th-order model. At 50 frames/s, the total bit rate is 2400 bits/s. The speech from pitch-excited coders sounds synthetic with some loss of intelligibility and of the speaker's identity. Furthermore, these coders are generally more sensitive to background acoustic noise than waveform coders.

The bit rate can be decreased further by taking advantage of the nonstationary character of speech and transmitting parameters only when they have changed sufficiently from the previous transmission. At the receiver, the parameters are interpolated linearly and the synthesis filter coefficients are updated every 10 ms. Using this variable-frame-rate (VFR) technique [19], the transmission rate can be reduced from 2400 bits/s to about 1600 bits/s with no loss in speech quality.

Another reduction in bit rate without quality loss can be achieved by transmitting only 6 instead of 10 LARs during unvoiced sounds. This result takes advantage of the fact that, perceptually, unvoiced spectra need not be represented in as much detail as voiced spectra.

6.2 Optimal Scalar Quantization

Above, the LARs were quantized assuming that the individual components were independent. Further reduction in bit rate can be achieved by taking advantage of the correlation among parameters. The method used is analogous to transform coding.

Let the set of p parameters be represented by a random vector $\underline{x}$, which we would like to quantize to a total of b bits. Further, assume that we wish to quantize $\underline{x}$ such that the mse is minimized. Then the optimal scalar quantizer consists of three steps: Parameter decorrelation, bit allocation, and scalar quantization.

1. Parameter Decorrelation: Let Q be the matrix whose columns are the eigenvectors of the covariance matrix C of the random vector $\underline{x}$. The new parameter vector

$$\underline{y} = Q^T\underline{x} \tag{58}$$

will have p uncorrelated components.

2. Bit Allocation: The second step is to allocate the given b bits among the p components of $\underline{y}$. This is done in an analogous manner to bit allocation in transform coding using (52) with N=p and v_i = the variance of component y_i.

3. Scalar Quantization: The third and final step is to perform the scalar quantization of each component y_i using b_i bits as allocated in the previous step. Here one simply uses a Max quantizer designed for each component.

At the receiver, the quantized vector $\hat{\underline{x}}$ can be retrieved from the quantized vector $\hat{\underline{y}}$ by an inverse transformation

$$\hat{\underline{x}} = Q\,\hat{\underline{y}}. \tag{59}$$

When compared to independent scalar quantization of LAR parameters, optimal scalar quantization achieves the same quantization error with a saving of only 3 bits per frame [43].

6.3 Vector Quantization (VQ) [29,30,32]

In optimal scalar quantization we took advantage of the correlation among parameters to reduce the bit rate. Here we reduce the bit rate further by taking advantage of the statistical dependence among parameters beyond correlation. The vector $\underline{x}$ is quantized as a single vector in p-dimensional space.

In vector quantization, the space of all LAR vectors is partitioned into $L = 2^b$ regions, where b is the number of bits to quantize each vector. The partition $P=\{C_i;\ 1\leq i\leq L\}$ defines L regions or clusters C_i, each with a representative template vector $\underline{z}_i$ which is chosen as the center of mass of cluster C_i, using some distance measure $d(\underline{x},\underline{z}_i)$.

The first problem in vector quantization is to determine the clusters C_i and the corresponding templates $\underline{z}_i$ from a given amount of training data. This can be accomplished by well-known iterative clustering techniques. The templates are chosen to minimize the total quantization error between the training data and the templates, using the distance d as a measure of quantization error.

Once the templates are determined, a vector $\underline{x}$ is quantized by determining the closest template, which usually entails an exhaustive search. If b=30 bits, for example, then one must be able to compute 2^{30} distances every frame. Also, one must be able to store that many templates. Because of computational and storage requirements, vector quantization is usually used for b<15 bits.

We mentioned earlier that vector quantization takes advantage of statistical dependence beyond correlation. For example, vector quantization would not be expected to be of much help if the components of $\underline{x}$ are jointly Gaussian. Optimal scalar quantization would be sufficient because once the parameters are decorrelated, they also become independent and vector quantization would not be able to take advantage of any remaining dependence beyond correlation. (For very large b, vector quantization would still be expected to increase the S/Q by about 2 dB over optimal scalar, but for smaller values of b, such a gain does not in fact take place.) But speech parameters are not Gaussian and they do exhibit dependence beyond correlation. Fig. 13 shows a plot of the LAR mse in dB versus the number of bits b for vector and optimal scalar quantization [30]. The number of LAR parameters used is p=14. Note that optimal scalar quantization has the same error with 15 bits as vector quantization with 10 bits: a saving of 5 bits. This difference between scalar and vector quantization would not be expected to increase significantly at higher bit rates for the following reason. Note that, in Fig. 13, the scalar curve becomes linear after 10 bits with a slope of -6 dB per bit per parameter, which implies that the data in the 2^{10} regions must be uncorrelated and fairly uniformly distributed. A similar behavior would be expected to take place for the vector case at a point where the parameters within each of the clusters become independent, after which the slope of the curve should also become -6 dB per bit per parameter. From that point on, the difference between scalar and vector quantization would remain constant for higher bit rates. Therefore, the advantages of vector quantization become less significant as the bit rate increases. Because of the large increase in computational and storage requirements of vector quantization with bit rate, we conclude that vector quantization is worthwhile only for relatively low bit rates, on the order of one bit per parameter.

One method that reduces the computational load drastically is by performing a binary search instead of an exhaustive search. In this method, the partitions are formed by binary division, that is, the space is first divided into two regions, then each region in turn is divided into two, until the desired number of regions is attained. Quantization of a vector $\underline{x}$ is then achieved in like manner by finding which of the first two regions it belongs to, then which of the two

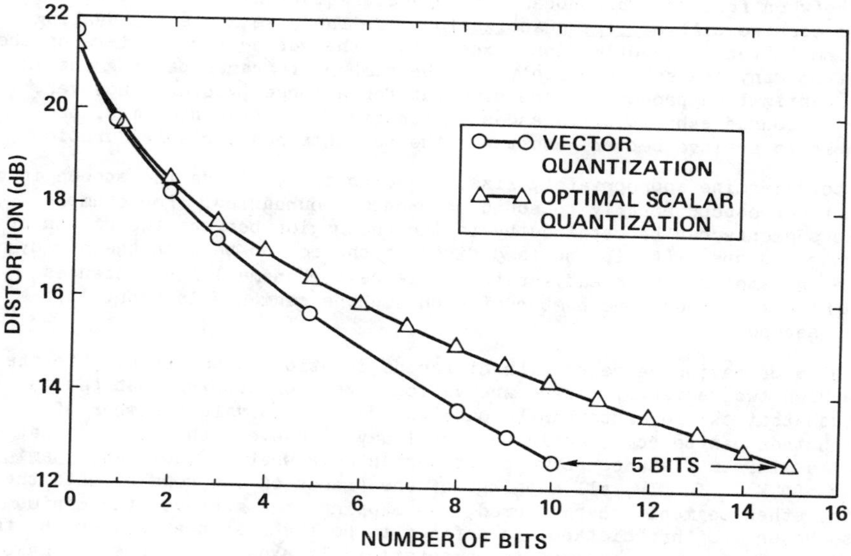

FIGURE 13. Performance comparison between vector quantization and optimal scalar quantization.

subregions it belongs to, etc. This binary search requires the computation of only 2b distances as compared to 2^b distances in an exhaustive search. Using a mse criterion on LARs , binary search requires only 0.5 bits more than exhaustive search at b=10 bits. Given the vast reduction in computation, binary search is certainly adequate for most applications. We note, though, that binary search doubles the required storage compared to exhaustive search.

While binary search is an effective way to reduce the computations in vector quantization, there is no comparable method that reduces the storage requirements. In fact, at this time, storage appears to be the limiting factor in vector quantization.

Most vector quantizers to date employ a bit rate of b=10 bits per frame or less and total bit rates < 1000 bits/s. Quite intelligible speech has been achieved for bit rates of 300-400 bits/s when trained on a single speaker [30].

It is important to note that vector quantizers require a large amount of training. The number of training vectors should be at least 50 times the number of templates L. Also, the training must include samples from different acoustic environments and different speakers. In general, vector quantizers are not as robust as scalar quantizers under environmental and channel noise conditions. Therefore, they should be used only at very low data rates where scalar quantizers are simply inadequate.

6.4 Segment Quantization [33]

Vector quantization, as described above, takes advantage of the dependence among parameters in a single frame. One can extend the same notion to a segment of speech comprising several frames, so that one can capitalize on the additional dependence between frames. The process of parameter quantization based on a segment of speech we call <u>segment quantization</u>. In fact, we are still dealing with a problem in vector quantization, except that the vector now consists of the parameters from many frames. Certainly, if the number of frames per segment is fixed, the quantization process is the same but for a longer vector. However, because speech sounds exhibit differences in duration at different times, one should be able to achieve better results if the segments had variable duration.

One method to determine appropriately sized segments is to divide the speech into segments that correspond roughly to sound or phoneme boundaries. For example, diphone-like segments can be obtained by taking the region between two points of minimum spectral slope. (A diphone is defined as the region between the steady-state of one phoneme and the steady-state of the next phoneme.) In sentences spoken at a moderate speed, the number of such diphone segments is about 11 segments per second.

A major problem of having segments with different duration is how to compute the distance between two segments. There are various types of warping that can be performed such that two segments can be made to have an equivalent number of frames, which then can be compared in the usual way. However, the warping that results in the minimum distance between two segments is what is known as <u>dynamic time warping</u> (DTW). In DTW, all mappings of the frames of one segment onto the frames of the other segment are performed; the mapping that achieves the minimum distance is chosen and that distance is defined to be the distance between the two segments. Clearly, such a procedure is computationally expensive. One can make use of <u>dynamic programming</u> to reduce the computational load somewhat [34].

Before we quantize the input speech into segments, we need a codebook of segments. One could go through a clustering procedure, as in vector quantization, but the problem now is more difficult because the codebook is expected to be larger and the distance computations are much more time consuming because of DTW. Fortunately, one can obtain a suboptimal but reasonable codebook by what is known as random quantization or random selection. Here, the entries in the codebook are chosen at random from a large number of training segments. This method achieves reasonable results because the number of parameters in a segment is large (126 parameters on the average if we assume 9 frames/segment and 14 parameters/frame). We know from information theory that if the components of a large vector are identically distributed and independent Gaussian random variables, a random selection of codewords achieves close to optimal performance. Because the conditions of this result are not met for speech, one would not expect performance that is optimal, but we do expect reasonable performance.

In an experiment using b=13 bits or 2^{13} segments in the codebook for a single speaker, corresponding to 15 minutes of speech, and transmitting one value each for a gain adjustment, a pitch adjustment, and a duration adjustment, a bit rate of about 200 bits/s is achieved. The resulting speech was fairly intelligible in context.

If one attempts to reduce the bit rate down into the 100 bits/s range, then one enters the domain of speech recognition.

7. SPEECH SYNTHESIS [1,2,40]

Fig. 2 shows the basic model used for speech synthesis. Depending on the application, the excitation and filter parameters are obtained in various ways. One way is to synthesize the speech from coded and stored parameters. Thus, instead of the transmission channel in Fig. 1, a storage device is used to store the coded parameters. Those parameters are recalled at a later time and are used to synthesize the speech. A digital voice message system which stores the speech message and plays it back upon request would make use of such a scheme. The coding method used depends on the relative cost of coding, storage, and transmission channel bandwidth (between storage and synthesis) on the one hand, and the speech quality desired on the other hand.

For some applications, it is desirable to be able to change the speed of speech playback or to change the voice characteristics, such as the general intonation (pitch, duration, and intensity). For such applications, one invariably uses a pitch-excited synthesis scheme that can perform all these changes in a straightforward manner.

One important application where the speed and intonation are usually specified independently of spectral information is text-to-speech synthesis. The objective here is to convert ordinary text to speech. The phoneme sequence for each word is stored in a dictionary, along with stress information. Another component of the system determines the general intonation of the text by analyzing its syntactic (grammatical) structure. The sequence of phonemes, along with stress and intonation information, are sent to a phonetic synthesizer, which synthesizes the speech.

Two popular types of phonetic synthesis are diphone synthesis and synthesis-by-rule. Diphone synthesis [35] is based on the concatenation of stored diphone segments, where each diphone spans the region from the middle of phoneme to the middle of the next phoneme. The diphone segments are usually stored in terms of LPC parameters. If one assumes the existence of approximately 50 phonemes in English, then the total number of diphones that need to be stored is 2500 diphones.

Synthesis-by-rule systems invariably use a formant synthesizer to synthesize the speech. (A formant is a resonance of the vocal tract.) Formant values for different phonemes and formant movements to describe coarticulatory effects of phonemes on each other are described by a fairly extensive set of rules. Such coarticulatory effects are taken care of automatically in a diphone synthesizer because they are inherently contained in the stored diphone segments.

Diphone synthesis has a few set of rules and requires a substantial amount of storage while synthesis-by-rule has a large set of rules and requires relatively little storage. Generally, diphone synthesis produces somewhat more natural-sounding speech than synthesis-by-rule.

8. SPEECH RECOGNITION

We mentioned in Section 6.4 that, in attempting to code speech at rates close to 100 bits/s, one enters the domain of speech recognition. In fact, one method of obtaining that data rate would be to recognize the sequence of phonemes automatically and transmit the identity of the phonemes. At the receiver, a phonetic synthesizer would synthesize the speech. One attempt to actually do all this used a codebook of diphone templates and attempted to "recognize" the phonetic sequence by the dynamic time-warping method mentioned in Section 6.4. Unfortunately, the recognition accuracy was not high enough to achieve intelligible speech output. In the long term, phonetic recognition with high accuracy (>80%) would still be a desirable goal because it does not place restrictions on what can be said. In the short term, however, we must limit the input to a speech recognition system if high recognition accuracy is needed.

Most high-performance speech recognition systems today are <u>isolated-word recognition</u> systems [36]. There is usually a relatively small, fixed number of words that are allowed. Each word is spoken in isolation. The system is trained by gathering spoken samples of the words and storing one or more templates for each word, or a statistical description of those templates. The word templates are stored as a sequence of spectral descriptions, either as LPC parameters, filter bank outputs, or other features such as zero crossing rates. The recognition is then performed by matching an input word to the stored set of templates and choosing the word template with the minimum distance to the input. Systems that employ some form of dynamic time-warping in computing the distance perform the best. Most isolated-word recognition systems to date are <u>speaker dependent</u>, i.e., the system is trained on each speaker separately. <u>Speaker-independent</u> systems are trained on the speech from many speakers. Today, only one speaker-independent system is available in the market, and it can handle up to twenty words. Speaker-dependent systems can usually handle tens of words with high accuracy. The better systems achieve 99% recognition for a vocabulary of 20 words.

More recently, several systems have been developed for <u>connected-word recognition</u> [36], where the speaker is allowed to speak without pausing between words. All systems to date have been speaker-dependent and the recognition rates have been substantially lower than in isolated-word recognition. High performance is often achieved by limiting the structure or syntax of allowable phrases and sentences. Most typically the grammar is specified in terms of a finite-state network. Emanating from each state of the network is a set of arcs representing the set of allowable words at that point in the network. A measure of difficulty of a recognition task specified by a given syntax can be given by the <u>branching factor</u>, defined by

$$F = 2^H$$

where

$$H = \frac{\text{Entropy of allowable sentences}}{\text{Average number of words per sentence}}$$

is the average entropy per word. F is essentially a measure of the average number of allowable words at each node of the network. The higher F is, the more difficult is the recognition task. Most high performance systems tend to have a value of $F<10$ and often less than five.

Still in the laboratory are continuous recognition systems. Such systems do not store word templates but rather use smaller elements such as phonemes as the basic unit of recognition. Therefore, in principle, one would be able to handle new words by simply specifying the phoneme sequence in each word. An experimental system at IBM, when trained on a single speaker, performs close to 90% correct word recognition on a task with a 1000-word vocabulary and branching factor F=24 [37]. The system is based on several tiers of hidden Markov models, starting from the spectral and phoneme levels to the word and syntax levels. Recognition is then performed by finding the sequence of words with the highest conditional probability, given the speech input.

Beyond continuous recognition is automatic speech understanding [38], whereby the system not only recognizes the sequence of words uttered, but also understands what was said, in the sense that it can act on each sentence by accessing a specially designed knowledge data base. While some work has been done in this area, further work awaits advanced methods in language understanding, phonetic recognition, and high-speed computers, for the computations needed are 100-1000 times what can be reasonably expected from most of today's computers.

9. SPEECH HARDWARE [39]

Recent developments in VLSI technology have made it possible to implement a number of speech processing functions on one or a few chips. The most famous speech synthesis chip perhaps was the Texas Instruments chip developed for their Speak and Spell learning toy. The synthesizer was a pitch-excited LPC synthesizer. A large assortment of synthesis chips by other manufacturers followed and now those chips can be found in many consumer products from toys to cars.

There are now on the market several text-to-speech systems which take ordinary text as input and produce speech with acceptable quality as output.

More basic to the communications industry was the development of what is known as a codec (coder/decoder) on a chip. A codec performs the functions of A/D and D/A conversion, along with anti-aliasing filters. Codecs are available which provide PCM coding or log PCM with A-law or μ-law coding.

At lower data rates, there are chips that provide 16 kbits/s or 32 kbits/s CVSD coding. Unfortunately, 16 kbits/s CVSD speech is quite noisy.

The most revolutionary development in signal processing technology for speech applications, I believe, is the appearance of signal processing microprocessors. NEC's SPI μPD7720 and TI's TMS320 are examples of such microprocessors. What makes such microprocessors important is that, using two or three of them on a small board, one can program a significant number of state-of-the-art speech processing algorithms in real time. A number of advanced speech coding algorithms, including APC, ATC, and LPC, have already been implemented in real-time using these chips. More speech processing algorithms will likely be implemented using such microprocessors, including speech recognition systems.

In addition to fairly general-purpose signal processing microprocessors, more special-purpose VLSI chips are being designed for specific applications. For example, there are a number of efforts that are devoted to designing chips that can do dynamic time warping for use in speech recognition systems.

The February 1983 issue of the IEEE Transactions on Acoustics, Speech, and Signal Processing, Part II [39], is devoted to integrated circuits for speech. The papers in that issue give the reader a representative view of current directions in speech hardware design.

GLOSSARY

ADM	Adaptive DM
ADPCM	Adaptive DPCM
APC	Adaptive Predictive Coding
APCM	Adaptive PCM
APC-NS	APC with Noise Shaping
ATC	Adaptive Transform Coding
BBC	Baseband Coding
CVSD	Continuously-Variable Slope Delta modulation
DM	Delta Modulation
DPCM	Differential PCM
DTW	Dynamic Time Warping
LAR	Log Area Ratio
LP	Linear Prediction
LPC	Linear Predictive Coding
PCM	Pulse Code Modulation
S/Q	Signal-to-Quantization-Noise Ratio
VQ	Vector Quantization

REFERENCES

1. Flanagan, J.L., Speech Analysis Synthesis and Perception, 2nd ed., Academic Press, New York, 1972.

2. Rabiner, L.R., and Schafer, R.W., Digital Processing of Speech Signals, Prentice-Hall, New Jersey, 1978.

3. Waveform Quantization and Coding, ed. N.S. Jayant, IEEE Press, New York, 1976.

4. Data Compression, ed. L.D. Davisson and R.M. Gray, Dowden, Hutchinson & Ross, Penn., 1976.

5. Berger, T., Rate Distortion Theory, A Mathematical Basis for Data Compression, Prentice-Hall, New Jersey, 1971.

6. Jayant, N.S., and Noll, P., Digital Coding of Waveforms, Principles and Applications to Speech and Video, Prentice-Hall, New Jersey, forthcoming.

7. Max, J., Quantizing for Minimum Distortion, IRE Trans. Info. Theory, vol. IT-6, pp. 7-12, March 1960.

8. Makhoul, J., Linear Prediction: A Tutorial Review, Proc. IEEE, vol. 63, pp. 561-580, 1975.

9. Atal, B.S., and Hanauer, S.L., Speech Analysis and Synthesis by Linear Prediction of the Speech Wave, J. Acoust. Soc. Amer., vol. 50, pp. 637-655, 1971.

10. Markel, J.D., and Gray, A.H., Jr., Linear Prediction of Speech, Springer-Verlag, New York, 1976.

11. Itakura, F., and Saito, A., A Statistical Method for Estimation of Speech Spectral Density and Formant Frequencies, Electron. Commun. Japan, vol. 53-A, pp. 36-43, 1970.

12. Atal, B.S., and Schroeder, M.R., Adaptive Predictive Coding of Speech Signals, Bell Syst. Tech. J., vol. 49, pp. 1973-1986, 1970.

13. Makhoul, J., and Berouti, M., Adaptive Noise Spectral Shaping and Entropy Coding in Predictive Coding of Speech, IEEE Trans. Acoustics, Speech, and Signal Processing, vol. ASSP-27, pp. 63-73, 1979.

14. Atal, B.S., and Schroeder, M.R., Predictive Coding of Speech Signals and Subjective Error Criteria, IEEE Trans. Acoustics, Speech, and Signal Processing, vol. ASSP-27, pp. 247-254, 1979.

15. Viswanathan, R., Russell, W., Higgins, A., Berouti, M., and Makhoul, J., Speech-Quality Optimization of 16 kb/s Adaptive Predictive Coders, IEEE Int. Conf. Acoustics, Speech, and Signal Processing, Denver, pp. 520-525, 1980.

16. Viswanathan, R., and Makhoul, J., Quantization Properties of Transmission Parameters in Linear Predictive Systems, IEEE Trans. Acoustics, Speech, and Signal Processing, vol. ASSP-23, pp. 309-321, 1975.

17. Gallagher, R.G., Information Theory and Reliable Communication, John Wiley, New York, 1968.

18. Huffman, D.A., A Method for the Construction of Minimum-Redundancy Codes, Proc. IRE, vol. 40, pp. 1098-1101, 1952.

19. Viswanathan, V.R., Makhoul, J., Schwartz, R.M., and Huggins, A.W.F., Variable Frame Rate Transmission: A Review of Methodology and Application of Narrowband LPC Speech Coding, IEEE Trans. Commun., vol. COM-30, pp. 674-686, 1982.

20. Zelinski, R., and Noll, P., Adaptive Transform Coding of Speech Signals, IEEE Trans. Acoustics, Speech, and Signal Processing, vol. ASSP-25, pp. 299-309, 1977.

21. Tribolet, J.M., and Crochiere, R.E., Frequency Domain Coding of Speech, IEEE Trans. Acoustics, Speech, and Signal Processing, vol. ASSP-27, pp. 512-530, 1979.

22. Berouti, M., and Makhoul, J., An Embedded-Code Multirate Speech Transform Coder, IEEE Int. Conf. Acoustics, Speech, and Signal Processing, Denver, pp. 356-359, 1980.

23. Makhoul, J., A Fast Cosine Transform in One and Two Dimensions, IEEE Trans. Acoustics, Speech and Signal Processing, vol. ASSP-28, pp. 27-34, 1980.

24. Fehn, H.G., and Noll, P., Multipath Search Coding of Stationary Signals with Applications to Speech, IEEE Trans. Commun., vol. COM-30, pp. 687-701, 1982.

25. Atal, B.S., Predictive Coding of Speech at Low Bit Rates, IEEE Trans. Commun., vol. COM-30, pp. 600-614, 1982.

26. Viswanathan, V.R., Higgins, A.L., and Russell, W.H., Design of a Robust Baseband LPC Coder for Speech Transmission Over 9.6 kbit/s Noisy Channels, IEEE Trans. Commun., vol. COM-30, pp. 663-673, 1982.

27. Gold, B., Digital Speech Networks, Proc. IEEE, vol. 65, pp. 1636-1658, 1977.

28. Makhoul, J., Viswanathan, R., Schwartz, R., and Huggins, A.W.F., A Mixed-Source Model for Speech Compression and Synthesis, J. Acoust. Soc. Am., vol. 64, pp. 1577-1581, 1978.

29. Buzo, A., Gray, A.H. Jr., Gray, R.M., and Markel, J.D., Speech Coding Based Upon Vector Quantization, IEEE Trans. Acoustics, Speech, and Signal Processing, vol. ASSP-28, pp. 562-574, 1980.

30. Roucos, S., Schwartz, R., and Makhoul. J., Vector Quantization for Very-Low-Rate Coding of Speech, IEEE Global Telecommun. Conf., Miami, FL, 1982.

31. Forney, G.D., Jr., The Viterbi Algorithm, Proc. IEEE, vol. 61, pp. 268-278, 1973.

32. Wong, D.Y., Juang, B.H., and Cheng, D.Y., Very Low Data Rate Speech Compression with LPC Vector and Matrix Quantization, IEEE Int. Conf. Acoustics, Speech, and Signal Processing, Boston, pp. 65-68, 1983.

33. Roucos, S., Schwartz, R.M., and Makhoul, J., A Segment Vocoder at 150 b/s, IEEE Int. Conf. Acoustics, Speech, and Signal Processing, Boston, pp. 61-64, 1983.

34. Sakoe, H., and Chiba, S., Dynamic Programming Algorithm Optimization for Spoken Word Recognition, IEEE Trans. Acoustics, Speech, and Signal Processing, vol. ASSP-26, pp. 43-49, 1978.

35. Schwartz, R., Klovstad, J., Makhoul, J., Klatt, D., and Zue, V., Diphone Synthesis for Phonetic Vocoding, IEEE Int. Conf. Acoustics, Speech, and Signal Processing, Washington, D.C., 1979.

36. Rabiner, L.R., and Levinson, S.E., Isolated and Connected Word Recognition - Theory and Selected Applications, IEEE Trans. Commun., vol. COM-29, pp. 621-659, 1981.

37. Bahl, L.R., Jelinek, F., and Mercer, R.L., A Maximum Likelihood Approach to Continuous Speech Recognition, IEEE Trans. Pattern Analysis and Machine Intelligence, vol. PAMI-5, pp. 179-190, 1983.

38. Trends in Speech Recognition, ed. W.A. Lea, Prentice-Hall, New Jersey, 1980.

39. IEEE Trans. Acoustics, Speech, and Signal Processing, vol. ASSP-31, No. 1, Part II, Feb. 1983.

40. Witten, I.H., Principles of Computer Speech, Academic Press, New York, 1982.

41. Speech Enhancement, ed. J.S. Lim, Prentice-Hall, New Jersey, 1983.

42. Atal, B.S., and Remde, J.R., A New Model of LPC Excitation for Producing Natural-Sounding Speech at Low Bit Rates, IEEE Int. Conf. Acoustics, Speech, and Signal Processing, Paris, France, pp. 614-617, 1982.

43. Viswanathan, V.R., Berouti, M., Higgins, A., and Russell, W., A Harmonic Deviations Linear Prediction Vocoder for Improved Narrowband Speech Transmission, IEEE Int. Conf. Acoustics, Speech, and Signal Processing, Paris, France, pp. 610-613, 1982.

6

Signal Processing in the Arab World

M. MRAYATI

We present the language dependent applications of signal processing. A huge effort of research on the Arabic language is needed in order to adapt these applications to it. We emphasize these applications in order to draw attention inside the Arab World to their growing importance. A brief description of the structure of these applications is given to indicate the type of research that can be undertaken in relation to the Arabic language. Finally a survey is presented of several laboratories in the Arab World that are currently active in the signal processing field.

Table of Contents

I. INTRODUCTION

Signal processing is one of the sciences established in the twentieth century. Its importance and potentials are not realized yet. Numerous and versatile applications are being introduced rapidly, especially with the IC technological impact. The wide use of mini and micro computers together with microprocessors, and the commercialization of signal processing chips such as the TMS 320 of Texas instruments and the μ PD 7720 of NEC, make digital signal processing easily accessible.

Some of the important applications of signal processing are language dependent and I will concentrate mainly on such aspects of signal processing. For example : Arabic speech signal processing and Arabic character recognition are particular to the Arab world. The Arabic language is spoken by more than 150 million people and if it is to be present in the actual modern life of communication, computer, automated office ... etc, a lot of research is needed. Moreover, it is quite evident that if the interfacing between the "processing machine" in its large sense and the Arabic language, both spoken and written, is achieved, we would see better development of management, industry, science and education in the Arab World. One aim of this chapter is to alert the scientists of the Arab World working in the field of signal processing to the applications which are particular to the Arab World and the importance of these applications.

Language dependent applications on speech signal processing (analysis, synthesis and recognition) will be mentioned. The need for basic data and rules on Arabic phonetics, language statistics, and syntax is emphasized and some of these data will be presented. We shall be describing very briefly these applications and pointing out the parts that are language dependent.

II. APPLICATION OF ONE DIMENSIONAL SIGNAL PROCESSING

One of the main one-dimensional signals is speech. Many significant applications involving speech processing are being introduced currently. Further more, some of these applications are important for development. The importance of speech processing systems stems from the fact that speech represents man's most powerful communication medium because it is [1] :

- more convenient and interferes less with other activities,
- independent of any factors affecting sight and reach,
- capable of use around corners, obstacles and in the dark,
- compatible with the normal telephone system,
- capable of providing an extra channel for multimodal operations (being therefore particularly suitable for urgent "break-in" message and for "hands-full" operations such as baggage handling).

In many applications, where information is to be presented in its speech form, the possibility of using the native language of the user is a prerequisite for their social and economical acceptance. In this section, some of these applications involving Arabic speech synthesis, analysis or recognition are enumerated. However, no consideration will be given to speech processing which is not language dependent such as analysis-synthesis telephony (vocoders), speech coding, underwater speech communication ... etc.

II.1 Speech Synthesis

In Table 1 we give a brief description of synthesis methods [2] where we have added their language dependency (L.D.) As we can easily see, it is possible and interesting to generate synthesized speech using digital storage of sentences, which is non-language dependent (N.L.D.) However this would be very expensive in systems having a large number of sentences; and even impossible for unlimited vocabulary system. All other synthesis methods are language dependent. In order to point out the needed coding parameters, Figure 1 shows a configuration for each of the synthesizers mentioned in table 1, see [3] and [4]. The main applications are:

1- Computer spoken-output [3] - [4] :

Figure 2 represents a block diagram of a system producing computer synthesized speech. Its use in public information systems, in PTT systems, audio response units ... etc, has already started [2]. Examining Figure 1 and 2 we can notice that for the installation of such systems using the Arabic language, research is needed in the following specific domains:

a- Description of the phonetic system.

b- Rules for orthographic-phonetic transcription.

c- Data and rules on micro and macro-melody.

d- Rules for the generation of prosodic features.

e- Dictionaries or rules for the generation of parameter values corresponding to the adopted type of synthesizer, e.g.,
- Formant values and rules for their transitions
- Diphoneme Formant values
- Spectral analysis of diphonemes for channel vecoder control
- Articulatory parameters or their corresponding A(x) values
- Diphoneme reflection-coefficient values for LPC synthesizers.

It is worth mentioning that speech synthesis by rule, from unlimited phonetic input, using formant synthesizers is already commercialized for several languages. Furthermore, speech synthesis by diphonemes using LPC or channel vocoders is

Table 1 :Speech Synthesis Methods and their Language Dependency

Synthesis Technique		Digital memory or tape recorder	Formant synth.	Channel vocoder synthesis	Linear Predictive coding synth.	Synthesis by vocal tract simulator
Coding Parameters		Signal	3-variable formants+nasality+amplitude+pitch	12-14 channels + excitation	Reflection coefficients+pitch	articulatory parameters
Quantity of information needed for synthesis		analog signal or 64 kb/s PCM 32 kb/s D-Mod	1000 b/s	2400 b/s	1000-2400	Final goal is 50 b/s
Output-Speech intelligibility and naturalness		Excellent	Good	Good	Very Good	Expected to be Good
Hardware Complexity		Simple	Medium	Medium	Medium	Complex
Synthesis method and language dependency	Sentence	N.L.D	N.L.D	N.L.D	N.L.D	N.L.D
	Word	L.D	L.D	L.D	L.D	L.D
	Diphoneme	not possible	V.L.D	V.L.D	V.L.D	V.L.D
	Rules	not possible	V.L.D	not possible	V.L.D	V.L.D

L.D. : Language dependent, N.L.D. : Non-language dependent, V.L.D. : Very language dependent

operational in a few languages. As for Arabic speech synthesis, an effort is being made to acheive it using diphonemes and LPC synthesizers (see section IV below).

2- Machine-to-man communication:

Vocal output of systems and equipment such as sophisticated alarm systems, automated announcement systems, voice-interactive training systems, etc, are being commercialized in several languages. The synthesis methods used here are mainly by sentence using one of the techniques mentioned in Table 1 . An example of synthesis by sentence using a formant synthesizer in Arabic will be demonstrated (tape).

3- Aids for the handicapped e.g. reading machine for blind [5] [6] :

The blind, who have lost the vast amount of written information that is an essential part of our existence, could partly overcome this severe situation by using a system that combines an optical character recognition unit with the synthesis system described in Figure 2 .

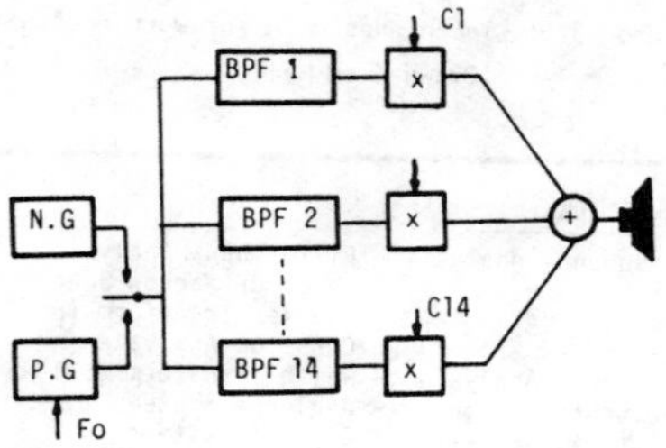

Channel Synthesizer

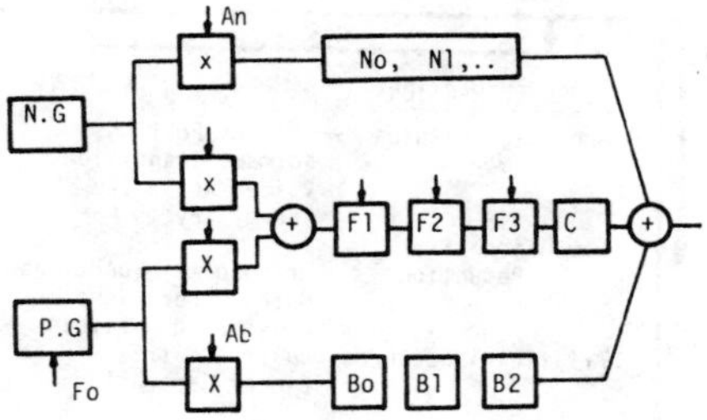

Formant Synthesizer

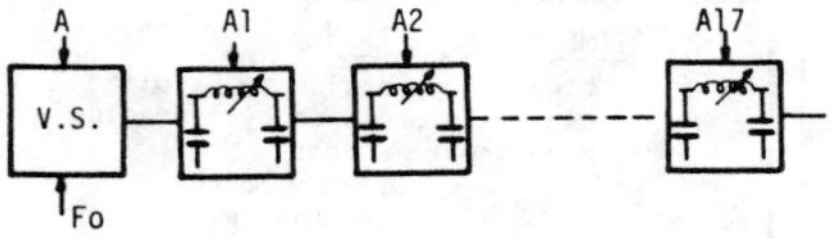

Vocal tract Simulator

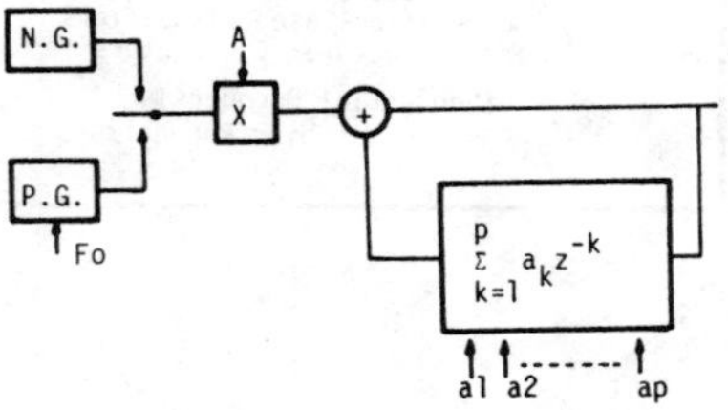

L.P. Synthesizer

Figure 1 : The four main speech synthesizer configurations showing type of parameters needed for each.

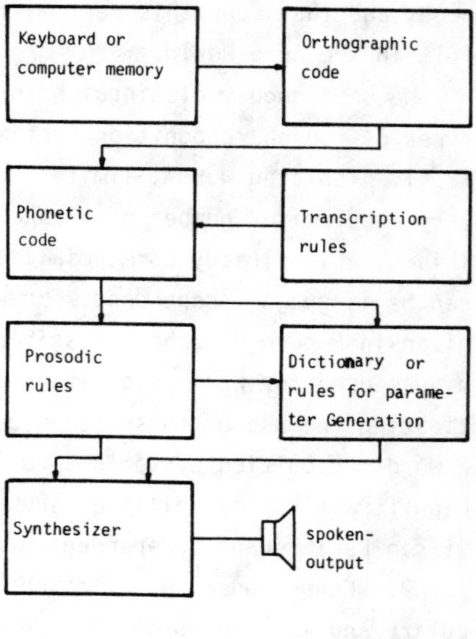

Figure 2 : Block diagram of a computer vocal response system

In order to use such a system in Arabic, we need to know, in addition to the important data mentioned above, a recognition algorithm for Arabic characters, see sections II.3 and III .

II.2 Speech Analysis and Recognition

Direct speech input is already successful in several fields, notably parcel sorting, inspection and quality control, airline baggage handling and special data capture, ... etc [1]. Many other applications are proposed. This demonstrates that it meets a significent need, and that soon this need will be felt in the Arab World.

Figure 3 : Techniques used for word recognition (Drawn from [1])

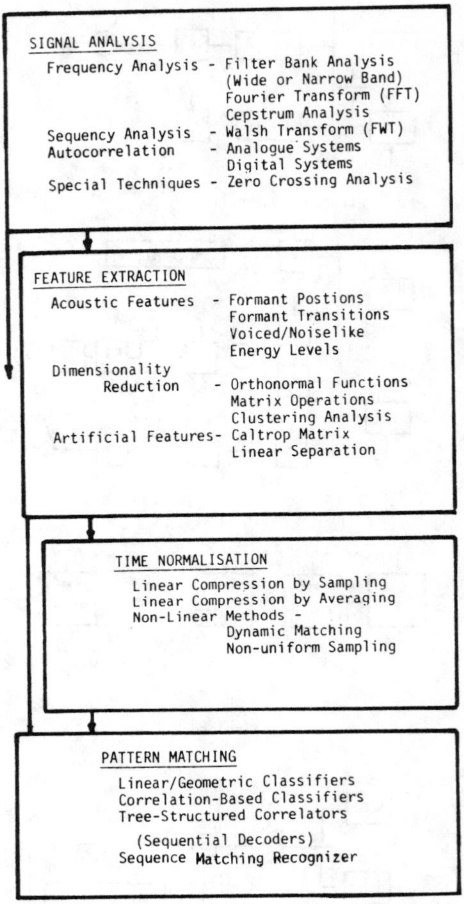

We can roughly distinguish three types of speech recognition systems:

1. "Isolated word", limited vocabulary and limited number of speakers. This type is already commercialized and can be language independent. The diagram given in Figure 3 , shows possible techniques used for word recognition. Investigating the use of these techniques in a word recognition system having learning (training) capabilities shows that it can be language independent.

2. "Connected word", limited vocabulary and limited number of speakers. This could be language independent for simple cases [7].

3. "Continuous speech", large vocabulary. This is definitely language dependent. This type uses phonemes as the basic unit of recognition and needs, in addition to phonetic distinctive features, knowledge of linguistic cues and constraints. This knowledge is essential to solve problems such as the following, [8] :

- Co-articulation effects between words and within them.
- The importance of a word in the message affects its stress and intonation and hence its acoustics realization.
- Difficulties of word boundaries detection.
- Variation in acoustic signals produced by different speakers.
- Variation in vocal tract size from speaker to another.
- Variation of speaking rate.

Klatt and Stevens [9] tried to perform a phonetic transcription from some spectrograms of continuous speech. They succeeded in transcribing only 33 per cent of the phonemes completely correctly. For a further 40 per cent they acheived a correct partial transcription. When they used, in addition to acoustic evidence, knowledge of vocabulary, syntax and semantics of the utterance, they manage to identify 96 per cent of the word. Figure 4 shows two hypothetical speech recognition systems [8], and Figure 5 represents a system proposed by Shigenaga et. al.[10]. It is quite evident that to realize such systems in Arabic, we need data and knowledge on:

- The phonetic system and phonetic-acoustic distinctive features,
- Phonological rules,
- Morphological rules and statistics,
- Lexical rules and statistics,
- Syntactical rules (predict and evaluate subject word at specific points in an utterance),
- Semantic knowledge,
- Prosodic features.

Elementary data and a few references on some of these topics will be given in section II.3. However, a lot of research has to be done on Arabic phonetics and computational lingistics if we want to have (in several years to come) Arabic speech recognition systems. These systems are important because of the many possible applications.

The number and type of possible speech recognition applications is limited more by the imagination of the system designer. The following list shows some posibilities of the very many that exist[1]:

1- Inspection systems:

Such systems could be language independent because of the limited vocabulary and limited number of users. Examples of

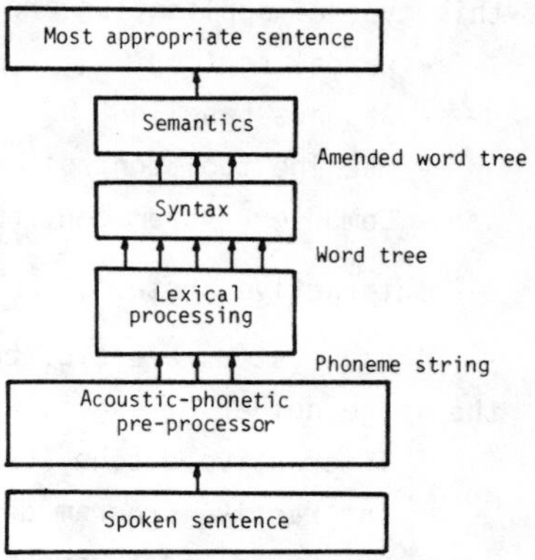

An hierarchical continuous speech recognition system (C.S.R.S.)

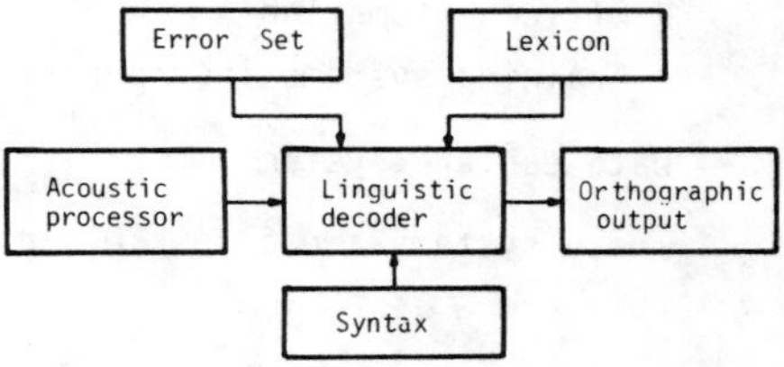

Fig. 4 System organization for C.S.R.S.

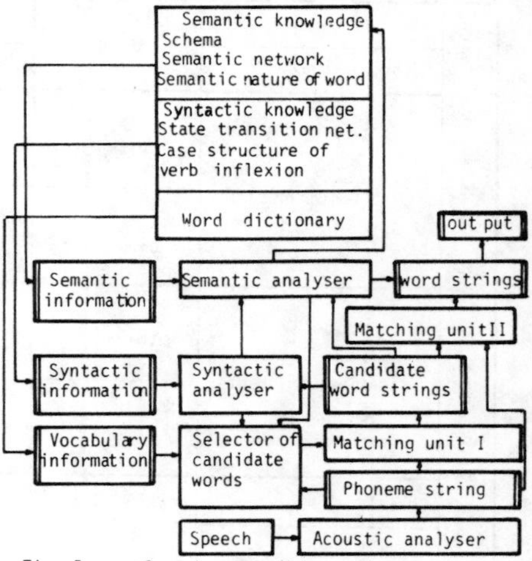

Fig. 5 Speech recognition system

this type of application are the following:

- Parcel sorting
- Baggage handling
- Machine tool control
- Computer system control (ships, aircraft ...)

2- Interactive systems:

These systems are L.D. because of the possibly large vocabulary needed and the large number of users. Examples of applications are as follows:

- Interactive design (CAD)
- Interactive program development
- Spoken input of computers (each telephone = a terminsl)
- Data base inquiry (eg: airlines ...)
- Office automation
- Training systems incorporating voice interaction

3- Data capture systems:

These systems are also L.D. and some applications are the following:

Table 2 : Literal Arabic Phonetic System

Vowels	Short Vowels	Voiced	fatḥa a, kasra i, ḍamma u	
	Long Vowels		Alif ā, yā3-maddiyah, wāw-maddiyah ū	
Consonants	Stop Consonants	Voiced	Pharyngealized	ḳ ḍ ṭ
				3 ǰ d b
		Unvoiced	Pharyngealized	
				k t
	Fricative Consonants	Voiced	Pharyngealized	ẓ
				ε ǧ z ḏ
		Unvoiced	Pharyngealized	ṣ
				h ḥ x š s ṯ f
	Nasals	Voiced	n m	
	Glides		Pharyngealized	ḷ
				l
			r	
	Semivowels		w y	

Vocal cords ⟶ Lips

Relative position of articulation

- Quality assurance and control (component inspection)
- Accounting systems

II.3 Arabic Language Phonetics

Some essential and preliminary data on Arabic phonetics is available. The Arab phonetic system is shown in Table 2 for literal Arabic. On the other hand Tables 3 and 4 show Arabic vowel classification and their formant frequencies respectively. Detailed study of Arabic phonology can be found in references [11] and [12]. The Arabic phonemes with their names and transcription are given in Table 5 . Finally, one reference dealing with these data in a comprehensive fashion will appear in the Proceedings of the Arab School on Science and Technology which will be held in Rabat - Moroco on 26/9/1983 entitled "Applied Arabic Linguistics and signal Data Processing".

Table 3 : Arabic Vowel Classification

Tongue hump position / Degree of constriction	front	central	back
High	yā3-maddiyah ī kasra i		wāw-maddiyah ū damma u
Low		Alif ä fatha a	

Table 4 : Formant Frequencies of Arabic Vowels

Formant / Vowel	F_1	F_2	F_3
i	290	2200	2700
ī	285	2200	2700
u	290	800	2150
ū	285	775	2050
a	600	1500	2100
ā	675	1200	2150

III. Application of 2.D.Signal Processing

Most of applications of 2.D.signal processing are not particular to the Arab World except Arabic character recognition. Character recognition is of growing importance and should be undertaken for the Arabic language. Several problems are peculiar to Arabic script. These are:

Table 5 : Arabic Phonemes

Phoneme	Name	Transcription
ا	alif	ā
و	wāw-Maddiyah	ū
ي	yā3-Maddiyah	ī
ـَ	fatha	a
ـُ	ḍamma	u
ـِ	kasra	i
ب	bā3	b
ت	tā3	t
ث	ṯā3	ṯ
ج	ǰīm	ǰ
ح	ḥā3	ḥ
خ	xā3	x
د	dāl	d
ذ	ḏāl	ḏ
ر	rā3	r
ز	zīy	z
س	sīn	s
ش	šīn	š
ص	ṣāḍ	ṣ
ض	ḍāḍ	ḍ
ط	ṭā3	ṭ
ظ	ẓā3	ẓ
ع	εayn	ε
غ	ğayn	ğ
ف	fā3	f
ق	ḳāf	ḳ
ك	kāf	k
ل	lām	l
م	mīm	m
ن	nūn	n
ه	hā3	h
و	wāw-līn	w
ي	yā3-līn	y
ء	hamza	3

1- Multiple representations of the same character depending on its position in the word.
2- Connectivity of characters.
3- Similarity of characters.
4- Variable-width characters.
5- Short vowels and other letter markers are written over or under characters (Al-Shakl).
6- Overlapping subwords.
7- The absence of standards for type font throughout the Arab World.
8- The absence of established standards for character codes.

Solutions to some of these problems have been proposed by several researchers or companies. For example Mr. AL-AGHDAR GAZAL. Director of "The Study and Research Institute for Arabization" Rabat-Moroco, proposed a type-font of Arabic characters having fixed width. He proposed, as well, to write the markers after the concerned character instead of over or under it. ADLER company also commercialized a typewriter with fixed width Arabic characters. However, work in these fields is still very limited if not rare. The following references are examples of such works: [13] [14]

Finally, a project for Arabic character recognition is being studied at the SSRC in Damascus. It is to be realized using a T.V. camera interfaced to a Vax-11/780. The resolution used is 512 pixel/line and 625 lines/ frame. Image enhancement is done by hardware circuitry. Primitive features for Arabic characters are defined as segments and segment-orientation. A statistical study will lead to specific rules for each character. These rules will be used in the classification operation. An algorithm for feature extraction has been tried for several characters and ecouraging results were obtained. We also propose to use character statistics to speed the recognition algorithm.

Finally, we give below a few applications:

- Automatic computer input of documents.
- Material handling (mail, baggage ...).
- Text communication with high speed and narrow BW.
- Office automation.

IV. Teaching and Research on Signal Processing in the Arab World

Signal processing was very recently introduced in some universities and research centers in the Arab World. No text books on this subject, as far as we know, exist in Arabic except one from Damascus University written by Mr. H.Abou-Al-Nour. The main laboratories known to us, active in the field are the following, with references made to the fields of interest where available:

- Laboratoire d'Electronique et d'Etude des Systemes Automatiques. Faculte des sciences . RABAT - MAROC .
  .L.P.C., Synth. by diphoneme.
- Kuwait Institute for Scientific Research. P.O.Box 24885 . KUWAIT .
  .Computer input of Arabic text.
- Scientific Studies and Research Center. P.O.Box 4470 . DAMASCUS - SYRIA .
  .Arabic phonetics, Formant synthesizer, LPC simulation, Diphoneme synthesis, Arabic character recognition.
- Institute de Linguistique et de phonetique . Ben Aknoun - ALGER .
  .Applied linguistics , Phonetics.
- Organisme National de la Recherche Scientifique . Route de Dely-Ibrahim-Ben-Aknoun . ALGER .
- National Computer Center . BAGHDAD .

References

1 Warren, J.H., Man-Machine communication using speech, EMI Central Research Laboratories.

2 Cartier, M. and Genin, J., La Reponse Vocale: Realisations Actuelles et Prespectives d'Avenir, Vol.54, Nol, 1974.

3 Mrayati, M., Guerin, B. and BOE, L.J., La Synthese de la parole, Images de la Physique, Supp. No 16 du courrier du CNRS, PARIS . Juin 1975.

4 ST-Bonnet, M. and Mrayati, M., La Synthese par Diphone, Bulletin de l'Institute de Phonetique de Grenoble, Vol. III, PP. 73-74, 1974.

5 Cooper, F.S., Gaintenby, J.H, Mattingly, I.G. and Uneda, N., Reading Aids for

the Blind: a special Case of Machine to Man Communication, IEEE Trans. Audio and Electroacoust., Vol. 17, No.4, 1969.

6 Suen, C.Y. et al., Talking Machines for the Blind Colloque International sur les Capteurs Biomedicaux, Tome: 2, PP. 543-544, Paris, Nov.

7 Rabiner, L.R., and Levinson, S.E., Isolated and Connected Word Recognition-Theory and Selected Applications, IEE Trans. Commun., Vol. COM - 29, PP.621-659, 1981 .

8 Ainsworth, W.A., Mechanisms of Speech Recognition, Pergamon Press, 1976 .

9 Klatt, D.H. and Stevens, K.N., On the Automatic Recognition of Continuous Speech: Implications from a Spectrogram-Reading experiment. IEEE Trans. Audio Electroacoust. AU-21,210 .

10 Shigenaga, M.m Sekiguchi, Y. and Yagisaway T., Speech Recognition System for Spoken Japanese Sentences, 11' ICA, Paris, 1983 .

11 Al-Ani, S.H., Arabic Phonology-An Acoustical and Physiological Investigation, Mouton, 1970.

12 Djafar, M.M., Grammaire de l'Arabe-Documents de Linguistique Quantitative, Dunod, Paris, 1973 .

13 Ali, M., Computer Recognition of Arabic Script, Proc. Seminar on the use of Computers in Universities, Baghdad, Nov., 1979 .

14 Parhami, B. and Taraghi M., Automatic Recognition of Printed Farsi Texts. Pattern Recognition , 1980 .

7

Signal Processing in Communications

JOHN G. PROAKIS

Digital signal processing techniques are widely used in transmission and reception of digital information. In this chapter we present a number of digital signal processing techniques and adaptive filtering algorithms which are applicable to the demodulation of digital communication signals. In particular, we consider digital implementation of a modulator, a demodulator, and a decoder, adaptive equalization algorithms for telephone channels and radio channels, and adaptive filtering of narrowband interference in direct sequence spread spectrum signals.

TABLE OF CONTENTS

1. INTRODUCTION

Digital signal processing plays an important and fundamental role in transmission and reception of information. Even when the information is in analog form, e.g., an analog signal such as speech, we have seen that it is common practice to convert the signal into a digital form and to transmit the information digitally. Conversion of the digital signal back to an analog form is performed at the receiving terminal of the communication system.

The widespread use of digital signal processing techniques in communications is due in a large part to the availability, low cost and high reliability of LSI (large-scale integrated) and VLSI (very large-scale integrated) circuits. As a consequence, many of the signal processing functions or operations that are encountered in modulation and demodulation of communication signals are presently performed by digital means, whereas, five to ten years ago, many of these same operations were performed in analog form.

In this presentation, we discuss a number of topics dealing with the application of digital signal processing techniques and algorithms in digital communications. We begin by describing a model of a digital communications system. The elements of such a system are illustrated in the block diagram shown in Figure 1.

The information source generates messages that are to be transmitted to the receiver. In general, the characteristics of the message depend on the type of information source that produces it. For example, the message may be a speech signal, or a video signal, or a signal obtained by optically scanning a picture. We refer to such signals as analog signals and the sources that produce them as analog sources.

In a digital communications system, the messages produced by a source are first converted to digital form, usually into a sequence of binary digits. Ideally, we would like to represent the source output (message) by as few binary digits as possible. That is, we seek an efficient representation that results in little or no redundancy. The process of converting the output of an analog source into a sequence of binary digits is called source encoding. For example, Makhoul [1] considers the encoding of signals from a speech source into binary digits.

In some cases the information source may be digital in nature and, hence, there is no need for a source encoder. For example, data stored on a magnetic tape or on a disk is already in the form of a sequence of binary digits.

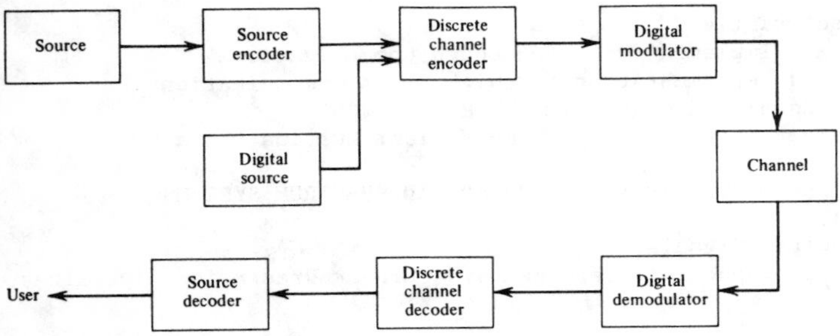

FIGURE 1. Model of a digital communications system.

The sequence of binary digits from the source encoder is transmitted through the channel to the intended receiver. The real channel may be either a pair of wires, a coaxial cable, an optical fiber, a radio channel, a satellite channel, or some combination of these media. Such channels are basically waveform channels and, hence, they cannot be used to transmit the sequence of binary digits directly. What is required is a device that converts the digital information sequence into waveforms that are compatible with the characteristics of the channel. Such a device is called a digital modulator, or simply, a modulator. The modulator is a part of a larger device called a channel encoder, which serves a second function, the need for which becomes clear when we consider the characteristics of real channels.

In general, no real channel is ideal. First of all, channels may have nonideal frequency response characteristics. In addition, there are noise disturbances and other interference that corrupt the signal transmitted through the channel. For example, there may be cross talk (interference) from signals being transmitted on adjacent channels. There is the thermal noise generated in the electronic equipment, such as amplifiers and filters used in the transmitter and in the receiver. There may also be noise and interference of the impulsive type, caused by switching transients in wire line channels and thunderstorms on radio channels. Finally, there may be intentional jamming of the signal transmitted over the channel. Such disturbances corrupt the transmitted signal and cause errors in the received digital sequence.

To overcome such noise and interference and, thus, to increase the reliability of the data transmitted through the channel, it is often necessary to introduce in a controlled manner some redundancy into the binary sequence from the source. The introduction of redundancy in the information sequence for the purpose of combatting the detrimental effects of noise and interference in the channel is the second function of the channel encoder. The redundancy introduced at the transmitter aids the receiver in decoding the desired information-bearing sequence. For example, a (trivial) form of encoding of the binary information sequence is simply to repeat each binary digit m times, where m is some positive integer. More sophisticated (nontrivial) encoding involves taking k information bits at a time and mapping each k-bit sequence into a unique n-bit sequence, called a code word. The amount of redundancy introduced by encoding the data in this manner is measured by the ratio n/k. This is also the ratio by which the channel bandwidth must be increased to accommodate the added redundancy in the information sequence. The reciprocal of this ratio, namely k/n, is called the rate of the code or, simply, the code rate.

An alternative to providing added redundancy in the information sequence as a means of overcoming the channel disturbances is to increase the power in the transmitted signal. Since the addition of redundancy implies the need to increase the channel bandwidth, there is a trade-off between transmitted power and channel bandwidth. In channels in which bandwidth is limited or expensive and power is available to overcome the channel degradations, no redundancy is employed. Consequently, not all digital communications systems employ a channel encoder that introduces redundancy in the information sequence. On the other hand, when such redundancy is introduced, it is both convenient and appropriate to view the channel encoder as consisting of two parts, a discrete channel encoder and a digital modulator. The former is discrete at both its input and its output. The latter has a discrete input but its output consists of waveforms.

To elaborate on the function performed by the modulator, suppose the information is to be transmitted 1 bit at a time at some uniform rate R bits/sec. The modulator may simply map the binary digit 0 into a waveform $s_1(t)$ and the binary digit 1 into a waveform $s_2(t)$. In this manner, each bit from the channel encoder is transmitted separately. We call this binary modulation. Alternatively, the modulator may transmit k information bits at a time by using $M = 2^k$ distinct waveforms $s_i(t)$, $i = 1,2,\ldots,M$, one waveform for each of the 2^k possible k-bit sequences. We call this M-ary modulation. We note that a new k-bit sequence enters the modulator every k/R seconds. Hence the amount of time available to transmit one of the M waveforms corresponding to a k-bit sequence is k times the time period in a system which uses binary modulation.

At the receiving end of the communication system, the digital demodulator processes the channel-corrupted transmitted waveform and reduces each waveform to a single number that represents an estimate of the transmitted information symbol (binary or M-ary). The processing performed by the demodulator consists of matched filtering or cross correlating the received channel corrupted signal with a replica of the M possible transmitted signal waveforms. The values of the M matched filter outputs or cross correlation operations provide a basis for making a decision on each received signal waveform. Generally, the decision is made in favor of the signal waveform resulting in the largest correlation. For example, when binary modulation is used, the demodulator may process the received waveform and decide on whether the transmitted bit is a 0 or a 1. In such a case we say that the demodulator has made a binary decision. As one alternative, the demodulator may make a ternary decision. That is, it decides that the transmitted bit is either a 0 or 1 or it makes no decision at all, depending on the apparent quality of the received signal. When no decision is made on a particular bit, we say that the democulator has inserted an erasure in the demodulated data. Using the redundancy in the transmitted data, the decoder attempts to fill in the positions where erasures occurred. Viewing the decision process performed by the demodulator as a form of quantization, we observe that binary and ternary decisions are special cases of a demodulator that quantizes to Q levels, where $Q \geq 2$. In general, if the digital communications systems employs M-ary modulation, where $m = 0,1,\ldots,M-1$ represent the M possible transmitted symbols, each corresponding to $k = \log_2 M$ bits, the demodulator may make a Q-ary decision, where $Q \geq M$. In the extreme case where no quantization is performed, $Q = \infty$.

When there is no redundancy in the transmitted information, the demodulator must decide which of the M waveforms was transmitted in any given time interval. Consequently $Q = M$ and, since there is no redundancy in the transmitted information, no discrete channel decoder is used following the demodulator. On the other hand, when there is redundancy introduced by a discrete channel encoder at the transmitter, the Q-ary output from the demodulator is fed to the decoder, which attempts to reconstruct the original information sequence using the redundancy contained in the code.

A measure of how well the demodulator and decoder perform is the frequency with which errors occur in the decoded sequence. More precisely, the average probability of a bit error at the output of the decoder is a measure of the performance of the demodulator-decoder combination. In general, the probability of error is a function of the code characteristics, the types of waveforms used to transmit the information over the channel, the transmitter power, the characteristics of the channel, i.e., the amount of noise, the nature of the interference, etc., and the method of demodulation and decoding.

As a final step, when an analog output is desired, the source decoder accepts the output sequence from the channel decoder and, from knowledge of the source encoding method used, attempts to reconstruct the original signal from the source. Due to channel decoding errors and possible distortion introduced by the source encoder and, perhaps, the source decoder, the signal at the output of the source decoder is an approximation to the original source output. The difference or some function of the difference between the original signal and the reconstructed signal is a measure of the distortion introduced by the digital communications system illustrated in Figure 1.

In the following sections we describe the use of digital signal processing techniques in digital communications over telephone channels and fading multipath channels and in digital communications by means of spread spectrum signals. First, we give a brief description of basic modulation and coding techniques.

1.1 Digital Modulation Techniques

Basic modulation techniques that are commonly used for transmitting the information are phase-shift keying (PSK), amplitude-shift keying (ASK), which is also called pulse amplitude modulation (PAM), frequency-shift keying (FSK) and combined PSK-PAM, which is also called QAM.

In PSK, the signalling waveforms may be expressed as

$$s_m(t) = A \cos(2\pi f_c t + \theta_m), \quad m = 1,2,\ldots,M \quad 0 \leq t \leq T \tag{1}$$

where f_c is the carrier frequency, T is the signal duration (1/T is the symbol rate) and θ_m, $m = 1,2,\ldots,M$ represent the M possible phases that carry the digital information. For example, with $M = 2$, $\theta_1 = 0$, $\theta_2 = \pi$ and we have binary PSK. With $M = 4$, $\theta_1 = 0$, $\theta_2 = \pi/2$, $\theta_3 = \pi$, $\theta_4 = 3\pi/4$ and, hence, we have four-phase PSK where each signal phase carries two bits of information. In general, with $M = 2^k$ signal phases, we may transmit k bits in a symbol duration of T seconds and, thus, we achieve a bit rate of $R = (\log_2 M)/T = k/T$ bits per second. In a commonly used variant of PSK, called differential PSK (DPSK), the information is conveyed to the receiver by phase changes between successive signal intervals.

In ASK or PAM, the signalling waveforms may be expressed as

$$s_m(t) = A_m \cos 2\pi f_c t, \quad m = 1,2,\ldots,M \quad 0 \leq t \leq T \tag{2}$$

where the signal amplitude A_m carries the information. For example, with $M = 2$, $A_1 = A$ and $A_2 = -A$ and we have binary ASK. Note that this is identical to binary PSK. With $M = 4$, $A_1 = 3A$, $A_2 = A$, $A_3 = -A$, $A_4 = -3A$, where A is an amplitude scale factor which determines the average transmitted power.

The signalling waveforms for FSK may be expressed as

$$s_m(t) = A \cos 2\pi f_m t, \quad m = 1,2,\ldots,M \quad 0 \leq t \leq T \tag{3}$$

where $\{f_m\}$ are the M possible frequencies. For $M = 2$, we may select $f_1 = f_c + \Delta f/2$ and $f_2 = f_c - \Delta f/2$, where Δf is the frequency spacing, i.e., $\Delta f = f_1 - f_2$, and f_c is the center (carrier) frequency. Usually, Δf is selected to be equal to 1/T where T is the signal interval. This selection of Δf yields orthogonal waveforms. Similarly, for $M > 2$, the spacing between successive frequencies is also selected as 1/T.

Finally, the signalling waveforms for combined PSK-ASK, or QAM, may be expressed as

$$s_{mn}(t) = A_m \cos(2\pi f_c t + \theta_n), \quad m = 1,2,\ldots,N_1, \quad n = 1,2,\ldots,N_2 \quad 0 \leq t \leq T \tag{4}$$

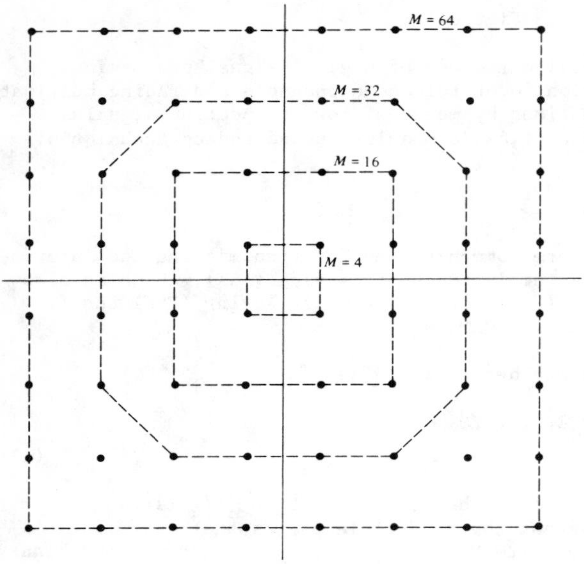

FIGURE 2. QAM signal constellation.

where $M = N_1N_2$. Figure 2 illustrates QAM signal point constellations for M = 4 (except for a phase rotation of 45°, this constellation is identical to four phase PSK), M = 16, M = 32 and M = 64.

The choice of a digital modulation technique depends on a number of factors, including the available channel bandwidth W, the amount of transmitter power, the desired data rate R in bits per second and the channel characteristics such as noise and other disturbances. When the channel simply adds thermal noise, which is characterized as an additive white Gaussian noise (AWGN) process, the selection of an appropriate modulation technique is relatively straightforward. This may be accomplished by comparing the modulation techniques described above on the basis of the normalized data rate R/W (bits per second per Hertz of bandwidth) versus the signal-to-noise ratio (SNR) per bit required to achieve a given probability of error. Figure 3 illustrates such a comparison for a fixed symbol error probability of 10^{-5} with AWGN. Also shown in Figure 3 is the curve for the channel capacity formula, originally derived by Shannon [2], which is

$$\frac{C}{W} = \log_2\left(1 + \frac{C}{W}\frac{E_b}{N_0}\right) \quad , \tag{5}$$

where C is the capacity in bits per second of the AWGN channel having a bandwidth W, E_b is the energy per information bit and N_0 is the power spectral density of the additive noise.

The graph is subdivided into two regions, R/W > 1 and R/W < 1. Efficient modulation techniques for operating in a bandwidth-limited environment (R/W > 1) are PSK, DPSK, PAM and QAM. Of these, PAM and QAM are the most efficient. In fact, the selection of (power) efficient QAM constellations is a problem that has

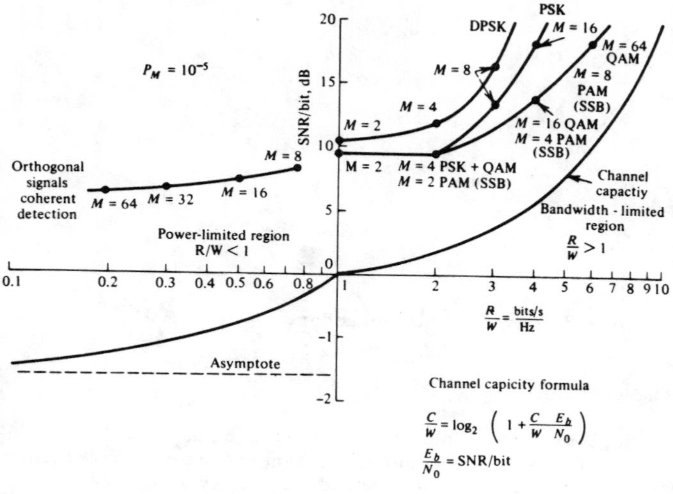

FIGURE 3. Comparison of modulation methods.

received considerable attention in the technical literature [3],[4]. In general, the rectangular signal constellations shown in Figure 2, although they are not optimum, yield relatively good performance and are easily generated. On the other hand, when $R/W < 1$, we are operating in a power-limited environment, where modulation techniques such as orthogonal FSK and coded waveforms are appropriate.

1.2 Channel Coding and Decoding Techniques

As previously indicated channel coding is a means of providing redundancy in the information sequence to be transmitted, which can be used by the receiver to overcome the effect of noise and other interference encountered in transmission. Two types of linear encoding are commonly used in practice.

One is block encoding, in which blocks of k information bits are encoded into corresponding blocks of n bits ($n > k$). Each block of n bits from the encoder constitutes a code word contained in a set of $M = 2^k$ possible code words. The code rate, defined as the ratio k/n and denoted as R_c, is a measure of the amount of redundancy introduced by the encoder. Thus, the bit rate at the output of the block encoder is R/R_c. The generation of the code words in a linear block code is accomplished digitally, usually by means of a shift register.

The second type of encoding is convolutional encoding. An example of such an encoder implemented as a shift register is shown in Figure 4. For every bit entering this encoder, there are three output bits. Thus, the code rate $R_c = 1/3$. The adders shown in Figure 4 perform modulo-2 addition. The number of stages in this shift register is $L = 3$. This parameter is called the constraint length of the convolutional code.

The binary digits from the encoder are fed into the modulator which maps the data stream into signal waveforms. PSK, FSK, PAM or QAM are commonly used signalling waveforms. At the receiving end, these waveforms are first processed by the

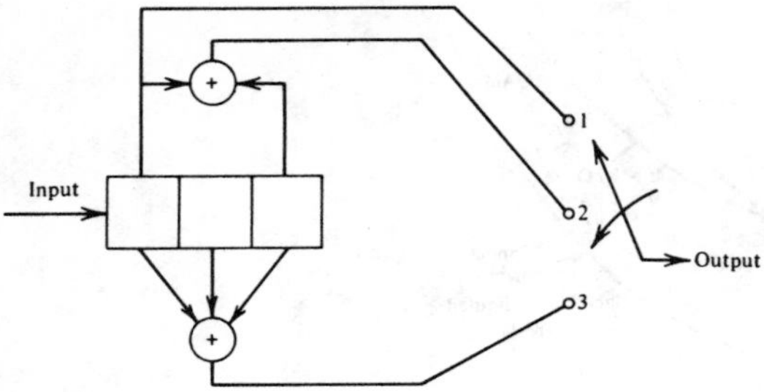

FIGURE 4. An $L = 3$, $k = 1$, $n = 3$ convolutional encoder.

demodulator and then by the decoder.

The demodulator is usually implemented either as a parallel bank of filters matched to the transmitted waveforms or as a parallel bank of cross correlators. The output of these filters or correlators is sampled at the symbol rate and quantized. For example, with binary signalling waveforms we may quantize the output to one bit, say either zero or one. We say that the demodulator has made a hard decision on each bit. The resulting sequence of detected bits is fed into the decoder which attempts to reconstruct the original information sequence by using the redundancy introduced by the encoding process. Since the decoder operates on the hard decisions made by the demodulator, the decoding process is termed hard-decision decoding. In general, the filter or correlator output may be quantized to Q levels. The decoder, in turn, makes use of the additional information contained in the finer quantized samples to recover the information sequence with a higher reliability than that achieved with hard decisions. We refer to the resulting decoding as soft-decision decoding.

Efficient decoding algorithms exist for both block codes and convolutional codes [5]. Implementation of both hard- and soft-decision decoding is performed digitally by means of either dedicated hardware or software on a programmable signal processor. Recent investigations on decoding algorithms for error control have been focused on the similarity between such decoding algorithms and discrete Fourier transforms, FIR filters and cyclic convolutions, topics familiar to people working in digital signal processing [6].

A detailed discussion of decoding algorithms for block and convolutional codes is beyond the scope of this presentation. The interested reader is referred to the books of Viterbi and Omura [7] and Proakis [5] for tutorial treatments and to the vast technical literature for additional information.

2. DIGITAL COMMUNICATIONS OVER TELEPHONE CHANNELS

2.1 Channel Characteristics

Telephone channels may be characterized as analog band-limited linear filters with a specified passband. We shall consider specifically voice-band channels for which the passband is 300 to 3000 Hz. We may express the frequency response of the channel as

$$C(f) = A(f)\, e^{j\theta(f)} \qquad (6)$$

where A(f) is the amplitude response characteristic and θ(f) is the phase response characteristic. Another characteristic, related to θ(f), is the envelope delay defined as

$$\tau(t) = -\frac{1}{2\pi}\frac{d\theta(f)}{df} \tag{7}$$

The frequency response characteristics of telephone channels have a significant impact on the design of signals for modulation and on signal processing techniques for demodulation. These topics are discussed below.

An ideal or nondistorting channel is defined as one for which the amplitude A(f) is constant for all frequencies in the passband and the phase θ(f) is a linear function of frequency for frequencies within the passband. On the other hand, if A(f) is not constant within the passband, we say that the signal transmitted through the channel suffers amplitude distortion. Similarly, if the envelope delay is not a constant within the passband we say that the signal transmitted through the channel suffers delay distortion.

As a result of the amplitude and delay distortion caused by the nonideal channel frequency response characteristic C(f), a succession of pulses transmitted through the channel at rates comparable to the channel bandwidth are smeared to the point that they are no longer distinguishable as well-defined pulses at the receiving terminal. Instead, they overlap and, thus, we have intersymbol interference. As an example of the effect of delay distortion on a transmitted pulse, Figure 5a illustrates a band-limited pulse having zeros periodically spaced in time at points labeled $\pm T$, $\pm 2T$, etc. If information is conveyed by the pulse amplitude, as in PAM for example, then one can transmit a sequence of pulses each of which has a peak at the periodic zeros of the other pulses. However, transmission of the pulse through a channel modeled as having a linear envelope delay characteristic τ(f) [quadratic phase θ(f)] results in the received pulse shown in Figure 5b having zero crossings that are no longer periodically spaced. Consequently a sequence of successive pulses would be smeared into one another and the peaks of the pulses would no longer be distinguishable. Thus the channel delay distortion results in intersymbol interference. It is possible to compensate for the nonideal frequency response characteristic of the channel by use of a filter or equalizer at the demodulator. Figure 5c illustrates the output of a linear equalizer which compensates for the linear distortion in the channel.

In addition to linear distortion, signals transmitted through telephone channels are subject to other impairments, specifically nonlinear distortion, frequency offset, phase jitter, impulse noise and thermal noise.

The degree to which one must be concerned with these channel impairments depends on the transmission rate over the channel and the modulation technique. For rates below 1800 bits/sec ($R/W < 1$), one can choose a modulation technique, for example FSK, which is relatively insensitive to the amount of distortion encountered on typical telephone channels from all the sources listed above. For rates between 1800 and 2400 bits/sec ($R/W > 1$), a more bandwidth-efficient modulation technique such as four-phase PSK is usually employed. At these rates some form of compromise equalization is often employed to compensate for the average amplitude and delay distortion in the channel. In addition, the carrier recovery method is designed to compensate for the frequency offset. The other channel impairments are not that serious in their effects on the error rate performance at these rates. At transmission rates above 2400 bits/sec ($R/W > 1$), bandwidth-efficient modulation techniques such as QAM, PAM, and PSK are employed. For example, at rates from 2400 to 9600 bits/sec, special attention must be paid to linear distortion, frequency offset, and phase jitter. Linear distortion is

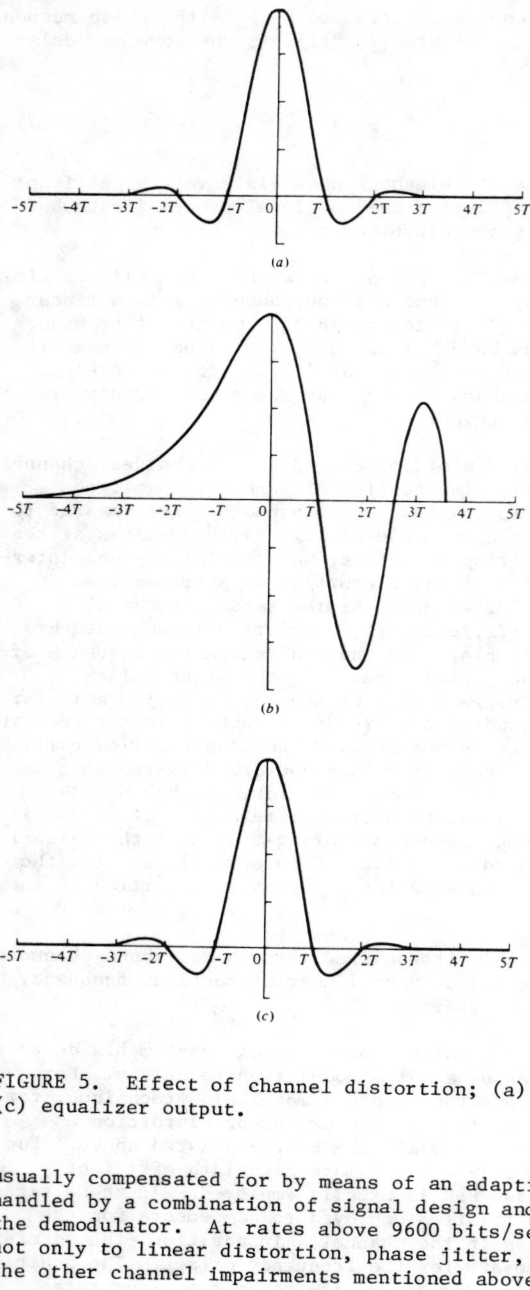

FIGURE 5. Effect of channel distortion; (a) channel input, (b) channel output, (c) equalizer output.

usually compensated for by means of an adaptive equalizer. Phase jitter is handled by a combination of signal design and some type of phase compensation at the demodulator. At rates above 9600 bits/sec, special attention must be paid not only to linear distortion, phase jitter, and frequency offset, but also to the other channel impairments mentioned above.

2.2 Bandwidth Considerations and Signal Design

In the following discussion we assume that the band-limited channel has an equivalent lowpass frequency response characteristic which is zero for $|f| > W$. Since any bandpass channel has an equivalent lowpass representation, there is no loss in generality by imposing this condition. In this section, we assume that the channel is ideal.

Suppose the equivalent lowpass channel through which the signal pulses are transmitted is band-limited to W Hz. Then, the signal pulses must also be band-limited to the same bandwidth. In general, the sampling theorem allows us to represent a band-limited signal pulse as

$$x(t) = \sum_{n=-\infty}^{\infty} x\left(\frac{n}{2W}\right) \frac{\sin 2\pi W\left(t - \frac{n}{2W}\right)}{2\pi W\left(t - \frac{n}{2W}\right)} \tag{8}$$

where $x(n/2W)$ are samples of the signal taken at a rate of 2W samples per second and W is the signal bandwidth. The band-limited signal pulse $x(t)$ is related to its voltage spectrum via the inverse Fourier transform, i.e.,

$$x(t) = \int_{-W}^{W} X(f)\, e^{j2\pi ft}\, df \tag{9}$$

and, hence, the samples are

$$x\left(\frac{n}{2W}\right) = \int_{-W}^{W} X(f)\, e^{j2\pi fn/2W}\, df \tag{10}$$

Now, suppose we transmit such band-limited pulses at a rate of 2W pulses per second. This particular choice of pulse rate is called the Nyquist rate. If we let $T = 1/2W$, then $x(t)$ becomes

$$x(t) = \sum_{n=-\infty}^{\infty} x(nT) \frac{\sin \pi(t - nT)/T}{\pi(t - nT)/T} \tag{11}$$

and a sequence of such pulses, modulated by the information sequence $\{I_n\}$, is expressed as

$$y(t) = \sum_{n} I_n x(t - nT) \tag{12}$$

Note that $y(t)$ is also a band-limited signal. Sampling $y(t)$ periodically at $t = kT$ yields

$$y(kT) = \sum_{n} I_n x(kT - nT) = x(0)\, I_k + \sum_{n \neq k} I_n x(kT - nT) \tag{13}$$

If we select $x(t)$ such that

$$x(kT) = \begin{cases} 1, & k = 0 \\ 0, & k \neq 0 \end{cases} \tag{14}$$

then $y(kT) = I_k$, so that the kth sample of $y(t)$ yields the kth information symbol. Thus, we have no intersymbol interference from a band-limited pulse that satisfies the condition given by (14). This pulse is

$$x(t) = \frac{\sin \pi t/T}{\pi t/T} \tag{15}$$

and its voltage spectrum is the rectangular characteristic

$$X(f) = \begin{cases} T, & |f| \leq \frac{1}{2T} \\ 0, & |f| > \frac{1}{2T} \end{cases} \tag{16}$$

The graphs of x(t) and X(f) are illustrated in Figure 6.

There are two basic problems associated with this pulse shape. One is the problem of realizing a pulse having the rectangular spectral characteristic X(f) given above. That is, X(f) is not physically realizable. The other problem is concerned with the fact that the tails in x(t) decay as 1/t. Consequently a mistiming error in sampling results in an infinite series of intersymbol interference components. Such a series is not absolutely summable and, hence, the sum of the resulting interference does not converge.

In order to avoid these problems, we restrict the transmission rate to $1/T < 2W$ symbols/sec and attempt to determine if pulses exist that satisfy the condition of no intersymbol interference.

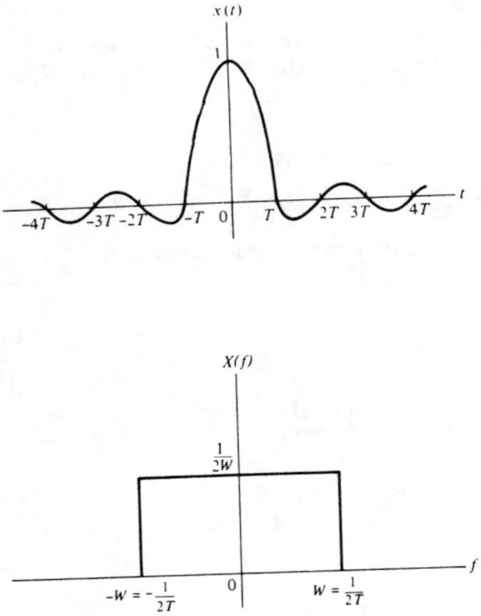

FIGURE 6. Band-limited signal for no intersymbol interference.

Now

$$x(t) = \int_{-W}^{W} X(f)\, e^{j2\pi ft}\, df \tag{17}$$

We wish to sample x(t) at a rate 1/T so that

$$x(kT) = \int_{-W}^{W} X(f)\, e^{j2\pi fkT}\, df \qquad k = 0, \pm 1, \pm 2, \ldots \tag{18}$$

Since W > 1/2T, x(kT) can be expressed as

$$x(kT) = \sum_{n=-N}^{N} \int_{(2n-1)/2T}^{(2n+1)/2T} X(f)\, e^{j2\pi fkT}\, df$$

$$= \sum_{n=-N}^{N} \int_{-1/2T}^{1/2T} X(f + \frac{n}{T})\, e^{j2\pi fkT}\, df \tag{19}$$

where N is the integer [2TW]. Let us define

$$X_{eq}(f) = \sum_{n=-N}^{N} X(f + \frac{n}{T}) \quad , \qquad |f| \leq \frac{1}{2T} \tag{20}$$

Then, by making use of the relation given by (20) in (19), we obtain

$$x(kT = \int_{-1/2T}^{1/2T} X_{eq}(f)\, e^{j2\pi fkT}\, df \tag{21}$$

Now the condition for no intersymbol interference, i.e., x(kT) = 0 for k ≠ 0, x(0) = 1, requires that

$$X_{eq}(f) = \begin{cases} T, & |f| \leq 1/2T \\ 0, & |f| > 1/2T \end{cases} \tag{22}$$

Figure 7 shows an example of a spectrum that has a bandwidth W > 1/2T and $X_{eq}(f)$ that results in no intersymbol interference. It is apparent that when X(f) has odd symmetry about the frequency 1/2T, then $X_{eq}(f)$ has the rectangular spectral characteristic as shown in Figure 7. Therefore it is possible to design band-limited pulses that give no intersymbol interference. Practically, if one selects the rate so that

$$W < \frac{1}{T} < 2W$$

it is possible to design a variety of pulses that have good spectral characteristics and are free of intersymbol interference on an ideal channel.

A pulse that has found wide use in digital transmission on band-limited channels such as telephone lines is one that has the raised cosine spectral characteristic,

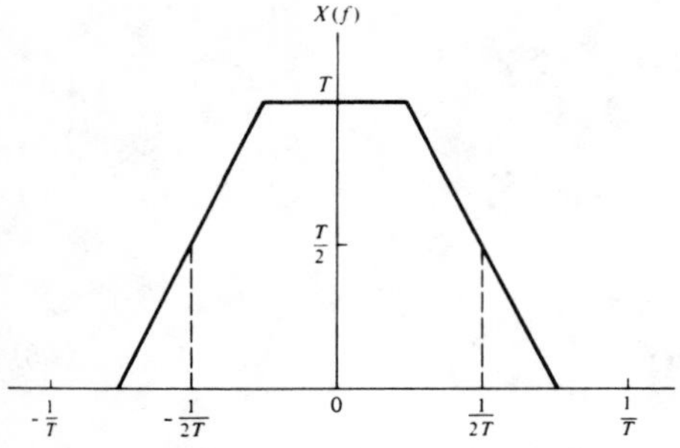

FIGURE 7. A signal spectrum which results in no intersymbol interference.

$$X(f) = \begin{cases} T, & 0 \leq |f| \leq (1 - \beta)/2T \\ \frac{T}{2} \left[1 - \sin \pi T(f - \frac{1}{2T}) \beta\right], & (1 - \beta)/2T \leq |f| \leq (1 + \beta)/2T \end{cases} \tag{23}$$

where β is called the rolloff parameter. The pulse $x(t)$ having this spectrum is

$$x(t) = \frac{\sin \pi t/T}{\pi t/T} \frac{\cos \beta\pi t/T}{1 - 4\beta^2 t^2/T^2} \tag{24}$$

The tails of the pulse decay as $1/t^3$. Hence a mistiming error in sampling leads to a series of intersymbol interference components that converge. Figure 8 illustrates the spectral characteristic $X(f)$ and the pulse $x(t)$ for several values of β.

By taking a slightly different approach, it is possible to design band-limited signal pulses that do not use any excess bandwidth. Suppose we impose the condition that the pulse rate be 2W pulses per second but we remove the constraint that there be no intersymbol interference at the sampling instants. Thus, we may use any pulse that has the general representation given by (8). In other words, there is a class of infinitely many, physically realizable pulses that satisfy this condition. The intersymbol interference caused by such pulses is deterministic or "controlled" and, hence, it can be taken into account at the demodulator, where the information is recovered from the received signal. The signal pulses that are represented by (8) have been called partial response signals [5],[8],[9],[10].

The simplest types of the partial response signals are obtained by allowing two of the samples $x(n/2W)$ to be non-zero. One special case is the duobinary pulse [9] which is specified in the time domain as

$$x\left(\frac{n}{2W}\right) = \begin{cases} 1, & n = 0,1 \\ 0, & \text{otherwise} \end{cases} \tag{25}$$

and in the frequency domain as

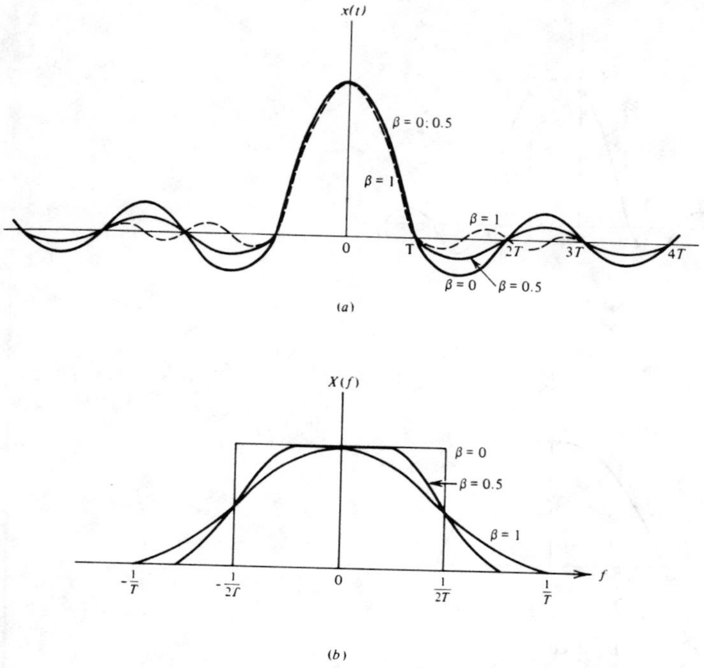

FIGURE 8. Pulses with a raised cosine spectrum.

$$X(f) = \begin{cases} \frac{1}{2W}\,(1 + e^{-j\pi f/W}) = \frac{1}{W}\, e^{-j\pi f/2W} \cos\frac{\pi f}{2W}, & |f| \le W \\ 0, & |f| > W \end{cases} \tag{26}$$

The pulse x(t) and the magnitude of the frequency characteristic X(f) are shown in Figure 9. Note that the frequency characteristic of the duobinary pulse does not possess sharp discontinuities and, as a result, it can be more easily approximated by a physically realizable filter.

A second special case, called a modified duobinary signal, is specified in the time domain by

$$x\left(\frac{n}{2W}\right) = \begin{cases} 1, & n = -1 \\ -1, & n = 1 \\ 0, & \text{otherwise} \end{cases} \tag{27}$$

and in the frequency domain as

$$X(f) = \begin{cases} \frac{1}{2W}\,(e^{j\pi f/W} - e^{-j\pi f/W}) = \frac{j}{W} \sin\frac{\pi f}{W}, & |f| \le W \\ 0, & |f| > W \end{cases} \tag{28}$$

The pulse x(t) and the magnitude of the frequency characteristic X(f) for the modified duobinary signal are shown in Figure 10. Again, note the smoothness of

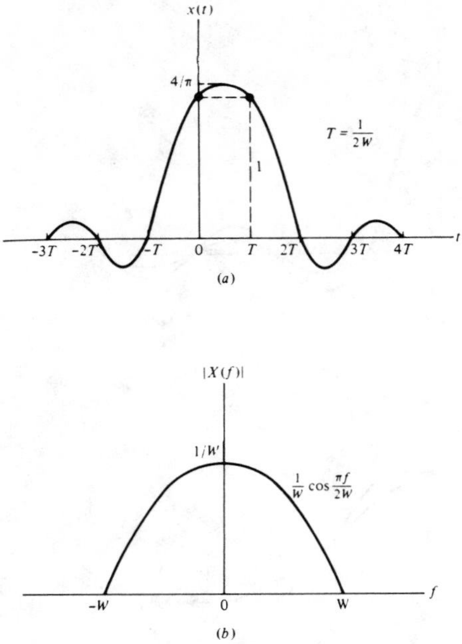

FIGURE 9. Duobinary signal.

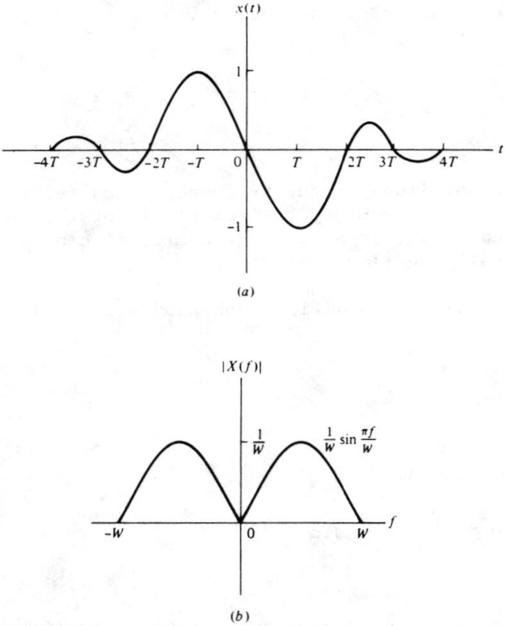

FIGURE 10. Modified duobinary signal pulse.

the function X(f). Also note that X(0) = 0, a characteristic that makes this signal especially suitable for SSB transmission on a carrier due to the ease with which the two sidebands can be separated by filtering.

One can generate other interesting pulse shapes at will [8]. However, as the number of nonzero weighting coefficients is increased, the problem of unravelling the "controlled" intersymbol interference becomes more cumbersome and impractical. Moreover, the recovery of the data sequence becomes more complex as the number of nonzero coefficients x(n/2W) is increased beyond two.

2.3 Application of Digital Signal Processing Techniques in Modem Implementation

The signal processing operations performed in a modulator and demodulator (modem) usually involve the generation of the signal waveforms, frequency translation, filtering and equalization. At the modulator, the digital signalling waveforms are usually synthesized at baseband and translated in frequency to a carrier frequency that is appropriate for the channel. In the case of voice-band telephone channels, an appropriate carrier frequency is approximately 1800 Hz. Appropriate modulation techniques for telephone channels are PSK, PAM (ASK) and QAM. In other words, the signal-to-noise ratio (SNR) is sufficiently high to support information of transmission in the band-limited region R/W > 1 of Figure 3.

The demodulation of the received signal is usually accomplished by first translating the signal to baseband (or, perhaps, some intermediate frequency) followed by filtering and adaptive equalization when needed. In order to perform the frequency translation coherently, the carrier component must be regenerated at the demodulator. This is usually accomplished in one of several ways employing phase-locked loops as described in [11],[12]. In addition to carrier recovery, there is also the problem of timing recovery or synchronization which is needed to keep the demodulator in synchronism with the modulator clock.

A number of techniques have been devised for deriving a timing signal from the received signal. For example, a block diagram of a time synchronizer, appropriate for QAM, is illustrated in Figure 11. The in-phase and quadrature components resulting from the multiplications are each filtered by a narrow bandpass filter tuned to frequency 1/2T, where 1/T is the pulse (symbol) rate. The two filtered outputs are squared (rectified), summed, and then filtered by a narrowband filter tuned to the clock frequency 1/T. The resulting sinusoidal waveform is the

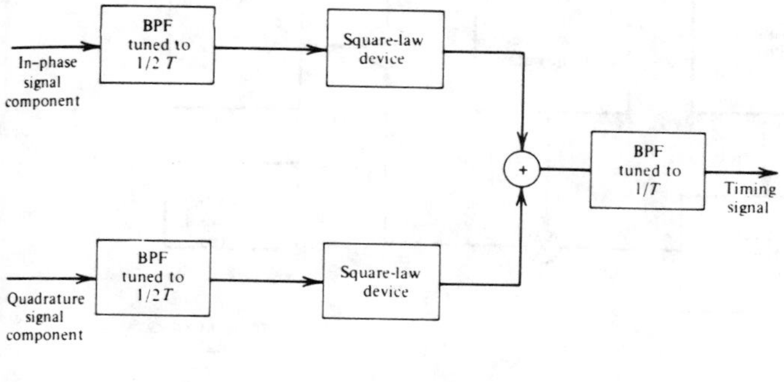

FIGURE 11. Block diagram of a synchronizer for QAM.

appropriate clock signal for sampling the outputs of the demodulation filters to recover the information.

Traditionally, the signal processing functions in a modem were performed in analog form, employing analog filters for rejecting out-of-band noise and signal harmonics, and multipliers or mixers for frequency translation. In recent years, the trend has shifted to the point where, today, high-speed voice band modems are implemented digitally either on a programmable digital signal processor or by means of hard-wired digital circuits utilizing LSI and VLSI technology. Either approach is more economical than a comparable analog implementation and provides high precision in the signal processing functions. A programmable digital signal processor provides an additional degree of flexibility in the implementation that is unparalleled by any other approach.

Figure 12 illustrates a block diagram of a digital modulator for synthesizing digitally the signalling waveforms for PSK ($M \geq 4$) or QAM. The desired pulse shaping of the in-phase and quadrature signal components is performed by two identical FIR filters. For telephone channel transmission, the impulse response of the FIR filters will be a sampled version of the pulse having either a raised cosine spectral characteristic or, perhaps, one of the partial response signals such as the duobinary pulse or the modified duobinary pulse. In most implementations of these filters, it is advantageous from a noise performance viewpoint to split the spectral characteristic of the desired pulse equally between the modulator and the demodulator. Hence, if $X(f)$ is the voltage spectrum of the desired pulse, the FIR filter at the modulator is designed to have the spectral shape $\sqrt{X(f)}$ and the corresponding FIR filter at the demodulator is designed to have the spectral shape $\sqrt{X(f)}$. Thus, the cascade of the two FIR filters has the desired spectral shape $X(f)$.

A sampling rate of $4/T$ (where $1/T$ is the symbol rate) is usually adequate for practical applications. Moreover, the pulse must be truncated to accommodate an FIR implementation. We observe that truncating the impulse response results in a modified (distorted) spectral characteristic from the desired spectrum $X(f)$.

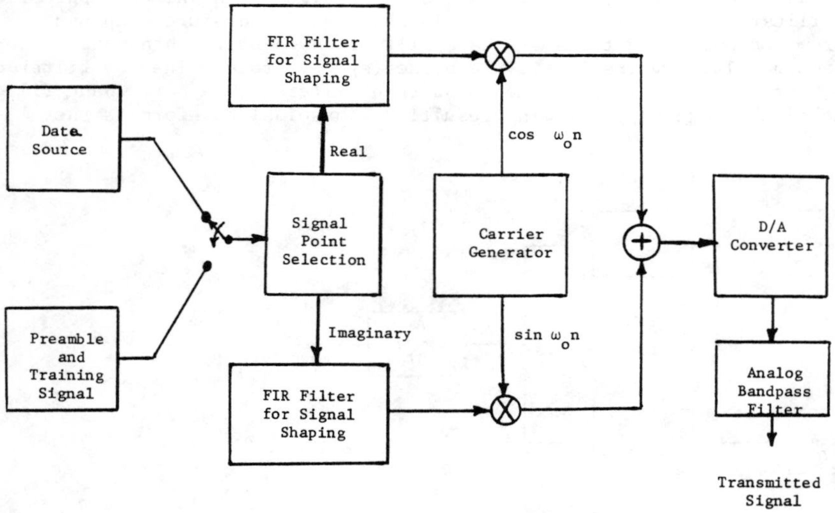

FIGURE 12. Digital implementation of modulation waveforms.

Usually, the frequency distortion of the desired spectrum is relatively small if we truncate the impulse response at $t \geq 10T$. In such a case the FIR filter has a minimum of 40 taps. Note, however, that the filter must have linear phase and, hence, the number of distinct tap weights is reduced by a factor of two.

Following the signal pulse shaping, the outputs of the FIR filters are translated in frequency by multiplication with the sine and cosine of a sampled version of the carrier frequency. These operations are easily performed digitally using a table of sine and cosine functions stored in a read only memory (ROM). After the in-phase and quadrature components of the signal are added digitally, the resultant signal samples are fed to a digital-to-analog (D/A) converter and further filtered at bandpass prior to transmission over the channel. Thus, the signal at the modulator is synthesized entirely in digital form and, then, it is converted to analog form for transmission over the analog channel.

At the demodulator, shown in Figure 13, the received analog signal is passed first through an analog bandpass filter which serves to limit the out-of-band noise. Following the analog filter is an automatic gain control (AGC) loop and, then, an analog-to-digital (A/D) converter. The AGC loop is slow-acting and serves to keep the signal level within the range of the A/D converter. The signal processing following the A/D converter is digital and, hence, it can be implemented in either a hard-wired fixed processor or in a programmable digital processor. The latter is preferable in most practical applications because of the flexibility afforded by a programmable machine when changes in the signal processing are desired.

The sampled bandpass signal is converted to baseband through multiplication by the sine and cosine of the carrier frequency, followed by lowpass filtering. As indicated previously, the pair of FIR lowpass filters are identical to the FIR filters used in the signal generation at the modulator.

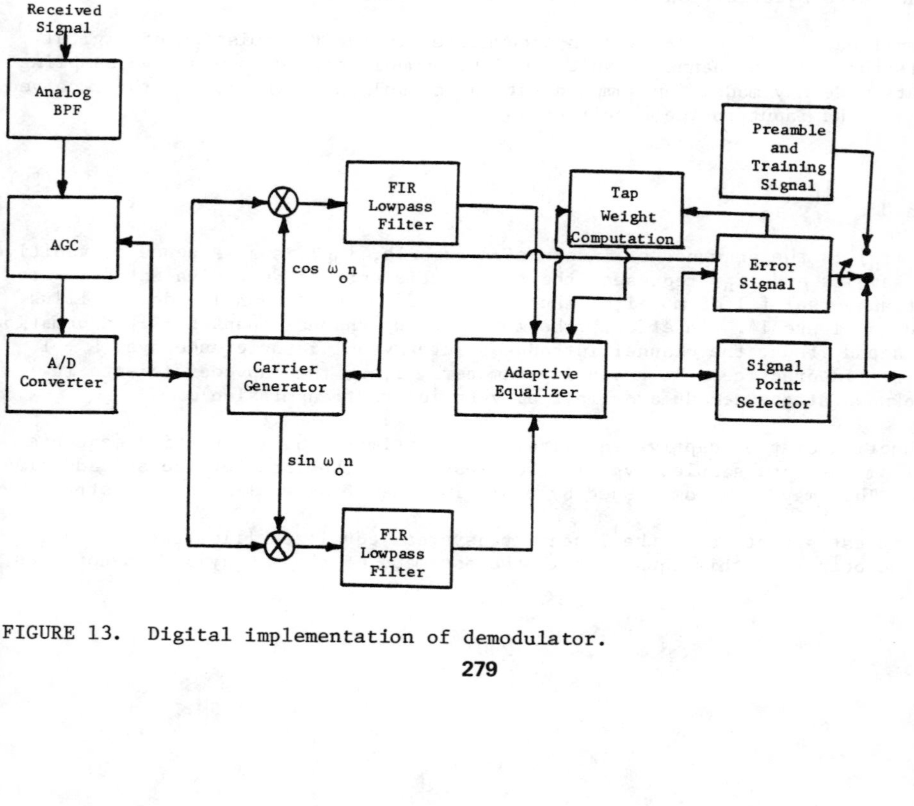

FIGURE 13. Digital implementation of demodulator.

The FIR filters are operating at some multiple of the pulse rate, say, 4/T for instance. Their output is down sampled to a rate 1/T for further processing by the adaptive equalizer. In some cases, adaptive equalizers are designed to operate on samples taken at a faster rate, such as 2/T or 3/T, for the purpose of easing synchronization requirements [13],[14]. Such equalizers are called fractionally-spaced equalizers.

For our purposes, we will assume that time synchronization has been achieved, so that the equalizer operates on samples taken at the symbol rate 1/T. Functionally, the equalizer operates digitally on a group of complex-valued (in-phase and quadrature) samples at a time and outputs an estimate of the transmitted symbol once every T seconds. The estimate is compared with the M possible transmitted symbols in the symbol selector and a decision is made in favor of the symbol that is closest in distance to the estimate.

Initially, the equalizer is trained by transmitting a known sequence of symbols over the channel. This sequence is called a preamble or a training sequence. The difference between the known sequence and the estimate at the output of the equalizer is an error signal that is used to adjust the equalizer parameters so as to minimize the mean square value of the error. Once the equalizer is trained, its decisions are sufficiently reliable so that the error signal is generated by taking the difference between the decision at the output of the symbol selector and the estimate of the equalizer. In addition to using the complex-valued error signal in the adjustment of the equalizer parameters, the error signal may also be used to track (digitally) the phase of the carrier of the incoming signal as indicated in Figure 13.

In the following section we present some practical adaptive equalizer structures and describe algorithms for adjusting the equalizer parameters adaptively.

2.4 Adaptive Equalization of Intersymbol Interference

The amplitude and delay distortion encountered in the transmission of digital signals through the channel results in intersymbol interference at the sampling instants. We may model the sampled (at the symbol rate) version of the received signal at the input to the equalizer as

$$y_k = \sum_{n=0}^{L} f_n I_{k-n} + \eta_k \qquad (29)$$

where $\{I_n\}$ is the sequence of information symbols, $\{\eta_k\}$ is a sequence of additive noise samples and $\{f_n\}$ represent the tap coefficients of an equivalent discrete-time transversal filter model of the channel [5]. This channel model is illustrated in Figure 14. In effect, the model of the channel spans L + 1 information symbols and, thus, the channel introduces intersymbol interference over L + 1 symbols. In practice, the channel parameters, i.e., the tap coefficients $\{f_n\}$, are unknown at the demodulator initially (prior to transmission).

The function of the adaptive equalizer is to estimate the information sequence $\{I_n\}$ from the data samples $\{y_k\}$ in the presence of the interference and additive noise. This may be accomplished by employing one of several equalizer structures.

The simplest structure is the linear transversal equalizer illustrated in Figure 15. The output of this equalizer is the estimate of the kth symbol, denoted as

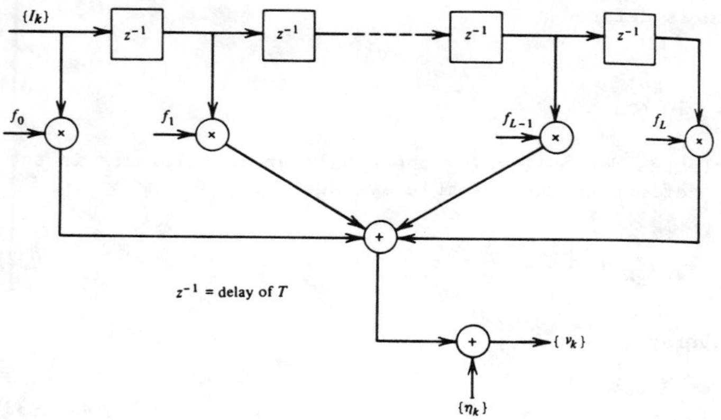

FIGURE 14. Discrete-time model of channel with intersymbol interference.

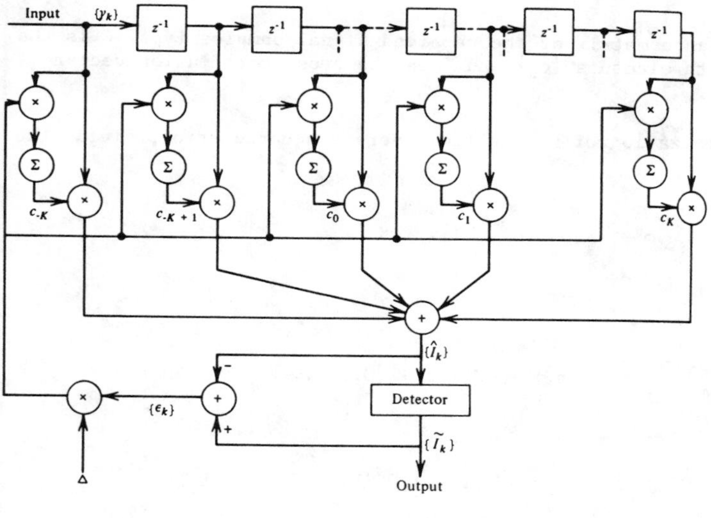

FIGURE 15. A linear adaptive equalizer based on the mean square error criterion.

$$\hat{I}_k = \sum_{j=-K}^{K} c_j \, y_{k-j} \quad , \tag{30}$$

where $\{c_j\}$ are the (2K + 1) tap weight coefficients of the equalizer. The symbol selector compares the estimate with the possible set of transmitted symbols and selects the symbol I_k that is closest in distance to $\hat{I}_k$. If the decision is correct, the selected symbol $\tilde{I}_k = I_k$, where I_k denotes the actual transmitted symbol.

The error signal sequence is defined as

$$\varepsilon_k = I_k - \hat{I}_k = I_k - \sum_{j=-K}^{K} c_j \, y_{k-j} \tag{31}$$

The criterion usually selected for optimizing the equalizer coefficients is the mean square error (MSE), defined as the ensemble average

$$J = E|\varepsilon_k|^2 = E|I_k - \sum_{j=-K}^{K} c_j \, y_{k-j}|^2 \tag{32}$$

or the equivalent time average,

$$\mathcal{E} = \frac{1}{N}\sum_{k=1}^{N} |\varepsilon_k|^2 = \frac{1}{N}\sum_{k=1}^{N} |I_k - \sum_{j=-K}^{K} c_j \, y_{k-j}|^2 \tag{33}$$

The minimization of either J or $\mathcal{E}$ with respect to the coefficients $\{c_j\}$ leads to a set of linear equations of the form

$$\underline{\Gamma}\ \underline{C} = \underline{\xi} \tag{34}$$

where $\underline{\Gamma}$ is the covariance matrix of the received signal samples $\{y_k\}$, $\underline{C}$ is the coefficient vector with elements $\{c_k\}$ and $\underline{\xi}$ is the cross correlation vector between I_k and $\{y_{k-n}\}$.

For example, the minimization of $\mathcal{E}$, the time average squared error, yields the equations

$$\sum_{j=-K}^{K} c_j \, \Gamma_{mj} = \underline{\xi}_m \,, \qquad m = -K,\ldots k-1,0,1,\ldots,K \tag{35}$$

where*

$$\Gamma_{mj} = \frac{1}{N}\sum_{k=1}^{N} y_{k-j} \, y_{k-m}^{*} \quad , \qquad m,j = -K,\ldots,-1,0,1,\ldots,K$$

$$\xi_m = \frac{1}{N}\sum_{k=1}^{N} I_k \, y_{k-m}^{*} \quad , \qquad m = -K,\ldots k-1,0,1,\ldots,K$$

The solution for the optimum coefficients is

$$\underline{C}_{opt} = \underline{\Gamma}^{-1}\ \underline{\xi} \tag{36}$$

The matrix inversion indicated in (36) is a computationally inefficient method for determining the optimum equalizer coefficients $\underline{C}_{opt}$. Instead, an iterative

---

*An asterisk denotes complex conjugation.

algorithm based on the method of steepest descent is commonly used [5]. This algorithm is

$$\underline{\hat{C}}_{m+1} = \underline{\hat{C}}_m - \Delta\underline{\hat{G}}_m = \underline{\hat{C}}_m + \Delta\varepsilon_m \underline{Y}_m^* \tag{37}$$

where $\underline{\hat{C}}_m$ denotes the estimate of the equalizer coefficients at the mth iteration, $\underline{\hat{G}}_m = -\varepsilon_m \underline{Y}_m^*$ is an estimate of the gradient vector, Δ is a step size parameter which is selected to ensure convergence of the algorithm, and $\underline{Y}_m$ is the vector of received signal samples that make up the estimate $\hat{I}_m$, i.e., $\underline{Y}_m = (y_{m+K},\ldots,y_m,\ldots,y_{m-K})$. The algorithm in (37) is the basic least mean square (LMS) error algorithm for recursively adjusting the tap weight coefficients of the equalizer [15]. It is illustrated in the linear equalizer shown in Figure 15.

The basic algorithm given by (37) and some of the possible variations of it have been incorporated into many commercial adaptive equalizers. The variations of the basic algorithm are obtained by using only sign information contained in the error signal ε_k and/or in the components of $\underline{Y}_k$. Hence three possible variations are

$$\begin{aligned} c_{(k+1)j} &= c_{kj} + \Delta \operatorname{sgn}(\varepsilon_k)\, y_{k-j}^* & j &= -K,\ldots,-1,0,1,\ldots,K \\ c_{(k+1)j} &= c_{kj} + \Delta\varepsilon_k \operatorname{sgn}(y_{k-j}^*) & j &= -K,\ldots,-1,0,1,\ldots,K \\ c_{(k+1)j} &= c_{kj} + \Delta \operatorname{sgn}(\varepsilon_k)\operatorname{sgn}(y_{k-j}^*) & j &= -K,\ldots,-1,0,1,\ldots,K \end{aligned} \tag{38}$$

where sgn(x) is defined as

$$\operatorname{sgn}(x) = \begin{cases} 1 + j & \operatorname{Re}(x) > 0,\ \operatorname{Im}(x) > 0 \\ 1 - j & \operatorname{Re}(x) > 0,\ \operatorname{Im}(x) < 0 \\ -1 + j & \operatorname{Re}(x) < 0,\ \operatorname{Im}(x) > 0 \\ -1 - j & \operatorname{Re}(x) < 0,\ \operatorname{Im}(x) < 0 \end{cases} \tag{39}$$

Clearly the third algorithm in (38) is the most easily implemented, but it gives the smallest rate of convergence relative to the others.

It is well-known [5] that the performance of the linear equalizer is poor when the channel frequency response possesses one or more deep spectral nulls. Wire line channels are not plagued by this problem and, hence, the linear equalizer is adequate. However, in many radio channels multipath propagation effects often result in frequency selective fading with deep spectral nulls. In such a case, a more powerful equalizer structure is required.

The decision feedback equalizer, illustrated in Figure 16 is one such structure [16],[17]. As shown, it consists of two delay lines, a feedforward section and a feedback section. The input to the first section is the sequence of received signal samples $\{y_k\}$. The input to the second section is the sequence of previously detected symbols. In this configuration, the intersymbol interference caused by previously detected symbols is eliminated from the estimate of the received symbol at any given instant. Thus, the estimate of the symbol at the output of the equalizer is

$$\hat{I}_k = \sum_{j=-K_1}^{0} c_j\, y_{k-j} + \sum_{j=1}^{K_2} c_j\, \tilde{I}_{k-j} \tag{40}$$

where $\hat{I}_k$ is an estimate of the kth information symbol, $\{\tilde{I}_{k-1},\ldots,\tilde{I}_{k-K2}\}$ are the

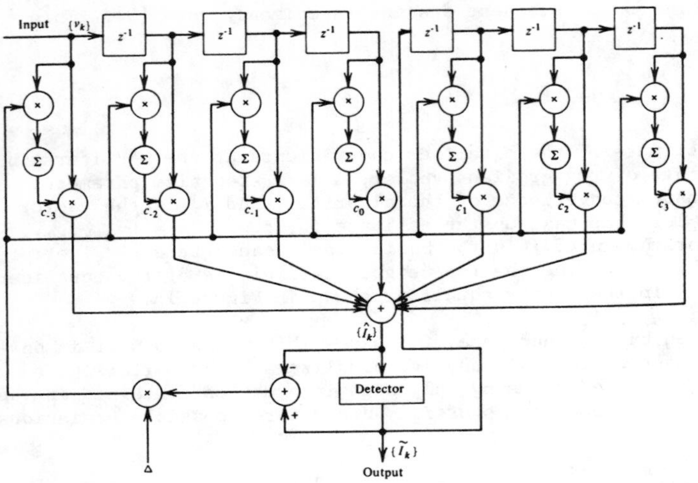

FIGURE 16. An adaptive decision-feedback equalizer.

previously detected symbols, c_j, $j = -K_1,\ldots,0$, are the tap coefficients of the feedforward section and c_j, $j = 1,\ldots,K_2$, are the tap coefficients of the feedback section of the equalizer.

Minimization of the mean square value of the error $\varepsilon_k = I_k - \hat{I}_k$ again leads to the set of linear equations for the $\{c_j\}$ as shown in [5]. Proceeding as in the case of the linear equalizer, we use the steepest-descent (gradient) algorithm for adjusting the coefficients recursively as

$$\underline{\hat{C}}_{m+1} = \underline{\hat{C}}_m + \Delta\varepsilon_k \underline{V}_m^* \tag{41}$$

where $\underline{V}_m = \{y_{m+K_1},\ldots,y_m, \tilde{I}_{m-1},\ldots,\tilde{I}_{m-K_2}\}$.

In general, the decision-feedback equalizer outperforms the linear equalizer, because the former, in effect, eliminates the intersymbol interference caused by the tail of the channel pulse response. In other words, once a decision is made on an information symbol, the intersymbol interference caused by that symbol on subsequently received symbols is completely eliminated.

Although the decision-feedback equalizer performs well in the presence of deep spectral nulls in the channel frequency response, it is not the optimum equalizer from a statistical detection viewpoint. The optimum equalization algorithm is one that performs maximum likelihood sequence estimation (MLSE) of the received symbols. The MLSE criterion is efficiently implemented by employing the Viterbi algorithm, which is described in some detail in [5],[18],[19]. The MLSE criterion has a complexity which grows exponentially with the time dispersion introduced by the channel, as compared to the linear growth for the DFE. Consequently, MLSE is not widely used in channel equalization applications, whereas the DFE is used frequently in practice.

The above treatment on adaptive equalization algorithms represents developments in equalization through about 1975. During the past five to seven years, a number of new developments in adaptive filtering and equalization have occurred that have generated considerable interest. Specifically, we mention the work on recursive least squares Kalman and lattice algorithms, which are particularly attractive for applications in time-variant multipath channels. These algorithms and filter structures are particularly attractive in equalizing time-variant multipath channels and are treated in the following section.

3. DIGITAL COMMUNICATIONS OVER FADING MULTIPATH CHANNELS

3.1 Channel Characteristics

In communication over some radio channels, such as high frequency (HF) ionospheric channels and tropospheric scatter channels, and undersea acoustic channels, the transmitted signal propagates to the receiver via several paths. This type of time-dispersive propagation is called multipath. The path delays vary randomly with time, thus, resulting in a received signal having an amplitude and phase that vary with time. The amplitude variations in the received signal are called fading, and it is a direct consequence of the time-variant characteristics of the multipath propagation [5].

A time-variant multipath channel is modeled as a linear, time-variant filter having an equivalent lowpass impulse response $c(\tau;t)$ or, equivalently, a time-variant transfer function $C(f;t)$, which is the Fourier transform of $c(\tau;t)$, i.e.,

$$C(f;t) = \int_{-\infty}^{\infty} c(\tau;t)\, e^{-j2\pi f\tau}\, d\tau \tag{42}$$

Both the impulse response $c(\tau;t)$ and the transfer function $C(f;t)$ are treated as random processes in the t-variable and characterized by the corresponding autocorrelation functions, which are defined as

$$\phi_c(\tau_1,\tau_2;\Delta t) = \frac{1}{2} E[c^*(\tau_1;t)\, c(\tau_2;t + \Delta t)] \tag{43}$$

$$\phi_C(f_1,f_2;\Delta t) = \frac{1}{2} E[C^*(f_1;t)\, C(f_2;t + \Delta t)] \tag{44}$$

In many cases, the attenuation and phase shift of the channel associated with a path delay τ_1 is uncorrelated with the attenuation and phase shift associated with a path delay τ_2. This is usually called uncorrelated scattering and it implies that

$$\phi_c(\tau_1,\tau_2;\Delta t) = \phi_c(\tau_1;\Delta t)\, \delta(\tau_1 - \tau_2) \tag{45}$$

$$\phi_C(f_1,f_2;\Delta t) = \phi_C(\Delta f;\Delta t) \tag{46}$$

where $\Delta f = f_2 - f_1$. With $\Delta t = 0$, the autocorrelation function $\phi_c(\tau) \equiv \phi_c(\tau;0)$ is called the multipath intensity profile and provides a measure of the time dispersion or multipath spread of the channel, which is denoted as T_m. Also $\phi_c(\tau)$ is related to $\phi_C(\Delta f) \equiv \phi_C(\Delta f;0)$ by the Fourier transform relation

$$\phi_C(\Delta f) = \int_{-\infty}^{\infty} \phi_c(\tau)\, e^{-j2\pi\Delta f\tau}\, d\tau \tag{47}$$

The autocorrelation $\phi_C(\Delta f)$ provides a measure of the frequency coherence of the channel. As a consequence of the Fourier transform relationship between $\phi_c(\Delta f)$ and $\phi_c(\tau)$, the coherence bandwidth of the channel, denoted as B_c, is approximately equal to the reciprocal of the multipath spread, i.e.,

$$B_c \simeq \frac{1}{T_m} \tag{48}$$

When the bandwidth W of the transmitted signal is small compared to the coherence bandwidth, the channel is said to be frequency nonselective, or flat fading. On the other hand, if $W >> B_c$, the channel is said to be frequency selective and the multipath is resolvable with a time resolution of 1/W.

A measure of the rapidity of the fading is provided by the autocorrelation function $\phi_c(\Delta t) \equiv \phi_c(0;\Delta t)$ and its Fourier transform

$$S_c(\lambda) = \int_{-\infty}^{\infty} \phi_c(\Delta t)\, e^{-j2\pi\lambda\Delta t}\, d\Delta t \tag{49}$$

The function $S_c(\lambda)$ is a power spectrum that gives the signal intensity as a function of the Doppler frequency λ. We call $S_c(\lambda)$ the Doppler power spectrum of the channel. The range of values over which $S_c(\lambda)$ is essentially nonzero is called the Doppler spread B_d of the channel. Since $S_c(\lambda)$ and $\phi_c(\Delta t)$ are a Fourier transform pair, the reciprocal of B_d is a measure of the coherence time of the channel, i.e.,

$$T_c \simeq \frac{1}{B_d} \tag{50}$$

Finally, the double Fourier transform of the correlation function $\phi_d(\Delta f;\Delta t)$ results in another two variable function, called the scattering function of the channel, i.e.,

$$S(\tau;\lambda) = \int_{-\infty}^{\infty}\int_{-\infty}^{\infty} \phi_c(\Delta f;\Delta t)\, e^{-j2\pi\lambda\Delta t}\, e^{j2\pi\tau\Delta f}\, d\Delta t\, d\Delta f \tag{51}$$

The scattering function provides us with a measure of the average power output of the channel as a function of the time delay τ and the Doppler frequency λ.

As an example, the scattering function measured on a 240 km tropospheric scatter channel is shown in Figure 17. The signal used to probe the channel had a time resolution of 0.1 μs. Hence, the time delay axis is quantized in increments of 0.1 μs. From the graph we observe that the multipath spread T_m = 0.7 μs. On the other hand, the Doppler spread, which may be defined as the 3-dB bandwidth of the power spectrum for each signal path, appears to be different for each signal path. In particular, in one signal path it is less than 1 Hz while in some other paths the Doppler spread is several hertz. We may select the largest of these 3-dB bandwidths associated with the different multipath components and call that the Doppler spread of the channel.

In the following two subsections we present two examples where digital signal processing techniques are employed in the demodulation and/or decoding of digital communication signals transmitted over the channel. In both cases we assume that there is available a channel with bandwidth W, which exceeds the coherent bandwidth B_c of the channel. We consider two possible approaches in the design of signals for reliable digital communications. The first is appropriate for the case $R/W << 1$ and the second is appropriate when $R/W \geq 1$.

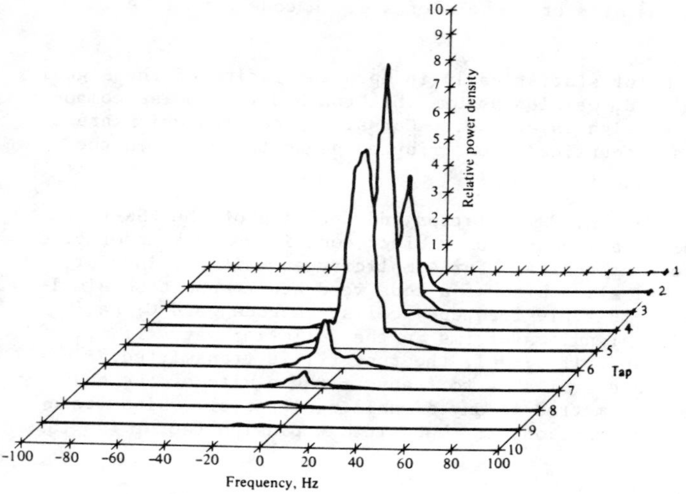

FIGURE 17. Scattering function of a medium-range tropospheric scatter channel.

3.2 Modem Design and Implementation for Coded Communications

One approach to modem design for the case in which $R/W \ll 1$ is to subdivide the channel bandwidth into a number of subchannels, each having a bandwidth that is much smaller than the coherence bandwidth of the channel. In such a case, each subchannel is frequency nonselective. To combat the detrimental effects of fading due to multipath, we may use frequency diversity in conjunction with M-ary FSK. Toward this end, the available time-bandwidth space is subdivided into a number of cells. The time duration T associated with each cell is selected to be much greater than the multipath spread T_m, i.e., $T \gg T_m$. Consequently, the intersymbol interference due to the multipath spread becomes negligible. Cross-talk between adjacent frequency cells due to Doppler spread in the channel is rendered negligible by providing sufficient separation in frequency (frequency guard band). The signal that occupies a cell consists of a sinusoidal pulse of duration T and frequency equal to the center frequency of the cell.

With this subdivision of the time-bandwidth space, coding provides an efficient means for obtaining diversity in transmission of the information over the channel [5],[20]. Block codes or convolutional codes or concatenated codes (one code superimposed on another code) may be used.

To elaborate, suppose that a binary code of rate $R_c = k/n$ (either a block code or a convolutional code) is employed. The code bits are transmitted by binary FSK. To obtain the maximum benefit from coding, we require that the FSK tone corresponding to a particular code bit fades independently from the FSK tone corresponding to any other code bit. This is possible if the tones or chips corresponding to the coded bits are separated either in time by an amount exceeding the coherence time T_c or in frequency by an amount exceeding the coherence bandwidth B_c of the channel. This requirement may result in inefficient utilization of the available time-frequency space with the existence of a large number of unused cells. To reduce the inefficiency, a number of code words may be interleaved either in time or in frequency or both, in such a manner that the waveforms corresponding to the bits or symbols of a given code word fade independently.

In addition to the assumption of statistically independent fading of the signal components of a given code word, we also assume that the additive noise components corrupting the received signals are white Gaussian processes which are statistically independent and identically distributed among the cells in the time-frequency space.

In order to explain the modulation, demodulation and decoding of the FSK-type (orthogonal) signals, consider a linear binary block code in which k information bits are encoded onto a block of n bits. For simplicity and without loss of generality, let us assume that all n bits of a code word are transmitted simultaneously over the channel on multiple frequency cells. A code word $\underline{C_i}$ having bits $\{c_{ij}\}$ is mapped into FSK signal waveforms in the following way. If $c_{ij} = 0$, the tone f_{0j} is transmitted, and, if $c_{ij} = 1$, the tone f_{1j} is transmitted. This means that 2n tones or cells are available to transmit the n bits of the code word, but only n tones are transmitted in any signalling interval. Since each code word conveys k bits of information, the bandwidth expansion factor for this approach is 2n/k.

The demodulator for the received signal must separate the signal into 2n spectral components corresponding to the available tone frequencies at the transmitter. This spectral decomposition is accomplished by sampling the signal and computing the discrete Fourier transform (DFT) by means of the Fast Fourier transform (FFT) algorithm. For codes having a large block length, the FFT algorithm provides an efficient means for performing the demodulation. The FFT outputs corresponding to the 2n possible transmitted frequencies are squared and appropriately combined for each code word to form the $M = 2^k$ decision variables corresponding to the $M = 2^k$ possible code words. The code word corresponding to the maximum of the decision variables is selected. Alternatively, if hard-decision decoding is employed on each pair of frequencies corresponding to a bit in a code word, the optimum maximum likelihood decoder selects the code word having the smallest Hamming distance relative to the received code word. Thus, the demodulator and the decoder can be implemented totally in digital form.

Although the discussion above assumed the use of a block code, a convolutional encoder can be easily accommodated.

3.3 Modem Design and Implementation for High Speed Communications

In this discussion we assume that the desired transmission rate R is comparable to or exceeds the channel bandwidth W, i.e., $R/W \geq 1$. In this case we use the entire signal bandwidth for serial transmission at a pulse rate of approximately 1/W pulses per second. Since the channel is assumed to be frequency selective, it may be characterized by the tapped delay line model shown in Figure 18 where the tap spacing between successive taps is 1/W and the total time spanned by the delay line is the multipath spread T_m. The time-variant tap weights $c_i(t)$, $1 \leq i \leq L$, represent the fading characteristics of the channel associated with the ith delay.

Since the transmission rate R exceeds $1/T_m$, it follows that the symbol duration $T << T_m$ and, hence, the received symbols are corrupted by intersymbol interference. Consequently, some form of adaptive equalization is needed to achieve reliable communication. Since the time-variant multipath often results in one or more spectral nulls in the frequency response characteristics of the channel, a linear equalizer yields poor performance. On the other hand, nonlinear equalization techniques such as decision-feedback equalization and maximum-

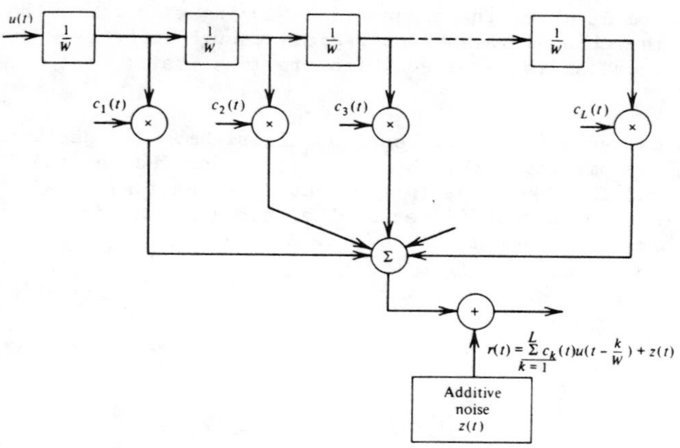

FIGURE 18. Tapped delay line model of frequency-selective channel.

likelihood sequence estimation result in relatively good performance. From an implementation viewpoint, a decision-feedback equalizer (DFE) is probably the simplest. The structure of the DFE may be in the form of a fixed size filter with feedforward and feedback taps as previously shown in Figure 16, or it may be in the form of a lattice filter with a variable number of stages as described below.

Algorithms for adjusting the equalizer coefficients to track the time variations in the channel may range from the simple LMS algorithm which converges relatively slowly, to the recursive least squares Kalman algorithm or recursive least squares lattice algorithms which converge relatively fast as will be illustrated. The LMS algorithm has proven to be adequate in tracking the time variations in tropospheric scatter channels where the time variations of the channel are extremely slow relative to the data rate or the symbol rate. For example, the data rate on a tropospheric scatter channel may be 10^7 bits per second and the fading rate or Doppler spread may be of the order of 10 Hz. Consequently, the bit rate to fade rate ratio is of the order of 10^6 bits, which implies that there is a considerable amount of time to track the channel.

In contrast, the LMS algorithm is unable to closely track the time variations in an HF channel where the data rate may be of the order of 5000 bits per second and the fade rate is one Hertz. Consequently, we must use a more sophisticated tracking algorithm for equalizing the channel adaptively. The following section describes the use of the Kalman algorithm [21] and its implementation in an HF modem.

3.4 An Adaptively Equalized High Speed HF Modem

A beyond-line-of-sight (BLOS) HF channel is a time-variant multipath channel with a multipath spread of up to 5 milliseconds and a fade rate of approximately 1 Hz. For a nominal 3 kHz bandwidth allocation, it is possible to transmit at a symbol rate of about 2400 samples per second by using shaped pulses such as a pulse having the raised cosine spectrum described in Section 2.2. The signal-to-noise ratio in the received signal is sufficiently high to support octal PSK transmission and possibly $M = 16$ phase PSK.

Due to the relatively fast rate at which the channel is changing with time, the LMS algorithm cannot follow the channel variations rapidly enough to be used in an equalizer. Instead, an algorithm is required which provides faster convergence.

The recursive least squares Kalman algorithm is such an algorithm. Its derivation is somewhat lengthy and will be omitted here. However, it is derived in detail in [5], pp. 412-417. In brief, the equations in the recursive least squares Kalman algorithm for adjusting the tap coefficients of either a linear or a decision-feedback equalizer are as follows:

Compute Equalizer Output:

$$\hat{I}_m = \underline{V}'_m \underline{C}_{m-1} \tag{52}$$

Compute the Error Signal:

$$\varepsilon_m = I_m - \hat{I}_m \tag{53}$$

Compute the Kalman Gain Vector:

$$\underline{K}_m = \frac{\underline{P}_{m-1} \underline{V}_m}{w + \underline{V}'_m \underline{P}_{m-1} \underline{V}_m} \tag{54}$$

Update the Error Covariance Matrix:

$$\underline{P}_m = \frac{1}{w} [\underline{P}_{m-1} - \underline{K}_m \underline{V}_m \underline{P}_{m-1}] \tag{55}$$

Update the Equalizer Coefficient Vector:

$$\underline{C}_m = \underline{C}_{m-1} + \underline{K}_m \varepsilon_m \tag{56}$$

In the above equations, $\hat{I}_m$ is the estimate of the desired symbol I_m, $\underline{P}_m$ is the error covariance matrix, $\underline{V}_m$ is the data vector with elements equal to the contents of the equalizer at the mth iteration, and w is a constant selected to be in the range (0,1), to provide exponential weighting into the past. In the case of the DFE, $\underline{V}_m$ consists of samples of the received signal and prior symbol decisions.

In comparison, the LMS algorithm for updating the equalizer coefficient vector is

$$\underline{C}_m = \underline{C}_{m-1} + \Delta \underline{V}_m \varepsilon_m \tag{57}$$

Figure 19 illustrates the initial convergence rate of these two algorithms for a channel with fixed parameters, $f_0 = 0.26$, $f_1 = 0.83$, $f_2 = 0.26$, and a linear equalizer which has 11 taps. All the equalizer coefficients were initialized to zero. The LMS algorithm was implemented with $\Delta = 0.02$. A larger value of Δ will increase the convergence rate but will yield more output noise [5]. The superiority of the Kalman algorithm is clearly evident.

In spite of its superior tracking performance, the Kalman algorithm described above has two disadvantages. One is its complexity. The second is its sensitivity to round off noise that accumulates due to the recursive computations.

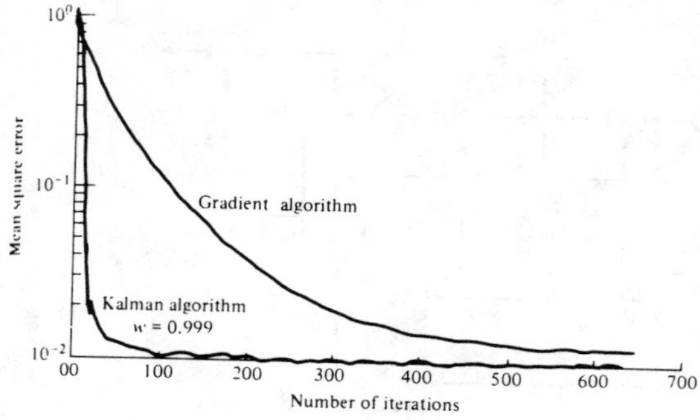

FIGURE 19. Comparison of convergence rate for the Kalman and gradient algorithms.

The latter may cause instabilities in the algorithm.

The number of computations or operations (multiplications and divisions are the most time-consuming) in each iteration of the Kalman algorithm is proportional to N^2. Most of these operations are involved in updating $\underline{P}_m$. This part of the computation is also susceptible to round off noise. To remedy this problem, algorithms have been developed which avoid the computation of $\underline{P}_m$ directly [22]. The basis of these algorithms lies in the decomposition of $\underline{P}_m$ in the form

$$\underline{P}_m = S_m \underline{\Lambda}_m \underline{S}'_m \tag{58}$$

where $\underline{S}_m$ is a lower triangular matrix and $\underline{\Lambda}_m$ is a diagonal matrix. Such a decomposition is called a Cholesky factorization or a square root factorization. In a square-root algorithm, $\underline{P}_m$ is not computed. Instead, the time updating is performed on $\underline{\Lambda}_m$ and $\underline{S}_m$ as described in [22],[23].

The square-root Kalman algorithm was incorporated recently in a modem developed by GTE to transmit data at high speed on a 3 kHz HF channel. The modulation technique used was eight-phase PSK with a symbol rate of 2400 symbols per second, i.e., a bit rate of 7200 bits per second. The modem was implemented digitally on a programmable processor. The Kalman updating was performed in a digital signal processor especially designed to match the requirements of the computation. Sixteen bit arithmetic was used in the implementation of the square-root Kalman equalizer algorithm. A transversal DFE was implemented on the programmable processor. The equalizer has 23 feedforward taps and 12 feedback taps. The square root Kalman algorithm for adjusting the equalizer coefficients has been described in [23]. Signal shaping at the transmitter was achieved by a pulse having a raised cosine spectral shape with a 50% rolloff factor. This shaping was accomplished by means of a 19-tap FIR digital filter. In short, the modulator and demodulator were implemented in software in a form similar to that described in Section 2.3.

The modem was successfully tested on several HF channel paths. To handle deep fades, it employs a low data rate feedback channel to signal the transmitter when a deep fade occurs. When this occurs, the transmitter switches from a data transmission mode to transmitting a training sequence which the equalizer uses to re-acquire the signal once it comes back from a deep fade. When the SNR reaches an acceptable level the receiver signals the transmitter via the feedback

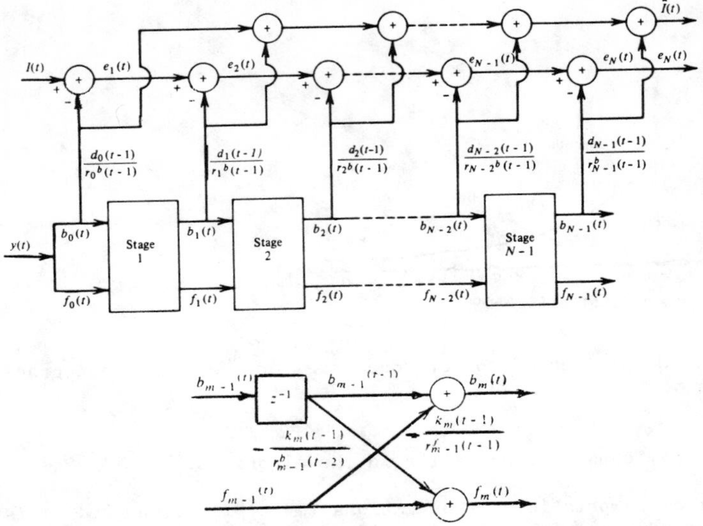

FIGURE 20. Least squares lattice linear equalizer.

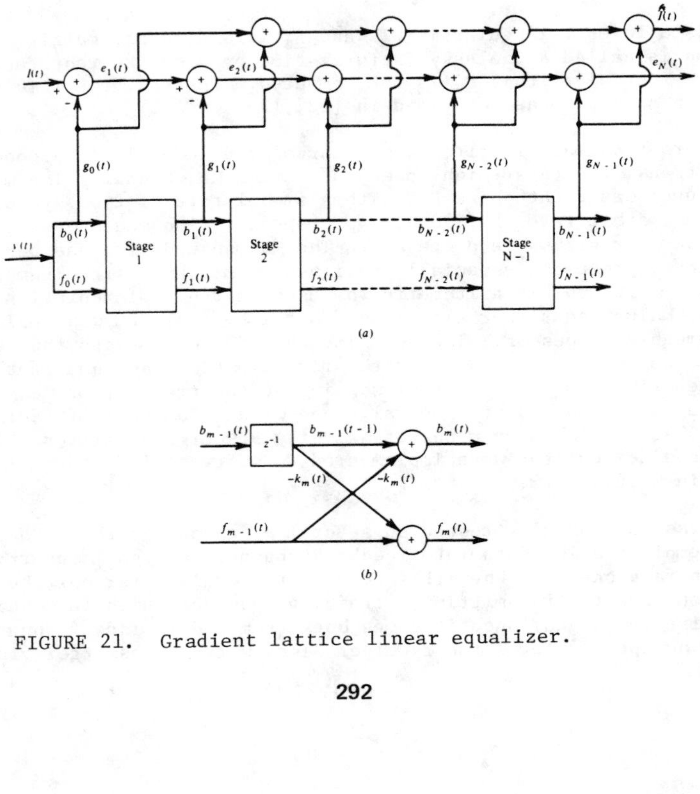

FIGURE 21. Gradient lattice linear equalizer.

channel to switch back to the data transmission mode.

3.5 Lattice Equalizer Structures for Time-Variant Multipath Channels

The linear and decision-feedback equalizers that were presented in Section 2.4 are realized as transversal filter structures. The LMS and the recursive least squares Kalman algorithms are simply two different algorithms for adjusting the equalizer coefficients.

There is another structure, called a lattice equalizer, that may be used to perform equalization. Such a structure is obtained as described in [5],[24],[25],[26],[27], by adopting the least squares criterion to perform forward and backward prediction on a set of data samples at the receiver. The derivation is lengthy and, hence, it will be omitted from this presentation. The complete derivation may be found in the references [5],[24]-[27].

The structure of a linear lattice equalizer, which is equivalent in performance and convergence rate to a tapped-delay-line linear equalizer with coefficients adjusted according to the least squares Kalman algorithm, is shown in Figure 20. The input data is denoted as $y(t)$, the desired symbol is $I(t)$, and the estimate is denoted as $\hat{I}(t)$, where t is discrete in time, i.e., $t = 0,1,2,\ldots$. The terms $f_k(t)$ and $b_k(t)$ represent the forward and backward k^{th} order prediction errors, respectively; $r_k^f(t)$ and $r_k^b(t)$ represent the minimum forward and backward least squares errors, respectively. The estimate $\hat{I}(t)$ is a weighted sum of the backward prediction errors. The weighting coefficients in the linear sum are determined recursively from the received signal samples to minimize the time-averaged (exponentially) weighted error.

A slightly simpler version of the least squares lattice is the gradient lattice equalizer, which is shown in Figure 21. In this structure, the weighting coefficients in the linear sum, denoted as $g_k(t)$, are no longer adjusted on the basis of the least squares criterion, but instead, are adjusted according to a gradient algorithm. However, the convergence rate of the gradient lattice is not degraded significantly compared to the optimum least squares lattice. This point is illustrated in Figure 22, which shows the convergence of the optimum least squares lattice algorithm, the gradient lattice algorithm, and the LMS (gradient) algorithm, where the latter utilizes a transversal structure [28].

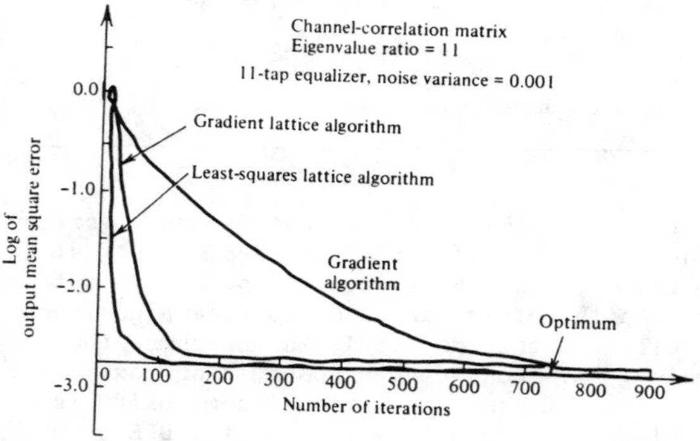

FIGURE 22. Comparison of the convergence rate of lattice equalizers and the gradient (LMS) algorithm [28].

A useful byproduct of the derivation of the least squares lattice algorithm is an efficient computation of the Kalman gain vector, which can be used in the transversal equalizer structures. The resulting algorithm is called the fast Kalman algorithm [29],[5],[25].

One important advantage of the lattice equalizer structures is that their computational complexity increases linearly with N, the number of equalizer parameters. By comparison, the Kalman algorithm requires a computational complexity proportional to N^2. A second important advantage of the lattice structures is their relatively low sensitivity to round off noise. Yet another important advantage is the ease with which one can add or delete stages in the lattice filter without affecting the parameters of the previous lattice stages. This characteristic is a consequence of the orthogonality properties of the least squares lattice filter as shown by Makhoul [30].

The DFE with coefficients adjusted according to the least squares Kalman algorithm also has a lattice-form realization. In contrast to the linear lattice, however, the DFE lattice consists of both one-dimensional lattice stages and two-dimensional lattice stages as has been indicated in [31]. This structure is illustrated in Figure 23. The suitability of this structure for adaptive equalization of such channels as HF has been confirmed by means of digital computer simulations.

We conclude this section by comparing the computational complexity of the various DFE equalizer structures as measured in terms of the number of multiplications and divisions. The results are illustrated in Table 1. For comparison, we also show the complexity of the LMS algorithm, although on some channels such as HF, it does not track sufficiently fast.

TABLE 1. Comparison of Computational Complexity of DFE Structures.

Algorithm	Number of Operations
Gradient Transversal DFE	$2N + 1$
Fast Kalman DFE	$20N + 1$
Kalman DFE	$2.5N^2 + 4.5N$
Square Root Kalman DFE	$1.5N^2 + 6.5N$
Gradient Lattice DFE	$13K_1 + 33K_2 - 36$
LS Lattice DFE	$18K_1 + 39K_2 - 39$

$N = K_1 + K_2$ = total number of taps

We observe that the Kalman, the square-root Kalman and the fast Kalman algorithms are especially suitable for tapped-delay-line equalizer structures. The fast Kalman is the most efficient of the rapidly convergent algorithms, but it suffers from sensitivity to round off noise. On the other hand, the lattice algorithms are highly stable and perform very well. Of the two lattice algorithms, the gradient lattice shows the most promise. Its convergence speed is approximately that of the optimum least squares lattice but its computational complexity is lower. In contrast, the computational complexity of the transversal DFE structures grows rapidly with N^2. For $N \leq 10$, the lattice, the Kalman and the fast Kalman algorithms require roughly comparable computational complexity. However, for $N > 10$, the advantage is in favor of the lattice structures. Consequently,

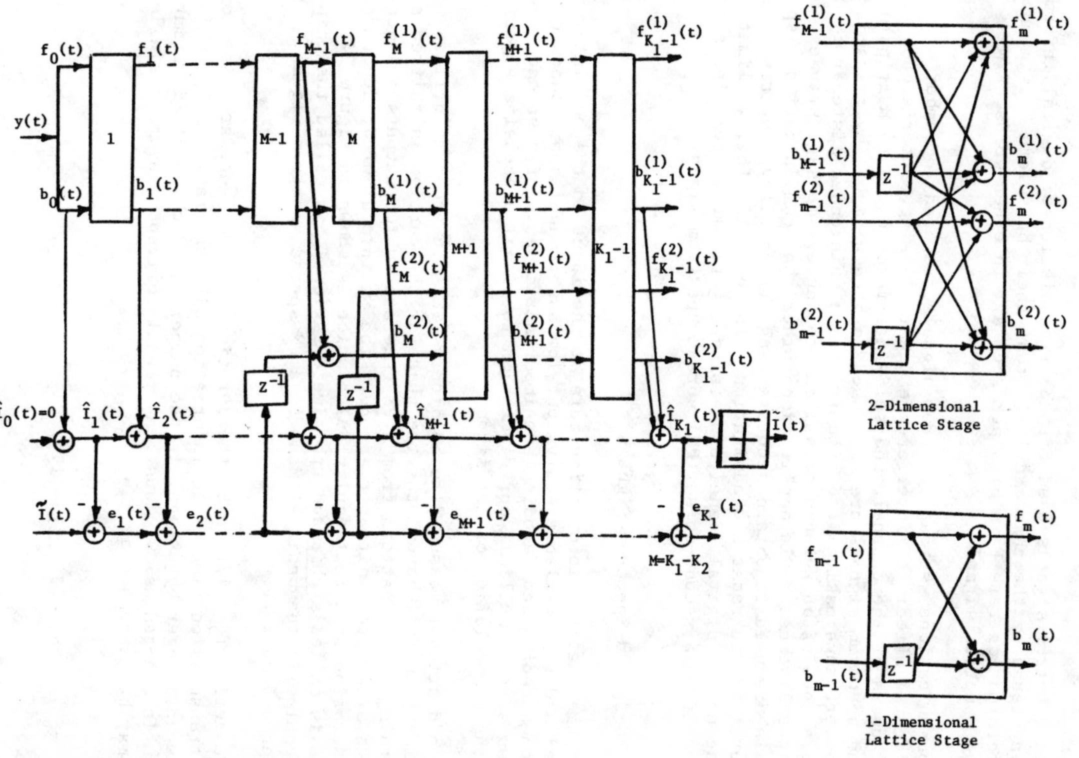

FIGURE 23. Least squares lattice DFE.

they will probably be used in the next generation of adaptively equalized modems for HF communications.

4. EXCISION OF NARROWBAND INTERFERENCE IN PN SPREAD SPECTRUM SYSTEMS

Spread spectrum signals used for the transmission of digital information are distinguished by the characteristic that their bandwidth W is much greater than the information rate R in bits per second. In other words, $R/W \ll 1$ for a spread spectrum signal.

In communications, spread spectrum signals are used to combat or overcome the detrimental effects of interference due to jamming, interference arising from other users of the channel, and self-interference due to multipath propagation. Spread spectrum signals are also used to hide a signal by transmitting it at a low power, thus making it difficult for an unintended listener to detect it in the presence of background noise, and to achieve message privacy in the presence of other listeners. In applications other than communications, spread spectrum signals are used to obtain accurate range (time delay) and range rate (velocity) measurements in radar and navigation. In the following, we limit our discussion to the use of digital signal processing algorithms in removing narrowband interference from a pseudo-noise (PN) spread spectrum signal.

4.1 PN Spread Spectrum Signals

We assume that the information rate at the input to the modulator is R bits per second and the available equivalent lowpass channel has a bandwidth W Hz. Although coding and decoding are important aspects of a spread spectrum communication system, we limit our discussion to signal processing algorithms appropriate for the demodulation of the signal.

The modulation is assumed to be binary PSK. In order to utilize the entire available channel bandwidth, the phase of the carrier is shifted pseudo-randomly according to the pattern from a PN generator which outputs a sequence of binary digits at a rate W bits per second. The reciprocal of W, denoted as T_c, defines the duration of a rectangular pulse which is shown in Figure 24. This rectangular pulse is called a <u>chip</u> and its time duration T_c is called the <u>chip interval</u>. This pulse is the basic element in a PN spread spectrum signal. The pulse energy is denoted as E_c.

If we define $T_b = 1/R$ to be the duration of a rectangular pulse corresponding to the transmission time of an information bit, the ratio T_b/T_c, which is assumed to be an integer by design, is the number of PN chips per information bit. This ratio is denoted as L. Thus, the signal corresponding to a single information bit may be expressed as

$$s(t) = \sum_{i=1}^{L} (2b_i - 1)\, g(t - iT_c) \tag{59}$$

where b_i is the binary-valued (0,1) sequence from the PN generator.

The demodulator for this signal usually takes the form of a filter matched to the chip waveform followed by correlation with a replica of the PN sequence. The effect of this demodulation is to collapse the bandwidth of the spread spectrum signal to the bandwidth R of the information signal. Simultaneously, any narrowband interference in the received signal is spread in bandwidth to W Hz. Thus, the interference is rendered equivalent to a lower-level noise with a

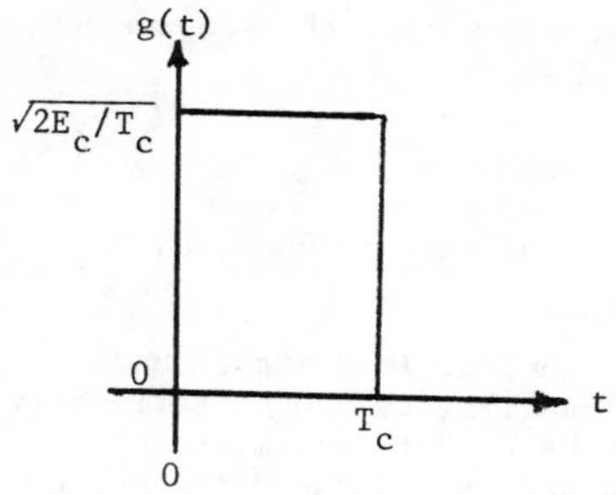

FIGURE 24. A PN chip.

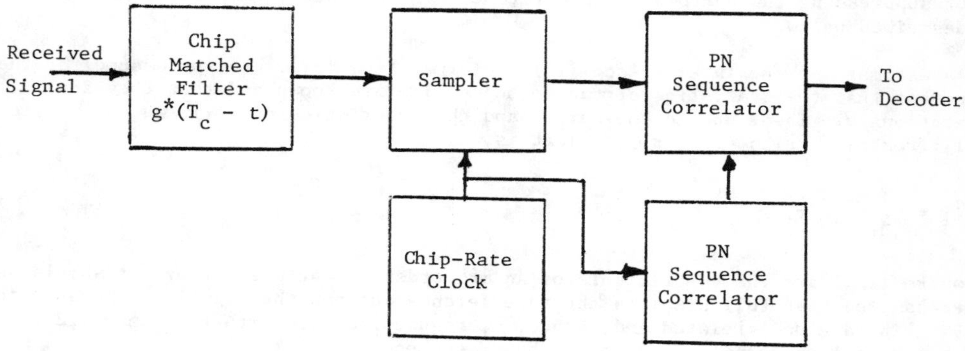

FIGURE 25. Conventional demodulator for PN spread spectrum signal.

relatively flat spectrum. In this manner, the spread spectrum system gains an advantage over the narrowband interference. This potential advantage over narrowband interference is just the ratio $L = T_b/T_c$, which is called the <u>processing gain</u>. Figure 25 illustrates a block diagram of the demodulator.

4.2 Excision of Narrowband Interference Using Linear Prediction Algorithm

The interference immunity of a PN spread spectrum communication system corrupted by narrowband interference can be further improved by filtering the signal prior to cross correlation, where the objective is to reduce the level of the interference at the expense of introducing some distortion on the desired signal. This filtering can be accomplished by exploiting the wideband spectral characteristics of the desired PN signal and the narrowband characteristic of the interference, as described below.

To be specific, we consider a demodulator in which the received signal is first passed through a filter matched to the chip pulse g(t). The output of this filter is synchronously sampled every T_c seconds to yield a signal component and an additive noise term. The additive noise is assumed to consist of two components, one corresponding to a broadband noise (usually thermal noise) and the other to narrowband interference. Consequently, we express the samples at the output of the chip matched filter as:

$$r_j = s_j + i_j + n_j \tag{60}$$

where s_j denotes the signal component, i_j the narrowband interference, and n_j the broadband noise.

The interference component i_j can be estimated from the received signal and suppressed by passing it through the linear transversal filter. Computationally efficient algorithms based on linear prediction may be used to estimate the interference. Basically, in this method the narrowband interference is modeled as having been generated by passing white noise through an all-pole filter. Linear prediction is used to estimate the coefficients of the all-pole model. The estimated coefficients specify an appropriate noise-whitening all-zero (transversal) filter through which the received signal is passed for the purpose of suppressing the narrowband interference. Linear prediction algorithms are described below.

An estimate of the interference $\{i_j\}$ is formed from $\{r_j\}$. Let us assume for the moment that the statistics of the sequence $\{i_j\}$ are known and that they are stationary. Then, due to the narrowband characteristics of $\{i_j\}$, we can predict i_j from $r_{j-1}, r_{j-2}, \ldots, r_{j-m}$. That is,

$$\hat{i}_j = \sum_{n=1}^{m} a_{mn}\, r_{j-n} \tag{61}$$

where $\{a_{mn}\}$ are the coefficients of an mth-order linear predictor. It should be emphasized that (61) predicts the interference but not the signal s_j, because the PN chips are uncorrelated and, hence, s_j is uncorrelated with r_{j-n}, $n = 1,2,\ldots,m$, when m is less than the length of the PN sequence.

The coefficients in (61) are determined by minimizing the mean-square error between r_j and $\hat{i}_j$, which is defined as

$$\varepsilon(m) = E[(r_j - \hat{i}_j)^2] = E[(r_j - \sum_{n=1}^{m} a_{mn}\, r_{j-n})^2] \tag{62}$$

Minimization of $\varepsilon(m)$ with respect to the predictor coefficients leads to the familiar set of linear equations, namely,

$$\sum_{n=1}^{m} a_{mn}\, \phi(k - n) = \phi(k) \qquad k = 1,2,\ldots,m \tag{63}$$

where

$$\phi(k) = E(r_j\, r_{j+k}) \tag{64}$$

is the autocorrelation function of the received signal $\{r_j\}$.

The solution of (63) for the coefficients of the prediction filter requires knowledge of the autocorrelation function $\phi(k)$. In practice, the autocorrelation function of i_j and, hence, r_j is unknown and it may also be slowly varying in time. Consequently we must consider methods for obtaining the predictor coefficients directly from the sequence $\{r_j\}$. This may be accomplished in a number of ways. Three different methods are described below. In all cases, the predictor coefficients are obtained by using a block of N samples of $\{r_j\}$.

The first method is simply based on the direct estimation of $\phi(k)$ from the block of N samples. The estimate of $\phi(k)$ is

$$\hat{\phi}(k) = \sum_{n=0}^{N-k} r(n)\, r(n+k) \quad , \qquad k = 0,1,\ldots,m \tag{65}$$

The estimate $\hat{\phi}(k)$ may then be substituted in place of $\phi(k)$ and the Levinson-Durbin algorithm can be used to solve the equations efficiently [32],[33].

The second method that may be used for obtaining the prediction coefficients is the Burg algorithm [34],[35]. Basically the Burg algorithm may be viewed as an order-recursive least-squares algorithm in which the Levinson recursion is used in each iteration. The performance index used in the Burg algorithm is

$$\varepsilon_B(m) = \sum_{j=m+1}^{N} [f_m^2(j) + b_m^2(j)] \tag{66}$$

where $f_m(j)$ and $b_m(j)$ are the forward and backward errors in the mth-order predictor, which are defined as

$$f_m(j) = r_j - \sum_{k=1}^{m} a_{mk}\, r_{j-k}$$

$$b_m(j) = r_{j-m} - \sum_{k=1}^{m} a_{mk}\, r_{j-m+k} \tag{67}$$

The predictor coefficients a_{mk} for $1 \leq k \leq m-1$ are forced to satisfy the Levinson-Durbin recursion, namely,

$$a_{mk} = a_{m-1\,k} - a_{mm}\, a_{m-1\,k-1} \quad , \qquad k = 1,2,\ldots,m-1 \tag{68}$$

As a consequence of this constraint, the forward and backward errors satisfy the recursive relations

$$f_m(j) = f_{m-1}(j) - a_{mm}\, b_{m-1}(j-1)$$

$$b_m(j) = b_{m-1}(j-1) - a_{mm}\, f_{m-1}(j) \tag{69}$$

$$f_0(j) = b_0(j) = r_j$$

The relations in (69) are substituted into (66) and $\varepsilon_B(m)$ is minimized with respect to a_{mm}. The result of this minimization is

$$a_{mm} = \frac{2 \sum_{j=m+1}^{N} f_{m-1}(j)\, b_{m-1}(j-1)}{\sum_{j=m+1}^{N} [f_{m-1}^2(j) + b_{m-1}^2(j-1)]} \quad , \qquad m \geq 1 \tag{70}$$

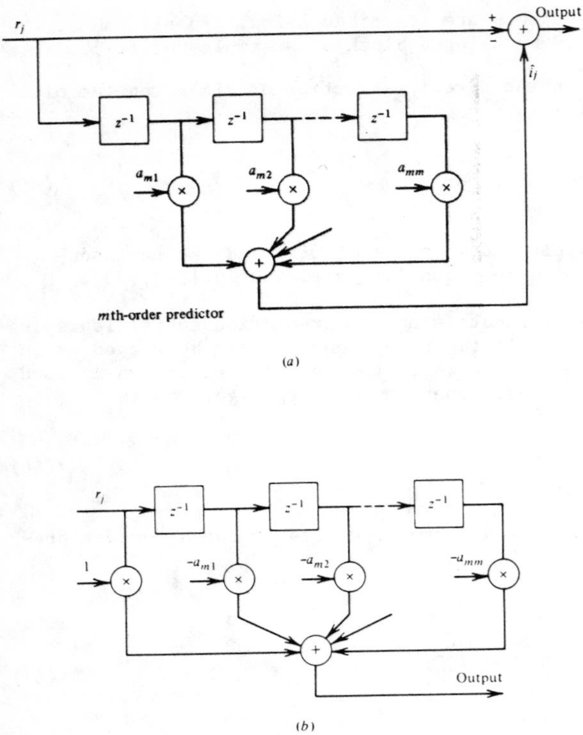

FIGURE 26. Equivalent structures for the interference suppression filter.

The relations given in (69) and (70) along with the Levinson-Durbin recursion for $\{a_{mk}\}$, $1 \leq k \leq m-1$, given by (68), constitute the Burg algorithm for obtaining the prediction coefficients directly from the data.

The Burg algorithm is basically a least-squares lattice algorithm with the added constraint that the predictor coefficients satisfy the Levinson recursion. As a result of this constraint, an increase in the order of the predictor requires only a single parameter optimization at each stage. In contrast to this method, we may use an unconstrained least-squares algorithm to determine the predictor coefficients. Specifically, this approach is based on minimizing (66) over the entire set of predictor coefficients. This minimization yields the set of equations

$$\sum_{k=1}^{m} a_{mk}\,\phi(n,k) = \phi(n,0) \quad , \qquad n = 1,2,\ldots,m \tag{71}$$

where

$$\phi(n,k) = \sum_{j=m+1}^{N} (r_{j-k}\, r_{j-n} + r_{j-m+k}\, r_{j-m+n}) \tag{72}$$

A computationally efficient algorithm for solving the equations in (71) has been given by Marple [36].

Once the prediction coefficients are determined by any one of the methods described above, the estimate i_j of the interference, given by (61), is subtracted from r_j and the difference signal is processed further in order to extract the digital information. Consequently the filter for suppressing the interference has the transfer function

$$A_m(z) = 1 - \sum_{k=1}^{m} a_{mk} z^{-k} \tag{73}$$

where z^{-1} denotes a unit (one-chip) delay. Two equivalent structures for this filter are shown in Figure 26.

As an example of the filter characteristics let us consider the suppression of a narrowband interference which is confined to 20% of the spectral band occupied by the PN spread spectrum signal. The average power of the interference is 20 dB above the average power of the signal. The average power of the broadband noise is 20 dB below the average power of the signal. Figure 27 illustrates the frequency characteristics of the filter with transfer function given by (73) when the coefficients of a fourth-order predictor are obtained by means of the three algorithms described above.

We observe only minor differences in the spectral characteristics of the filters. These results were generated by Monte Carlo simulation on a digital computer. Good estimates of the coefficients were obtained with as few as 50 samples (PN chips) and predictors that varied in length from fourth to fifteenth order. In general, the Burg algorithm and the unconstrained least-squares algorithm yield

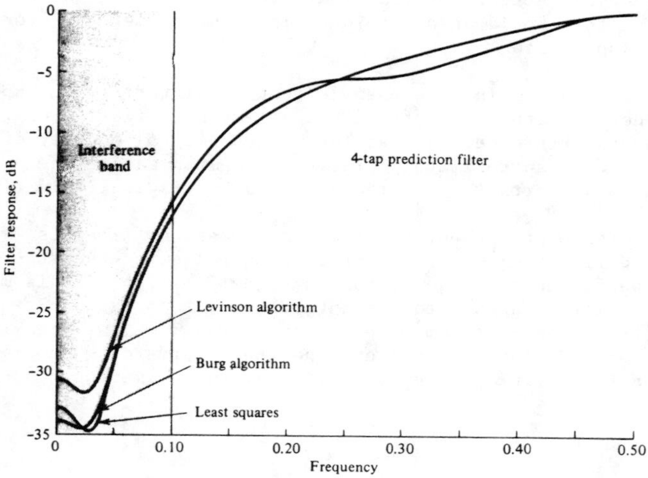

FIGURE 27. Frequency response characteristics for the interference suppression filter.

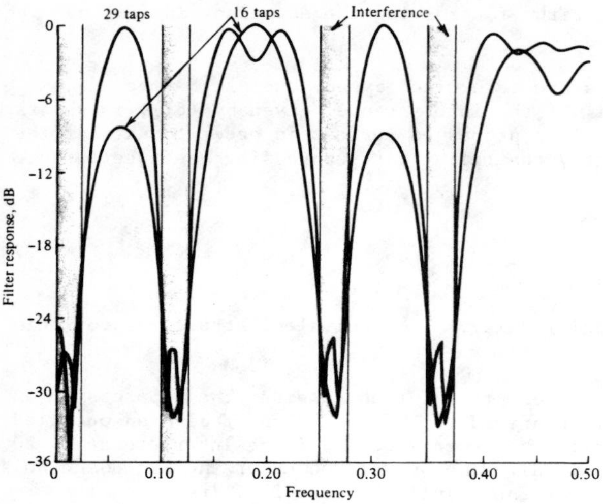

FIGURE 28. Frequency response characteristics of 16- and 29-tap filters for four bands of interference.

good estimates of the predictor coefficients provided the number of signal samples is at least twice the length of the prediction filter.

Figure 28 illustrates an example in which narrowband interference occupying 20% of the signal band is split equally into four frequency bands. The spectral characteristics of a 16- and 29-tap filter are shown in this example. It is apparent that the 29-tap filter has better spectral characteristics. In general, the number of taps in the filter should be about four times the number of interference bands for adequate suppression.

It is apparent from the results given in this example that the interference suppression filter acts as a notch filter. In effect, it attempts to flatten or whiten the total noise plus interference, so that the power spectral density of these components at its output is approximately flat. While suppressing the interference, the filter also distorts the desired signal by spreading it in time.

Since the noise plus interference at the output of the suppression filter is spectrally flat, the matched filtering or cross correlation following the suppression filter should be performed with the distorted signal [37],[5]. This may be accomplished by having a filter matched to the interference suppression filter, i.e., a discrete-time filter with impulse response $\{-a_{mm}, -a_{mm-1}, -a_{m1}, 1\}$ followed by the PN correlator. In fact, we can combine the interference suppression filter and its matched filter into a single filter having an impulse response

$$h_0 = -a_{mm}$$

$$h_k = -a_{mm-k} + \sum_{n=0}^{k-1} a_{mm-n}\, a_{mk-n}\,, \qquad 1 \leq k \leq m-1 \tag{74}$$

$$h_m = 1 + \sum_{n=1}^{m} a_{mn}^2$$

$$h_{m+k} = h_{m-k} \qquad 0 \le k \le m$$

The combined filter is a linear phase (symmetric) transversal filter with $K = 2m + 1$ taps. The impulse response may be normalized by dividing every term by h_m. Thus the center tap is normalized to unity.

The effectiveness of the interference suppression filter can be assessed by comparing the signal-to-noise ratio (SNR) at the output of the spread spectrum demodulator with and without the suppression filter. The SNR without the suppression filter is [37],[5],

$$SNR_{n0} = \frac{E_c L}{N_0 + \phi_{ii}(0)/2E_c} \tag{75}$$

and with the inclusion of the suppression filter it becomes

$$SNR_0 = \frac{E_c L}{N_0 \sum_{k=0}^{K} h_k^2 + \frac{1}{2E_c} \sum_{k=0}^{K} \sum_{n=0}^{K} h(k)\, h(n)\, \phi_{ii}(k - n) + 2E_c \sum_{k=0}^{K/2-1} (2 - \frac{k}{L})\, h_k^2} \tag{76}$$

The ratio of the SNR in (76) to the SNR in (75) represents the improvement in the performance due to the use of the interference suppression filter. This ratio, denoted by η, is

$$\eta = \frac{N_0 + \phi_{ii}(0)/2E_c}{N_0 \sum_{k=0}^{K} h_k^2 + \frac{1}{2E_c} \sum_{k=0}^{K} \sum_{n=0}^{K} h(k)\, h(n)\, \phi_{ii}(k - n) + 2E_c \sum_{k=0}^{K/2-1} (2 - \frac{k}{L})\, h_k^2} \tag{77}$$

This ratio is called the improvement factor resulting from interference suppression [38]. It may be plotted against the normalized SNR per chip without filtering, defined as

$$\frac{SNR_{n0}}{L} = \frac{E_c}{N_0 + \phi_{ii}(0)/2E_c} \tag{78}$$

The resulting graph of η versus SNR_{n0}/L is universal in the sense that it applies to any PN spread spectrum system with arbitrary processing gain for a given E_c, N_0, and $\phi_{ii}(0)$.

As an example, the improvement factor in (decibels) is plotted against SNR_{n0}/L in Figure 29 for a single-band of equal-amplitude randomly phased sinusoids covering 20% of the frequency band occupied by the PN spread spectrum signal. The interference suppression filter consists of a nine-tap suppression filter which corresponds to a fourth-order predictor. These numerical results indicate that the notch filter is very effective in suppressing the interference prior to PN correlation. As a consequence, the jamming margin of the PN spread spectrum system is increased by the improvement factor.

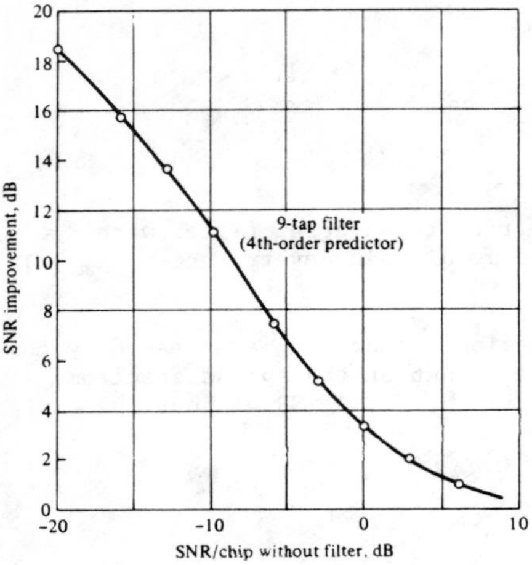

FIGURE 29. Improvement factor for interference suppression filter.

The estimation and suppression of narrowband interference in PN spread spectrum signals is just one example of the use of digital signal processing algorithms and techniques in spread spectrum communications. Digital signal processing techniques are also widely used in frequency-hopped (FH) spread spectrum communications. In addition, digital processing techniques are prevalent in the implementation of modulators and demodulators for PN, FH and hybrid PN-FH spread spectrum systems.

5. CONCLUDING REMARKS

In this brief treatment we have attempted to illustrate via several examples the use of digital signal processing algorithms and techniques in digital communications over telephone channels, fading multipath radio channels and additive noise channels corrupted by narrowband interference. Specifically, we cited applications of signal processing techniques in the implementation of modulators, demodulators, and in the use of digital signal processing algorithms for adaptive equalization and in narrowband interference suppression.

There are numerous other aspects of communications where digital signal processing techniques have been applied. For example, there is the broad area of source encoding/decoding, which includes speech processing and image processing. There are also other applications in telephone channel transmission, including the design and implementation of digital filter banks for frequency division multiplexing and in the design and implementation of echo cancellers.

Digital signal processing techniques are expected to play an important role in future developments of new communications services, such as satellite communications, fiber optic communications, and mobile communications. They will also have a significant impact in the development of modern packet switching and digital switching systems.

The rapid development of relatively inexpensive integrated digital circuits and digital computers has been primarily responsible for the widespread use of digital signal processing techniques in communications. In addition to the development of inexpensive digital hardware, we should also mention the new developments in sampled-analog integrated circuits, especially charge-coupled devices (CCD). Programmable CCD's are now available which are capable of performing FIR filtering, correlation and convolution at clock (sampling) rates exceeding 10 MHz. These new sampled-analog devices serve to complement the digital integrated circuit technology and insure the continued and expanding use of digital signal processing techniques in communications.

REFERENCES

1. Makhoul, J., *Speech Processing,* this volume.

2. Shannon, C. E., A Mathematical Theory of Communication, *Bell System Tech. J.,* vol. 27, pp. 379-423, July 1948, and pp. 623-656, October 1948.

3. Simon, M. K. and Smith, J. G., Hexagonal Multiple Phase-and-Amplitude-Shift Keyed Signal Sets, *IEEE Trans. Communications,* vol. COM-21, pp. 1108-1115, October 1973.

4. Thomas, C. M., Weidner, M. Y., and Durrani, S. H., Digital Amplitude-Phase Keying with M-ary Alphabets, *IEEE Trans. Communications,* vol. COM-22, pp. 168-180, February 1974.

5. Proakis, J. G., *Digital Communications,* McGraw-Hill, New York, 1983.

6. Blahut, R. E., Transform Techniques for Error Control Codes, *IBM Journal of Res. and Devel.,* vol. 23, pp. 299-315, May 1979.

7. Viterbi, A. J. and Omura, J. K., *Principles of Digital Communication and Coding,* McGraw-Hill, New York, 1979.

8. Lucky, R. W., Salz, J., Weldon, E. J., Jr., *Principles of Data Communication,* McGraw-Hill, New York, 1968.

9. Lender, A., The Duobinary Technique for High-Speed Data Transmission, *1963 IEEE Winter General Meeting.*

10. Kretzmer, E. R., Generalization of a Technique for Binary Data Communication, *IEEE Trans. Comm. Tech.,* vol. COM-14, pp. 67-88, February 1966.

11. Stiffler, J. J., *Theory of Synchronous Communications,* Prentice Hall, Englewood Cliffs, NJ, 1971.

12. Lindsey, W. C., *Synchronization Systems in Communications,* Prentice Hall, Englewood Cliffs, NJ, 1972.

13. Ungerboeck, G., Fractional Tap-Spacing Equalizer and Consequences for Clock Recovery in Data Modems, *IEEE Trans. Communications,* vol. COM-24, pp. 856-864, August 1976.

14. Qureshi, S. U. H. and Forney, G. D., Jr., Performance and Properties of a T/2 Equalizer, *Nat'l Telecom. Conf. Record,* Los Angeles, CA, December 1977.

15. Widrow, B., Adaptive Filters, I: Fundamentals, Stanford Electronics Laboratory, Stanford University, CA, Tech. Report No. 6764-6, December 1966.

16. Austin, M. E., Decision-Feedback Equalization for Digital Communication over Dispersive Channels, MIT Lincoln Laboratory, Lexington, MA, Tech. Report No. 437, August 1967.

17. Monson, P., Feedback Equalization for Fading Dispersive Channels, *IEEE Trans. Inform. Theory,* vol. IT-17, pp. 56-64, January 1971.

18. Forney, G. D., Jr., Maximum-Likelihood Sequence Estimation of Digital Sequences in the Presence of Intersymbol Interference, *IEEE Trans. Information Theory,* vol. IT-18, pp. 363-378, May 1972.

19. Magee, F. R. and Proakis, J. G., Adaptive Maximum-Likelihood Sequence Estimation for Digital Signaling in the Presence of Intersymbol Interference, *IEEE Trans. Information Theory,* vol. IT-19, pp. 120-124, January 1973.

20. Pieper, J. F., Proakis, J. G., Reed, R. R., and Wolf, J. K., Design of Efficient Coding and Modulation for a Rayleigh Fading Channel, *IEEE Trans. Information Theory,* vol. IT-24, pp. 457-468, July 1978.

21. Godard, D., Channel Equalization using a Kalman Filter for Fast Data Transmission, *IBM J. Res. and Devel.,* pp. 267-273, May 1974.

22. Bierman, G. J., *Factorization Methods for Discrete Sequential Estimation,* Academic, New York, 1977.

23. Hsu, F. M., Square Root Kalman Filtering for High-Speed Data Received on Fading Dispersive HF Channels, *IEEE Trans. Information Theory,* vol. IT-28, pp. 753-763, September 1982.

24. Kailath, T., *Linear Estimation for Stationary and Near-Stationary Processes,* this volume.

25. Cioffi, J. M., and Kailath, T., Fast Recursive-Least-Squares, Transversal Filters for Adaptive Filtering, *IEEE Trans. Acoustics, Speech and Signal Processing,* 1984.

26. Carayannis, G., Manolakis, D., and Kalouptsidis, N., Fast Kalman Type Algorithms for Sequential Signal Processing, *Proc. 1983 International Conference on Acoustics, Speech and Signal Processing,* pp. 186-189, Boston, MA, April 1983.

27. Satorius, E. H. and Alexander, S. T., Channel Equalization using Adaptive Lattice Algorithms, *IEEE Trans. Communications,* vol. COM-27, pp. 899-905, June 1979.

28. Satorius, E. H. and Pack, J. D., Application of Least Squares Lattice Algorithms to Adaptive Equalization, *IEEE Trans. Communications,* vol. COM-29, pp. 136-142, February 1981.

29. Falconer, D. D., and Ljung, L., Application of Fast Kalman Estimation to Adaptive Equalization, *IEEE Trans. Communications,* vol. COM-26, pp. 1439-1446, October 1978.

30. Makhoul, J., A Class of All-Zero Lattice Digital Filters: Properties and Applications, *IEEE Trans. Acoustics, Speech and Signal Processing,* vol. ASSP-26, pp. 304-314, August 1978.

31. Ling, F. and Proakis, J. G., Generalized Least Squares Lattice and Its Application to DFE, *Proc. 1982 IEEE Int. Conf. on Acoustics, Speech and Signal Processing,* Paris, France, May 1982.

32. Levinson, N., The Wiener RMS (Root Mean Square) Error Criterion in Filter Design and Prediction, *J. Math. Phys.,* vol. 25, pp. 261-278, 1947.

33. Durbin, J., Efficient Estimation of Parameters in Moving-Average Models, *Biometrika,* vol. 46, parts 1 and 2, pp. 306-316, 1959.

34. Burg, J. P., Maximum Entropy Spectral Analysis, *Proc. 37th Meeting of the Society of Exploration Geophysicists,* 1967; also reprinted in *Modern Spectrum Analysis,* D. G. Childers (ed.), pp. 34-41, IEEE Press, New York, 1978.

35. Ulrych, T. J. and Bishop, T. N., Maximum Entropy Spectral Analysis and Autoregressive Decomposition, *Rev. Geophys. and Space Phys.,* vol. 13, pp. 183-200, February 1975.

36. Marple, L., A New Autoregressive Spectrum Analysis Algorithm, *IEEE Trans. Acoustics, Speech and Signal Processing,* Vol. ASSP-28, pp. 441-454, August 1980.

37. Ketchum, J. W. and Proakis, J. G., Adaptive Algorithms for Estimating and Suppressing Narrow-Band Interference in PN Spread-Spectrum Systems, *IEEE Trans. on Communications,* vol. COM-30, pp. 913-924, May 1982.

38. Hsu, F. M. and Giordano, A. A., Digital Whitening Techniques for Improving Spread Spectrum Communications Performance in the Presence of Narrowband Jamming and Interference, *IEEE Trans. on Communications,* vol. COM-26, pp. 209-216, February 1978.

8

Radar/Sonar Signal Processing

H. J. WHITEHOUSE

ABSTRACT

Selected aspects of radar/sonar signal processing are presented from the perspective of real-time implementation. The range equation is introduced to show that signal detection depends primarily on the received signal energy and is thus independent of the structure of the transmitted waveform. The ambiguity function is introduced as the output of a matched filter receiver for a point target in range and doppler. The issues of waveform selection and matched filter processing complete the discussion of temporal signal processing. Spatial signal processing is introduced with a discussion of phased array beamforming. Adaptive array techniques are described as an example of advanced spatial processing. Synthetic aperature radar (SAR) and inverse synthetic aperature radar (ISAR) are considered as examples of imaging radars.

CONTENTS

INTRODUCTION

Radar and active sonar have many signal processing characteristics in common. In this paper some of these common areas will be examined. By necessity only a few of the many possible signal processing questions will be examined and it is hoped that the reader will consult the bibliography at the end of this paper for a selection of textbooks, conference proceedings, and reprint volumes which provide valuable additional material.

From the point of view of the engineer designing the signal processing for either a radar or a sonar system the primary target to be detected, airplane or submarine, is approximately the same size. Thus, if we choose a common wavelength, the number of cycles of the carrier signal which illuminates the target will be similar.

For example, if the wavelength of the carrier signal is chosen to be 0.1 meters, and the velocity of propagation of the radar signal is assumed to be 300,000 kilometers per second, then the carrier frequency will be 3.0 GigaHertz. Using the same wavelength of the carrier signal, 0.1 meters, and assuming a velocity of propagation of the sonar signal of 1.5 kilometers per second, the corresponding carrier frequency for the sonar signal will be 15 kiloHertz.

Thus it is seen that the most striking difference between radar and sonar signals is their carrier frequencies which is a direct result of the 200,000 ratio in the velocity of propagation of the signals. Two more subtle differences are the ways in which velocity and angular position are measured.

Most typical radar targets have velocities one to two orders of magnitude greater than typical sonar targets, however, even with this difference in target velocity the larger ratio of the velocities of propagation leads to the implementation of velocity measurement in radar by multiple pulse techniques and in sonar with single pulse techniques.

Similarly, most radars have a maximum range greater than the maximum range of a corresponding type of sonar. However, again the large ratio of electromagnetic velocity to acoustic velocity leads to different implementations of angular measurement, particularly the measurement of azimuth. In many radars azimuth is measured on a multiple pulse basis by mechanically scanning the transmitting/receiving antenna.This is possible because of the high pulse repetition frequency, PRF, that results from the large velocity of propagation of radar signals. By way of contrast the slow velocity of propagation of sonar signals lead to the introduction of electronic scanning and phased array receivers much earlier than they were introduced in radar.

Although we have discussed some of the more important differences between radar and sonar and their implications on the way these systems are implemented there are many more differences between the systems which we will not be able to discuss. Some of these which could be considered are:

(1) radar can transmit and receive polarized signals while sonar can't;

(2) radars other than high-frequency, HF, radars are not significantly influenced by refraction while all sonars are;
(3) radars have a different form of frequency dependent attenuation than sonars.

FUNDAMENTAL CONCEPTS

Range Equation

We start our study of radar/sonar signal processing by examining the range equation. The range equation is the elementary expression which relates the detectability of an isolated point target of specified cross-sectional area, σ, to a signal-to-noise ratio at the output of the receiver.

For historical reasons the range equation in radar is usually presented as the product of a sequence of factors related to the physical design of the antennas and the slant range to the target. In the case of sonar the equation is usually presented as a sum of logarithmic terms with an implicit range term which includes both the loss due to spreading of the radiated signal and the loss due to attenuation at the center frequency of the

The radar equation is

$$Pr = \frac{Pt\ Gt\ Gr\ \lambda^2\ \sigma}{(4\pi)^3\ R^4} \tag{1}$$

where Pr is the received power, Pt is the transmitted power, Gt is the transmitting antenna gain, Gr is the receiving antenna gain, λ is the wavelength, σ is the target cross-section, and R is the slant range to the target.

From detection theory we know that for coherent detection in white noise of power density, No, the received energy divided by the noise power density should be compared to a threshold which depends upon both the probability of detection and the probability of false alarm. Thus if we assume that the transmitted signal has constant envelope we may rewrite the radar equation as

$$\frac{E}{No} = \frac{Pt\ \tau\ Gt\ Gr\ \lambda^2\ \sigma}{(4\pi)^3\ k\ T\ R^4} \tag{2}$$

where E is the energy in the returned signal after integrating for time, τ, k is Boltzmann's constant, and T is the receiver noise temperature in degrees Kelvin.

It should be noted that although the system parameters of the radar as well as the range to the target and the noise background determine the detection performance the form of the transmitted signal does not enter into the expression. Thus the signal can be chosen for other considerations such and range and doppler resolution. The choise of signals is a major part of radar/sonar signal processing and good discussions can be found in Cook and Bernfeld, Rihaczek, and Vakman.

Ambiguity Function

Woodward showed that the return from a point target can be characterized by a two dimensional function called the ambiguity function, which expresses the receiver response in range and doppler offset to the transmitted signal. The ambiguity function and several equivalent representations are shown in Fig. 1.

$$A_{12}(\tau,f) = \int g_1(t)\, g_2^*(t+\tau)\, e^{-i2\pi ft}\, dt$$

GROUPING OF TERMS	METHOD
$\int \left[g_1(t)\, G_2^*(u)\, e^{-i2\pi ut}\right] e^{-i2\pi(ft+u\tau)}\, dt\, du$	2-D FOURIER TRANSFORM
$\int \left[g_1(t)\, g_2^*(t+\tau)\right] e^{-i2\pi ft}\, dt$	τ-SLICE
$e^{-i\pi f^2} \int \left[g_1(t)\, e^{-i\pi t^2}\right]\left[g_2^*(t+\tau)\right]\left[e^{i\pi(t-f)^2}\right] dt$	TIME-INTEGRATING
$\int \left[g_1(t)\, e^{-i2\pi ft}\right]\left[g_2^*(t+\tau)\right] dt$	SPACE-INTEGRATING

FIGURE 1. Crossambiguity calculations.

Inspection of the expression for the cross-ambiguity function shows it to be a Fourier transformation of the product of two signals. If a matched filter such as that shown in Fig. 2 is used a part of the receiver then the ambiguity function is seen to result naturally as the receiver output for a point target when the expotential is interpreted as a doppler mismatch.

Thus, from linear filter theory, we may interpret the output of a radar/sonar receiver as the convolution of the point target ambiguity function response with the target's range-doppler distribution. From these considerations it is readily seen that the selection of the transmitted signal is not a simple task. Furthermore, it is known that the volume under the ambiguity is conserved when the signal form is changed. That is, although different signals will have different ambiguity functions it is not possible to have exact knowledge of range without corresponding ambiguity in measuring doppler. Conversely, exact knowledge of doppler will result in ambiguity in range.

WAVEFORM SELECTION

Range Resolution

The time resolution of a transmitted waveform may be found by evaluating the ambiguity function for zero doppler frequency. The resulting matched filter output is the auto-correlation function of

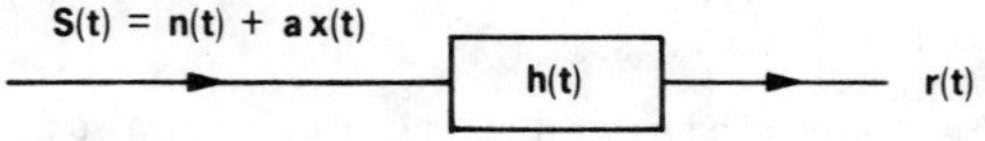

$$h(t) = x(-t)$$

$$r(t) = \left[S * h\right](t) = \int_{-\infty}^{\infty} S(\tau)\, x(\tau - t)\, d\tau$$

FIGURE 2. Matched filter.

the transmitted waveform. To convert to range resolution multiply the time duration of the auto-correlation function by the velocity of propogation divided by two. Thus small range resolution can be acheived with wide bandwidth signals. However, other aspects of the signal should also be considered. In order to maximize the efficiency of the transmitter the envelope of the transmitted signal should be constant. In addition, to resolve two closely separated targets the time sidelobes of the transmitted signal's auto-correlation function must be uniformly small. A well known example of such signals are the Barker codes [1] for which a length seven example is 1,1,1,-1,-1,1,-1 with corresponding auto-correlation function -1,0,-1,0,-1,0,7,0,-1,0,-1,0,-1. Unfortunately, the longest known Barker code is of length 13 and many other codes have been developed subsequently.

Since the velocity of propagation of a radar signal is about 300,000 kilometers/second a range resolution of 1.0 meters implies a bandwidth of about 150 MHz. Thus a 10 meter radar range resolution implies a bandwidth of about 15 MHz. At a sonar velocity of propagation of 1.5 kilometers/second a range resolution of 1.0 meters implies a corresponding bandwidth of only 750 Hz. Therefore, a 10 meter sonar resolution corresponds to only 75 Hz of bandwidth. The implication for real-time signal processing is that 1.0 meter range resolution in sonar can be achieved easily with digital techniques but that a corresponding 1.0 meter range resolution in radar is a difficult digital design.

Doppler Resolution

The doppler resolution of a transmitted waveform may be found by evaluating the ambiguity function for zero time delay. The resulting signal is found to be the Fourier transform of the magnitude squared of the transmitted signal. Thus the doppler resolution increases with increasing coherent integreation time. The doppler shift caused by the component of target motion along the line of sight given as $2vf/c$ where v is the radial velocity of the target, c is the velocity of propagation, and f is the carrier frequency. For $v = 300$ m/sec, $c = 300{,}000$ km/sec, and $f = 1.0$ GHz the doppler shift is 2kHz. For $v = 15$ m/sec, $c = 1.5$ km/sec and $f = 1.0$ kHz/s the doppler shift is 20 Hz. Thus multiple pulse are ususlly used for radar signals while a single pulse can be used for an active sonar signal.

Simultaneous Range-Doppler Resolution

By independently choosing the bandwidth and time duration of the transmitted signal both the range resolution and the doppler resolution of the signal may be specified. However, due to the volume constrant on the ambiguity function simultaneous unambiguious range and doppler resolution implies high sidelobes and thus is undesirable in a dense targe environment.

Since the range resolution and the doppler resolution correspond to slices through the ambiguity function at zero doppler and zero range specific signals can be found which achieve both range and doppler resolution but with ambiguity. These signals which a called "chirp" waveforms since the frequency usually changes in a linear manner have a well defined diagonal ridge for their ambiguity function. These waveforms are well described in the book "Principles of High-Resolution Radar" by A. Rihaczek.

Even though these waveforms have low sidelobes other than on the ridge the ambiguity of the ridge presents some difficulties. If the error in estimated range caused by an unknown doppler can be tolerated than the waveform can be used directly. If the doppler and the range of the target must be simultaneously estimated than a pair of chirps with opposite chirps can be used. Even in a dense target environment such as ground mapping these waveforms are often used because of the simple signal processing and the small velocity of ground targets. If a more accurate reconstruction of a dense range-doppler environment is required the chirp waveform can be used along with tomographic reconstruction principles of signal processing [2].

WAVEFORM PROCESSING

Modern radar/sonar signal processing must be flexible in order to accommodate a wide variety of waveforms. If the signal bandwidth isn't too large then digital signal processing using the Fast Fourier Transform (FFT) may be attractive. Current digital multipliers and accumulators operate at clock rates of 5 to 20 MHz. If parallel inplementation of complex arithmetic and pipeline techniques are used then corresponding signal bandwidths can be processed in real time. However, if larger bandwidths must be processed then either analog signal processing or parallel digital processing must be used.

Transversal Filters

Figure 3 shows a generic transversal filter convolver. The output y(t) = [x * h](t), where h(t) = {W1,W2,...Wn}. When used as a matched filter the impulse response, h(t), is the time reversed and complex conjugated transmitted signal. Although the transversal filter can be implemented directly with digital circuits the number of multiplications required is the square of the number of taps and indirect implementations via the fast Fourier transform (FFT) would usually be preferred. However, the transversal filter realization is especially attractive with analog implementations such as charge coupled devices (CCDs) and surface acoustic wave (SAW) devices. Also, the analog implementations often provide a reduction in size, weight, and power consumption over corresponding digital hardware.

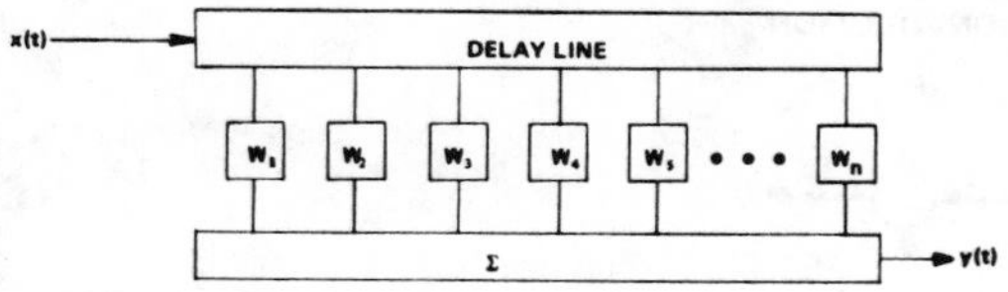

FIGURE 3. Transversal filter convolver.

Discrete Fourier Transforms

Figure 4 shows a Fourier transform implementation of convolution. The implementation is indirect by means of a forward Fourier transform, multiplication by the Fourier transform of the filter's impulse response, and an inverse Fourier transform. To realize a matched filter the filter's transfer function H(f) is replaced by complex conjugate of the Fourier transform of the transmitted signal.

These two techniques offer simplicity of implementation without flexibility, or flexibility at the expense of digital complexity. If it were possible to implement the Fourier transform with time invariant linear filters then it would be possible to have both flexability and simplicity of implementation. The Fourier transform, however, is not a shift invariant operation. If the input is delayed then the output Fourier transform is a phase shifted version of the undelayed Fourier transform. However, there is an algorithm for the computation of the Fourier transform called the Chirp-Z-Transform (CZT) in which the major computational tasks can be done with time invariant linear filters. The derivation of the algorithm is described in the companion paper on Signal Processing Technology in this volume.

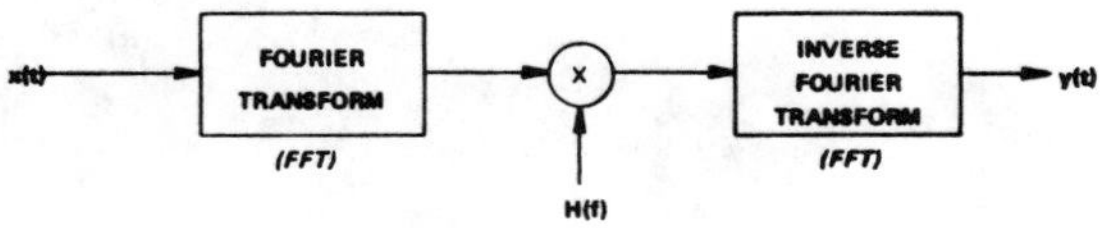

FIGURE 4. Fourier transform convolver.

The Chirp-Z-Transform implements either the continuous or the discrete Fourier transform. The CZT consists of multiplying the input signal by a continuous or discrete complex chirp, filtering the resulting signal with either a continuous of a discrete chirp filter, and post multiplying the resulting complex signal by another complex chirp. The discrete version of the algorithm is shown in Fig. 5.

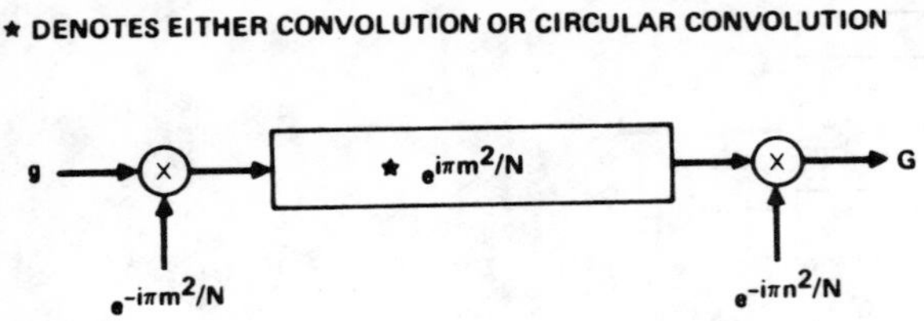

FIGURE 5. Chirp-z-transform implementation of the DFT.

If three of these CZT implementations are combined as in Fig. 6a then an arbitrary cross comvolution can be computed. However, several of the multipliers are redundant. These can be eliminated and the result is shown in Fig. 6b. If the Fourier transform of the second function is known or can be computed ahead of time then the configuration of Fig. 6c can be used. Please note the similarity to the FFT convolver shown in Fig. 4.

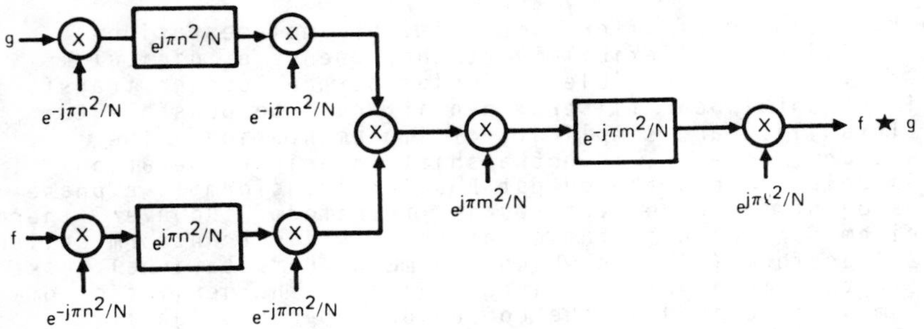

FIGURE 6a. Implementation of cross convolution - Direct Implementation.

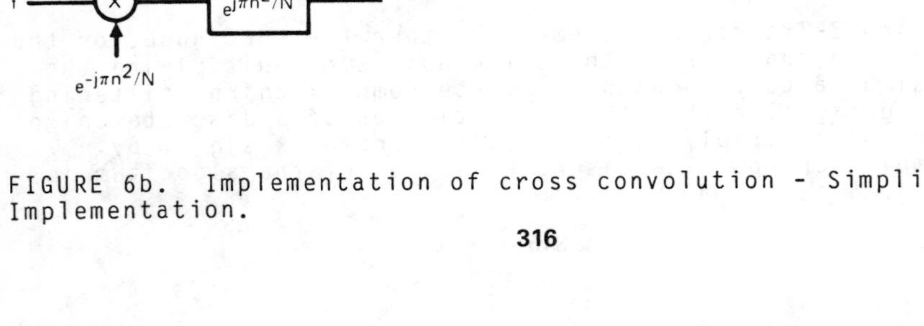

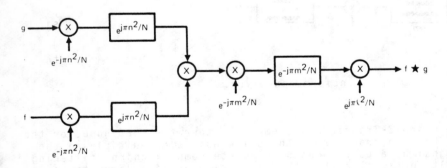

FIGURE 6b. Implementation of cross convolution - Simplified Implementation.

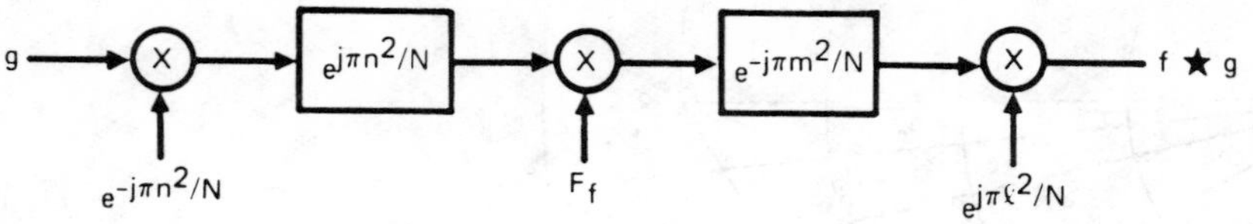

FIGURE 6c. Implementation of cross convolution - Matched Filter configuration with the Fourier transform of f stored in a Read Only Memory (ROM).

Although it may seem extravagant to compute a simple convolution via such an indirect method as through the use of the CZT algorithm this method may be preferable if severe narrow band jamming is present and the signal to be detected is broadband. Under these conditions the introduction of a nonlinear element such as a limiter in the signal path just before the reference function multiplier may result in significant improvement in performance.

However, in some applications, constraints on size, weight, and power may make analog signal processing attractive. In particular, if a doppler invariant waveform such as a linear frequency modulated chirp is used then Surface Acoustic Wave (SAW) filters make attractive matched filters in applications where more complicated filters cannot be implemented.

In summary, what has been shown is that even for complicated time varying transmitted waveforms it is possible to build matched filters using a combination of multipliers and time invariant linear filters with fixed impulse responses.

SENSOR PROCESSING

Although many radars use rotating dish antennas, sonar systems and some new radars use electronic phased arrays. The processing of these arrays has many characteristics in common with the processing of time series. In particular, if the antenna elements form an equally spaced line array, then it can be shown that the spatial Discrete Fourier Transform (DFT) of the sensor outputs at a given frequency is equivalent to a conventional beamformer at the same frequency.

Phased Array Beamforming

If the signal propagation medium is assumed to be homogeneous and isotropic the required signal processing for beamforming is completely determined by the geometrical structure of the array and

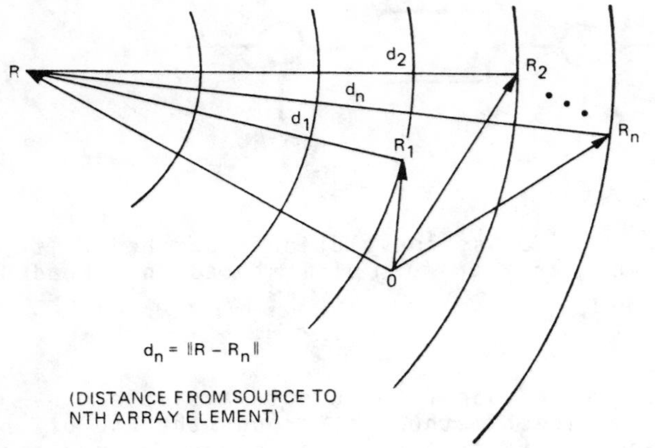

FIGURE 7a. Propagation geometry for an arbitrary array.

the location of possible sources. An arbitrary array propagation geometry is shown in Fig. 7a. Assuming no refraction and equal attenuation to all of the array elements the signal component at the nth array element is

$$s_n(t) = s(t - ||R - R_n||/c) \tag{3}$$

The desired beamformer output is the sum of the contributions from each of the individual array elements and is of the form

$$b(t) = \sum_n s_n(t + ||R - R_n||/c) = \sum_n s_n(t - T_n) \tag{4}$$

The expression for the norm may be expanded to give

$$||R - R_n|| = ([R - R_n, R - R_n])^{1/2} = (||R||^2 - 2[R, R_n] + ||R_n||^2)^{1/2} \tag{5}$$

which in the Fresnel approximation is

$$||R - R_n|| \approx ||R|| - [R, R_n]/||R|| + ||R_n||^2/(2\,||R||) \tag{6a}$$

$$T_n \approx -||R||/c + [U, R_n]/c - ||R_n||^2/(2c\,||R||), \qquad U = R/||R|| \tag{6b}$$

The three terms in the above equation correspond respectively to a gross delay to the origin of the array, a steering delay, and a focussing delay. Ignoring the gross delay to the origin of the array the beamformer output in the jth look direction is

$$b_j(t) = \sum_n s_n(t - T_{nj}) \tag{7a}$$

$$T_{nj} = -(1/c)(-[U_j, R_n] + ||R_n||^2/(2\,||R||)) \tag{7b}$$

In sonar signal processing the fractional bandwidth of the received signal may be sufficient to justify the wide bandwidth interpretation of the above time domain beamformer. However, in radar signal processing the fractional bandwidth of the received signal is always sufficiently small to justify a frequency domain intrepretation of beamforming. Expressing s(t) and b(t) interms of their Fourier transforms

$$s_n(t) = \int S_n(f)\, e^{i2\pi ft}\, df \tag{8a}$$

$$b_j(t) = \int B_j(f)\, e^{i2\pi ft}\, df \tag{8b}$$

Then

$$s_n(t - T_{nj}) = \int S_n(f)\, e^{-i2\pi f T_{nj}}\, e^{-i2\pi ft}\, df \tag{9}$$

Assuming all sources are in the Fresnel region or the far field the frequency-domain beamformer is given by the following expressions

$$B_j(f) = \sum_n S_n(f)\, \underbrace{e^{-i2\pi[U_j,R_n](f/c)}}_{\text{Steering Phase Shift}}\, \underbrace{e^{i2\pi(f/c)(\|R_n\|^2/2\|R\|)}}_{\text{Focussing Phase Shift}} \tag{10a}$$

$$B_j(f) = \sum_n S_n(f)\, e^{-i2\pi[U_j,R_n]/\lambda}\, e^{i\pi\|R_n\|^2/(\lambda\|R\|)} \tag{10b}$$

where $U_j = R/\|R\|$ is a unit vector in the desired look direction. Inspection of equations 10 show that the focussing phase shifts are independent of the steering direction and the steering phase shifts and independent of the focussing required for sources in the Fresnell region. Usually the sources can be assumed to be all in the far filed and the focussing phase shifts may be omitted. For an array of diameter D the assumption of far field is justified if the distance R to the source satisfies $R \gg D^2/\lambda$, where $\lambda = c/f$ is the wavelength corresponding to the frequency f.

The equations of the time-domain and frequency-domain beamformer may both be interpreted as matrix multiplication. Thus parallel processing architectures such as systolic arrays and wavefront processors could be used for digital phased array beamforming. However, since the processing is so computationally intensive for an arbitrary configuration of receiving elements either regular array geometries or arrays of sub-arrays are often used instead.

The equation of the time-domain and frequency-domain beamformer may both be interpreted as matrix multiplication. Thus parallel processing architectures such as systolic arrays and wavefront processors could be used for digital phased array beamforming. However, since the processing is so computationally intensive for an arbitrary configuration of receiving elements either regular array geometries or arrays of sub-arrays are often used instead.

For many regular geometries, frequency-demain beamforming either provides a reduction in the required number of multiplications through the use of the fast Fourier transform (FFT), or the equivalent use of a large number of simultaneous miltiplies through the use of transversal filters and the chirp-Z-transform (CZT) algorithm. Consider the line array geometry illustrated in Fig. 7b. The beamformer output is

$$G(a) = \sum_n g_n e^{-i2\pi f X_n (\cos a)/c} \qquad (11)$$

since $[R_n, U] = X_n \cos a$.

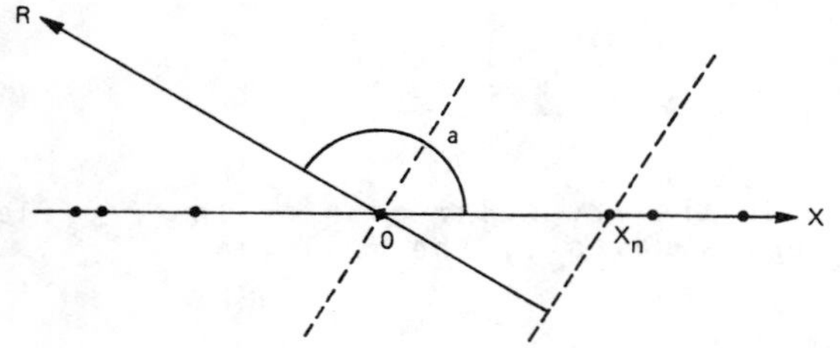

FIGURE 7b. Line array beamforming.

For the special case of a uniformly spaced line array, the beamformer equation reduces to

$$G(a) = \sum_n g_n e^{-i2\pi n[fd(\cos a)/c]} \qquad X_n = nd \qquad (12)$$

which is equilivant to a discrete Fourier transform (DFT) as is shown in the following equation where the angle θ_k satisfies the expression $\cos\theta_k = (k/N)(\lambda/d)$.

$$G_k = \sum_{n=0}^{N-1} g_n e^{-i2\pi nk/N} \qquad (13)$$

Thus a two-dimensional temporal-spatial Fourier transform can form beams in a set of nonuniformly spaced look directions

$$G(\theta) = \sum_{n=0}^{N-1} g_n e^{-i2\pi(nd\cos\theta)/\lambda} \qquad \theta_k = \cos^{-1}(k\lambda/Nd) \qquad (14)$$

This type of signal processing is equivalent to decomposing the incoming signal field into a frequency, wave-number spectrum. Thus the extension of this technique to uniformly spaced two-dimensional arrays is straight forward. Williams [3] describes this technique when the DFT is replaced by the FFT.

Figure 8 shows an example of frequency-wave number beamforming. The output of each sensor is Fourier transformed using either the FFT or CZT algorithm. The results of this initial processing is stored in a temporary memory for subsequent processing. If the initial data was stored in row format then it is read out in column format and processed a second time with a Fourier transform algorithm. In other words a two-dimensional Fourier transform is used on the signal.

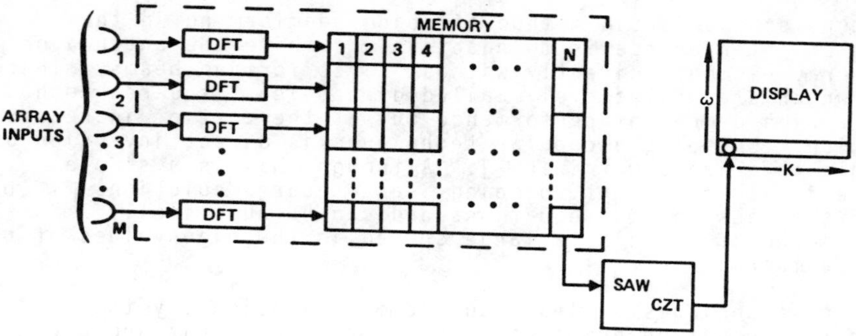

FIGURE 8. Frequency-wavenumber beamformer.

If this beamformer is to work in real time a whole column of data must be transformed in the second dimension while a single Fourier coefficient is being computed in the first dimension. Thus it may be desirable to use two different technologies to implement the transformations in the two directions. In the example shown the first DFTs could be implemented digitally while the second DFT is shown as being implemented using the CZT algorithm with a surface acoustic wave (SAW) device.

Adaptive Array Techniques

As the signal environment for an antenna or sensor array becomes more complex the possibility of intentional and unintentional interference increases. Conventional sensor processing techniques such a beamforming do not control the location of the nulls in the sidelobes of the resulting beam patterns. The first approach to solving this problem is to use auxiliary beams which can be steered in the direction of the interference and then to subtract the interference from the main beam response. Although this technique has been used it requires the the auxiliary beams have nulls in the direction of the desired signal otherwise that signal will be reduced along with the interference.

The gains and phases of the auxiliary antennas have to be continually adjusted to compensate for changes in interference. Thus a number of adaptive techniques have been developed which in either analog or digital form minimize the interference power in the desired beam direction [4]. From a signal processing perspective the simplest of the algorithms are the gradient descent algorithms which iteratively solve a least squares problem. The difficulty with these algorithms are that even for a stationary interference environment the algorithms have to iterate to a solution and in a dynamic environment the convergence rate of the algorithm may be slow due to a large spread in the eigenvectors of the corresponding covariance matrix. Inspite of these difficulties these algorithms are often used since there are simple digital implementations available.

A more sophisticated method of doing beamforming in the presence of interference is to adaptively combine the antenna or sensor array elements directly without first forming beams. These methods which are collectively called direct least squares methods provide a high degree of performance but at the cost of increased computation. One of the popular techniques is direct inversion of the sample covariance matrix [4]. Although this is a simple technique it is equilivant to doing least squares solutions [5] by means of the Gauss normal equations and can be numerically illconditioned if there is a large spread in the eigenvalues of the covariance matrix.

Modern techniques of least square minimization try to avoid altering the condition number of the covariance matrix and are equilivant to square root methods of solving Kalman filters. However, as the numerical accuracy of the least square beamforming method increases so does the computational complexity of the algorithms. It is thus important to develop new high accuracy numerically stable algorithms and architectures if digital signal processing is going to be applied to radar and sonar phased arrays.

One method of solution is to apply digital parallel pipeline computational techniques to the adaptive beamforming problem [6]. Systolic arrays and wavefront processors which are composed of many nearest neighbor interconnected processing elements appear to be well suited to these types of computational problems. Several of these architectures are described elsewhere in this volume.

As a result of the concurrent advances in numerical linear algebra, parallel architectures for computation, and VLSI hardware implementations, it is no longer necessary for the signal processing engineer to use suboptimal computations when optimal computations will provide improved performance. Direct least squares minimization and iterative eigensystem calculation can be considered for future signal processing systems. The use of the singular value decomposition (SVD) and the generalized SVD make possible the use of single precision computation in many applications where double precision computation was previously required. Although much research still needs to be done in applying these new techniques to situations with time varying statistics, it is clear that the ability to implement computational macros in silicon and to connect them together without conflicts for access to resources will make possible many new and important advances in signal processing.

IMAGING RADARS

There are two primary classes of imaging radars, Synthetic Aperature Radars (SAR) and Inverse Synthetic Aperature Radars (ISAR). Fig. 9 shows the typical geometry of a SAR flight path. The airplane tries to fly a straight trajectory and in radar looks orthogonal to the flight path. This SAR geometry is sometimes referred to a "side looking radar."

The object of the signal processing for the SAR is to form an image of the terrain with the same resolution in cross range as in range. In order to acheive this objective a pulse compression signal such as a linear fm chirp is transmitted. Rihaczek shows that the range resolution of this signal is inversely proportional to the signals bandwidth. Thus it is sufficient to use a fixed pulse compression matched filter for each succesive radar pulse.

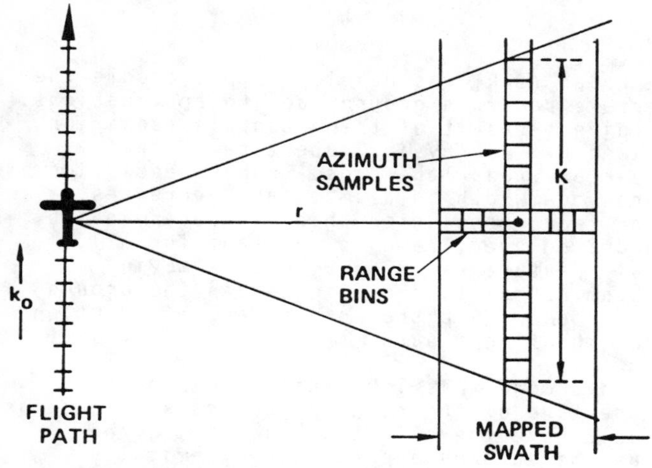

SAR processor computes:

$$g_{r,k_o} = \sum_{k=\frac{-K}{2}+1}^{\frac{K}{2}} h_{r,k}\, f_{r,k_o+k}\,, \qquad k_o = \frac{K}{2}, \frac{K}{2}+1, \ldots$$

where $h_{r,k}$ is of the form $e^{\frac{ia k^2}{r}}$

and $f_{r,\ell}$ is the received signal from range bin r on the ℓth pulse.

FIGURE 9. SAR geometry.

Cross range resolution is more difficult to obtain. Since it is desired to have constant cross range resolution then the coherent integration time in cross range must be range dependant. In addition, if pulse compression techniques are to be employed in the cross range direction then the radar pulses must be coherent from pulse to pulse.

Only a few books have been written on SAR radar technology and the reader is encouraged to consult a journal reprint volume such as Kovaly's. In these papers the history of SAR is chronicled from its beginning using optical processing to the current time when both analog and digital technology is used.

In SAR the dificult processing is the cross range signal which is different for each range. From a point fixed on the ground the doppler signal starts when the radar first illuminates the ground point and continues through zero doppler when the plane is at its point of closest approach. It then continues until the radar stops illuminating the point. The doppler history of a point on the ground is thus determined by the beamwidth of the radar, the slant range to the point and the geometry of the flight path which can be inertially measured.

Synthetic Aperture Radar

Synthetic aperture radar (SAR) was developed to overcome the limitations of real aperture radar in ground mapping applications. These limitations are a direct result of the antenna's radiation pattern which is determined by the physical aperture of the antenna. With a real aperture radar the cross- range resolution is determined by the antennas beamwidth in azimuth and decreases as the range increases. The range resolution which is determined by the bandwidth of the radar, however, remains constant for all ranges. Thus when a real aperture radar is used on a moving platform with its main beam normal to the flight path the ground near the flight path is mapped with better cross-range resolution than the ground away from the flight path.

One possible way to overcome this limitation would be to dynamically change the aperture of the radar as the radar signal is being received. This would have the effect of increasing the cross-range resolution as the antenna aperture is dynamically increased. However, there is a signal-to-noise penalty associated with this mode of operation since the radar would receive energy from only a portion of the illuminated ground at long range since the transmitting aperture is constant. Despite these limitations this method of cross-range resolution improvement has been used successfully in a side looking sonar.

Alternatively, the received radar signal could be decomposed into frequency cells and each frequency cell processed separately. If the received signal comes entirely from illumination restricted to one side or the other of the flight path then the doppler shift induced in the radar return by the motion of the platform can be used to improve the effective cross-range resolution of the radar. This method of processing is sometimes referred to as doppler beam sharpening and it is equivalent to using doppler to interpolate the response of the receiver beamwidth.

Two observations should be made about doppler beam sharpening. The first is that the signal-to-noise ratio decreases more rapidly with range since there is no corresponding increase of antenna aperture with range. The second is that the cross-range resolution is still range dependent since the doppler filters simply interpolates the antenna beam pattern. However, if the doppler filters were to become narrower as the range increased then the cross-range resolution could be made independent of range.

Since the bandwidth of a finite duration sinusoid decreases in direct proportion to the increase in the duration of the sinusoid, range dependent doppler bandwidth implies range dependent doppler filtering. If matched filtering is used in order to maximize the signal-to-noise ratio of the doppler filter then the duration of the matched filter's impulse response will increase as the range increases. Thus for a radar moving at a constant velocity the distance over which the antenna is sampling the returning signal will depend on range. For this reason radars based on range dependent doppler filtering are called synthetic aperture radars (SARs). A good tutorial review paper on SAR is [7] by Kiyo Tomiyasu.

If the real antenna of the radar is oriented so that the boresite of the antenna is normal to the direction of motion of the radar and if the real beamwidth of the radar is not too large then the doppler shift introduced by the motion of the radar will be linear as the radar passes the scatter. These radars are often called side looking radars (SLRs) and were the first types of SARs since the signal processing is simpler than for SARs with the boresite of the antenna squinted in the direction of motion of the radar. However, the principles of the signal processing are the same for both types of radars and the operation of the SLR is easier to understand.

In the previous discussion it has been assumed that the change in the range to the radar scatter does not change by more than a fraction of a wave length during the integration of the radar return. This type of radar is called an unfocused SAR and although its performance in cross-range resolution is better than that of a conventional radar cross-range resolution still depends on range. If the phase of the radar returns are corrected when the range to the radar scatter changes by more than a fraction of a wavelength then the radar is called a focused SAR. It is this type of SAR which has cross-range resolution independent of range.

These results can be summarized as follows. If the slant range to the target is denoted by R, the velocity of propagation of the radar wave by c and the frequency of the carrier by f, then the following relations hold for cross-range resolution:

$$\text{Resolution(conventional)} = cR/fD$$

$$\text{Resolution(unfocused)} = 0.5(cR/fD)^{\frac{1}{2}}$$

$$\text{Resolution(focused)} = D/2$$

where D is the horizontal aperture of the antenna. Although the expression for the resolution of the conventional radar, as given, goes to zero as the range goes to zero the expression is only valid in the far field of the antenna for ranges greater than R(minimum) = D(Df/c) where (Df/c) is the horizontal dimension of the antenna measured in wavelengths. In the near-field the minimum conventional resolution is D. Thus the focused SAR has a potential resolution one-half of that of the conventional radar at its best.

Signal processing considerations

In the previous section the signal processing aspects of the SAR have been simplified in order to appreciate the fundamental considerations of SAR operation. In actual practice the integration time of the doppler filters is much greater than the interpulse period of the radar. Thus many radar transmissions must be accumulated before doppler filtering. It is therefore necessary that each radar pulse be phase coherent with its predecessors. It also means that many radar returns must be stored in memory before initial azimuth processing can begin. Therefore the signal processing of a SAR is inherently two-dimensional. For these reasons the first SAR processors stored the radar returns on film and the film was subsequently processed using an optical processor. Even the Seasat spaceborne SAR which was put into orbit in June 1978 used optical processing for full coverage with digital processing used only for selected areas of observation.

For real-time processing either analog or digital techniques can be used. The simplest pulse modulation technique to use to achieve range modulation is to frequency modulate the radar carrier with a linear sweep. Then, if the radar is a SLR the received signal will be simultaneously frequency modulated in both range and azimuth. The corresponding radar return from a point scatterer will be a Fresnel zone plate which can be reconstructed using anamorphic optics. If the signal is to be electronically processed then range compression can be done while the radar signal is at an intermediate frequency (IF) and the azimuth compression can be done at baseband. An appropriate combination of processors might be a surface acoustic wave (SAW) [8] matched filter at IF and a charge coupled device (CCD) [9] matched filter at baseband. Alternatively, digital processing using the fast Fourier transform (FFT) [10] could be employed after analog to digital (A/D) conversion of the inphase and quadrature components of the baseband signal.

If the SAR has only modest resolution and is flown at aircraft altitudes then the signal processing is separable as indicated above. That is, the Fresnel zone plate formed by a point scatterer is an ellipse with its minor axis in the range direction and its major axis in the azimuth direction and azimuth compression is independent of range compression. If the SAR has high resolution or if it is flown at orbital altitudes then the signal processing may not be separable. That is, the Fresnel zone plate formed by a point scatterer may be a distorted ellipse and a two-dimensional matched filter may be required. If this is the case the signal processing is significantly more difficult. The Seasat optical processor is able to compensate for some "range curvature" distortion of the Fresnel zone plate.

A more recent SAR radar is the Shuttle imaging radar-A (SIR-A). This radar was flown on the second flight of the Columbia shuttle in November 1981. This radar is similar to the seasat radar but with lower resolution, 38 m for SIR-A versus 25 m for seasat. The SIR-A radar acquired a total of 7 1/2 hours of SAR data which was recorded on an on-board optical recorder. The total area covered was about 10 million square kilometers, and the film was processed on an optical processor after the flight.

Another aspect of SAR processing has to to do with the nature of the noise in the reconstructed image. Since the radar is a source of narrow bandwidth coherent illumination the reconstructed images even for high radio frequency (RF) signal-to-noise ratio will exhibit a spatial noise much like that of a visible laser image. In order to overcome this noise multiple looks of the radar are averaged together by the processor. For seasat 4 looks are used while for SIR-A 6 looks are used.

Figure 10a shows the areas of the world which have been mapped as of January 1, 1983. Also included are the areas of SEASAT SAR coverage. SIR-A images can be obtained by requesters within the United States at the following address:

National Space Science Data Center
Code 601.4
NASA-Goddard Space Flight Center
Greenbelt, Maryland 20771

Scientists outside the United States should direct their requests to:

World Data Center A
Rockets and Satellites
Code 601
NASA-Goddard Space Flight Center
Greenbelt, Maryland 20771

Seasat SAR images are available from:

National Oceanic and Atmospheric Administration
Satellite Data Services Division
Room 100, World Weather Building
Washington, D.C. 20233

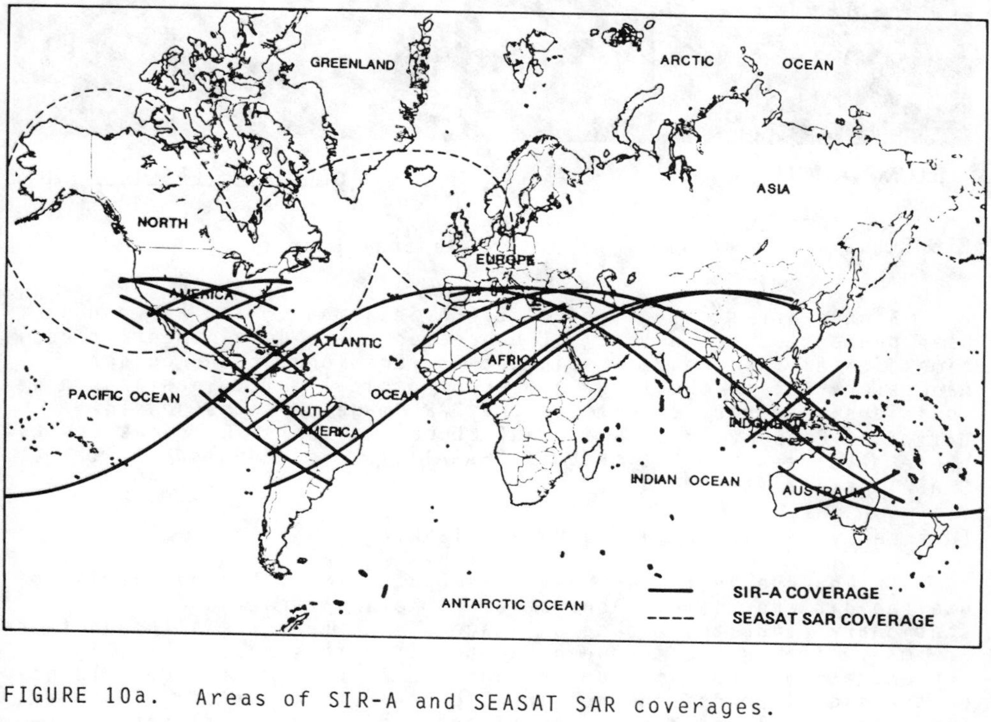

FIGURE 10a. Areas of SIR-A and SEASAT SAR coverages.

Figure 10b is a SIR-A image of an area of coastal Belize in Central America and the adjacent ocean and islands. The area has a heavy cover of tropical vegetation and thus the SIR-A image shows an even, diffuse radar return that is modulated by slope changes. The mainland coast is characterized by marsh, lagoons, barrier bars and offshore barrier reefs.

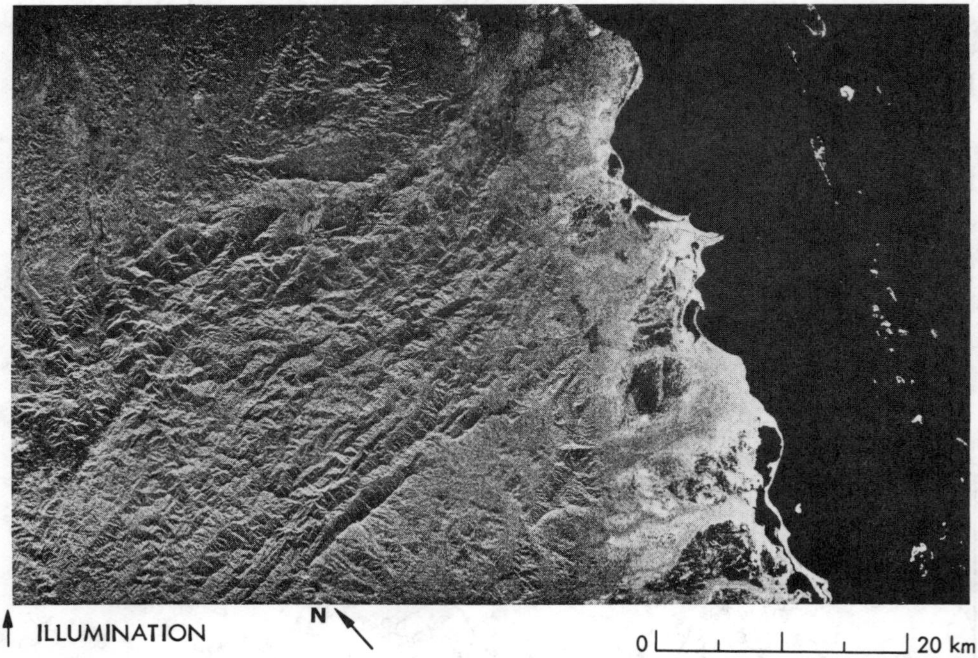

FIGURE 10b. SIR-A image of Belize Central America.

Figure 10b shows a typical SIR-A SAR image. The image has been processed with an optical processor. The development of real-time digital SAR processors is not as far along as the optical processors due to the large number of digital multiplications and additions necessary to compute one SAR image. The MacDonald-Dettwiler Company in Canada manufactures digital SAR processors and these have been used to process SEASAT images inaddition to the SEASAT optical processor.

Inverse Synthetic Aperature Radar (ISAR)

In inverse synthetic aperature radar (ISAR) the object is to use the differential motion between a moving target and a stationary radar to produce an image of the target. Although there are many similarities between SAR and ISAR this section will concentrate on the differences since these have a strong influence on the signal processing required to generate an image. Consider first the simpliest case of an aircraft in straight flight past a stationary radar. This case most closely corresponds to a SAR geometry. However, in the case of the SAR the moving radar platform could be instrumented so that departures from a straight path could be compensated for before trying to form an image. In the case of the ISAR departures of the target aircraft from a straight path have to be measured first and then the data from the radar reprocessed inorder to form an image. If a high resolution image is required there may be more than one step in the estimation and correction of the target motion.

C. C. Chen discusses some of the techniques which may be used for ISAR motion compensation in his University of Southern California thesis "Imaging with Radar Returns", August 1978. In this thesis he developes a method of estimating the range motion of the target from averaged range histories of the targets motion. Figures 10a and 10b show a typical range history and the corresponding reconstructed image.

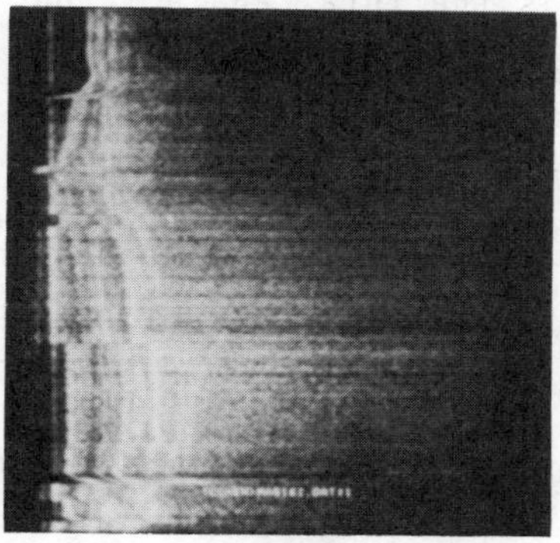

FIGURE 11a. Log magnitude of every 16th pulse return.

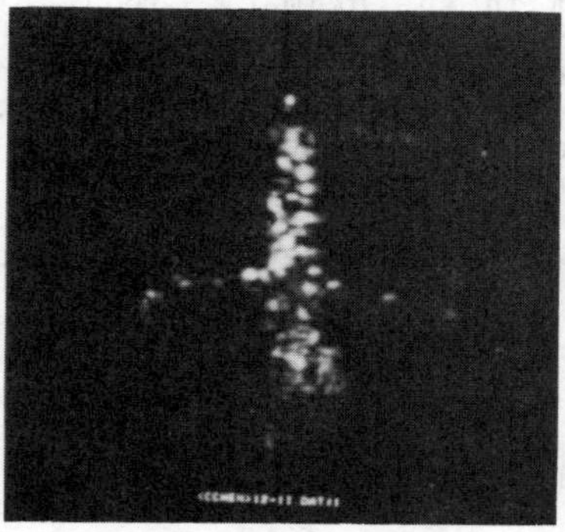

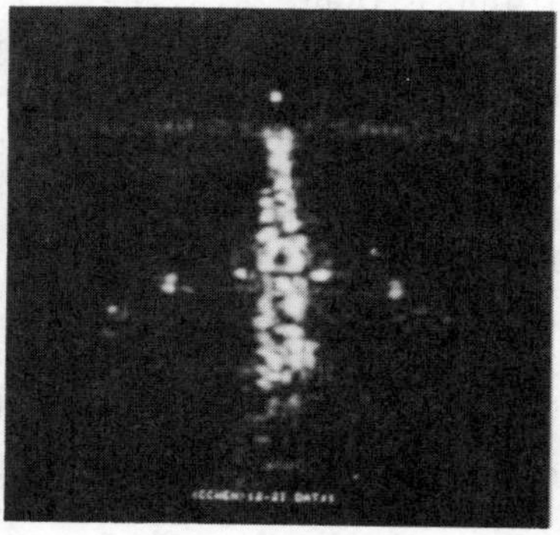

FIGURE 11b. Aircraft radar images. a) 1st 2.5 seconds or 256 signatures (approx. 4.5 degrees aspect change). b) 2nd 2.5 seconds.

Now consider the case of the target aircraft executing an arbitrary flight path. The yaw component of the compensated aircraft motion will generate the same image as the SAR, however, the pitch and roll motions of the target will generate different images in the reconstruction algorithm and the image will be distorted unless corrections are introduced for the unwanted motion components. In order to simplify the discussion only yaw motion will be considered at this point. Since the straight line motion of the target was only used to provide an equilivant yaw motion it simplifies the discussion to consider the target as a rotating rigid body.

In this simplified discription the role of the signal processing is to estimate the magnitude of the rotation and to construct a range-doppler image of the illuminated scattering distribution of the target. Since the doppler or cross-range resolution of the signal is proportional to the duration of the signal and more cross-range resolution is required for slowly rotating targets it follows that it is the total angle through which the target has rotated and the frequency, i.e. doppler sensitivity, of the radar that determine the quality of the reconstructed image. For typical radar frequencies the total angle of rotation needed is about 1 to 3 degrees.

If approximately equal resolution is required in range and cross-range then both large signal bandwidth and long coherent integration times will be required. It is thus difficult to satisfy the assumption of constant rotation rate as well as difficult to implement the necessary transmitted signal. For these reasons the use of new techniques of ISAR imaging should be considered. One of the most promising of these new techniques is to model the SAR or ISAR image formation process by analogy with tomographic reconstruction [2].

CONCLUSION

Radar and active sonar signal processing are both computation intensive. Two types of signal processing dominate the computational load. In the time domain the primary computation is matched filtering or ambiguity function calculation. In the spatial domain the computation is beamforming and interferrence rejection. Although there are large differences in the bandwidths which must be processed in radar and sonar the effective computational loads are similar. In radar the velocity of propagation is high, therefore, the pulse repetition rate is usually high and only one beam look direction needs to be processed each transmission. However, the processing of this one look direction often taxes the capability of a digital processor and analog circuits are often used both for the temporal signal processing and for the spatial processing in phased array systems.

In active sonar the propagation velocity is so slow that all look directions must be processed simultaneously. This resulted in phased array technology being applied to sonar many years before being applied to radar. However, the computational load induced by the simultaneous computation in many look directions makes up for the simplicity of the computations in a given look direction.

Digital signal processing is well matched to the computational load of sonar although analog processing may be used when size, weight, or power constraints limit the application of digital techniques.

However, in radar the time processing of a single beam output often exceeds the capacity of a simple digital processor and the use of analog surface acoustic wave (SAW) pulse compression filters is often considered. If digital signal processing is going to be used for beamforming and interference cancellation then new parallel signal processing techniques such a systolic array and wavefront matrix processors will be needed in the future.

REFERENCES

[1] R. H. Barker , "Group Synchronizing of Binary Digital Systems," in Communication Theory, Edited by Jackson, W. Butterworths Scientific Publications, 1953, pp. 273-287.

[2] M. Bernfeld, CHIRP Doppler Radar, Proceedings of the IEEE, vol. 72, no. 4, April 1984, pp. 540-541.

[3] J. R. Williams, Fast Beam Forming Algorithm, Journal of the Acoustical Society of America, Vol 44, pp.1454-1455, 1968.

[4] R.A. Monzingo and T.W. Miller, Introduction to Adaptive Arrays, John Wiley & Sons, New York, 1980

[5] C.L. Lawson and R.J.Hanson, Solving Least Squares Problems, Prentice-Hall, 1974

[6] J.M. Speiser and H.J. Whitehouse, "Parallel Processing Algorithms and Architectures for Real-Time Signal Processing", SPIE Vol. 298, Real-Time Signal Processing IV (1981), paper 298-01

[7] K. Tomiyasu , "Tutorial review of synthetic aperture radar with application to imaging of the ocean surface." Proc IEEE, vol. 66, pp. 563-583, 1978.

[8] H. Matthews, Editor, Surface wave filters: Design, Construction and Use. New York: Wiley, 1977.

[9] C. H. Sequin and M. F. Tompsett, Charge Transfer Devices. New York: Academic Press, 1975.

[10] E. O. Brigham, The Fast Fourier Transform. Englewood Cliffs, NJ: Prentice-Hall, 1974.

BIBLIOGRAPHY

Radar texts

C.J.A Bird, Radar Precision and Resolution, John Wiley & Sons, New York, NY, 1974.

E. Brookner, Radar Technology, Artech House, Dedham, MA, 1977.

W.S. Burdic, Radar Signal Analysis, Prentice-Hall, Englewood Cliffs, NJ, 1968.

M.H. Carpentier, Radars:New Concepts, Gordon and Breach, New York, NY, 1968.

J. Constant, Introduction to Defense Radar Systems Engineering, Spartan Books, New York, NY, 1972.

C.E. Cook and M. Bernfeld, Radar Signals an Introduction to Theory and Application, Academic Press, New York, NY, 1967.

S.A. Hovanessian, Introduction to Synthetic Array and Imaging Radars, Artech House, Dedham, MA, 1980.

F.E Nathanson, Radar Design Principles Signal Processing and the Environment, McGraw-Hill, New York, NY, 1969.

D.R. Rhodes, Introduction to Monopulse, Artech House, Dedham, MA, 1980.

A.W. Rihaczek, Principles of High-Resolution Radar, McGraw-Hill, New York, NY, 1969.

M.I. Skolnik, Introduction to Radar Systems, McGraw-Hill, New York, NY, 1962.

M.I. Skolnik, Radar Handbook, McGraw-Hill, New York, NY, 1970.

D.E. Vakman, Sophisticated Signals and the Uncertainty Principle in Radar, Springer-Verlag, New York, NY, 1968.

P.M. Woodward, Probability and information theory, with Applications to Radar, Pergamon Press, Oxford, UK, 1953.

Journal article reprint volumes and Conference Proceedings

D.K.Barton,
Vol 1, Monopulse Radar,

Vol 2, The Radar Equation,

Vol 3, Pulse Compression,

Vol 4, Radar Resolution and Multipath Effects,

Vol 5, Radar Clutter,

Vol 6, Pulse Compression,

Vol 7, CW and Doppler Radar,
Artech House, Dedham, MA, 1974-1978.

IEEE International Radar Conference Proceedings,
1975:75 CHO 938-1 AES

1977:77CH1271-6 AES

1980:80CH1493-6 AES
The Institute of Electrical and Electronics Engineers, New York, NY.

J.J. Kovaly, Synthetic Aperture Radar, Artech House, Dedham, MA, 1976.

Radar/Sonar and Sonar texts

R. Benjamin, Modulation, Resolution and Signal Processing in Radar, Sonar and Related Systems, Pergamon Press, Oxford, UK, 1966.

C.W. Horton, Signal Processing of Underwater Acoustic Waves, US Government Printing Office, Washington D.C., 1969.

R.J. Urick, Principles of Underwater Sound, McGraw-Hill, New York, NY, 1975.

H.L. Van Trees, Detection, Estimation, and Modulation Theory, Part III, Radar/Sonar Signal Processing and Gaussian Signals in Noise, John Wiley and Sons, New York, NY, 1974.

Conference Proceedings

University of Birmingham, Proceedings of the Symposium on Signal Processing in Radar and Sonar Directional Systems, Institution of Electronic and Radio Engineers, London, UK, 1964.

9

Digital Image Processing: Problems and Methods

ANIL K. JAIN

A review of digital image processing problems and techniques is presented. Some recent advances in this field are blended with known successful algorithms to present an up-to-date perspective. Theoretical framework of different problems is discussed to explain the advantages and limitations of various techniques.

TABLE OF CONTENTS

Research supported in part by the Army Research Office under Grant DAG29-82-K-0077 and in part by ONR under Grant N00014-81-K-0191 as part of a Special Research Opportunity (SRO) project.

I. INTRODUCTION

Digital image processing refers to manipulation of two dimensional data sets by a digital computer. Fig. 1 shows the key steps in digital image processing. An input object, transparency, slide, photograph, or a chart is first digitized and stored as a matrix of binary words in the computer memory. This digitized image is subsequently processed and displayed on a high resolution television monitor. For display, the image is stored in a rapid access buffer memory which refreshes the monitor periodically to produce a visibly continuous display. A digital computer controls all the digitization, storage, processing, and display operations. Program inputs to the computer can be made through a terminal and the outputs are available on a terminal, television monitor, or a printer/plotter.

Digital image processing has found numerous applications, common examples of which are satellite image processing, space image applications, image transmission and storage applications, medical image processing, radar, sonar and acoustic image processing, robotics, etc. Although image processing applications are numerous and problems are many, for our purpose here we will consider the following basic classes of problems that occur in image processing.

1. Image Representation
2. Image Enhancement
3. Image Restoration
4. Image Data Compression
5. Image Reconstruction
6. Image Analysis

II. IMAGE REPRESENTATION

In image representation one is concerned with characterization of the quantity that each picture-element (also called pixel or pel), represents. An image could represent luminances of objects in a scene (pictures taken by an ordinary camera), the absorption characteristics of the body tissue (in x-ray imaging), the radar cross-section of a target (in radar imaging), the temperature profile of a region (in infrared imaging) or the gravitational field in an area (in geophysical imaging). In general any two dimensional function which bears information is an image. Image models give a logical or quantitative description of the properties of this function.

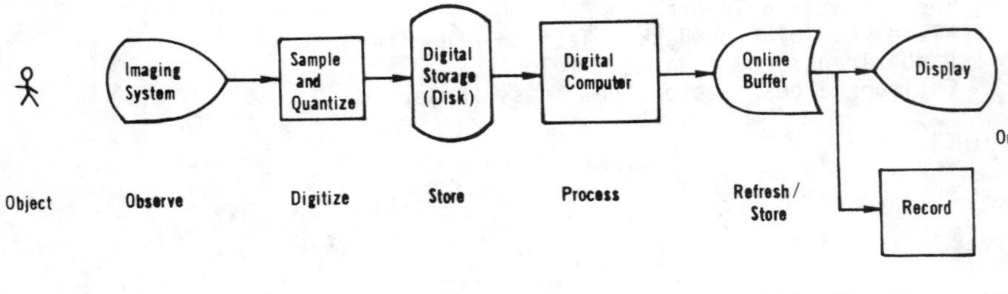

Figure 1: A Typical Digital Image Processing Sequence

Sampling and Quantization

The fundamental requirement for digital processing is that images be sampled and quantized. The sampling rate has to be large enough to preserve the useful information in an image and is determined by the bandwidth of the image. For example, the bandwidth of raster scanned common television signal is about 4 MHz. From the sampling theorem this requires a minimum sampling rate of 8 MHz. At 30 frames/s and a 512 line raster this means each image frame should contain approximately 512 x 512 pixels.

Image quantization is the analog to digital conversion of a sampled image to a finite number of gray levels. The number of bits required to represent the minimum detectable contrast of the human visual system is approximately six bits (=64 levels). Since contrast is logarithmically related to the luminance scale, roughly 8 bits (=256 levels) are sufficient for visual perception of monochrome images.

Series Representations or Image Transforms

A classical method of representing signals is by series expansion in terms of orthogonal basis functions. A common example is the Fourier series. For images, analogous representation via two dimensional orthogonal functions called basis images is possible. For sampled images the basis images are determined from unitary matrices generally called image transforms. If U denotes an NxN image then its A-transform is generally defined as a separable transformation

$$V = A\,U\,A^T \tag{1}$$

where A is an NxN unitary matrix i.e.,

$$A^{-1} = A^{*T} \text{ or } A\,A^{*T} = A^{*T}A = A^TA^* = I \tag{2}$$

Inverse transformation of (1) gives

$$U = A^{*T}VA^* \tag{3}$$

This can also be written as

$$U = \sum_{k,\ell=0}^{N-1}\sum v(k,\ell)a_k^{*T}a_\ell^* = \sum_{k,\ell=0}^{N-1}\sum v(k,\ell)B_{k,\ell} \tag{4}$$

where a_k^* is the kth row of A^* and

$$B_{k,\ell} = a_k^{*T}a_\ell^*\ ,\quad 0\leq k,\ell\leq N-1 \tag{5}$$

are called the basis images of the A-transform. Eqn. (4) gives an expansion of U in an N^2-dimensional vector space. The elements $v(k,\ell)$ are called the A-transform coefficients. The basis images $B_{k,\ell}$ from a complete orthonormal basis for an N^2-dimensional vector space.

Generally, the transform A is chosen to be what is called a fast transform. Such transforms have structural properties which lead to fast Fourier transform (FFT) type algorithms, i.e., a transformation of the type y=Ax for an Nx1 vector x can be performed in O(NlogN) operations. Therefore, the two dimensional operation of (3), which requires A-transformation of each column of U followed by the A-transformation of each row of the previous result, can be completed in

$O(N^2 \log N)$ operations. Examples of common fast unitary transforms are the discrete Fourier (DFT), Cosine (DCT), Sine (DST), and Walsh-Hadamard (WHT) transforms and are defined below.

$$\text{DFT:} \quad a(m,n) \triangleq \frac{1}{\sqrt{N}} \exp\left(-j\frac{2\pi mn}{N}\right), \quad 0 \le m,n \le N-1 \tag{6}$$

$$\text{DCT:} \quad a(m,n) \triangleq \begin{cases} \frac{1}{\sqrt{N}}, & m = 0, \; 0 \le n \le N-1 \\ \sqrt{\frac{2}{N}} \cos \frac{m(2n-1)\pi}{2N}, & 1 \le m \le N-1, \; 0 \le n \le N-1 \end{cases} \tag{7}$$

$$\text{DST:} \quad a(m,n) \triangleq \sqrt{\frac{2}{N+1}} \sin \frac{(m+1)(n+1)\pi}{N+1}, \quad 0 \le m \le N-1 \tag{8}$$

$$\text{WHT:} \quad a(m,n) \triangleq (-1)^{\sum_{i=0}^{p-1} m_i n_i}, \quad m_i, n_i = i^{th} \text{ binary digit (0 or 1)}, \; i=0,1\ldots,p-1 \text{ in the binary expansion of } m \text{ and } n \text{ respectively.} \tag{9}$$

Fig. 2 shows the basis images of the 8x8 Cosine and Hadamrd transforms.

Unitary transforms have three important properties which are useful in image processing.

1. ENERGY CONSERVATION: The energy of the transform coefficients is equal to the image energy, i.e.,

$$\sum_{m,n=0}^{N-1}\sum |u(m,n)|^2 = \sum_{k,\ell=0}^{N-1}\sum |v(k,\ell)|^2 \tag{10}$$

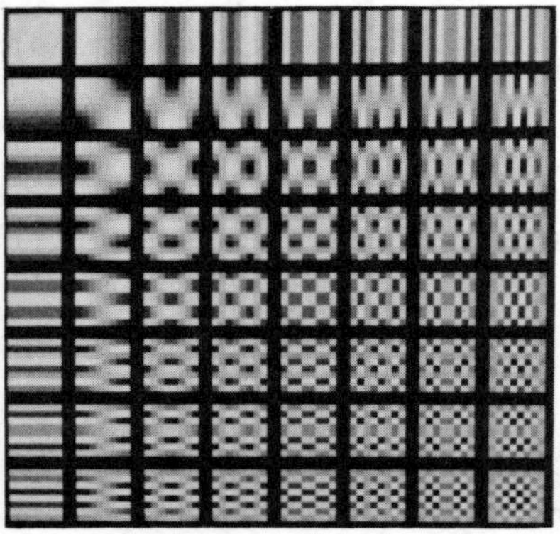

a) Basis Images of the 8 x 8 Two Dimensional Cosine Transform

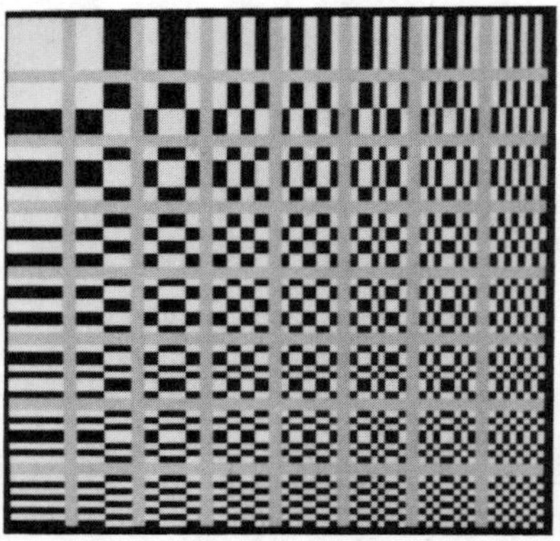

b) Basis Images of the 8 x 8 Two Dimensional Sequency Ordered Hadamard Transform

Figure 2: 8x8 Basis Images

2. ENERGY COMPACTION: Generally, a large fraction of the total energy is compressed in a relatively few transform coefficients. For example, it is quite typical for the above mentioned transforms to compress approximately 90% of the energy in 10% of the transform coefficients of television quality monochrome images.

3. DECORRELATION: Generally, the interpixel correlation of monochrome images is quite high (≈0.95 for 512x512 tv images). However, the transform coefficients tend to be highly decorrelated.

These properties make transform coefficients useful in enhancement, filtering, compression, and feature extraction problems. When an ensemble of images is considered, a transform called the Karhunen-Loeve (KL) transform becomes optimum with respect to the last two properties. It compresses maximum average (over the ensemble) energy in a given number of transform coefficients and decorrelates them perfectly. This means the KL transform gives the minimum mean square error in representing any image from the ensemble by a fixed number of basis images. The KL transform basis images are determined from the eigenfunctions of the image covariance function.

Although the KL transform is optimal, it is not fast in general. The large dimensionality associated with images mandates use of fast transform substitutes of the KL transform. For large image sizes (>256) modeled by stationary random fields there is a class of sinusoidal transforms [21] whose membership includes the DFT, DCT, DST, etc., which approaches the efficiency of the KLT. For smaller image blocks, the DCT is known to be a very good substitute for the KLT when the interpixel correlation is high [21]. This property of DCT makes it useful in block by block processing of images. The DST is another useful transform which approaches (or even exceeds!) the performance of the KLT in block by block processing of the images with overlapping block boundaries [22].

Fig. 3 shows different transforms of an image. The energy compaction property of the transforms is evident. Applications of image transforms will be discussed in the subsequent sections.

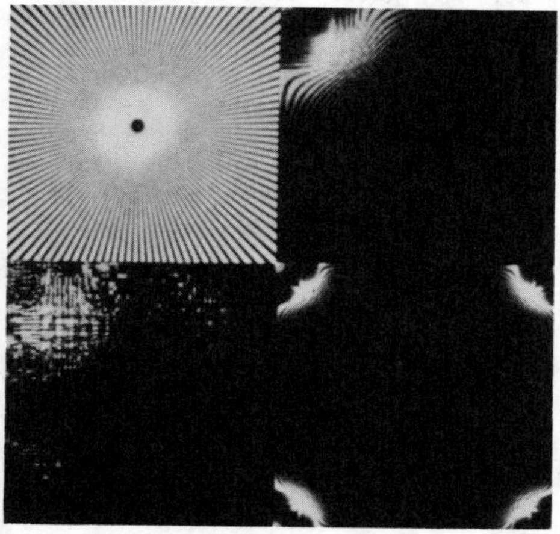

Figure 3: Top row: Original Image and its Cosine Transform. Bottom row: Hadamard and Discrete Fourier Transforms.

Stochastic Representations

Here an image is considered as a member of an ensemble. Generally, for reasons of dimensionality, the ensemble is described by its mean and covariance functions only. Often, one starts with the stationary case so that for an image u(m,n) its mean is constant and its covariance is shift invariant, i.e.,

$$E[u(m,n)] = \mu = \text{constant}$$
$$E[u(m,n)-\mu)(u(i,j)-\mu)] = r(m,n;i,j) = r(m-i,n-j) \tag{11}$$

The two-dimensional eigenvectors of r(m-i,n-j) give the KL transform of this ensemble. Two common covariance functions used in digital image processing are

1. Separable Covariance Model:

$$r(k,\ell) = \sigma^2 \rho_1^{|k|} \rho_2^{|\ell|} \tag{12}$$

2. Nonseparable Exponential Covariance Model:

$$r(k,\ell) = \sigma^2 \exp\{\sqrt{\alpha_1 k^2 + \alpha_2 \ell^2}\} \tag{13}$$

For $\alpha_1=\alpha_2$, (13) is called the isotropic covariance model. Stationary mean and covariance models are useful in image data compression problems such as transform coding, restoration problems such as Wiener filtering and in other applications where global properties of the ensemble are sufficient. A more effective use of these models is to consider them to be spatially varying slowly or piecewise spatially invariant. This requires estimation of short-term spectral properties of the image data.

A method of characterizing short-term or local properties of an image is to characterize the relationship of a pixel with its neighbors. For example, a difference equation forced by white noise, or some other random field with known power spectrum density, can be useful for representing the ensemble. Fig. 4 shows examples of three canonical forms of stochastic models where an image pixel is characterized in terms of its neighboring pixels. If the image were scanned from top to bottom incrementing from left to right, the model of Fig. 4a would be called a causal model. This is because the pixel A is characterized by pixels which lie in the 'past'. Extending this idea, the model of Fig. 4b is a non-causal model because the neighbors of A lie in the 'past' as well as the 'future' in both the directions. In Fig. 4c, we have a semicausal model because the neighbors of A are in the 'past" in the j-direction and are in the 'past' as well as 'future' in the i-direction. In these models ε_A is a random variable which is suitably defined to 'fit' the joint statistics of u_A and its neighbors.

In general these models can be written (for zero mean image ensembles) in the form

$$u(m,n) = \overline{u}(m,n) + \varepsilon(m,n)$$

$$\overline{u}(m,n) \triangleq \sum_{k,\ell \in \hat{S}_x}\sum a(k,\ell)u(m-k,n-\ell)\ ,\ x=1,2,3 \tag{14}$$

where $\hat{S}_1, \hat{S}_2, \hat{S}_3$ correspond to causal (C), semicausal (SC), and noncausal (NC) neighborhoods defined as

$$\hat{S}_1 \triangleq (k,\ell): \quad \{\ell \geq 1, \forall k\} U \{\ell = 0, k \geq 1\} \ , \ C$$

$$\hat{S}_2 \triangleq (k,\ell): \quad \{\ell \geq 1, \forall k\} U \{\ell = 0, \forall k \neq 0\} \ , \ SC \tag{15}$$

$$\hat{S}_3 \triangleq (k,\ell): \quad \{\forall (k,\ell) \neq (0,0\} \ , \ NC$$

The quantity $\overline{u}(m,n)$ represents the interaction phenomenon in the neighborhood of a pixel. One useful way of specifying $\overline{u}(m,n)$ is by requiring it to be the minimum mean square predictor of u(m,n) whose region of support is $\hat{S}_x$, x=1,2,3. This gives rise to the so-called causal, semicausal, and noncausal minimum variance representations (MVRs) of discrete random fields. The prediction filter coefficients a(m,n) can be uniquely determined from the associated orthogonality condition

$$E[\varepsilon(m,n)u(m-k,n-\ell)] = \beta^2 \delta(k,\ell) \ , \ (k,\ell) \in S_x$$

$$\beta^2 \triangleq \min E[\varepsilon^2(m,n)] \ , \ S_x \triangleq \hat{S}_x U(0,0) \ , \ x=1,2,3 \tag{16}$$

which yields the normal equations

$$r(k,\ell) - \sum_{m,n \in \hat{S}_x} \sum a(m,n) r(k-m, \ell-n) = \beta^2 \delta(k,\ell) \ , \ (k,\ell) \in S_x \ , \ x=1,2,3 \tag{17}$$

If $S(\omega_1, \omega_2)$ denotes the power spectrum density of the image ensemble, i.e.,

$$S(\omega_1,\omega_2) \triangleq \sum_{k=-\infty}^{\infty} \sum_{\ell=-\infty}^{\infty} r(k,\ell) \exp[-j(\omega_1 k + \omega_2 \ell)] \tag{18}$$

then for noncausl MVRs (17) can be solved easily for a(m,n) and β^2 as shown in Fig. 5. Note that all we need is the Fourier series of $1/S(\omega_1,\omega_2)$. For semicausal and noncausal models (17) can be reduced to

$$r_\ell(\omega_1) = \sum_{n=1}^{\infty} \hat{a}_n(\omega_1) r_{\ell-n}(\omega_1) + b(\omega_1)\delta(\ell) \ , \ \ell \geq 0 \tag{19}$$

where

$$r_\ell(\omega_1) \triangleq \sum_{k=-\infty}^{\infty} r(k,\ell) \exp(-j\omega_1 k), \ \hat{a}_n(\omega_1) \triangleq a_n(\omega_1)/a_0(\omega_1),$$

$$a_n(\omega_1) \triangleq \sum_{m=-\infty}^{\infty} a(m,n) \exp(-j\omega_1 m), \ n \geq 1$$

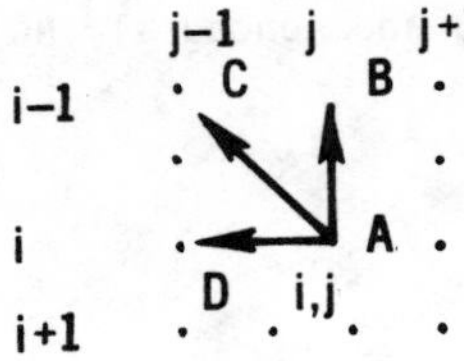

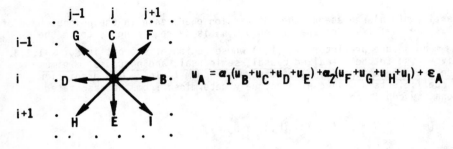

$$u_A = p_1u_B + p_2u_D + p_3u_C + \varepsilon_A$$

a) Causal Model

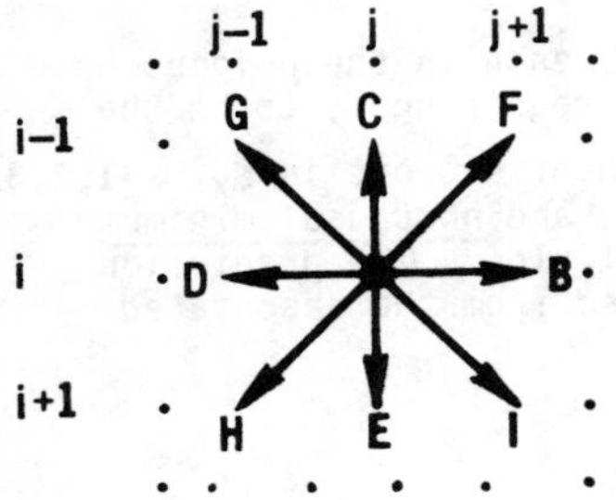

$$u_A = \alpha_1(u_B+u_C+u_D+u_E) + \alpha_2(u_F+u_G+u_H+u_I) + \varepsilon_A$$

b) Noncausal Model

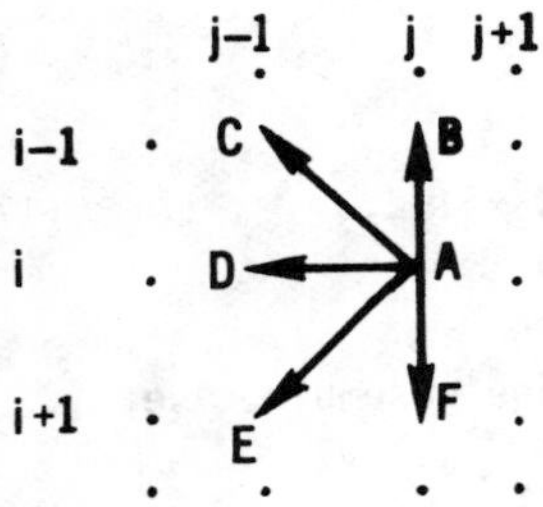

$$u_A = \alpha_1(u_B+u_F) + \alpha_2(u_C+u_E) + u_D + \varepsilon_A$$

c) Semicausal Model

Figure 4: Examples of Three Canonical Forms of Stochastic Models

$$\xrightarrow{1/S(\omega_1,\omega_2)} \boxed{\;F^{-1}\;} \longrightarrow r^+(m,n) \longrightarrow a(m,n) = -r^+(m,n)/r^+(0,0), \quad \beta^2 = 1/r^+(0,0)$$

Figure 5: Realization of Noncausal MVRs

$$b(\omega_1) = \begin{cases} \beta^2/|a_0(\omega_1)|^2, \; a_0(\omega_1) \triangleq 1 - \sum_{m=0}^{\infty} a(m,0)\exp(-j\omega_1 m), \quad \text{C MVRs} & \text{(20a)} \\ \beta^2 a_0(\omega_1), \; a_0(\omega_1) \triangleq 1 - \sum_{\substack{m=-\infty \\ m\neq 0}}^{\infty} a(m,0)\exp(-j\omega_1 m), \quad \text{SC MVRs} & \text{(20b)} \end{cases}$$

Given $r_\ell(\omega_1), \forall \ell$, (19) can be solved recursively by a Levinson type algorithm parametric in ω_1 to obtain $\hat{a}_n(\omega_1)$ and $b(\omega_1)$. For noncausal MVRs $b(\omega_1)$ must be factored, as in (20a), by a one-dimensional all pole filter (AR model) whose frequency response is $a_0(\omega_1)$. For semicausal MVRs, one only needs to calculate the Fourier series

$$1/b(\omega_1) \triangleq \sum_{m=-\infty}^{\infty} b^+(m)\exp(-j\omega_1 m) \tag{21}$$

which yields

$$\beta^2 = 1/b^+(0) \ , \ a(m,0) = -b^+(m)/b^+(0) \tag{22}$$

Once $a_0(\omega_1)$ and $\hat{a}_n(\omega_1)$ are known, the $a(m,n)$ are obtained as the Fourier series coefficients of $a_0(\omega_1)\hat{a}_n(\omega_1)$. Thus it is seen that two-dimensional causal MVRs require a two-stage spectral factorization, the semicausal MVRs require a one-stage spectral factorization and noncausal MVRs do not require any spectral factorization. Defining the prediction error filter PEF as

$$A(z_1,z_2) = 1 - \sum_{m,n \in S_x}\sum a(m,n)z_1^{-m}z_2^{-n} \ , \tag{23}$$

the power spectrum density of the prediction error $\varepsilon(m,n)$ can be shown to be

$$S_\varepsilon(\omega_1,\omega_2) = \begin{cases} \beta^2 \ , & \text{C-MVR} \\ \beta^2 A(e^{j\omega_1},\infty) \ , & \text{SC-MVR} \\ \beta^2 A(e^{j\omega_1},e^{j\omega_2}) \ , & \text{NC-MVR} \end{cases} \tag{24}$$

which yields

$$S(\omega_1,\omega_2) = \begin{cases} \beta^2/|A(e^{j\omega_1},e^{j\omega_2})|^2 \ , & \text{C-MVR} \\ \beta^2 A(e^{j\omega_1},\infty)/|A(e^{j\omega_1},e^{j\omega_2})|^2, & \text{SC-MVR} \\ \beta^2/A(e^{j\omega_1},e^{j\omega_2}) \ , & \text{NC-MVR} \end{cases} \tag{25}$$

The causal and semicausal prediction error filters can also be obtained by Wiener-Doob homomorphic spectral decomposition [23] using the ideas of [27] reported for causal filters. Causal realizations have been studied in the framework of linear prediction by Marzetta [26] also.

Finite Order Models

The models considered above have at least semi-infinite order and are of limited practical use as such. Finite order approximate realizations can be obtained by truncating the filter coefficients in such a way that the resulting systems are stable. For noncausal MVRs, it is sufficient to truncate the Fourier series of $1/S(\omega_1,\omega_2)$ such that the result remains positive, i.e.,

$$\sum_{m=-p}^{p} \sum_{n=-q}^{q} r^{+}(m,n)\exp[j(\omega_1 m+\omega_2 n)] > 0 \ , \ -\pi\leq\omega_1,\omega_2\leq\pi \tag{26}$$

For semicausal MVRs, stable, finite order, approximate realizations can be obtained by

i) running the Levinson recursions associated with (19) to a finite limit, say n=q;
ii) truncating the Fourier series of the associated reflection coefficients $\rho_n(\omega_1)$ such that the result remains bounded in magnitude by unity for all $-\pi\leq\omega_1\leq\pi$;
iii) truncating the Fourier series of (21) such that the result remains positive for all $-\pi\leq\omega_1\leq\pi$.

In the case of causal MVRs, stability is assured by the steps i) and ii) above and if $b(\omega_1)$ is rationalized by a finite order AR model spectrum. An alternate approach for causal models is to restrict the region of support of the reflection coefficients to be finite and design them such that they remain less than unity in magnitude and the causal prediction error is minimized [26]. This approach also guarantees (stable) finite order causal models because a finite region of support of reflection coefficients maps into another finite region of support for the prediction error filter coefficients (although the converse is not true).

If an MVR happens to be of finite order, then the predictor $\overline{u}(m,n)$ will have finite support, i.e.,

$$\overline{u}(m,n)\,|\,\{u(m-k,n-\ell),(k,\ell)\varepsilon\hat{S}_x\} = \overline{u}(m,n)\,|\,\{u(m-k,n-\ell),(k,\ell)\varepsilon\hat{W}_x\},\ x=1,2,3 \tag{27}$$

where corresponding to $\hat{S}_x$, we have

$$\hat{W}_1 = \{1\leq\ell\leq q,\ -p\leq k\leq p\}\ \cup\ \{\ell=0,\ 1\leq k\leq p\} \tag{28a}$$

$$\hat{W}_2 = \{0\leq\ell\leq q,\ -p\leq k\leq p;\ (k,\ell)\neq(0,0)\} \tag{28b}$$

$$\hat{W}_3 = \{-p\leq k\leq p,\ -q\leq\ell\leq q;\ (k,\ell)\neq(0,0)\} \tag{28c}$$

Then u(m,n) has a finite order MVR

$$u(m,n) = \sum_{(k,\ell)\varepsilon\hat{W}_x} a(k,\ell)u(m-k,n-\ell) + \varepsilon(m,n),\ x=1,2,3 \tag{29}$$

which satisfies the normal equation

$$r(k,\ell) = \sum_{(m,n)\varepsilon\hat{W}_x} a(m,n)r(k-m,\ell-n) + \beta^2\delta(k,\ell);\ (k,\ell)\ \varepsilon\ S_x \tag{30}$$

With $a(0,0) \triangleq 1$, and $W_x \triangleq \hat{W}_x U(0,0)$, x=1,2,3; the subset of (29) corresponding to $(k,\ell)\varepsilon W_x$ can be conveniently written in matrix notation as

$$R\,\underline{a} = \beta^2 \underline{1} \tag{31}$$

where R is the covariance matrix of $\{u(i,j),(i,j)\varepsilon W_x,\ x=1,2,3\}$ and $\underline{1}$ is a unit vector that takes unit value at a location i_0 corresponding to the (0,0) location in the window W_x. The specific structure of R, $\underline{a}$ and $\underline{1}$ for the three types of MVRs can be found in [24,25].

If R is positive definite, then the solution of (31) will yield the unique MVR parameters $\underline{a}$ and $\beta^2>0$, that will also satisfy (30).

The spectral density function (SDF) realized by the MVR of (29) will be similar to (25) except that now the prediction error filter has finite support. Note that while the causal and noncausal MVR spectra are all pole, the semicausal is of ARMA type. Finite order MVRs, thus correspond to a restrictive class of rational SDFs. This means for arbitrary SDFs, finite order MVRs, will not exist in general. Therefore, in practice, one has to settle for finite order MVRs which are approximations, in some sense, of the infinite order (irrational) realizations. Enforcing finite supports for the prediction regions, as in (28), requires the orthogonality condition for minimum variance prediction to be

$$E[\varepsilon(i,j)u(i-k,j-\ell)] = \beta^2\delta(k,\ell)\ ,\ (k,\ell)\varepsilon W_x\ ,\ x = 1,2,3 \tag{32}$$

which yields (30) except that (k,ℓ) now belong to W_x rather than S_x as in (17), unless the given samples happen to come from that finite order MVR - a rare possibility. The above equation yields precisely the same equation as (31) as far as the unknowns $\underline{a}$ and β^2 are concerned.

Any positive definite matrix R in (31) will yield a unique solution for the model parameters, but does not guarantee a stable model. Also, to calculate $\underline{a}$ and β^2 on W_x, covariances are needed from a window W_c, which is twice the size of W_x. Hence, even if the model is stable, the covariances it generates need not match those on W_c. In spite of this, the advantage of these linear prediction models is that only finite order linear equations have to be solved. These equations have a nearly block Toeplitz structure and efficient algorithms are available for calculation of model coefficients [24]. More importantly, recent results [24,25] have shown that if the given covariances are from an SDF that is positive analytic, solution of the linear equations (31) for successively larger model orders (p,q), yield prediction error filters $\{A_{p,q}(z_1,z_2)\}$ that converge uniformly in appropriate regions of the z_1-z_2 plane to unique limit PEFs that realize the given SDF exactly. From this it follows that after a finite model order (p,q), the PEFs will have no zeroes in the appropriate regions of the z_1-z_2 plane, leading to stable models. From this is has been shown that sufficiently well behaved power spectrum densities or the associated covariances can be matched arbitrarily closely by finite order, stable, MVRs. Moreover, one only needs to solve the finite number of equations (31), rewritten as

$$R_{p,q}\underline{a}_{p,q} = \beta^2_{p,q}\underline{1}_{p,q} \tag{33}$$

for increasing values of p,q until a stable filter with desired spectral match is found. It can also be shown that the above result implies that for a fixed q, as $p\to\infty$, (33) yields the maximum entropy solution for extrapolating a covariance sequence given on the semi-infinite strip $\{k,\ell:0<\ell<q,\ \forall k\}$. These results provide the basis and motivation for using MVRs for two-dimensional spectrum estimation and for designing finite order random field representations for images. Some results on convergence of causal filters have also been obtained by Delsarte et al. [28]. The foregoing theory, discussed at greater

length in [24,25], provides a unified treatment of the three canonical forms of linear prediction on a two-dimensional grid.

The above models are useful in developing algorithms which have different hardware realizations. For example, causal models can realize recursive filters which require small memory while yielding an infinite impulse response. On the other hand, noncausal models can be used to design fast transform based finite impulse response filters and image coding algorithms. Semicausal models can yield two dimensional algorithms which are recursive in one dimension and nonrecursive in the other.

III. IMAGE ENHANCEMENT

In image enhancement, the goal is to bring out or accentuate certain image features for subsequent image analysis or display. Examples are contrast and edge enhancement, pseudo-coloring, filtering, etc. The enhancement process itself does not increase the inherent information content in the data. It simply emphasizes certain specified image characteristics. Enhancement algorithms are generally interactive and application/user dependent.

Table 1 lists some of the common image enhancement techniques. Point operation methods map each gray level into another gray level by a zero memory transformation. Spatial operation techniques perform local operations such as averaging, finding median etc., over a local neighborhood of each pixel. In transform techniques a point operation filter is implemented on the transform coefficients of the image. In pseudocoloring a gray level image is mapped into a color image by assigning different colors to different gray levels. Fig. 6 shows some examples of image enhancement.

IV. IMAGE RESTORATION

This term refers to estimation of an image from observations with known degradations, e.g., motion blur, defocus, scanner nonlinearity, etc. The difference between image restoration and image enhancement is that the former techniques are formulated with respect to a quantitative observation model and an optimization criterion, while the latter techniques are designed on a subjective or heuristic basis because a quantitative formulation is not available.

Wiener Filtering

Fig. 7 shows the typical situation in image restoration. If the imaging system is linear and shift invariant, the image of an object can be expressed as

$$g(x,y) = \int\int_{-\infty}^{\infty} h(x-\alpha,y-\beta)f(\alpha,\beta)d\alpha d\beta + \eta(x,y) \tag{34}$$

where $\eta(x,y)$, $f(x,y)$ and $g(x,y)$ represent the additive noise, the object, and the observed image respectively, and $h(x,y)$ is called the point spread function (PSF). The image restoration problem is to find an estimate of $f(x,y)$ given the PSF, the blurred image, and certain properties of the noise process. A common restoration method is called Wiener filtering, which is given in the spatial frequency coordinates as

$$\hat{F}(\xi_1,\xi_2) = A(\xi_1,\xi_2)G(\xi_1,\xi_2) \tag{35}$$

Table 1: Image Enhancement Techniques

POINT OPERATIONS

Contrast Stretching: Gray levels lying in a small region are stretched to occupy a larger region.

Noise Clipping: Gray levels above or below certain known signal limits are clipped to their respective limits to reduce noise effects.

Window Slicing: Gray levels lying in a range are made fully bright and remaining are set to zero.

Histogram Equalization or Modeling: Gray levels are modified such that their resulting histogram is uniform or equal to a specified distribution.

Binarization: Gray levels are thresholded so that the output is 0 or 1.

SPATIAL OPERATIONS

Noise Smoothing: Each pixel is replaced by the average of pixel values in its neighborhood. This method works well when the noise is Gaussian.

Median Filtering: Each pixel replaced by its neighborhood median. It works best when the noise is binary.

Unsharp Masking: The image and its (digital) Laplacian are mixed in certain proportion e.g., output $v(m,n) = u(m,n) + \lambda g(m,n)$ where $g(m,n) = u(m,n) - 1/4[u(m-1,n)+u(m+1,n)+u(m,n-1)+u(m,n+1)]$ and λ is a positive parameter.

Lowpass Filtering: Weighted averaging of pixels over a window.

Highpass Filtering: Lowpass filtered image subtracted from the original.

Bandpass Filtering: Pixel-average over a large window minus pixel-average over a small window.

Magnification/Interpolation: Pixel replication, bilinear interpolation, etc.

TRANSFORM OPERATIONS

Filtering: $V = A\,U\,A^T$, $v^{\bullet}(k,\ell) = g(k,\ell)v(k,\ell)$, output $u^{\bullet}(m,n) = [A^{-1}v^{\bullet}(A^T)^{-1}]_{m,n}$; $g(k,\ell)$ is the A-transform domain filter. For example if $g(k,\ell)$ is zero when (k,ℓ) is far from the origin then it is a lowpass filter.

Root Filtering: $v^{\bullet} = |v|^{\alpha}\exp(j\theta_v)$, $0 < \alpha < 1$, θ_v is the phase of v.

Homomorphic Filtering: $v^{\bullet}(k,\ell) = W(k,\ell)\log v(k,\ell)\exp\{j\theta_v\}$, W = 2-D window.

PSEUDOCOLORING: Map each gray level into a distinct color.

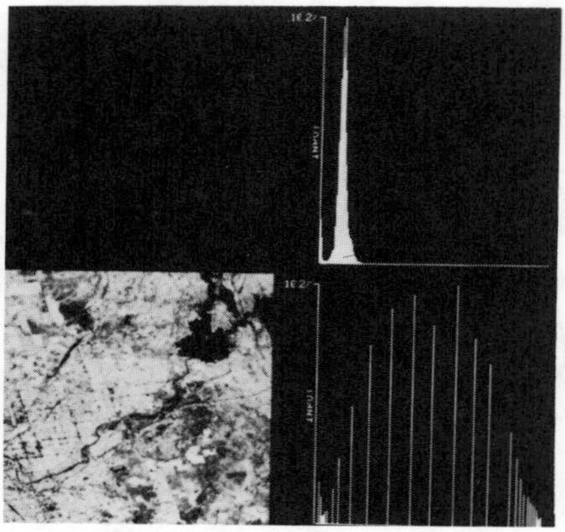

Histogram Equalization: Top Row: Input image, its histogram, Bottom row: Processed image and its histogram.

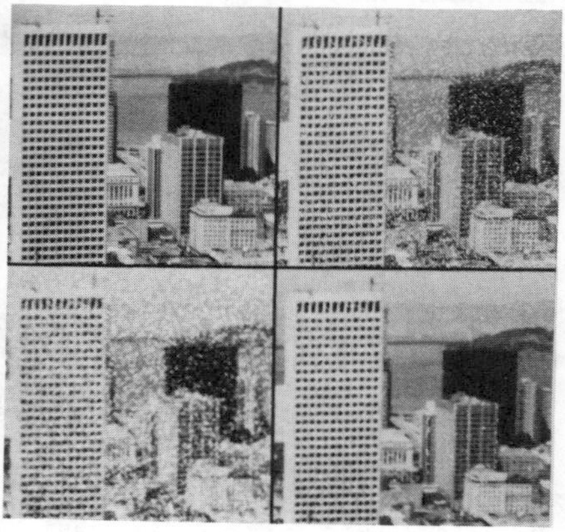

Noise Removal Via Median Filtering a) Original, b) with binary noise, c) spatial average, d) 3x3 median filtered.

a	b
c	d

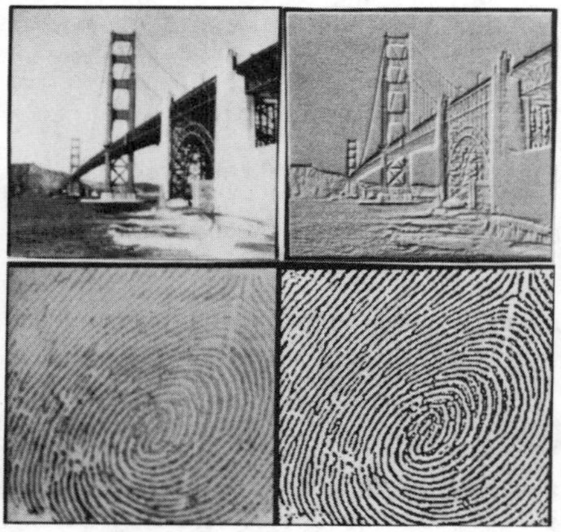

Highpass Filtering (right) of original images (left).

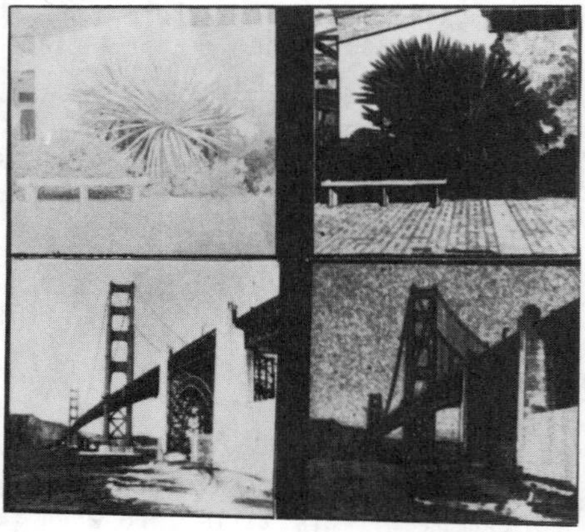

Contrast Ratio Mapping (right) of images on left. Note bricks on the patio and suspension cables on the bridge.

Figure 6: Image Enhancement Examples

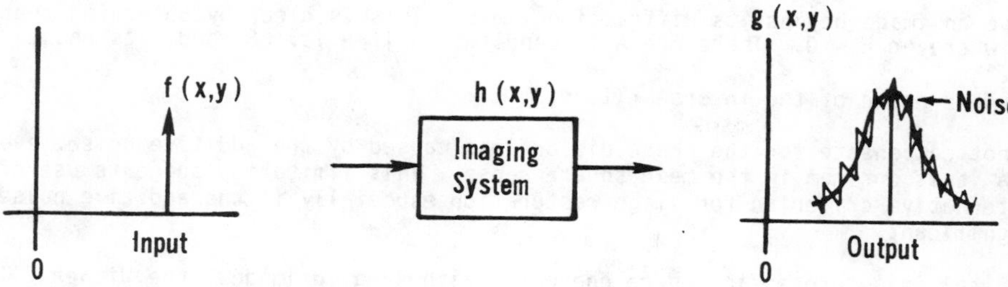

Figure 7: Blurring due to an Imaging System. Given the noisy and blurred image the image restoration problem is to find an estimate of the input image f(x,y).

$$A(\xi_1,\xi_2) = H^*(\xi_1,\xi_2)S_f(\xi_1,\xi_2)/[|H(\xi_1,\xi_2)|^2 S_f(\xi_1,\xi_2) + S_n(\xi_1,\xi_2)] , \quad (36)$$

where the upper case variables represent quantities in the Fourier transform domain, H is the frequency response of the imaging system, S_f and S_n are the power spectrum densities of the object and noise which are assumed to be uncorrelated and stationary. The restored image is given by the Fourier inverse of $\hat{F}$. The Wiener filter gives the best linear mean square estimate of the object from the observations. In the absence of noise, setting $S_n = 0$ gives the well-known <u>inverse filter</u>

$$A\big|_{S_n=0} \triangleq A^- = 1/H \quad (37)$$

This filter does not exist if H = 0 for any (ξ_1,ξ_2). However, if we take the limit $S_n \to 0$, then we obtain a filter A^+

$$\lim_{S_n \to 0} A \triangleq A^+ = \begin{cases} 1/H, & H \neq 0 \\ 0, & H = 0 \end{cases} \quad (38)$$

which is called the <u>pseudo-inverse filter</u>. Thus, if H is a lowpass filter (e.g., in lens aberration), the Wiener filter will act as a highpass filter if the noise at high frequencies is negligible.

On the other hand, for a perfect imaging system (H = 1) with additive noise, we get

$$A = S_f/(S_n+S_f) \quad (39)$$

$$= S_{nr}/(S_{nr}+1) \quad (40)$$

which is also called the <u>noise smoothing filter</u>. Here $S_{nr} \triangleq S_f/S_n$ is the signal to noise power ratio at frequency (ξ_1,ξ_2). For common images, the power spectrum takes larger values at lower frequencies. Hence, (40) acts like a lowpass filter if the additive noise is white.

The Wiener filter, in general, achieves a compromise between a lowpass filter for noise smoothing and a highpass filter for deblurring, which results in a bandpass filter. It has two other important characteristics. First, it cannot resolve an image beyond its diffraction limit. This is clear by observing that $A = 0$ wherever $H = 0$. Therefore A is bandlimited if H is. Second, its phase

$$\theta_A = -\theta_H = \text{phase of the inverse filter,} \tag{41}$$

does not compensate for the phase distortions caused by the additive noise, even though it is optimum in the mean square sense. This limitation suggests use of an alternative criterion for image restoration especially if the additive noise is significant.

In digital image processing where one works with sampled images, the Wiener filter is implemented by sampling the frequency response of (36) and using the FFT to estimate the inverse Fourier transform of $\hat{F}$. It can also be designed in different image transform domains by mapping the given filter impulse response matrix in the desired transform domain resulting in a realization shown in Fig. 8. For many fast transforms, the matrix $\tilde{A}$ often turns out to be highly sparse or nearly diagonal. For NxN images such an algorithm thus requires $O(N^2 \log N) + O(N^2)$ operations which is quite low compared to $O(N^3)$ for direct convolution methods.

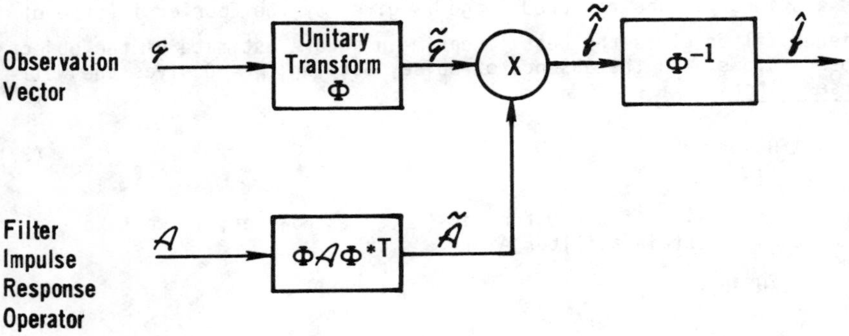

Figure 8. Application of Unitary Transforms for Image Restoration. This method is also called Generalized Wiener Filtering.

FIR Filters

Even with fast transforms the Wiener filter can be quite unsatisfactory. This is because the stationarity assumption of f(x,y) is often not a good one. Moreover, if the PSF is spatially varying, the Fourier technique cannot be used. An alternative is to design a finite impulse response (FIR) Wiener filter. It is useful for piecewise spatially invariant models and can be adapted to changes in the PSF and object statistics. Also, if the filter size is not very large, it can be implemented directly as a convolution operation, a facility which many modern digital image processing systems offer.

An FIR Wiener filter is of the form

$$\hat{u}(m,n) = \sum_{i,j \in W}\sum g(i,j)v(m-i,n-j) \tag{42}$$

$$W \triangleq \{-M<i,j<M\}$$

where v(m,n) is the observed image represented by the digitized model

$$\begin{aligned} v(m,n) &= h(m,n) \circledast u(m,n) + \eta(m,n) \\ &\triangleq \sum_{i,j}\sum h(m-i,n-j)u(i,j) + \eta(m,n) \end{aligned} \tag{43}$$

where $\eta(m,n)$ is additive white noise with variance σ_n^2. The optimum filter weights g(i,j), which minimize the mean square error, are obtained by solving the doubly Toeplitz system of equations

$$[\sigma_\eta^2 \delta(k,\ell) + r(k,\ell) \circledast a(k,\ell)] \circledast g(k,\ell) = h(-k,-\ell) \circledast r(k,\ell) \quad (k,\ell)\varepsilon W \tag{44}$$

$$a(k,\ell) \triangleq h(k,\ell) \circledast h(-k,-\ell)$$

Fig. 9 shows examples of FIR Wiener filtering of two images which were misfocused during acquisition. The restored images shown were obtained after trying several values of the unknown blur and noise parameters.

Other Methods

Several other image restoration methods such as least squares, constrained least squares, and spline interpolation methods can be shown to belong to the class of Wiener filtering algorithms. Other methods such as maximum likelihood, maximum entropy and maximum a-posteriori are nonlinear techniques which require iterative solutions. Finally, there are linear recursive filtering methods [30] which essentially implement the Wiener filter by sweeping the image recursively in two directions (forward and backward). Recursive filters are generally quite complex in design and implementation and are really useful only when the degradations are causal and spatially varying.

A compromise between transform based Wiener filters and the two dimensional recursive filters is obtained by the so-called semicausal filters. These filters are implemented by taking an image transform along one of the coordinates and performing recursive filter operations in the other coordinate. Such filters combine the advantages of recursive and transform based algorithms [30,32].

In practice image restoration techniques work best when the blur and noise levels are low to moderate. Difficulties mount very quickly when the blur and/or noise become large. Deblurring filters become very sensitive to noise if the blur is heavy. Likewise the noise smoothing filters tend to oversmooth the image edges when the noise is large. Adaptive filtering techniques which track the spatial variations in image statistics are useful when the noise is large but turn out to be very much image dependent. Techniques based on short term spectral estimation and adaptive FIR filtering have shown some success [24,29].

V. IMAGE DATA COMPRESSION AND CODING

The amount of data associated with visual information is so large that it often places impractical levels of computer memory requirements. Typical television images generate data rates exceeding eight million bytes per second. Other image sources generate data at even higher rates. Storage and/or transmission

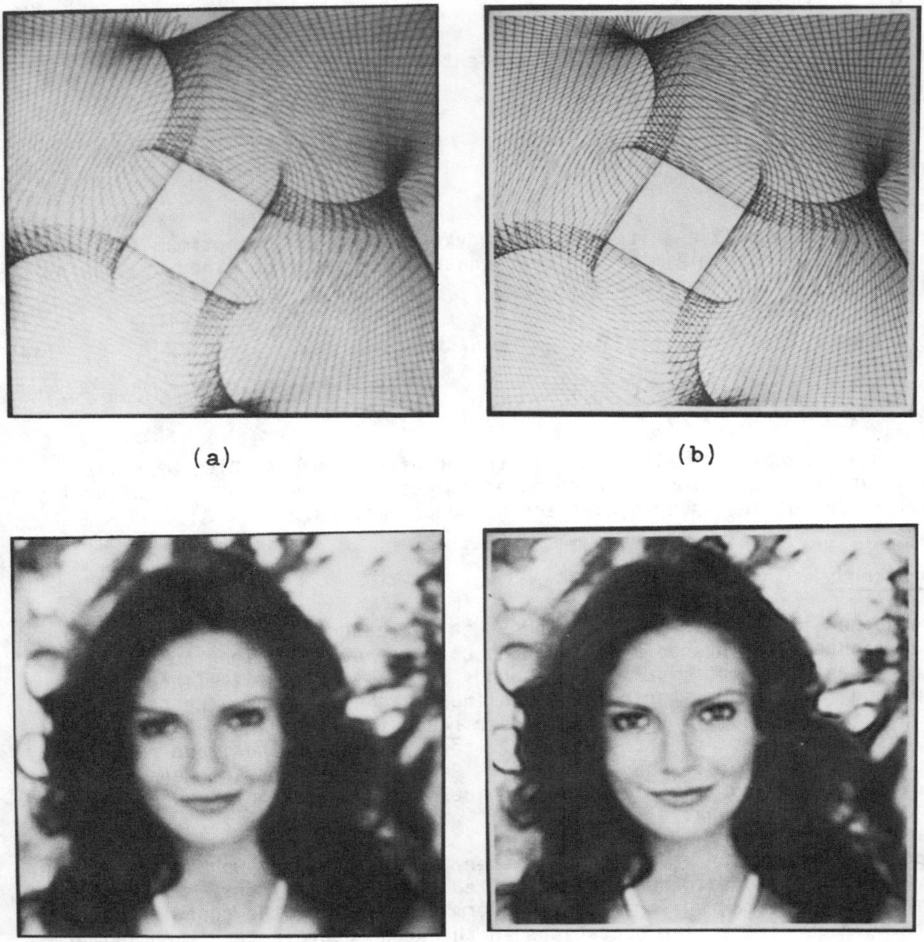

(a) (b)

(c) (d)

Figure 9: (a), (c): Images Digitized with Camera Misfocused; (b), (d): Images After FIR Wiener Filtering Using Fast Hardware Convolver.

of such data requires large capacity and/or bandwidth, which could be very expensive. Image data compression and coding technique reduce the number of bits required to store or transmit images without any appreciable loss of information (or quality). Image transmission applications are in broadcast television, remote sensing via satellite, aircraft, radar, sonar, teleconferencing, computer communications, facsimile transmission, etc. Image storage is required most commonly for educational and business documents, medical images used in patient monitoring systems, etc. Because of their wide applications, data compression and coding schemes are of great importance in digital image processing.

DPCM

Table 2 lists different types of image data compression techniques. Pixel by pixel coding techniques are based on efficient coding of binary sequences representing the image pixels. Predictive coding methods exploit redundancy in the data. Fig. 10 shows a common method of predictive coding called differential pulse code modulation (DPCM). Typically, the image is scanned line by line and a difference between a pixel and its predicted value from previously coded outputs is formed. This difference signal is quantized and coded for transmission or storage.

If the image is represented by a causal MVR, e.g.,

$$u(m,n) = \overline{u}(m,n) + \varepsilon(m,n) \tag{45a}$$

$$\overline{u}(m,n) = \sum_{i,j \varepsilon \hat{W}_1} \sum a(i,j)u(m-i,n-j) \tag{45b}$$

then at time k corresponding to pixel location (m,n), the predictor in the DPCM loop is generally chosen as

$$\overline{u}^{\bullet}(m,n) \triangleq \sum_{i,j \varepsilon \hat{W}_1} \sum a(i,j)u^{\bullet}(m-i,n-j) \tag{46}$$

and the difference

$$e(m,n) \triangleq u(m,n) - \overline{u}^{\bullet}(m,n) \tag{47}$$

is quantized and coded for transmission. Most images have high interpixel correlation so that the difference e(m,n) is generally small (except near edges) and can be coded accurately by a small number of bits. The reconstructed image is obtained as

$$u^{\bullet}(m,n) = \overline{u}^{\bullet}(m,n) + e^{\bullet}(m,n) \tag{48}$$

where $e^{\bullet}(m,n)$ is the reproduced value of e(m,n) after coding. From (47) and (48) it is easy to infer that

$$u(m,n) - u^{\bullet}(m,n) = e(m,n) - e^{\bullet}(m,n) \tag{49}$$

i.e., the error in the reproduced image would be small if error in quantization of e(m,n) is small.

In practice, for eight bit raw image data (such as television images), it is possible to reduce the data rate to about 2 to 3 bits/pixel without any perceivable degradation in the reproduced image. Predictive coding ideas can be extended to moving images also where three dimensional (inter-frame) prediction is done. This results in further compression and rates of about 1 bit/pixel are easily attained.

Table 2: Image Data Compression and Coding Problems

IMAGE DATA COMPRESSION AND CODING

- Pixel by Pixel Coding
  - PCM
  - Run Length Coding
  - Huffman Coding
  - Universal Coding
- Predictive Coding
  - Delta Modulation
  - DPCM
- Transform Coding
  - Zonal Coding
  - Thresholding
- Hybrid Coding
  - Transform/DPCM

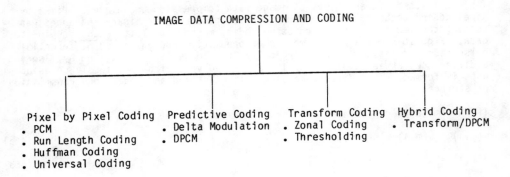

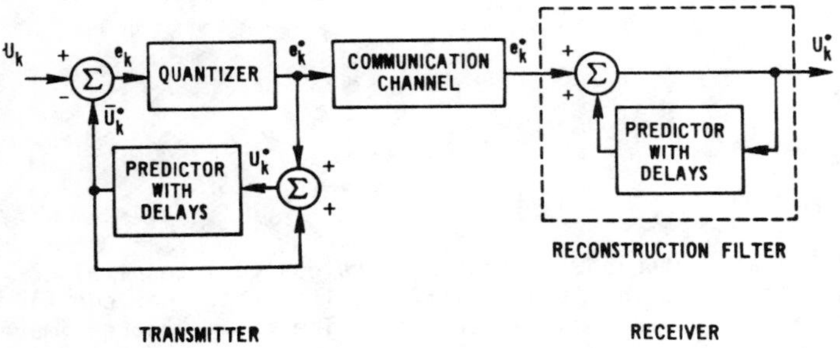

Figure 10: DPCM Method of Predictive Coding

Transform Coding

Compression can also be achieved by a unitary transform such that most of the information is packed into a small number of samples. This reduced set of samples is then quantized and coded to achieve compression. To reconstruct the image, the rejected samples are replaced by zeros and the inverse transform is taken. This is called transform coding (Fig. 11). For a given level of mean square distortion, the KL transform achieves the lowest bit rate (or highest compression) among all possible linear transforms. In practical transform coding, the image is generally divided into small blocks (typically 16x16) and each block is coded independently. For images with high interpixel correlation, the cosine transform turns out to be the closest (among the known fast transforms) to the KL transform. Fig. 12 shows an example of cosine transform coding which achieves a compression of 8.

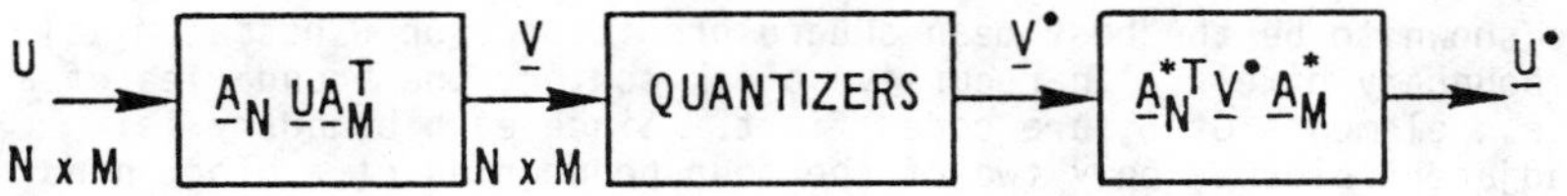

Figure 11: Two-Dimensional Transform Coding

a) Original 8 Bits/Pixel

b) Cosine Transform Coded 1 Bit/Pixel

Figure 12: Transform Coding Example

Recursive Block Coding

Conventional block by block transform coding yields visibly objectionable block boundaries especially at low bit rates (less than 1 bit/pixel). Recursive block coding is a technique which removes redundancy between successive blocks while reducing the block boundary effect. Its performance, both theoretically and experimentally, has been found to be superior to the conventional block by block KL transform coding. The theoretical basis of this method lies in the noncausal representations considered above. For example, suppose an image has the noncausal MVR

$$u(m,n) = \alpha[u(m-1,n) + u(m+1,n) + u(m,n-1) + u(m,n+1)] + \varepsilon(m,n) \tag{50}$$

where $\varepsilon(m,n)$ is a moving average field which is correlated with its nearest four neighbors only with correlation $-\alpha$. If an NxN block of u(m,n) is mapped into an N^2x1 vector, then (50) can be written in the form

$$\mathcal{A}u = b + e \tag{51}$$

where e represents the NxN block of $\varepsilon(m,n)$ and b is an N^2x1 vector which depends only on pixels lying on the outside boundary of the NxN image clock represented by u. Eqn. (51) yields a useful orthogonal decomposition

$$u = u^o + u^b$$
$$u^o \triangleq A^{-1}e\ ,\ u^b \triangleq A^{-1}b\ ,\ E[u^o(k)u^b(\ell)] = 0,\ \forall k, \forall \ell \tag{52}$$

where u^b can be shown to be the best mean square predictor (for Gaussian MVRs) of u given the boundary pixels. In recursive block coding, the boundaries of a given block, i.e., elements of b, are coded first. Since each boundary is shared by two adjacent blocks, only two of the four boundaries of a block need to be coded at each step since the other two boundaries are available from previous blocks. From the encoded boundary values, u^b is estimated and the residual u^o is transform coded. It can be shown that the residual vector u^o is uncorrelated from block to block - unlike conventional block by block transform coding where the successive blocks are correlated. The reproduced block u is simply the sum of the reproduced values of u^o and u^b. Fig. 13 shows the superiority of this algorithm over cosine transform coding.

Comparisons

Predictive coding techniques have generally low computational complexity which makes them attractive for hardware implementation. However, they are quite sensitive to channel noise (due to transmission or storage) as well as to changes in the statistical parameters of the image. On the contrary, transform techniques have a high computational complexity but are quite robust with respect to channel noise and image characteristics. For the same level of distortion, transform coding techniques achieve approximately twice the compression of predictive techniques. Other image data compression algorithms (Hybrid Coding) exist which use a combination of these two methods (Fig. 14). Such algorithms combine the advantages of DPCM and transform coding schemes and can be shown to follow from semicausal representations. Details of these and other image data compression algorithms can be found in [33,34].

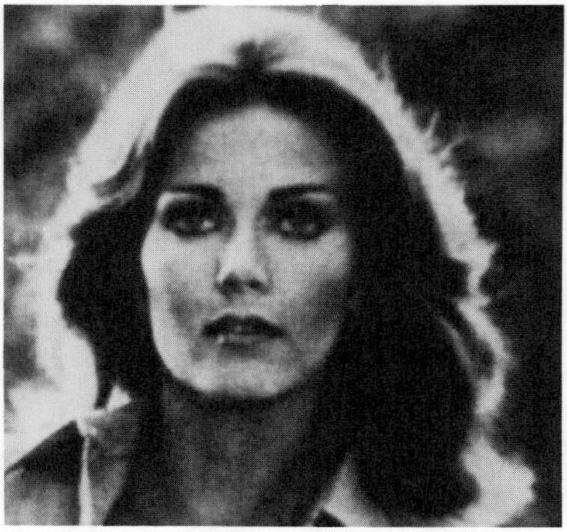

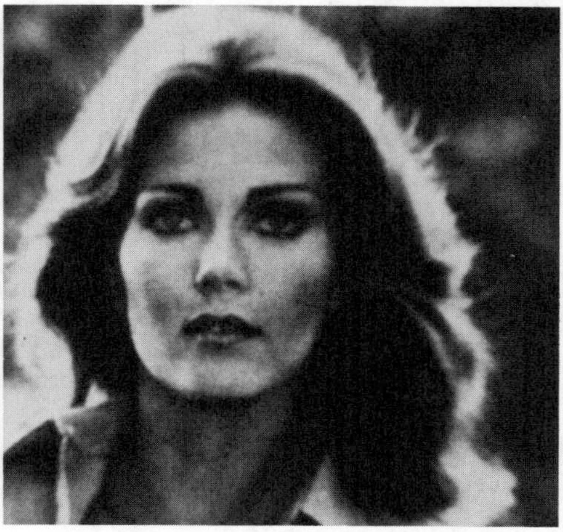

a) Cosine Transform Coded, MSE=.64%.

b) RBC, MSE = .49%.

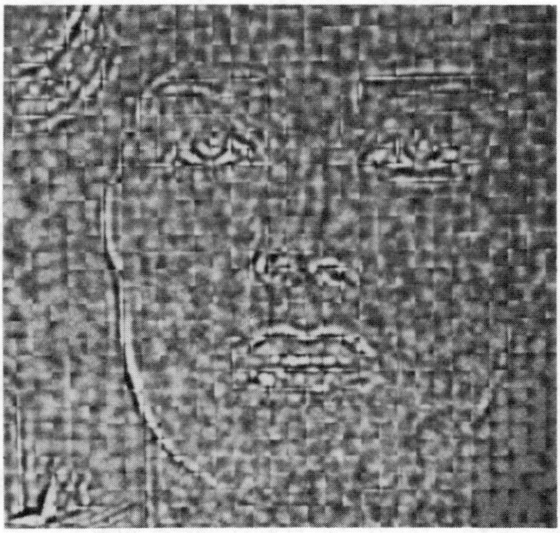

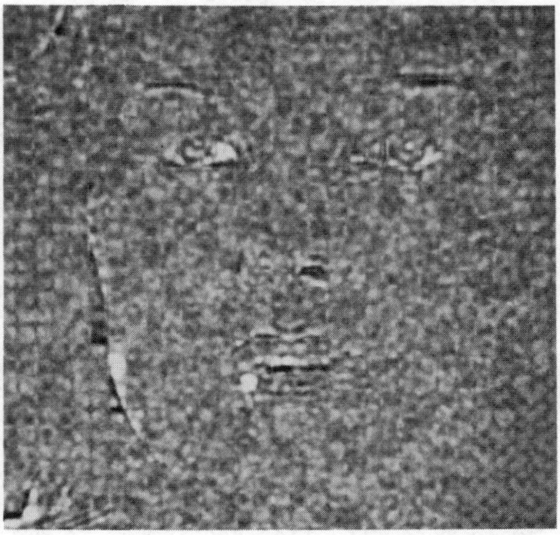

c) Magnified Error, Cosine Transform Coding

d) Magnified Error, RBC

Figure 13: Recursive Block Coding (RBC) using 8x8 blocks at 0.24 bits/pixel (entropy). This method is also called fast KLT coding [22,33].

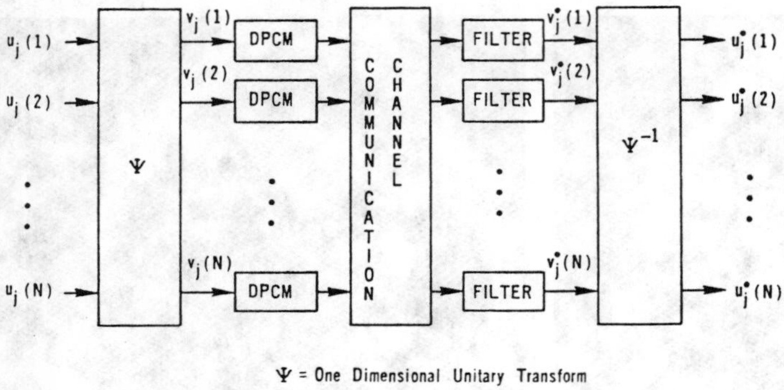

Figure 14: Hybrid Coding

VI. IMAGE RECONSTRUCTION FROM PROJECTIONS

The Radon Transform

A special class of image restoration problems is called image reconstruction from projections. It is important in medical imaging, non-destructive testing of assemblies, astronomy, geological exploration, radar imaging and in many other fields. It deals with the problem of estimating a two (or higher) dimensional object from several one dimensional projections. A projection is a shadowgram obtained by illuminating an object by penetrating radiation (Fig. 15). Mathematically it is modeled as a line integral of a function (e.g. absorption density in the case of x-rays) along the path of the radiation. For a function f(x,y), its projection at an angle θ, denoted by g(s,θ) (Fig. 16) is written as

$$g(s,\theta) = \int_S f(x,y)ds = \int\int_{-\infty}^{\infty} f(x,y)\delta(x\cos\theta+y\sin\theta-s)dxdy, \quad -\infty<s<\infty \ , \ 0\leq\theta<\pi \tag{53a}$$

This mapping is also called the <u>Radon transform</u> ($\mathcal{R}$) and is often abbreviated as

$$g(s,\theta) = [\mathcal{R}f(x,y)](s,\theta) \tag{53b}$$

The Radon transform is an operator which maps the spatial domain (x,y) to another domain (s,θ). Each point in (s,θ) corresponds to a line in (x,y). The image reconstruction problem is to determine the function f(x,y) from g(s,θ). In practice g(s,θ) is available for only a finite number of values of θ and one can only estimate f(x,y).

The Projection Theorem

A key result in Fourier theory which enables the reconstruction of f(x,y) from g(s,θ) is called <u>the projection theorem</u> which states that if

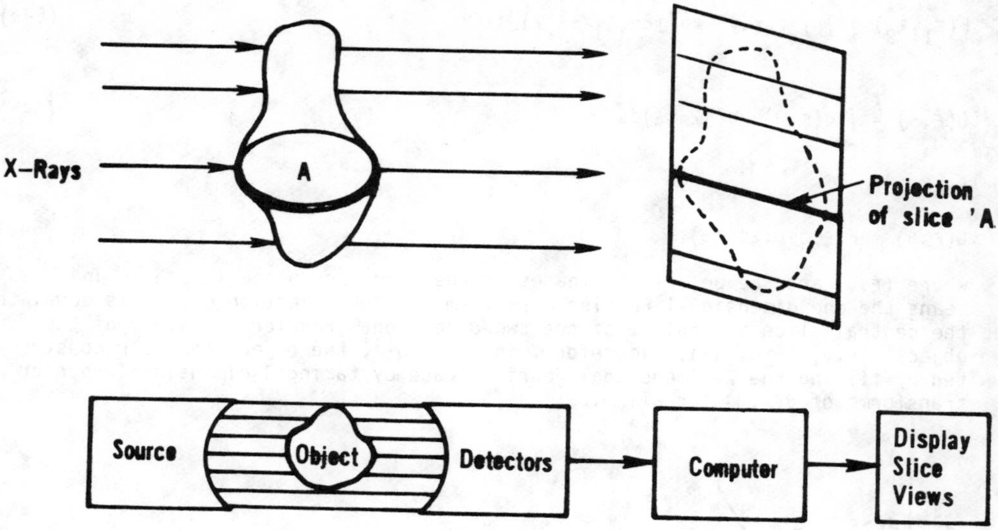

Figure 15: An X-Ray CT Scanning System

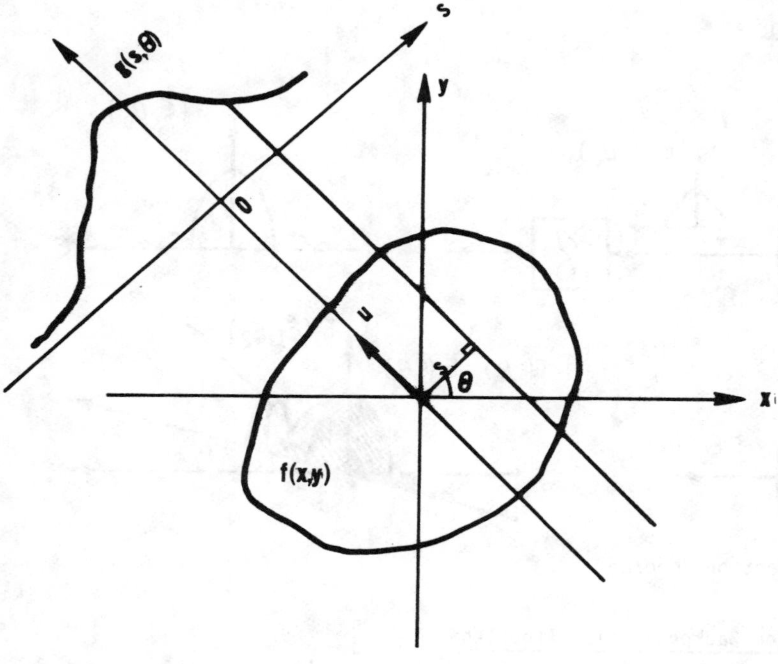

Figure 16: The Radon Transform, $g(s,\theta)$, of $f(x,y)$. It is the projection of $f(x,y)$ at an angle θ.

$$F(\xi_1,\xi_2) \triangleq \int_{-\infty}^{\infty}\int f(x,y)\exp[j2\pi(\xi_1 x+\xi_2 y)]dxdy \tag{54a}$$

$$G(\xi,\theta) \triangleq \int_{-\infty}^{\infty} g(s,\theta)\exp(j2\pi s\xi)ds \tag{54b}$$

then

$$G(\xi,\theta) = F(\xi\cos\theta,\xi\sin\theta). \tag{54c}$$

where (ξ,θ) are the polar coordinates in the frequency plane (ξ_1,ξ_2). This means the one dimensional Fourier transform of the projection $g(s,\theta)$ is equal to the central slice at angle θ of the two dimensional Fourier transform of the object $f(x,y)$ (Fig. 17). Therefore, in principle, the object can be reconstructed by filling the 2-dimensional Fourier space by taking 1-dimensional Fourier transforms of $g(s,\theta)$ for all θ.

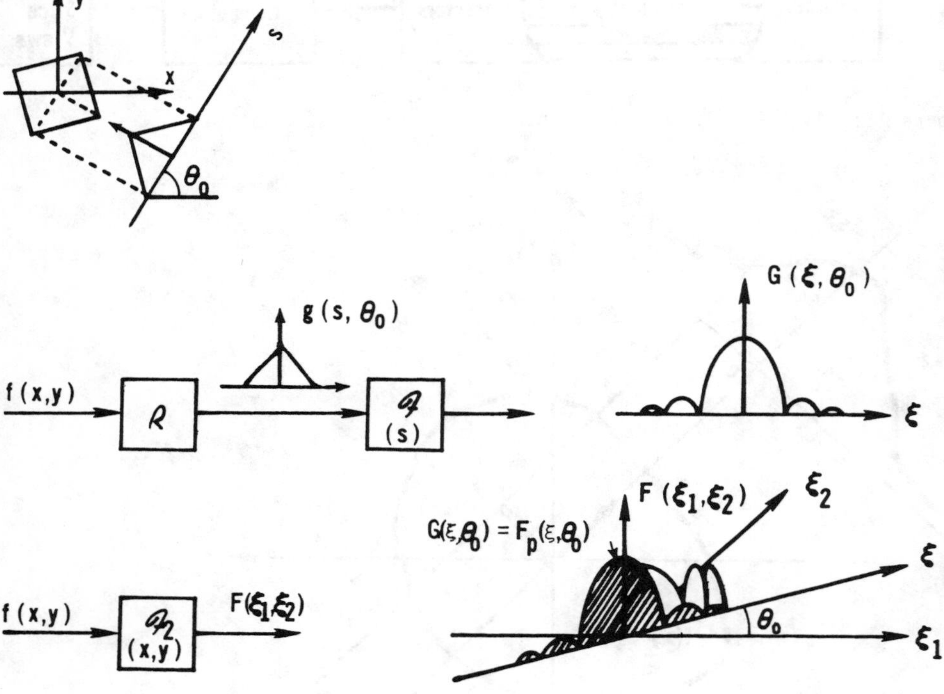

Figure 17: Projection Theorem

Filter/Convolution Backprojection Algorithm

A more useful and popular method of reconstruction is shown in Fig. 18. For each θ, the projection $g(s,\theta)$ is first filtered in the s-coordinate by a one dimensional filter whose frequency response is $H(\xi) = |\xi|$. The result $g(s,\theta)$ is then passed through a so-called the back projection operator B, defined as

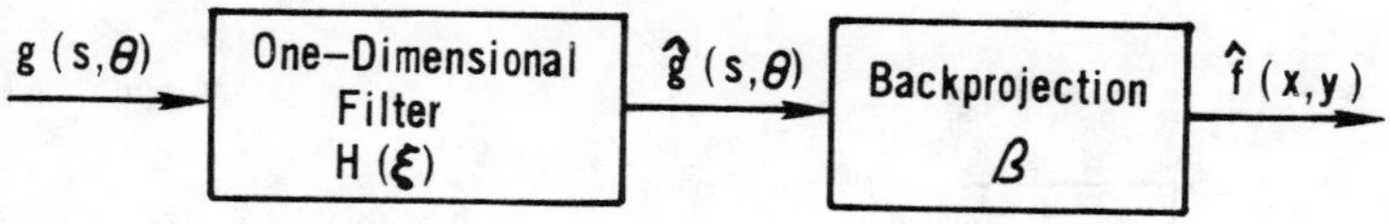

Figure 18: Convolution/Filter Backprojection Algorithm. $H(\xi) = |\xi|$ gives $\mathcal{R}^{-1}$; $H(\xi) = |\xi|W(\xi)$ gives practical algorithms.

$$[\mathcal{B}\hat{g}](x,y) = \int_0^{\pi} \hat{g}(x\cos\theta+y\sin\theta,\theta)d\theta \tag{55}$$

The result equals f(x,y). The filter $|\xi|$ plays a role similar to the inverse filter discussed before and is ill-conditional as such. In practice it is implemented as a bandlimited filter

$$H(\xi) = |\xi|W(\xi) \tag{56}$$

where $W(\xi)$ is a frequency-limiting window function. Two functions used often are

$$W(\xi) = \text{rect}(\xi d) \qquad \text{(Ram-Lak)} \tag{57}$$

$$W(\xi) = \text{sinc}(\xi d)\text{rect}(\xi d) \qquad \text{(Shepp-Logan)} \tag{58}$$

where

$$\text{rect}(x) = \begin{cases} 1, & |x|<1/2 \\ 0, & \text{otherwise} \end{cases} \tag{59}$$

and 1/2d is the bandwidth. The foregoing method, also called convolution/filter backprojection algorithm, has two noteworthy features; i) it requires only one-dimensional convolutions/filters, ii) the backprojection algorithm simply requires summation and interpolation operations.

Radon Transform of Random Fields

Recently the Radon transform theory has been studied for random fields and several interesting results have emerged. These results are summarized below.

Definitions:

1) Let $\tilde{\mathcal{R}}$ be defined as (Fig. 19) an operator

$$\tilde{\mathcal{R}} \triangleq \mathcal{H}^{1/2} \tag{60}$$

where $\mathcal{H}^{1/2}$ represents a filter whose frequency response is $|\xi|^{1/2}$.

2) $\tilde{g}(s,\theta) \triangleq \tilde{\mathcal{R}}f$ (61)

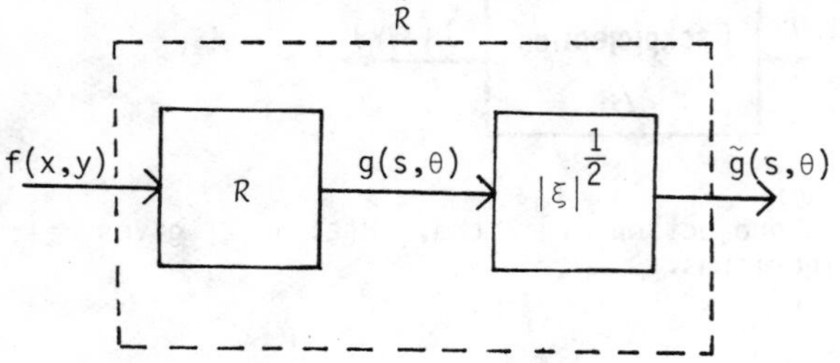

Figure 19: The $\tilde{R}$-Transform

3) Let f(x,y) be a stationary random field with power spectrum density $S(\xi_1,\xi_2)$ and autocorrelation function $r(\tau_1,\tau_2)$. In polar coordinates define

$$S_p(\xi,\theta) \triangleq S(\xi\cos\theta,\xi\sin\theta) \tag{62}$$

and let $r_p(s,\theta)$ denote the one-dimensional inverse Fourier transform of $S_p(\xi,\theta)$ w.r.t. ξ. From projection theorem it follows that

$$r_p(s,\theta) = Rr \tag{63}$$

Theorem 1: The operator $\tilde{R}$ is unitary, i.e.,

$$\tilde{R}^{-1} = \tilde{R}^* \tag{64}$$

and

$$\int\int_{-\infty}^{\infty} |f(x,y)|^2 dxdy = \int_0^{\pi}\int_{-\infty}^{\infty} |\tilde{g}(s,\theta)|^2 dsd\theta \tag{65}$$

Theorem 2: The operator $\tilde{R}$ is a whitening transform in θ for stationary random fields. In fact

$$r_{\tilde{g}\tilde{g}}(s,\theta;s',\theta') \triangleq E[\tilde{g}(s,\theta)\tilde{g}(s',\theta')] = r_p(s-s',\theta)\delta(\theta-\theta') \tag{66}$$

Corollary: The Radon transform is also a whitening transform in θ for stationary random fields.

Definition:

4) The one-dimensional power spectrum density of $\tilde{g}(s,\theta)$, at angle θ, is defined as the one-dimensional Fourier transform of $r_p(s,\theta)$ w.r.t. s, i.e.,

$$S_{\tilde{g}}(\xi,\theta) \triangleq F_1\{r_p(s,\theta)\} . \tag{67}$$

Theorem 3: (Projection Slice Theorem for Random Fields). The one-dimensional power spectrum density $S_{\tilde{g}}(\xi,\theta)$ of the $\tilde{R}$-transform of a stationary random field

$f(x,y)$ is the central slice at angle θ of the two-dimensional power spectrum density $S(\xi_1,\xi_2)$ of $f(x,y)$, i.e.,

$$S_{\tilde{g}}(\xi,\theta) = S_p(\xi,\theta)$$

$$= S(\xi\cos\theta, \xi\sin\theta) \tag{68}$$

Theorem 3 is noteworthy because it states that the central slice of the 2-D power spectrum of $f(x,y)$ is the 1-D power spectrum of $\tilde{g}(s,\theta)$, the $\tilde{}$-transform and not of $g(s,\theta)$, the Radon transform. Also, from the relationship between $g(s,\theta)$ and $\tilde{g}(s,\theta)$ (Fig. 18), it follows that

$$S_p(\xi,\theta) = S_{\tilde{g}}(\xi,\theta) = |\xi| S_g(\xi,\theta) \tag{69}$$

Optimum Reconstruction from Noisy Projections

The above results are useful in estimating two dimensional power spectrum of a random field from its one-dimensional projections, and also in optimum mean square reconstruction of an object from noisy and blurred projections. For instance, suppose the observed projections are modeled as

$$z(s,\theta) = \int_{-\infty}^{\infty} h(s-t,\theta) g(t,\theta) dt + \eta(s,\theta) \tag{70}$$

where $h(s,\theta)$ is the PSF that blurs the projections and $\eta(s,\theta)$ is additive noise independent of the object $f(x,y)$ and uncorrelated in θ, i.e.,

$$E[f(x,y)\eta(s,\theta)] = 0 \tag{71}$$

$$E[\eta(s,\theta)\eta(s',\theta')] \triangleq r_\eta(s-s',\theta)\delta(\theta-\theta') \tag{72}$$

$$r_\eta(s,\theta) \overset{\mathcal{F}_1}{\longleftrightarrow} S_\eta(\xi,\theta) \tag{73}$$

Then the optimum linear mean square estimate of $f(x,y)$ can be shown [37,38] to be given by the filter backprojection algorithm

$$\hat{f}(x,y) = \mathcal{B} A_\theta z \tag{74}$$

where A_θ is a filter whose frequency response is

$$A_p(\xi,\theta) \triangleq \frac{|\xi| H^*(\xi,\theta) S_g(\xi,\theta)}{|H(\xi,\theta)|^2 S_g(\xi,\theta) + S_\eta(\xi,\theta)} \triangleq |\xi| W(\xi,\theta) \tag{75}$$

where $W(\xi,\theta)$ is the Wiener filter for each projection considered independently. Note that now the filter $A_p(\xi,\theta)$ can vary with θ unlike the Ram-Lak or Shepp-Logan filters. In practice $S_g(\xi,\theta)$ may be estimated via one-dimensional spectral estimation techniques applied to the given projections. If a model for

$S(\xi_1,\xi_2)$ is available, then $A_p(\xi,\theta)$ can be rewritten, via (69), as

$$A_p(\xi,\sigma) = |\xi|H^* S_p/[|H|^2 S_p+|\xi|S_\eta] \tag{76}$$

Fig. 20 shows examples of reconstruction from noisy projections via the optimum stochastic filter described above and its comparison with the Shepp-Logan filter. Other comparisons and details may be found in [38].

VII. OTHER AREAS

There are several other areas in digital image processing which we have not covered. Notable among these are image analysis, color image processing, processing of facsimile images and image processing hardware and software systems.

Image analysis is concerned with the problems of obtaining measurements from images. It includes feature extraction, texture analysis, shape recognition, boundary analysis, scene matching, etc. Image analysis is important in a large number of applications such as robotics for industrial automation, medical diagnosis from images, target detection in radar imaging.

Color image processing deals with many of the foregoing problems (e.g., representation, enhancement, restoration, coding, etc.) except that the data is now three dimensional. Facsimile images are two level images such as half tones or printed documents, maps, etc. Representation and processing techniques of such images is quite different although the basic principles are the same.

With increasing number of applications of image processing, rapid advances are being made in the development of vision systems and image processing systems. Modern image processing systems include a digital-computer, an array processor, large capacity disks and tapes, color display monitors, scanners, film recorders, and computer terminals. Smaller systems for specific applications are evolving and are designed around microcomputers.

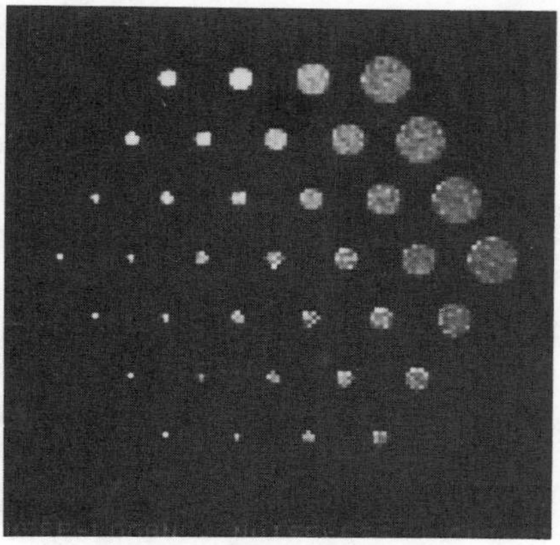

a) Shepp-Logan Filter, Noiseless Case

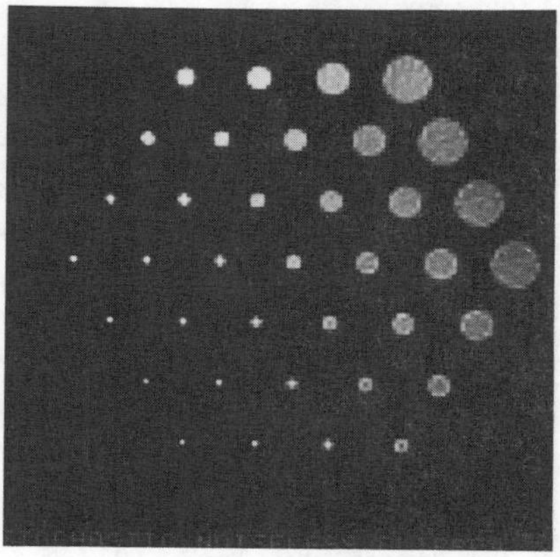

b) Stochastic Filter, Noiseless Case

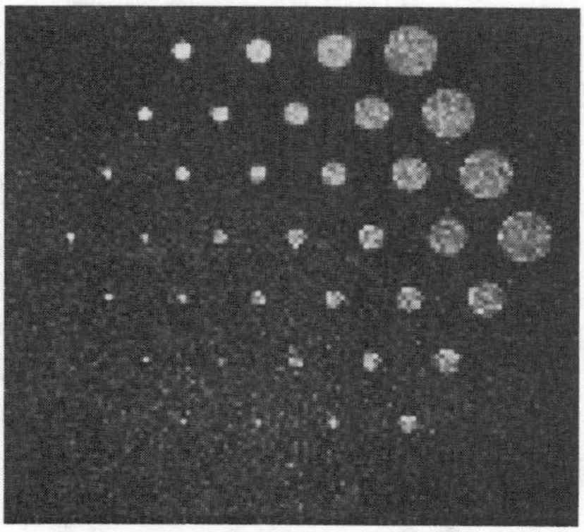

c) Shepp-Logan Filter, $\sigma_\nu^2=1$

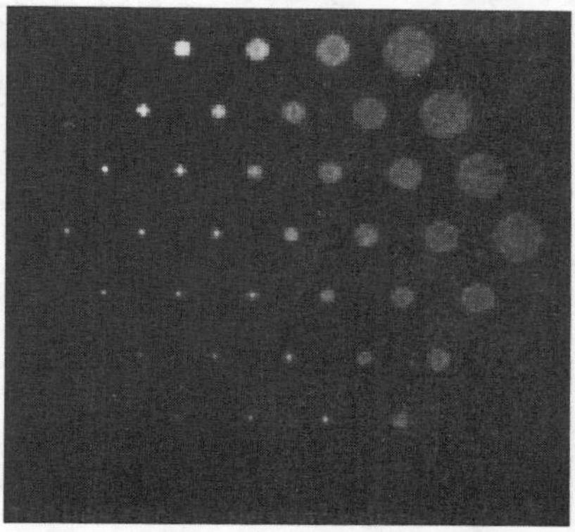

d) Stochastic Filter, $\sigma_\nu^2=1$

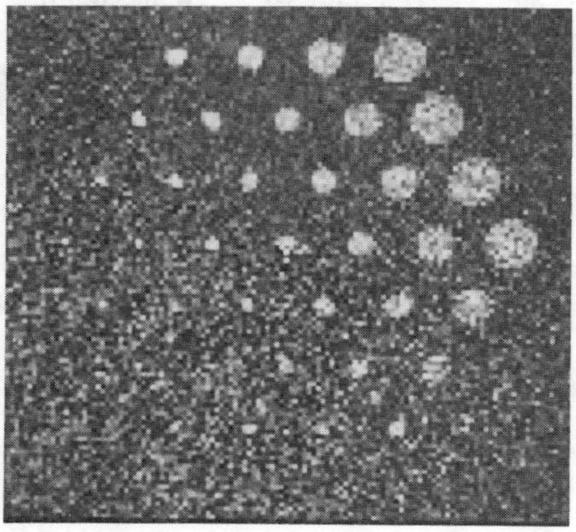

e) Shepp-Logan Filter, $\sigma_\nu^2=5$

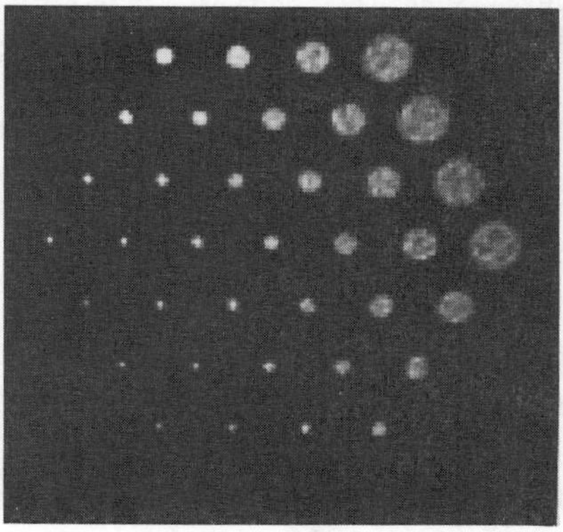

f) Stochastic Filter, $\sigma_\nu^2=5$

Figure 20: Image Reconstruction from 90 Projections Containing Additive Noise of Variance σ_ν^2.

BIBLIOGRAPHY

Section I

Several books exist which cover a variety of topics in digital image processing. These include the following.

1. Andrews, H.C., with contributions by W.K. Pratt and K. Caspari, Computer Techniques in Image Processing, Academic Press, New York, 1970.

2. Andrews, H.C. and B.R. Hunt, Digital Image Restoration, Prentice-Hall, Englewood Cliffs, N.J., 1977.

3. Gonzales, R.C. and P. Wintz, Digital Image Processing, Addison-Wesley, Reading, Mass., 1977.

4. Huang, T.S. and O.J. Tretiak, Eds., Picture Bandwidth Compression, Gordon and Breach, New York, 1972.

5. Huang, T.S., Ed., Topics in Applied Physics: Picture Processing and Digital Filtering, Vol. 6, Springer-Verlag, New York, 1975.

6. Lipkin, S. and A. Rosenfeld, Picture Processing and Psychopictorics, Academic Press, New York, 1970.

7. Pratt, W.K., Digital Image Processing, Wiley-Interscience, New York, 1978.

8. Rosenfeld and A.C. Kak, Digital Image Processing, Academic Press, New York, 1976.

9. Castleman, K.R., Digital Image Processing, Prentice Hall, Englewood Cliffs, N.J., 1979.

10. Hall, E.L., Computer Image Processing and Recognition, Academic Press, 1979.

11. Jain, A.K., Fundamentals of Digital Image Processing, (to appear).

Several special issues of technical journals and review papers have also appeared which have surveyed problems related to digital image processing. For examples, see

12. Special Issue on Redundancy Reduction, Proc. IEEE, Vol. 55, 3, March 1967.

13. Special Issue on Digital Communications, IEEE Commun. Tech., Vol. COM-19, 6, Part I, December 1971.

14. Special Issue on Digital Picture Processing, Proc. IEEE, Vol. 60, 7, July 1972.

15. Special Issue on Two-Dimensional Signal Processing, IEEE Trans. Computers, Vol. C-21, 7, July 1972.

16. Special Issue on Digital Image Processing, IEEE Computer, Vol. 7, 5, May 1974.

17. Special Issue on Digital Signal Processing, Proc. IEEE, Vol. 63, 4, April 1975.

18. Special Issue on Image Bandwidth Compression, IEEE Trans. Communications, Vol. COM-25, 11, November 1977.

19. Special Issue on Image Processing, Proc. IEEE, Vol. 69, 5, May 1981.

Section II

Sampling and Quantization of Images is discussed in several texts including [1,7,8,11]. For a comprehensive review see [33].

For detailed discussions on image transforms see [7,11] and

20. Ahmed, N. and K.R. Rao, Orthogonal Transforms for Digital Signal Processing, New York: Springer Verlag, 1975.

21. Jain, A.K., "A Sinusoidal Family of Unitary Transforms," IEEE Tans. Pattern Anal. Machine Intell., vol. PAMI-1, pp. 356-365.

22. Jain, A.K. and P.M. Farrelle, "Recursive Block Coding," Proc. 1982 Asilomar Conference, November 1982.

For stochastic image models and related details see

23. Jain, A.K., "Advances in Mathematical Models for Image Processing," Proc. IEEE, Vol. 69, No. 5, pp. 502-528, May 1981.

24. Ranganath, S., "Two Dimensional Spectral Factorization and Spectral Estimation with Applications in Image Processing," Ph.D. Dissertation, Department of Electrical and Computer Engineering, University of California, Davis, March 1983.

25. Ranganath, S. and A. K. Jain, "Two Dimensional Linear Prediction Models Part I: Spectral Factorization and Realization," (to appear in IEEE Trans. ASSP).

26. Marzetta, T. L., "A Linear Prediction Approach to Two Dimensional Spectral Factorization and Spectral Estimation," Ph.D. Dissertation, Dept. Elec. Eng. and Comput. Sci., MIT, Cambridge, MA, February 1978. Also see IEEE Trans. ASSP, Vol. ASSP-28, No. 6, pp. 725-733, December 1980.

27. Ekstrom, M. P. and J. W. Woods, "Two Dimensional Spectral Factorization with Applications in Recursive Digital Filtering," IEEE Trans. ASSP, Vol. ASSP-24, No. 2, pp. 115-128, April 1976.

28. Delsarte, P., Y. Genin, and Y. Kamp, "Half-Plane Toeplitz Systems," IEEE Trans. Inform. Theory, Vol. IT-26, No. 4, pp. 465-474, July 1980.

Sections III,IV

Image Enhancement and Restoration techniques are discussed in most texts on digital image processing cited, e.g. [1,3,7-11]. For FIR Wiener filters see [24] and

29. Jain, A.K. and S. Ranganath, "Image Restoration and Edge Extraction Based on 2-D Stochastic Models," Proc. 1982 ICASSP, Paris, France, 1982.

For recursive and semirecursive filtering of images see,

30. Jain, A. K., "A Semicausal Model for Recursive Filtering of Two Dimensional Images," IEEE Trans. Comput., Vol. C-26, pp. 343-350, April 1977.

31. Woods, J. W. and C. H. Radewan, "Kalman Filtering in Two Dimensions," IEEE Trans. Inform. Theory, Vol. IT-23, pp. 473-482, July 1977.

32. Jain, A. K. and J. R. Jain, "Partial Differential Equations and Finite Difference Methods in Image Processing, Part II: Image Restoration," IEEE Trans. Automat. Contr., Vol. AC-23, pp. 817-834, October 1978.

Section V

For comprehensive reviews of Image Data Compression Algorithms see

33. Jain, A.K., "Image Data Compression: A Review," Proc. IEEE, Vol. 69, No. 3, pp. 349-389, March 1981.

34. Netravali, A.N. and J.O. Limb, "Picture Coding: A Review," Proc. IEEE, Vol. 68, pp. 366-406, March 1980.

Section VI

A large amount of literature is available on image reconstruction. The fundamentals are available in [10,11] and

35. Herman, G.T., Image Reconstruction from Predictions: The Fundamentals of Computerized Tomography, New York, NY: Academic Press, 1980.

36. Proceedings of the IEEE, Special Issue on Computerized Tomography, March 1983.

For image reconstruction from noisy projections and related results are reported in

37. Jain, A.K. and S. Ansari, "Representation of Two Dimensional Random Fields Via Radon Transform," SIAM 1983 National Meeting, Denver, CO., June 1983.

38. Jain, A. K. and S. Ansari, "Radon Transform Theory for Random Fields and Optimum Image Reconstruction from Noisy Projections," (to appear in Proc. 1984 ICASSP, San Diego).

10

Signal Processing Technology

H. J. WHITEHOUSE

ABSTRACT

Selected aspects of signal processing technology are presented from the perspective of real-time hardware implementation. The transversal filter is introduced as the archetype time invariant linear filter. Surface acoustic wave devices and charge coupled devices are proposed for applications where size, weight, and low power consumption are more important than accuracy. The chirp-z-transform is introduced so that time variant filters can be implemented with time invariant linear filters. Systolic arrays are introduced so that advanced matrix operations can be implemented using digital integrated circuits. A systolic test bed is described for use as an algorithm development tool.

CONTENTS

INTRODUCTION

Signal processing is a continuously changing field. This paper will discuss selected aspects of real-time signal processing technology and their mathematical basis in linear system theory and numerical linear algebra. By exploiting alternative mathematical formulations of the basic algorithms of signal processing different archectures can be developed for their implementation.

Initally, in the early 1960's, only analog discrete component implementations were possible for real time signal processing. In the mid 1960's two events occured which changed signal processing: The development of surface acoustic wave (SAW) devices, and the publication of the fast Fourier transform (FFT) algorithm. The first provided analog integrated components, the second real time digital signal processing. By the early 1970's charge coupled devices (CCDs) were being developed and this technology made integrated analog circuits possible with silicon technology. Corresponding advances in digital integrated circuits (ICs) allowed the exploitation of the FFT algorithm and the development of array processors.

Although both the SAW and CCD technologies provided direct implementation of finite impulse response (FIR) filters it was the application of the chirp-z-transform algorithm which made possible the real time computation of the discrete Fourier transform (DFT). The CZT algorithm provided a means of low power implementation of the DFT for both video and intermediate frequency (IF) applications.

In the late 1970's a new architecture called a "systolic array" was introduced as a means of implementing advanced matrix operations such as matrix-matrix multiplication, matrix factorization, matrix inversion, singular value decomposition, and the solution of eigensystems. The systolic array provides an effective means of combining a large number of digital very large scale integration (VLSI) circuits with only nearest neighbor interconnections and a minimum of control overhead.

In the early 1980's there was a renewed interest in optical signal processing especially with optical processors which use incoherent illumination. In addition, a systolic array test bed was developed to evaluate the interaction of algorithms, architectures, and implementation for matrix signal processing.

SIGNAL PROCESSING

Signal Processing Perspective

Signal processing is a difficult concept to describe precisely yet most signal processsing applications are easily recognized. Signal processing often comprises transformations applied to measurements to facilitate their use by an observer or a computer. Such transformations will typically be performed for the purposes of signal-to-noise ratio enhancement, detection, parameter estimation, classification, data compression, demodulation, or reformatting for subsequent use.

The signals to be processed, whether in discrete or continuous representation, are usually processed homogeneously, in that all components of the signal vector are operated upon in a qualitatively similar manner. Furthermore, the computational operations in a signal processing system are ususally data-independent, permitting the use of regular processing structures and algorithms. Additionally, implementations can often be achieved which use fixed amounts of memory and have predetermined processing times.

Classical signal processing consists almost entirely of applying homomorphisms either to perform some intuitively justified 'ad hoc' operation, or as a step in likelihood ratio calculations using Gaussian statistics. Mathematically a homomorphism is a mapping from an algebraic structure to a second algebraic structure, which preserves one or more of the algebraic operations. For example, linear and bilinear transformations preserve vector addition and scalar multiplication of vectors while a logarithmic transformation maps a product into a sum.

Typical linear transformations in signal processing include: sampling, Fourier transforms, filtering, scaling, Hilbert transforms, fixed beamforming, quadrature demodulation, and Karhunen-loeve transforms. Typical bilinear transforms in signal processing include: crossconvolution, crosscorrelation, and crossambiguity functions. A typical logarithmic transformation is homomorphic filtering.

Modern signal processing, with conceptual roots beginning about 1950, extends the set of required operations via parametric models utilizing structural information about the signal or the noise. When appropriate, these models provide improved performance through reducing the number of alternative hypotheses or nuisance parameters which must be considered, thus permitting more powerful tests and better estimators to be constructed, in exchange for more 'a priori' information and more sophisticated computation. Required operations now include matrix multiplication, matrix inversion, solution of eigensystems, and the singular value decomposition.

Linear Time Invariant Filters

Linear filters are one of the most important forms of time invariant signal processing structures and their implementation has been studied for many years. There are three major classes of linear filters: longitudinal, recursive, and transversal. The ladder network is an example of a longitudinal filter, the linear feedback network is an example of a recursive filter, and the tapped delay line is an example of a transversal filter. The discussion of signal processing implementation will start with a review of the transversal filter.

A time-invariant linear filter can be characterized in the time domain by its impulse response h{t} which specifies the time behavior of the filter's output for a delta function input at time t = 0. The output y{t} of the time-invariant filter for an arbitrary input x{t} is the convolution of the filters impulse response h{t} with the input signal x{t}. Figure 1 shows the

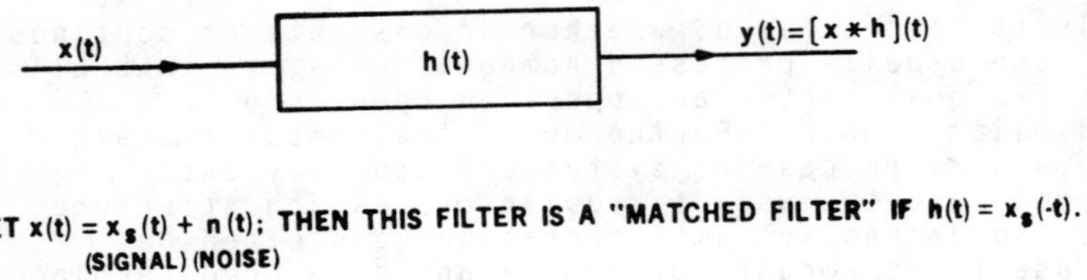
x(t)
h(t)
y(t) = [x * h](t)
LET x(t) = x_s(t) + n(t); THEN THIS FILTER IS A "MATCHED FILTER" IF h(t) = x_s(-t).
(SIGNAL) (NOISE)

FIGURE 1. Linear filter with fixed impulse response.

relationship between the input and output of the time-invariant filter with fixed impulse response. If the filter's impulse response is the time reversal of some expected signal plus additive noise then the filter is called a matched filter. These filters are used extensively for signal detection in radar, sonar, and communication systems.

Signal Processing Filters

In classical signal processing most of the filters are of the type called Infinite Impulse Response (IIR) filters. These filters correspond to a direct implementation of the filters' rational polynomial transfer function. The infinite impulse response is due to a recursive implementation of the zeroes of the denominator of the transfer function.

In modern signal processing most of the filters are of the type called Finite Impulse Response (FIR) filters. These filters correspond to a direct implementation of the filters impulse response approximated with a finite number of terms. These filters often arise as matched filters for radar and sonar signals. A direct transversal filter implementation using N taps requires N times N multiplications. An indirect implementation using the Discrete Fourier Transform (DFT) realized using the Fast Fourier Transform (FFT) algorithm requires approximately 2N times log base 2 of N multiplications. This difference in computation is one of the reasons for the FFTs popularity in audio signal processing.

The linear filter is the simplest signal processing task to implement using transversal filters. To implement a linear filter it is sufficient to choose the tap weights appropriately. If the desired impulse response is known and its sample values are available then the transversal filter tap weights are simply the sample values of the filter's impulse response.

The transversal filter was originally developed by H. E. Kallmann [1] for the equalization of television signals. The original filters developed in the late 1930's were constructed from contigously tapped electromagnetic transmission lines. However, the large propagation velocity of electromagnetic waves in ordinary transmission lines results in correspondingly large physical implementations of the transversal filter. A transversal filter is indicated schematically in Fig. 2a.

$$B \equiv (f_n \text{ rate})_{max}$$

$$T \equiv \text{integration time}$$

$$N \equiv \text{number of stages in filter}$$

$$N(\text{samples}) = T(\text{sec}) \cdot B(\text{samples/sec})$$

$$\text{multiplication rate} = NB$$

$$g_{n_o} = \sum_{n=1}^{N} f_{n_o + n} \, h_n$$

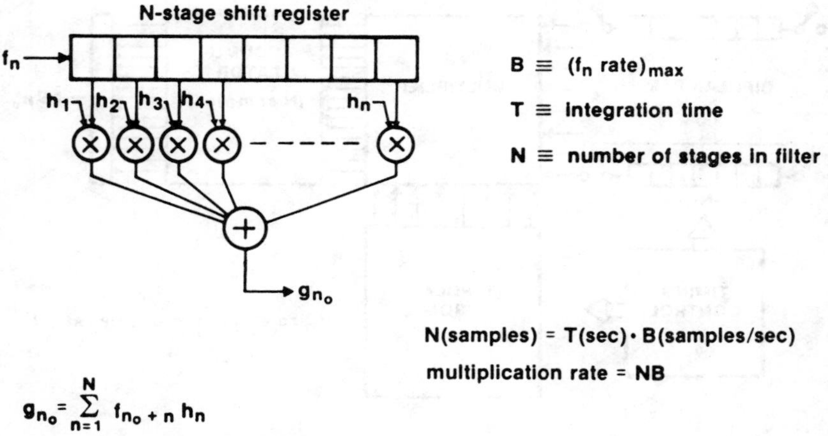

FIGURE 2a.   Transversal filter.

Although the original transversal filter of Kallmann used a continuous propagation medium the tapping mechanism was discrete. Individual tapps were made contiguously along the delay line at a spacing corresponding approximately to one half of the reciprocal of the bandwidth of the line. This Nyquist spacing is shown schematically in Fig. 2a by the vertical lines in the delay line. An alternative interpretation of Fig. 2a is that the delay line is digital and each cell of the delay line corresponds to a digital word which is clocked through the structure.

After the signals are tapped from the delay line they are multiplied by a weight, h, and summed together. The resulting filter, therefore, has an impulse response, h{n}, and the output is the convolution of the input signal with the impulse response h{n}. Figure 2b shows a digital implementation of a transversal filter. The input signal f{n} is quantized into a number of bits, typically 12 to 16 bits for audio signals, 6 to 8 bits for video signals.

The number representation for the signal is arbitrary provided it is compatible with the multiplier, however, sign magnitude and two's complement are popular representations. The digital signals then recirculate in virtual delay lines composed of Random Access Memory (RAM) and are multiplied successively by the weights which are stored in Programmable Read Only Memory (PROM). The number of multiplications which have to be performed to implement a length N filter is N times N.

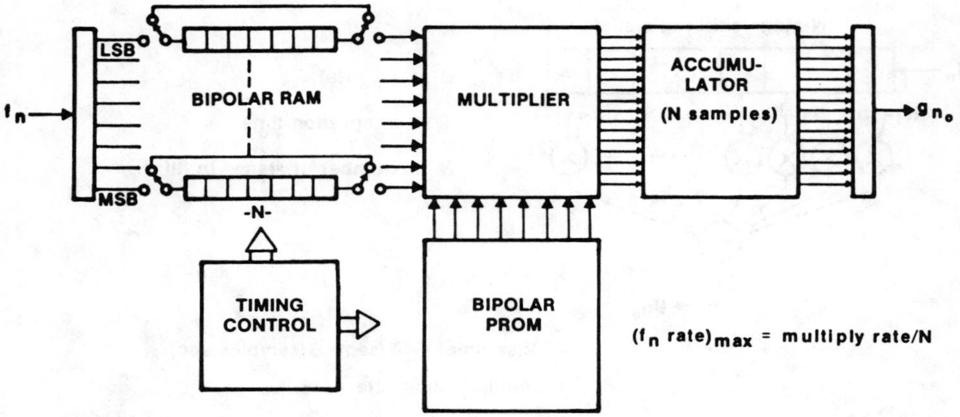

FIGURE 2b. A digital implementation of a transversal filter.

ANALOG SIGNAL PROCESSING

In many applications such as radar, sonar, communications, and image processing digital accuracy is not required. When analog accuracy is sufficient then real-time signal processing filters can be designed which significantly reduce the size, weight and power consumption of the signal processing electronics [2].

Surface Acoustic Wave (SAW) Devices

If an acoustic delay line is used instead of an electromagnetic delay line the physical size of the resulting filters can be significantly reduced since the velocity of sound is about 100,000 times slower than the velocity of light. Acoustic surface waves propagating on single crystals of piezoelectric material represent one form of the transversal filter which can be manufactured with photolithographic techniques similar to those used in the semiconductor industry. These surface acoustic wave (SAW) filters have potential application in radar signal processing. The signal processing application of acoustic transversal filters is described in [3].

Figure 3 shows a schematic representation of the operation of a SAW filter. The input transducer T1 and output transducer T2 are interdigited metalization on the surface of a piezoelectric crystal. The input transducer T1 acts as a transversal filter with impulse response T1 determined by the patterning of its interdigitations. Thus the surface wave generated is the convolution of the input transducer's impulse response. The second transducer T2 similarly acts as a transversal filter with impulse response T2. Since the surface wave generated by the first transducer is the input to the second transversal filter T2, the combined response of the filter is the convolution of the two individual impulse responses. However, if the physical pattern of the transducers T1 and T2 are considered instead of their impulse responses then the combined filter acts as though the two patterns were correlated together rather than convolved together. This is due to the fact that the impulse response of the first filter is the time reversal of its physical pattern while the impulse response of the second filter is a reproduction of its physical pattern. Thus to represent a Kallmann filter with a SAW device the imput transducer should be uncoded and the entire filters impulse response coded into the output transducer.

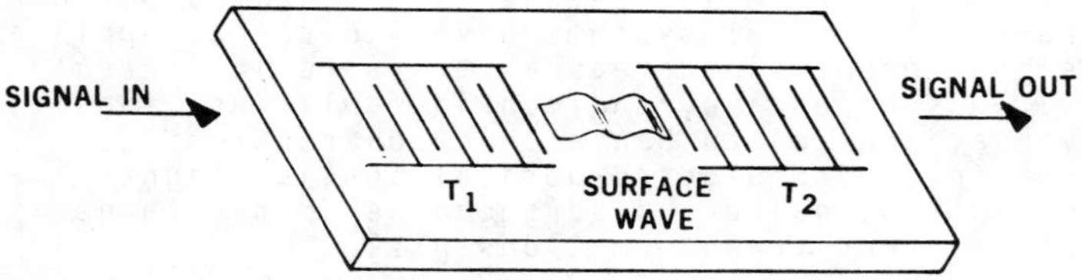

SURFACE WAVE= SIGNAL IN $\star$ T_1

SIGNAL OUT= SURFACE WAVE $\star$ T_2

SIGNAL OUT = SIGNAL IN $\star$ $T_1 \star T_2$

= SIGNAL IN $\star$ CROSS CORRELATION OF T_1 AND T_2

FIGURE 3. Surface acoustic wave filter.

Charge Coupled Devices (CCDs)

In 1970, five years after the introduction of SAW technology, a new analog technology was developed for audio through video frequencies. This technology is a semiconductor technique of Charge Coupled Devices. CCDs can be fabricated on standard semiconductor production lines. Unlike SAW devices where the support circuitry has to be external to the device, CCD can have their support circuitry fabricated on the same substrate. In addition, CCDs can operate at all frequencies from audio to video. The book <u>Charge Transfer Devices</u> [4] by Sequin and Tompsett provides background information on this important class of modern signal processing devices.

Optical Processing

Optical processing systems are of two major types: Those which use coherent optical illumination; and those which use incoherent optical illumination. Both types of systems are inherently two dimensional and thus are capable of processing large amounts of data. This allows a parallel processing capability of over 100,000 analog multiplications to be performed simultaneously in a single device.

Coherent optical systems have received considerable attention and many sophisticated systems have been operational for a number of years. These systems have been well documented and several survey papers and books have been written [5-10]. These systems are attractive since a two-dimensional Fourier transform relation exists between the amplitude distributions in the front and rear focal planes of the lens. By using this Fourier transform relationship filters can be constructed in the spatial frequency domain. Thus two-dimensional convolution and correlation can be achieved. These systems are used in applications which require spectrum analysis, pattern recognition, or image processing.

Incoherent optical systems may be lensless and do not have any natural Fourier transform relation between input and output planes. Nonetheless, incoherent optical systems have widespread application in a number of signal processing areas as they are in general simpler, smaller, less expensive, and considerably less sensitive to thermal and vibrational disturbance than coherent optical systems. When combined with electro-optical spatial light modulators incoherent optical processors can perform a large variety of discrete linear transformations [11].

A representative electro-optical processor is shown in Fig. 4. The system consists of three main components: A light-emitting diode (LED), a two-dimensional light modulator, and a two-dimensional charge coupled divice (CCD). The input signal (vector) modulates the LED which illuminates the spatial light modulator. The light which passes through the spatial light modulator is the product of the input signal with the spatial signal (matrix) represented by the modulator. The CCD integrates the product of the input signal with each row of the matrix and outputs another signal (vector). Thus the electro-optical processor is equivalent to a matrix-vector multiplier or to a linear filter.

FOURIER TRANSFORMATIONS

The discrete Fourier transform has been a staple signal processing technique since the popularization of the FFT algorithm in the sixties. However, until recently it has been difficult to implement the FFT in real time for signals with more than audio bandwidth. Even when high speed FFTs are implemented with bipolar multipliers they require substantial amounts of space, weight, and power.

In order to overcome these difficulties analog implementations of the DFT were sought for applications where size, weight, and power were at a premium. This lead to the development of Surface Acoustic Wave (SAW) and Charge Coupled Device (CCD)

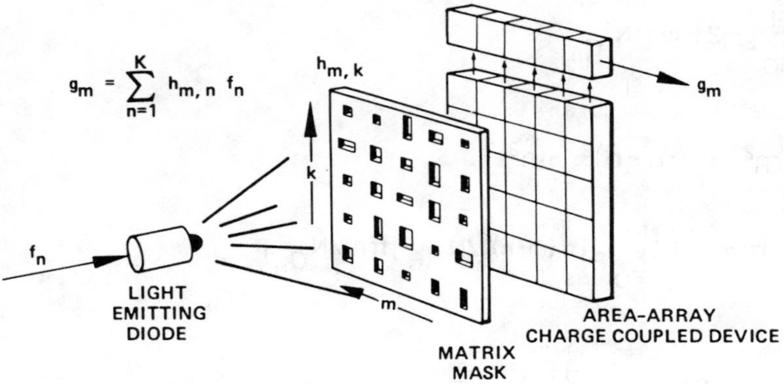

FIGURE 4. Two-Dimensional Optical Processor.

implementations of the DFT using the Chirp-Z-Transform (CZT) algorithm [12].

The Chirp-Z-Transform

The Fourier transform is difficult to implement with transversal filters since it cannot be directly implemented as a time invariant linear filter. In 1965 L. Mertz [13] wrote, "My personal speculation is that if dispersive delay lines or any other chirp filters become available for audio frequencies they will make elegant real time wave analyzers for rapid scanning Fourier transform spectrometers."

In the same year as Mertz was publishing his observation a new technology, SAW devices, was being developed [14]. These devices, which were made possible in large part by the development of practical photolitographic interdigital transducers, provided the possibility of real-time video frequency DFTs.

The chirp-z-transform algorithm is the direct result of a refactoring of the algebraic expression for the DFT after an identity subsitution. The derivation of the CZT algorithm is given in Fig. 5a. After removing terms from under the summation sign which do not depend on the summation variable the final expression results. This expression looks more complicated than the original expression for the DFT, however, it is amenable to a simple interpretation.

First the input signal is multiplied sample by sample by a complex chirp. The resulting signal is convolved with a linear time-invariant filter whose impulse response is a complex chirp. Then the resulting signal is multiplied again sample by sample by a complex chirp. The complex multipliers can be implimented as a combination of four real multipliers. The complex convolver can be implimented also as a combination of four real filters. The final

1) $$G_m = \sum_{n=0}^{N-1} e^{-j2\pi mn/N} g_n$$

2) $$-2mn = -m^2 + (m-n)^2 - n^2$$

3) $$\therefore G_m = e^{-j\pi m^2/N} \sum_{n=0}^{N-1} e^{j\pi (m-n)^2/N} \left(e^{-j\pi n^2/N} g_n\right)$$

where $m = 0, 1, 2, \ldots, N-1$

FIGURE 5a. Discrete Fourier transform CZT algorithm.

configuration is shown in Fig. 5b. If only the magnitude of the DFT is needed then the post multiplication is not required since the chirp signal has unit modulus.

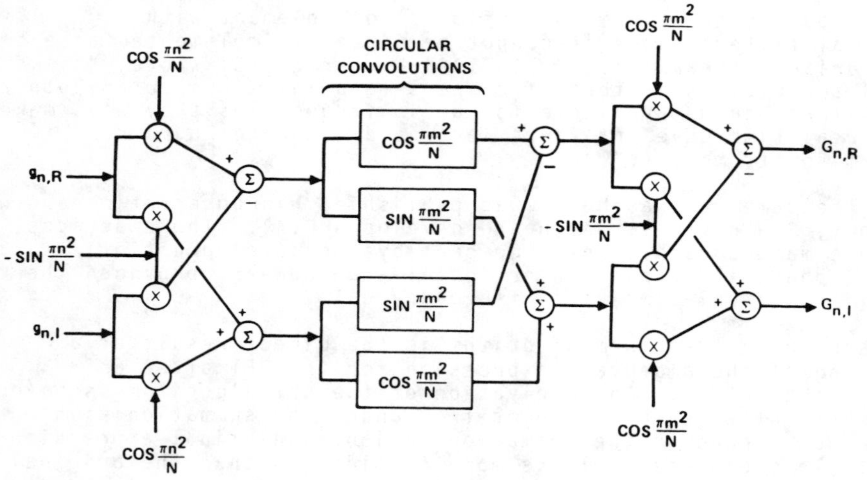

FIGURE 5b. DFT via CZT algorithm with parallel implementation of complex arithmetic.

Figure 6 illustrates a CCD chirp filter. The pattern is that of a cosine chirp and starts at the left of the illustration. The tap weighting can be visualized by looking for the break in every third electrode. The output of this CCD is formed by taking the difference of charge contribution from the upper part and the lower part of the split electrodes. Thus real numbers between +1 and -1 are represented by the varying position of the electrode break.

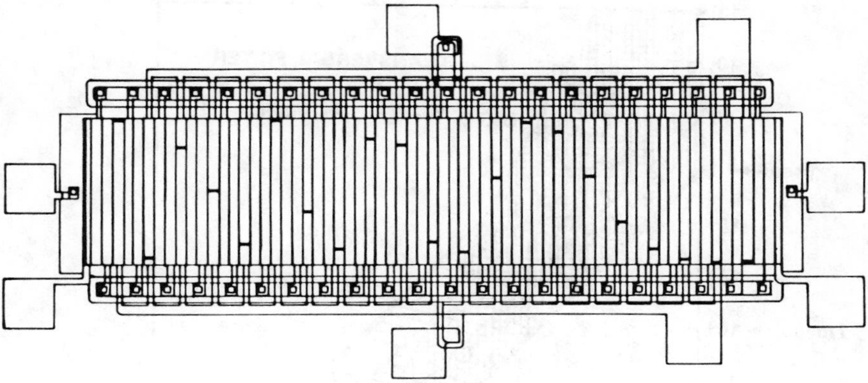

FIGURE 6. Charge coupled device (CCD) chirp filter.

Although both CCD and SAW devices process analog signals and are fabricated by photolithography there are several important differences. The CCD is a base band device while the SAW is a band-pass device. The CCD is a low frequency device while the SAW is a high frequency device. Both types of devices have limited time-bandwidth product due to aberrations in the propagation. For the CCD charge transfer inefficiency limits the number of taps while for the SAW attenuation and acoustic scattering from the electrodes limits the number of taps. However, for a fixed time-bandwidth product (number of taps) the CCD device usually is physically smaller and less expensive to fabricate since it can be fabricated on a semiconductor production line and SAW devices require special fabrication and packaging. Thus CCDs and SAWs are complimentary signal processing devices and both should be considered in application where size, weight, power and cost are the dominant factors in system performance.

With CCDs taking over signal processing at video frequencies SAW technology was expanded to higher frequencies. New SAW devices were developed with time bandwidth products in excess of one thousand. These devices called Reflective-Array Compressors (RACs) are continuous dispersive devices with bandwidth capabilites in to the hundreds of MegaHertz frequency range. The book Surface Wave Filters, Design, Construction, and Use, [15] edited by H. Matthews provides background information on this important class of signal processing devices.

Figure 7 shows a comparison of the performance of three types of transversal filter components: Digital, CCD, and SAW. From this comparison it can be observed that the three technologies have overlapping ranges of application. Thus the design engineer has flexibility in the choice of a technology for a particular application.

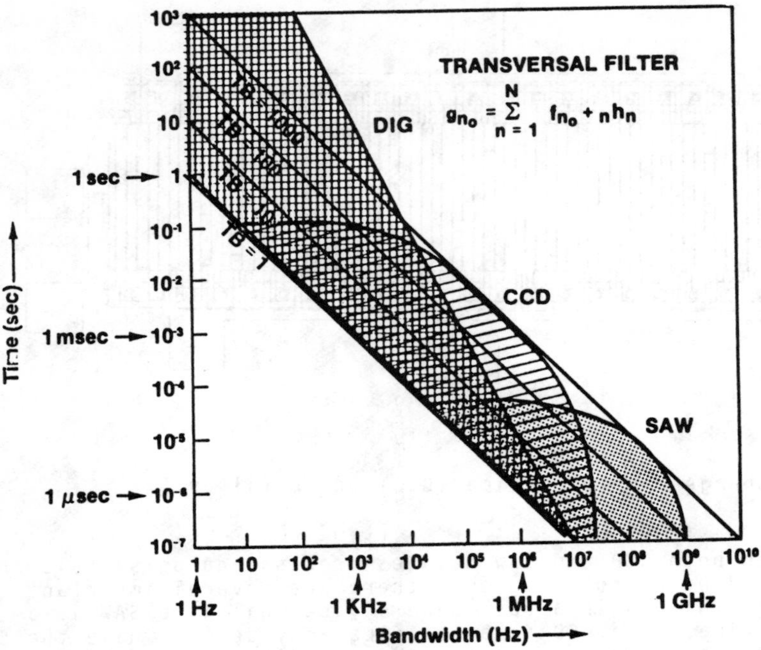

FIGURE 7. Transversal filter comparison.

Two-dimensional DFT

Two-dimensional signal processing requires two-dimensional convolutions, correlations, and Fourier transformations. These operations occur in image processing applications as well as in multiple sensor array systems such as phased array radar and sonar systems. If an indirect implementation of convolution and correlation is used then a two-dimensional Fourier transform is sufficient. Although the FFT algorithm can be directly extended to two dimensions there are often size, weight, and power constraints which favor analog implementations.

A real time two-dimensional DFT can be implemented using a combination of CCD and SAW components. The CCD's have small bandwidth and long integration time. The SAW's have large bandwidth and short integration time. When used in combination, these components provide the system designer with great flexability in implementing the necessary signal processing operations.

Consider a uniformly spaced line array of sensors as an example. In this example a Fourier transform of the output of the individual sensors provides the spectrum of the input signal or it can be used as the first step in an indirect implementation of linear filtering. At a fixed instant in time a Fourier transform of the multiple sensor outputs provides a representation of the spatial wave number structure of the radiation field incident on the sensors. Through the use of two-dimensional signal processing wide bandwidth beamforming and other operations can be implemented.

The advantage of using different technologies follows from the differences in the performance characterics of the individual technologies. If CCD CZT DFT's are used for the time processing of the individual sensor outputs then real time bandwidths from audio to video can be realized. If lower frequency sensors, such as seismic sensoars, are used then digital processing would probably be preferred since CCD filters have a minimum shift rate since they are dynamic storage elements.

MATRIX SIGNAL PROCESSING

The two major signal processing concepts examined thus far have been those of convolution and Fourier transformation. These operations have been considered from the point of view of operations on an array of samples, the sample values of the input vector. In recent years special digital hardware called an "array processor" has become popular as an add on device to minicomputers in order to provide the computationally intensive operations of convolution, correlation, and fast Fourier transformation.

Although analog signal processing with surface acoustic waves, charge coupled devices, and optical processing has been capable of surprising flexibility in the implementation of convolution, correlation, as well as Fourier and other transformations there is a continuing need for systems with higher dynamic range and greater flexibility than analog devices can provide. In addition, many of the algorithms of modern signal processing are not easily implemented with any of the current analog devices. Thus it is important to have a selection of parallel processing digital algorithms and the architectures for their real-time implementation.

It is conjectured that the computational requirements for many real-time signal processing tasks can be reduced to a common set of basic matrix operations. These include matrix-vector multiplication, matrix-matrix multiplication and addition, matrix inversion, the solution of systems of linear equations, least squares approximate solution of systems of linear equations, singular value decomposition (SVD) of matrices, the solution of eigensystems, the generalized singular value decomposition, and the solution of generalized eigensystems.

All of the matrix computations indicated above can be efficiently solved using a digital computer with a single arithmetic unit. The first five matrix operations may be computed non-iteratively and the others may be computed by a variety of iterative algorithms. There are libraries of numerically stable high quality numerical linear algebra software available under the following names: LINPACK [16] for linear equation solution and least squares problems, EISPACK [17] for eigensystem problems, and the publications of Van Loan [18,19] and Paige-Saunders [20] for the more recent generalized singular value decomposition.

Digital signal processing architectures underwent a revolutionary change in the mid 1960's with the development of the dedicated FFT processor, followed by an evolutionary change as further flexibility was added in the more general array processor. The requirements of modern signal processing together with the availability of VLSI technology are leading to another revolutionary change in signal processing architectures: The systolic [21] and wavefront [22] cellular processors with vector and matrix parallelism for matrix operations.

Signal Processing Operations

Much of present day signal processing is based on assumptions of wide-sense stationarity for noise and random signals, together with linear, time-invariant models for signal propagation. Since sinusoids are the eigenfunctions of shift-invariant kernels, the Fourier transform is important both for implementing convolutions and providing eigensystem expansions for spectrum analysis.

Recent signal processing algorithms tend to incorporate additional prior information concerning the signal structure and more realistic non-stationary noise models. This more realistic modeling will generally provide improved performance at the expense of a greater computational load. Maximum entropy spectrum analysis provides improved resolution compared to classical windowed spectrum analysis, when an all-pole spectal model is applicable, but requires the solution of a linear prediction problem. Adaptive cancellation of noise or interference by direct least squares techniques provides faster convergence [23], and hence better tracking behavior than gradient based adaptive filtering, but requires the orthogonal triangularization of a matrix or algorithms of comparable complexity. Recent beamforming/direction finding techniques provide improved resolution by incorporating a spatial multiple point source model for the signals, but require the solution of an eigensystem or generalized eigensystem problem at each resolved temporal frequency [24-25]. The common trend in these and many other recent signal processing algorithms is their extensive utilization of linear algebra operations, both conceptually and computationally. Real time signal processing will increasingly require the parallel hardware equivalent of LINPACK and EISPACK, together with the more recent generalized singular value decompositions of Van Loan [18,19] and Paige-Saunders [20].

Signal processing algorithms frequently require the solution of the eigensystem of an estimated covariance matrix, where the covariance matrix is estimated as the product of a data matrix and its transpose. Roundoff error in the formation of this matrix product can cause a significant, irreversible loss of accuracy. The computational wordlength requirement for such a problem can be reduced by about a factor of two by solving the eigensystems problem indirectly via computing the singular value decomposition of the data matrix. This is similar to the improvement in solving least squares problems via orthogonal triangularization of the data matrix, rather than forming the Gauss normal equations [26-28]. The "MUSIC" high resolution direction finding algorithm of R. Schmidt [29] requires the solution of a generalized eigensystem involving two estimated covariance matrices. For such computations, the generalized singular value decompositions of Van Loan and Paige-Saunders, permit a similar reduction in computational word length by applying a reduction directly to the data matrices.

Although in actuality array processors are neither arrays of processors nor simultaneous processors of arrays of numbers, these signal processing hardware and software additions provide a much needed increase in throughput for the General Purpose (GP) computer. Recently, a new class of signal processing hardware called the Systolic Array Processor (SAP) has been proposed [30]. This hardware consists of an array of processors simultaneously processing an array or vector of input signal. Since a whole vector of data may be processed simultaneously the SAP may be thought of as a matrix processor

Figure 8 shows a comparision of several popular parallel computation methods. The digital transversal filter is very effective at computing convolutions, correlations, and through the CZT algorithm the discrete Fourier transform. However, it has limited flexibility for computing matrix operations. The array processor, although more flexible is usually only a single real or complex multiplier and the necessary memory and software for algorithm execution one operation at a time. When more than one operation at a time is attempted as with a bus organized collection of microprocessors then interprocessor communication becomes a difficulty. Professor Minsky of MIT once speculated that although the hardware parallelism of bus organized microcomputers was increasing linearly with the number of microprocessors tied on the bus the computation throughput was only increasing logarithmically. This conjecture seams to be true in many simulations of bus organized microprocessors and is due to bus contention problems with the multiple processors trying to simultaneously access the bus.

GENERIC ARCHITECTURE	CHARACTERISTICS
Digital transversal filter	Limited to convolution, adaptive filtering
Array processors	Limited parallelism (typically one butterfly), Limited operations (FFT, convolution, pointwise mult)
Bus organized multiprocessors	Limited parallelism, difficult programming — due to communication architecture & conflicts for access to shared resources
Systolic (HT Kung, SY Kung)	Modular parallelism
	Synchronous data flow
	Simple control
	Sufficiently flexible for computation — intensive part of signal processing tasks: vector, matrix & sorting operations

FIGURE 8. Parallel computation comparison.

The problems just described have been substantially overcome by a new class of processing arrays developed by H.T. Kung of Carnige Mellon University and S.Y. Kung of the University of Southern California and called respectively systolic array and wavefront processor [21,22]. These signal processing concepts are the two-dimensional generalization of the digital transversal filter. These processors have both distributed computation and distributed memory. It is often possible to configure these processing arrays so that the interconnections between adjacent processors is to the nearest neighbors only.

Figure 9 describes several of the most important attributes of systolic arrays, a generic term which describes both the original systolic array of H.T. Kung and the more flexible wavefront processor of S.Y. Kung. A detailed description of these architectures is given by S.Y Kung in this volume.

- **An arrangement of processing elements efficiently using parallelism to achieve high-performance signal processing**
- **Generally characterized by**
  - **Identical processing elements**
  - **Interconnections between nearest neighbors**
  - **Synchronous data flow**
- **Well-matched to VLSI implementation**

FIGURE 9. Attributes of a systolic array.

Advanced Matrix Operations

A subset of the matrix operations which could form the basis of a real-time digital processing system are: the fast Fourier transform (FFT), matrix-matrix multiplication, orthogonal triangularization (QR factorization), and the singular value decomposition. With these operations it is possible to compute most of the matrix operations such as matrix inverse, linear least squares solutions, and the solution of eigensystems. If the subset is augmented with the generalized SVD then even the solution of the generalized eigensystem is possible.

However, no systolic array processor will have enough hardware resources for all possible sizes of matrices. It is therefore necessary to be able to partition the matrix operations. Matrix-Vector multiplication and matrix-matrix multiplication factor in a simple manner. The first problem occurs with matrix inversion. If the inverse of a partitioned matrix could be written as a partitioned matrix of the same size the problem would be solved. Fortunately this is true [31]. Figure 10a and Fig. 10b show the required computations.

$$P = \begin{bmatrix} A & B \\ C & D \end{bmatrix} \quad \text{where } P \text{ is } (2n, 2n) \text{ and } A, B, C, D \text{ are } (n, n)$$

$$P^{-1} = \begin{bmatrix} A & B \\ C & D \end{bmatrix}^{-1} = \begin{bmatrix} A^{-1} + A^{-1}B\Delta^{-1}CA^{-1} & -A^{-1}B\Delta^{-1} \\ -\Delta^{-1}CA^{-1} & \Delta^{-1} \end{bmatrix}$$

with $\Delta = D - CA^{-1}B$

FIGURE 10a. Inverse of a partitioned matrix.

Storage

A	$R = A^{-1}$	R	R	R	R	R	R + MT	R + MT
B	B	M = RB	M	M	M	M	M	$S = M\Delta^{-1}$
C	C	C	C	C	Q = CR	$T = \Delta^{-1}Q$	T	T
D	D	D	$\Delta = D - CM$	Δ^{-1}	Δ^{-1}	Δ^{-1}	Δ^{-1}	Δ^{-1}

Operation count
2 inversions
6 matrix multiplications or multiply-adds

FIGURE 10b. Computation of partitioned inverse matrix.

More complicated matrix operations such as orthogonal triangularization with Givens rotations [27], solving eigensystems, and finding singular value decompositions require even more powerful signal processing elements. The partitioning of these more complicated calculations is not currently resolved.

SIGNAL PROCESSING IMPLEMENTATIONS

Parallel Architectures

For applications in which the sampling rate approaches the computational cycle time, parallel architectures are required for real-time implementation of tasks requiring a number of operations proportional to the square or the cube of the number of points per data block, such as typical matrix operations [26]. Representative parallel architectures include the digital tapped delay line transversal filter, the array processor, the bus organized multiprocessor, and the systolic array. This paper emphasizes the systolic array because of its promising combination of characteristics for utilizing VLSI technology for real-time signal processing: modular parallelism with throughput directly proportional to the number of cells, simple control, synchronous data flow, local interconnects, and sufficient versatility for implementing the matrix operations needed for signal processing. It is suggested that the systolic array be viewed as an adjunct to the general purpose computer and the array processor rather than as a replacement for either, since the general purpose computer excels in unstructured computation and decision making and the array processor excels in fast Fourier transform computation.

The systolic array [21], introduced by H.T.Kung, is a regular geometric array of identical or nearly identical computational cells with common timing and control and synchronous data flow. Representative systolic architectures are shown in Fig. 11 for a variety of matrix computations. Systolic arrays may be viewed as distributed,space-time implementations of recursion relations. Occasionally the operation required to initiate or terminate a set of recursions will be slightly different from the intermediate recursive steps,necessitating one or more boundary cells which differ from the interior cells of the array. In addition for the operations of matrix multiplication and matrix inversion, constant efficiency may be maintained when a large matrix is partitioned to fit on a small systolic array, if memory is provided, conceptually orthogonal to the plane of the array, so that submatrices may be stored or loaded in a single memory cycle time [30].

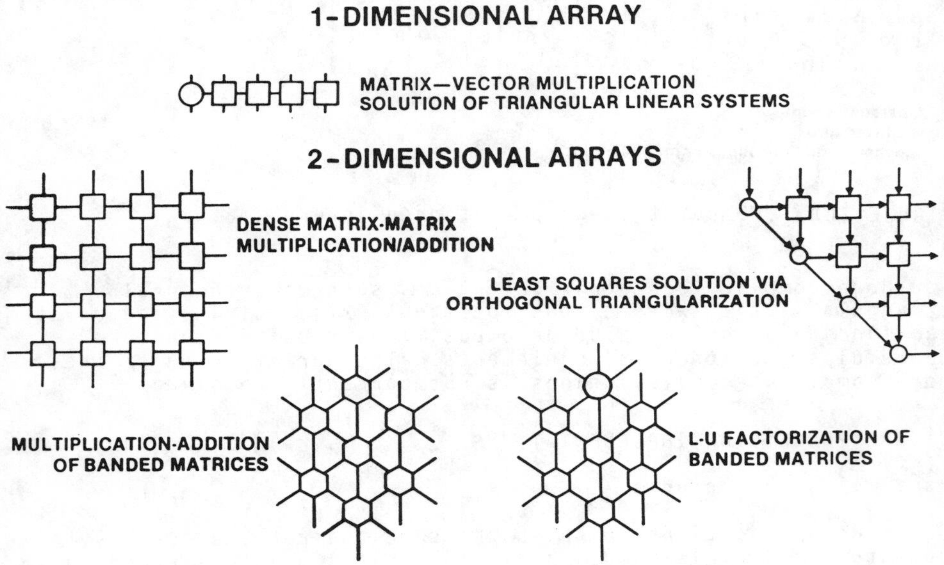

FIGURE 11. Systolic array solutions.

Two major systolic arrays have been developed to evaluate this new technology. TRW/ESL have developed a linear systolic array called Phoenix which incorporates 28 of the TRW fixed point multiplier/accumulators. The Naval Ocean System Center has developed a reconfigurable one/two dimensional systolic array with 64 floating point processing elements as a systolic architecture testbed.

Systolic Array Test Bed

The many systolic array architectures provide a method of implementing most of the important matrix computations of linear algebra. For each matrix computation the corresponding systolic array concept involves the use of many identical processing elements operating in parallel to achieve a high throughput. However, the individual cells in the array lattice are usually different for different matrix computations. Although the prospects for fabricating the individual cells of a systolic array as a single silicon chip are good the number of different cells required is still large. Fortunately, many of the design details of the individual algorithms such as data flow, control, numerical accuracy and input/output can be determined before fabrication through the use of a general purpose systolic test bed implemented with commercially available components.

In order to test the feasibility of a reconfigurable matrix processor a 64 element, (8-by-8), systolic array test bed was constructed [30]. The systolic array testbed was designed as a adjunct to a host computer in much the same manner as commercial vector array processors. The testbed consists of a cabinet containing an 8-by-8 array of systolic processor circuit boards, a mother board, and a rack for the host-interface electronics. The configuration is shown in Figure 12. In order to achieve flexibility in the use of the testbed, it was decided to use a microprocessor with both non-volatile and volatile memory in the construction of the systolic processing elements (SPEs).

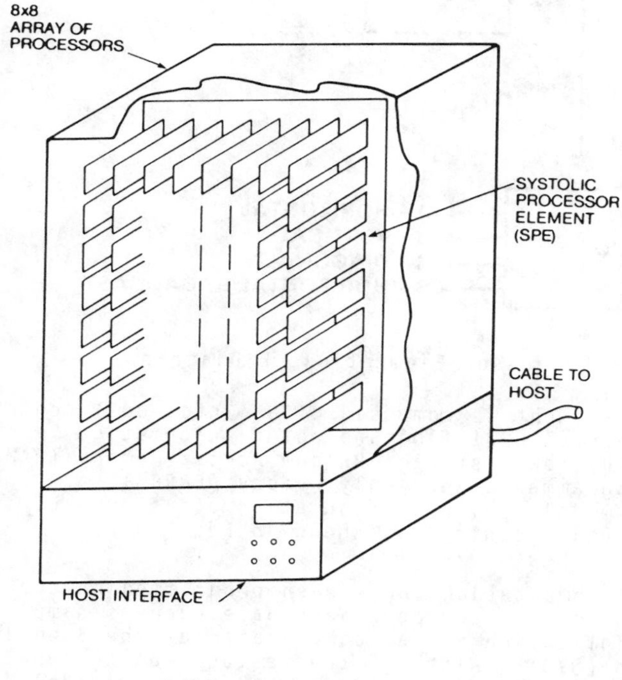

FIGURE 12. Systolic Test Bed.

Even though most of the processing elements of a systolic array are the same the boundary cells are often different and different matrix operations require different processing element cells. To overcome these problems a general purpose systolic processing element was constructed using commercially available Very Large Scale Integration (VLSI) circuits. The programmable systolic processor element is shown in Fig. 13. Because of the provision for non-volitle storage those cells which must perform different computations can be programmed with the appropriate instructions.

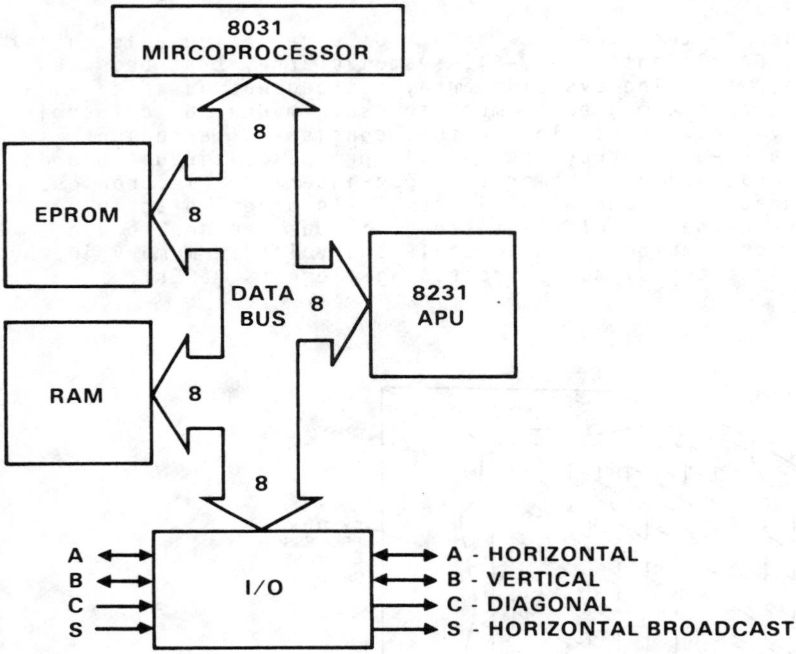

FIGURE 13. Systolic Processor Element Block Diagram.

The processing element is composed of an Intel 8031 control microprocessor, an Intel 8231 floating point Arithmetic Processing Unit (APU), data storage Random Access Memory (RAM), an Electrically Programmable Read Only Memory (EPROM), and an Input/Output (I/O) module. The configuration of a three-by-three systolic array including control is shown in Fig. 14.

Several matrix computations have been programmed on the systolic array testbed. These operations have been as simple as matrix-matrix multiplication and as complicated as the Singular Value Decomposition (SVD). Although the testbed was designed for flexibility instead of speed it has been a significant tool in the development of systolic architectures. An estimate of the complexity of the SPE is about 64,000 equilivant gates and thus it is reasonable to suggest that a single chip of this complexity could be constructed in the future.

The results obtained on an experimental eight-by-eight systolic test bed have shown the feasibility of systolic matrix processor addition to the computing resources of a minicomputer. It is felt that systolic architectures implemented via VLSI technology will make it feasible and desirable to provide real-time signal processing using recent algorithmic developments.

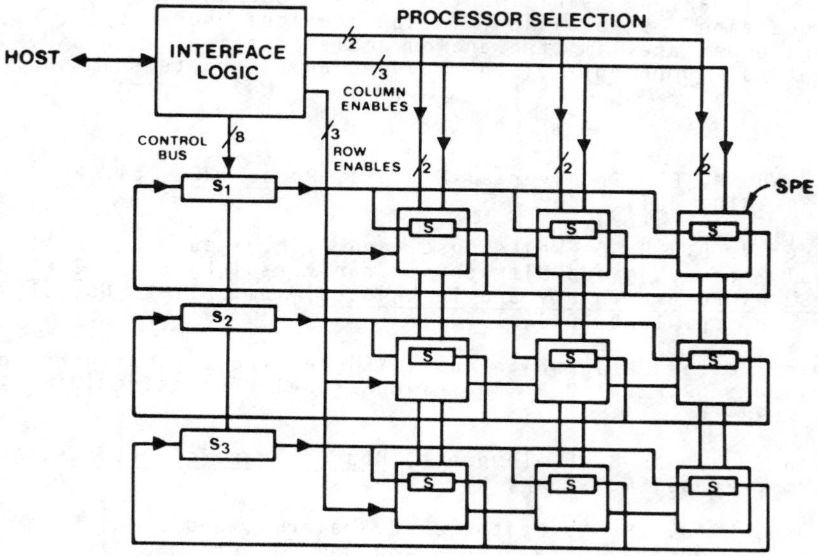

FIGURE 14. Systolic Array Control Structure.

CONCLUSIONS

Signal processing technology has developed concurrently with the development of new signal processing algorithms and architectures. Transversal filters using surface acoustic wave (SAW) technology were developed first for radar signal processing then charge coupled devices (CCD) for video and sonar signal processing. More recently with the development of very large scale integration (VLSI) digital circuits transversal filters with high accuracy are available across the range of applications from speech processing to radar processing.

Modern signal processing with its many matrix operations is currently being developed using VLSI circuits in systolic and wavefront processor architectures. As a result of the concurrent advances in numerical linear algebra, parallel architectures for computation, and VLSI hardware implementations, it is no longer necessary for the signal processing engineer to use suboptimal computations when optimal computations will provide improved performance. Although many advanced algorithms such as direct orthogonal triangularization, eigensystem solution, and the singular value decomposition (SVD) have systolic or wavefront implementations research is still needed to increase the number of algorithms which can be implemented.

The use of the singular value decomposition (SVD) and the generalized SVD make possible the use of single precision computation in many applications where double precision computation was previously required. Although much research still needs to be done in applying these new techniques to situations with time varying statistics, it is clear that the ability to implement matrix computation in silicon without conflicts for access to resources will make posssible many new and important advances in signal processing. In addition, in applications where space, weight, and power are limitted analog implementations with SAW, CCD, or optical components often provides an alternate solution.

REFERENCES

1. Kallmann, H. E., Transversal Filters, Proc. IRE, vol. 28, pp. 302-310, July 1940.

2. Squire, W. D., H. J. Whitehouse, and J. M. Alsup, Linear Signal Processing and Ultrasonic Transversal Filters, IEEE Trans. Microwave Theory and Techniques, vol. MTT-17 No. 11, November, pp. 1020-1040, 1969.

3. Bromley, K. and H.J. Whitehouse, "Signal Processing Technology Overview", SPIE Vol.298, Real-Time Signal Processing IV(1981), paper 298-14

4. Sequin C. H. and M. F. Tompsett, Charge Transfer Devices, Academic Press, New York, NY, 1975.

5. Cutrona, L. J., E. N. Leith, C. J. Palermo, and L. J. Procello, "Optical Data Processing and Filtering System." IRE Trans. Information Theory, vol. IT-6, pp. 386-, 1960.

6. Vander Lugt, A., "A Review of Optical Data-processing Techniques," Opt. Acta vol. 15, pp.1-, 1960.

7. Vander Lugt, A, "Coherent Optical Processing," Proc, IEEE vol. 62, 1300-, 1974.

8. Goodman, J. W., Introduction to Fourier Optics", McGraw-Hill New York, NY, 1968.

9. Shulman, A. R., Optical Data Processing, John Wiley & Sons, New York, NY, 1970.

10. Preston Jr., K., Coherent Optical Computers, McGraw-Hill, New York, NY, 1972.

11. Monahan, M. A., R. P. Bocker, K. Bromley, and A. Louie, "Incoherent Electro-optical Processing with CCD's," Optical Computing Conference, Washington, D.C., April 1975.

12. Whitehouse, H. J., and J. M. Speiser, "Linear Signal Processing Architectures," NATO Advanced Study Institute on Signal Processing with Emphasis on Underwater Acoustics, Portovenere, Italy, D. Reidel Publishing Co., Dordrecht, Holland, pp. 669-702, 1977.

13. Mertz, L., _Transformations in Optics_, John Wiley & Sons, New York, NY, 1965.

14. White R. M., and Voltmer, F. W., Direct Piezoelectric Coupling to Surface Elastic Waves, _Appl. Phys. Letters_, vol. 7, pp. 314-316, December 1965.

15. Matthew, H., _Surface Wave Filters, Design, Construction, and Use,_ John Wiley & Sons, New York, NY, 1977.

16. Dongarra, J.J. et al, "LINPACK Users' Guide", Society for Industrial and Applied Mathematics, Philadelphia, 1979.

17. Garbow, B.S. et al, "Matrix Eigensystem Routines-EISPACK Guide Extensions", Springer-Verlag, New York, NY, 1977.

18. Van Loan, C.F., "Generalizing the Singular Value Decomposition", _SIAM Journal on Numerical Analysis_, vol. 13, No.1, pp.76-83, March 1976.

19. Van Loan, C.F., "Generalized Singular Values, with Algorithms and Applications", Ph.D. Thesis, Univ. of Michigan, Ann Arbor, Michigan, 1973.

20. Paige, C.C. and M.A. Saunders, "Towards a Generalized Singular Value Decomposition", _SIAM Journal on Numerical Analysis_, vol. 18, No.3, pp.398-405, June 1981.

21a. Kung H. T., and C. E. Leiserson, Systolic Arrays (for VLSI), in _Sparse Matrix Proceedings 1978_, ed. I. S Duff and G. W. Stewart, pp 256-282, SIAM, Philadelphia, PA, 1979.

21b. Kung, H. T., Why Systolic Architecture?, _Computer_, 1981.

22a. Kung, S. -Y., VLSI Array Processor for Signal Processing, presented at _Conference on Advanced Research in Integrated Circuits_ held at MIT, Cambridge, Massachusetts, Jan. 1980.

22b. Kung, S. -Y., R. J. Gal-Ezer, and K. S. Arun, "Wavefront Array Processors: Architecture, Language, and Applications", Proceedings, Conference on Advanced Research in VLSI, edited by Paul Penfield, Jr., MIT, January 25-27, 1982.

23. Monzingo, R.A. and T.W.Miller, _Introduction to Adaptive Arrays_, John Wiley & Sons, New York, 1980.

24. Speiser, J.M., H.J. Whitehouse, and K. Bromley, "Signal Processing Applications for Systolic Arrays", Record of the 14th Asilomar Conference on Circuits, Systems, and Computers, held at Pacific Grove, California, 17-19 Nov., 1980, pp.100-104 IEEE Catalog No.80CH1625-3.

25. Whitehouse, H.J. and J.M. Speiser, "Sonar Applications of Systolic Array Technology", presented at IEEE EASCON, Washington, D.C., Nov.17-19, 1981.

26. Speiser, J.M. and H.J. Whitehouse, "Parallel Processing Algorithms and Architectures for Real-Time Signal Processing", SPIE vol. 298, Real-Time Signal Processing IV (1981), paper 298-01.

27. Gentleman, W.M. and H.T. Kung, "Matrix Triangularization by Systolic Arrays", SPIE vol. 298, Real-Time Signal Processing IV (1981), paper 298-03.

28. Lawson, C.L. and R.J. Hanson, Solving Least Squares Problems, Prentice-Hall, 1974.

29. Schmidt, R.O., "A Signal Subspace Approach to Multiple Emitter Location and Spectral Estimation", Ph.D. Dissertation, Dept. of Electrical Engineering, Stanford University, November, 1981.

30. Bromley, K., Symanski, J. J., Speiser J. M., and Whitehouse H. J., Systolic Array Processor Developments, in VLSI Systems and computations, ed. H. T. Kung, B. Sproull, and G. Steele, pp. 273-284, Computer Science Press, Rockville, MD, 1981.

31. Bodewig, E., Matrix Calculus, North-Holland Pub. Co., 1956.

11

VLSI Array Processor for Signal Processing

S. Y. KUNG

ABSTRACT

In modern signal processing, there are increasing demands for large-volume and high-speed computations. At the same time, the advent of VLSI has generated a visible impact on signal processing by offering almost unlimited computing hardware at extremely low cost. These factors combined have had major effect on the rapid up-grading of current signal processors, from the classical (one-dimensional) transversal filters to highly parallel (two-dimensional) array processors. This chapter first reviews the impact of the basic VLSI device technology and layout design on VLSI processor architectures. VLSI device technology offers a promising potential as well as creates some new design constraints. Thus systematic and well-defined hierachical design and CAD tools for simulation and verification are critical in reducing the design complexity. The solution to real-time signal processing hinges upon novel designs of high-speed, highly parallel array processors for the common primitives in signal processing, such as convolution, EET, and matrix operations, etc. This chapter presents the algorithmic and architectural footing for the evolution of the design of these VLSI oriented array processors. A brief review is first provided on the impact of MOS device technology and VLSI layout methodology on array processor design. With very large scale integration of systems in mind, the rest of the chapter focuses on locally interconnected computing networks. While the Systolic Array offers an elegant solution, avoiding global interconnection by effectively managing local data movements, its inherent global synchronization requirement may affect its practicality in ultra large scale systems. On the other hand, the asynchronous data-driven nature of the Wavefront Array offers a natural solution to this synchronization problem. Moreover, the wavefront array lends itself to a wavefront language (MDFL), drastically simplifying the description of parallel algorithms and paving a way to silicon compilation. We also note that, for synchronizable systems, a (synchronized version of) wavefront array will function just like a multi-rate systolic array and therefore may offer a higher throughput rate than a pure systolic array. Finally, two application examples of the wavefront array are discussed. One is on the wavefronts of Givens rotations with applications to least-square error solution and eigenvalue decomposition problems. The other is on solving Toeplitz systems, with extension to the solution of band Toeplitz systems.

---

[1] Research supported in part by the Office of Naval Research under contract N00014-81-K-0191; and by the National Science Foundation under grant ECS-82-12479

Table of Contents

1 Introduction

The ever-increasing demands for high-performance and real-time signal processing strongly indicate the need for tremendous computation capability, in terms of both volume and speed. Therefore, the realization of many modern signal processing methods depends critically on high speed computing hardware. The availability of low cost, high density, fast VLSI devices makes high speed, parallel processing of large volumes of data practical and cost-effective [38]. This presages major technological breakthroughs in real-time signal processing applications. On the other hand, it is quite obvious that the full potential of VLSI can be realized only when its application domains are discriminatingly identified. It is also noted that traditional computer architecture design considerations are no longer adequate for the design of highly concurrent VLSI computing processors.

For an example of new VLSI design principles, high layout and design costs suggest the utilization of a repetitive modular structure. Furthermore, the communication has to be restricted to localized interconnections, as communication in VLSI systems is very expensive in terms of area, power and time consumption [38]. A broad and fundamental understanding of the new impact presented by the VLSI technology hinges upon a cross-disciplinary research encompassing the areas of algorithm analysis, parallel computer design, system applications, VLSI layout methodology and device technology.

Our objective in this chapter is to introduce an integrated research approach aimed at incorporating the vast VLSI computational capability into modern signal processing applications. This should help shorten the design gap between signal processing theory/ algorithm and VLSI processor architecture/implementation.

2 Impact of VLSI on Array Processor Design

In this section, we first review the basic features of VLSI (MOS) circuitry. The building block concept as well as the scaling effects in VLSI design are briefly discussed. Finally, the major issues which influence the trend of VLSI system design are summarized.

2.1 MOSFET

MOS circuits are typically made of n-channel enhancement and depletion transistors (NMOS) or n-channel and p-channel enhancement transistors (CMOS). An n-channel transistor is made in a p-type substrate, whereas a p-channel transistor is made in an n-type substrate. In an n-channel transistor, the drain and source regions are created by n-type diffusions. The gate is made of a conductor (polysilicon) over a thin oxide covering the region between the drain and source diffusions. When the voltage of the gate is raised with respect to the drain, source and substrate voltages, electrons will be attracted to the surface of the substrate. Above a certain threshold the number of electrons will be so large that they form a conducting channel between the source and drain.

The basic MOS module is the inverter circuit. Figure 1 shows the basic NMOS and CMOS inverter circuits. Usually the transistor connected with the ground is referred to as the pulldown transistor, while the transistor connected with V_{dd} is referred to as the pullup transistor. The pullup transistor of the NMOS inverter is a depletion mode transistor, i.e. it will always be on. The pullup transistor of the CMOS inverter (cf. Fig. 1(b)) is a p-channel enhancement type; it will only be on when the voltage at the gate is low with respect to the voltages at the drain and source. When the inverters are properly designed, their outputs can be used to drive the input of the next inverter stage.

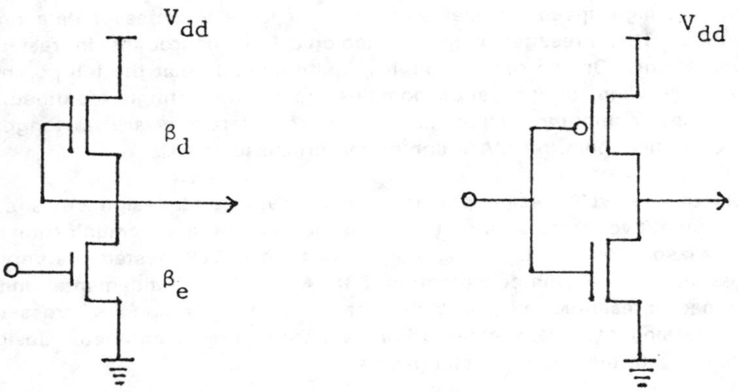

(a) NMOS-depletion load Static Inverter

(b) CMOS static Inverter

Figure 1: Basic NMOS and MOS inverter circuits

A major difference between NMOS and CMOS is the power dissipation. A CMOS inverter draws power only in a transient condition, due to the fact that normally only one of the transistors will be on. Thus the major power consumption is proportional to the switching frequency. An NMOS inverter, on the other hand, draws power whenever the pulldown transistor is conducting. One way of alleviating this problem is to use dynamic logic, which is however more difficult to design and requires more area. Consequently, CMOS is in general more preferable than NMOS.

2.2 Integrating Circuits into Chip

One strong point of VLSI design is the availability in the near future of a <u>hierarchical</u> and <u>multilevel</u> design method and the associated software packages. Usually we consider four description levels: (1) architectural, (2) register transfer, (3) logic/circuit, and (4) layout.

An upper level description should be an elegant and powerful abstraction of the more detailed implementation at the lower level. For instance, a control unit can be simply a box at the architectural level; one or more Finite State Machines (FSM) at the register transfer level; a PLA circuit description at the circuit level and a collection of rectangles at the layout level. At each of the description levels the cells are hierarchically specified to decrease the complexity of the description. In order to reduce the design costs a modular design approach is used; it is often less expensive to implement a more general module, that can be used in a number of different places than to implement a specific module that can be used only once.

The actual process of designing the layout of a VLSI chip is fairly well supported at the moment. Existing interactive layout editors and design rule checkers relieve the designer from most of the tedious work of specifying and checking the layout. An important aid in specifying the layout of a VLSI circuits is so-called sticks-diagrams. A stick-diagram specifies the topology of the circuit; the relative positions of the transistors and their interconnections. In a stick-diagram the transistors are symbolically depicted as the crossing of polysilicon and diffusion lines. A stick-diagram adequately models the functional behaviour, i.e. the logic gates and their interconnections of the circuit. However it does not allow the specification of certain capacitive effects, e.g. bootstrapping.

Building Blocks. At the circuit level we decompose the circuit in three major types of building-blocks. The basic memory module is the one bit register cell; the basic logic module is the AND-OR-INVERT gate, and the basic arithmetic module is the full adder. Basically a VLSI circuit consists of these three types of modules. A somewhat special, but widely used logic module is the PLA, (Programmable Logic Array). A PLA can be used to implement any set of Boolean equations, and, if combined with a state-register, even to implement a complete FSM. A PLA can be generated directly from a register transfer level specification. In general, with the increasing use of high-level design aids, we see more and more programs that are able to synthesize large portions of a VLSI circuit from a high-level (Register Transfer) specification. One specific example is the FLEX system that can generate small microcomputer circuits from a specification of the operations that it is supposed to perform.

The above sketched building block approach, when combined with high-level tools as silicon compilers/assemblers give the VLSI designer the flexibility and modularity needed to cope with the ever increasing complexity, in terms of number of transistors to integrate, of VLSI design.

2.3 Scaling

When the feature size is scaled down to the submicron level, millions of transistors may be integrated into one chip. In this section we assume that all of the dimensions, as well as the voltages and currents on the chip are scaled down by a factor of α. We note that, when scaling down a transistor with a factor α, the number of transistors that can be put together on a chip increases with a factor of α^2. We also note that the power dissipation per transistor decreases by a factor of α^2 , due to the fact that both the threshold voltage[2] and the supply voltage are scaled down by α. Finally, we note that the switching delay of transistor is also scaled down by α, due to the fact that the channel length is decreased with a factor of α.

Figure 2 depicts the effect of scaling down a conductor by a factor of α. Since the area of the conductor is decreased by a factor of α^2, and, simultaneously, the thickness of the oxide is decreased by a factor of α; therefore, the actual capacitance will decrease only by a factor of α. The resistance of the conductor might be expected to increase by a factor of α^2, however since the length of the resistor is in general also decreased by a factor of α,

---

[2]Due to the voltage compatibility constraints, we might not be able to scale down the supply voltage. In that case, the power dissipation of the chip would increase significantly, and we would be forced to use a technique with an inherent low power dissipation, such as CMOS

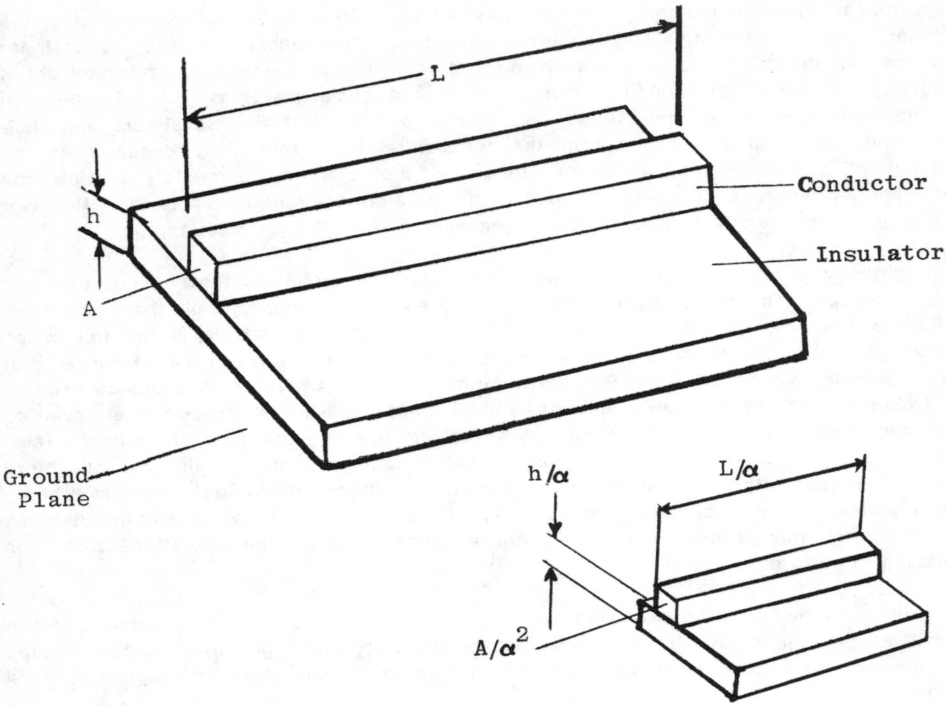

Figure 2: Scaling of VLSI circuitry

the actual increase of the resistance will be a factor of α. The result is that the actual RC time constant of an interconnection will not change, meaning that the actual interconnection delay will not scale down, but remain constant.

Side-effects of scaling

With increasing device densities, the side-effects of scaling become increasingly more critical.

Interconnect Delay. One effect is the increased importance of the interconnection delay: in the submicron technology, the interconnection delay will become greater than the switching delay of the logic gates. This is due to the fact that the line response time RC is unchanged by scaling while the switching speed of the logic gates is scaled down with a factor α. Moreover, if the average interconnection length is not scaled down with the same factor α, the interconnection delay may even increase[3], further aggravating the situation. When the delay time of the circuit depends largely on the interconnection delay (instead of the logic gate delay), minimal and local interconnections will become an essential factor for an effective realization of the VLSI circuits.

---

[3] It must be noted here that for typical VLSI circuits the average interconnection length does not scale down with a factor of α due to the fact that a large number of interconnections still have to cross the entire chip.

Electro-migration. Another side effect is that[4] the current density in the conductors will increase, with a factor of α. This will increase electro-migration effect, i.e. the movement of conductor atoms under the influence of electron bombardments, ultimately breaking the conductor lines. One way out of this problem is to scale the thickness of the conductor with a factor less then α. This however poses an extra challenge to the processing technology, particulary on the lithographic and etching processes.

2.4 Impact of Device Technology on VLSI architecture

Although VLSI provides the potential of tremendous increase in processing hardware, it imposes its own device constraints on the architectural design. Some of the key factors are communication, system clocking, modularity and versatility. Therefore, design of VLSI systems should be based on the potential and constraints imposed by the VLSI device technology and layout design. There are some influential issues in the VLSI system design, including: interconnection, system clocking, and modularity.

Interconnections. Interconnections in VLSI systems are implemented in a two-dimensional circuit layout with few crossover layers. Therefore, communication has to be restricted to localized interconnections, as communication is very expensive in terms of area, power and time consumption [38, 1, 23]. In addition, wiring in a localized interconnect network can be scaled down by the same factor as ω is reduced. Otherwise, the system performance could be severely limited by the RC-line delay.

Dynamic interconnections are often required for more general purpose computing. For signal processing array processors, however, mapping of activities onto a statically interconnected array of processors may prove more efficient without an undue loss of flexibility.

System Clocking. The timing framework is a very critical issue in designing the system. There are two opposite timing schemes: globally synchronous and locally synchronous timing approaches. In the synchronous scheme, there is a global clock network which distributes the clocking signals over the entire array. The clock skew incurred in global clock distribution has, however, increasingly become a nontrivial factor, causing unnecessary slowdown in clock rate. Under this circumstance, locally synchronous (but globally asynchronous) scheme may become more preferable. It involves no global clock, and information transfer is by mutual convenience and agreement between each processing element and its immediate neighbors. A key advantage of this scheme is that it is unaffected by the scaling of technology. Moreover, this scheme can be implemented by means of a simple handshaking protocol [30].

Modularity. Large design and layout costs suggest the utilization of a repetitive modular structure. To implement the modules we want to make optimal use of the device technology. Thus we have to identify the primitives that can be implemented efficiently.

Programmability. In addition, it is very important to make a processor programmable, so that the high design cost may be amortized over a broader market basis. There is however

---

[4]This is because, although the current is scaled down with a factor of α, the cross-sectional area is simultaneously scaled down by a factor of α^2.

a tradeoff between the flexibility of a processor and its hardware complexity. From a top-down point of view, one should exploit the fact that a great majority of signal processing algorithms possess a recursiveness and locality property, (cf. Sec. 4.2). Indeed, a major portion of the computational needs for signal processing and applied mathematical problems can be reduced to a basic set of matrix operations and other related algorithms [29]. This commonality can be identified and then exploited to simplify the hardware and yet retain most of the desired flexibility.

In summary, for the purpose of reducing design cost, repetitive modular structures and programmable processor elements are in general favorable. Moreover, interconnection tends to become very expensive in terms of area, power and time consumption; therefore, the burden of global interconnection and global synchronization should be alleviated via architectural design whenever possible.

3 VLSI Array Processors for Signal Processing

The realization of many modern signal processing methods depends critically on the availability and feasibility of high speed computing hardwares. In fact, until the mid-1960's most signal processing was performed with specialized analog processor because of the hardware complexity, power consumption, and lower speed of conventional digital systems. However, digital processors can provide better (and sometimes indispensable) precision, dynamic range, long-term memory and other flexibilities such as programmability, and expandability to accommodate changing requirements. More importantly, today's VLSI is offering a much greater hardware capacity, higher speed, and less power than, or at least compatible with, most leading analog devices.

A typical processor architecture will consist of input, output, memory, control unit, and ALU. Conventional digital processors use one or at most a few ALUs, and rely on memory to buffer the intermediate results. Even the so-called Floating Point Systems' AP-120B "Array Processor" consists of only one adder and one multiplier, assisted by smart software. Its maximum basic speed for signal processing operations is approximately 7 MOPS [24]. Today's VLSI technology can produce single chip multiplier and accumulator, e.g. the TRW MPY-16 multiplier chip. Moreover, the device speed also improves significantly as a consequence of feature size scaling. Nevertheless, all these advances alone will not be sufficient to match the rate of increase of the required processing speeds. In order to meet the stringent real-time signal processing requirement, novel algorithmic, architectural, and device technological solutions have to be resorted to. Thus the implementation of parallel signal processors may very well be the single most significant application of VLSI.

3.1 Signal Processing via Supercomputers

One of the main thrust of the VLSI technology is the realization of supercomputer systems capable of processing thousands of MOPS or MFLOPS. General purpose supercomputers have the key advantage of being able to respond to changes in the information streams in a complex task. It can redistribute the available hardware resources into variable type of processing units. It can also perform instruction set adaptions, activating certain dedicated instruction sets (and their corresponding hardware realizations) to perform special tasks. The major disadvantage of supercomputer, from a signal processing point of view, is the excessive supervisory overhead paid for the central control, scheduling, communication, etc. It often severely hampers the processing rates which are critical for real time processing requirement. In order to achieve such increases in throughput rate, the only effective solution appears to be massively concurrent processing. Evidently, for special applications, a (stand alone or peripheral) dedicated array processor will stand as a much faster and more cost-effective computing tool.

The roles of supercomputers and (special purpose) array processors may very well be complementary in several grand-scale applicational systems. For instances, we mention the computations in meteorology, geophysical exploration, and most military applications such as ballistic missile defense systems. In these cases, the (host) supercomputer will supervise central control, resource scheduling, database management and, most importantly, provide super-speed I/O interfaces with the (peripheral) array processors. The array processors will in turn speedily execute the (computing-bound) functional primitives such as FFT, digital filtering, correlation, and matrix multiplication/inversion and handle other possible computation bottleneck problems.

3.2 Special Purpose Array Processors

The integrated circuit technology has evolved rapidly from few transistors in the 1960's to tens of thousands of transistors per chip in today's technology. Very conceivably, in the next decade, this number will grow into millions of transistors in a single VLSI chip. Correspondingly, the basic VLSI signal processor primitives will also be rapidly upgraded. Specially, let us note that most signal processing algorithms share the common functional primitives such as inner product, outer product, correlation, convolution, FIR and IIR filters, DFT, FFT, etc. Very often these functional primitives need to be processed in very fast (real-time) processing rates and this represent the major computational bottleneck, need be handled by special hardware. Therefore, novel architectural designs of massively parallel array processors for these primitives are called for.

FFT Array Processor. To avoid loss of time due to memory write/fetch or the limited speed of "bus" carrying the information, special purpose hardwares often utilize direct interconnections, or smart exchange networks, connecting the PEs (Processing Elements). They are often concatenated in pipeline configuration, thereby increasing the overall computing speed [24]. A good example, is a fully pipelined FFT Array Processor, see Fig. 3, where butterfly-type interconnections (or perfect shuffle networks) are used for the communications between different stages of PEs.

Transversal Filtering. Apart from FFT computing, perhaps the most useful digital signal processing techniques are FIR (or convolution) digital filters. For example, a typical telephone echo canceller will use a 128 point transversal filter, with the tap weights derived from the correlation function between the sent signal and the received signal. In telecommunication networks, millions of cancellers are needed (one per 12 or 24 telephones) [48], thus digital echo canceller has been a most important application of VLSI digital signal processing.

An FIR digital filter with transfer function

$$H(z) = \sum_{i=1}^{M} b_i z^{-i}$$

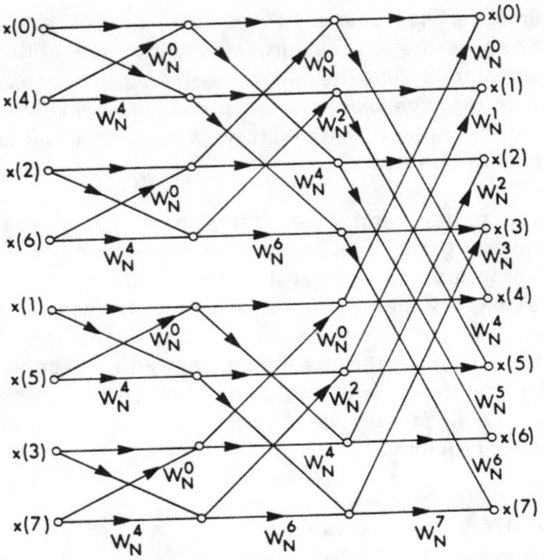

Figure 3: A fully pipeline/parallel FFT array processor configuration

is often implemented in a tapped-delay-line, sometimes called transversal filter [43], as shown in Fig. 4.

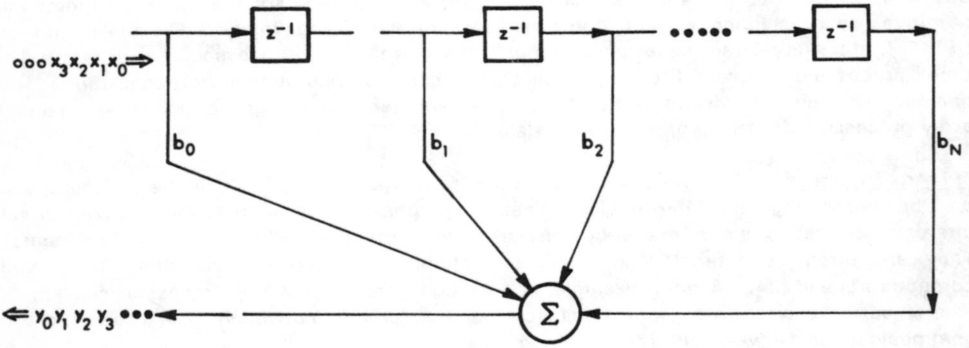

Figure 4: A transversal filter or tapped-delay-line design

Just like the FFT Array Processor, this conventional FIR filter design requires global communications (for the summing operation). Unlike the FFT processor, however, the FIR filter possesses an inherent property of "locality" of data flow, which may be exploited to simplify its VLSI implementation.

Our viable option seems to be that, for a carefully selected applicational domain (signal processing in this case), there may be significant commonality among the class of algorithms involved. This commonality can be identified and then exploited to simplify the hardware and yet retain most of the desired flexibility.

From a top-down point of view, one should exploit the fact that a great majority of signal processing algorithms possess a recursiveness and locality property, (cf. Sec. 3). Indeed, a major portion of the computational needs for signal processing and applied mathematical problems can be reduced to a basic set of matrix operations and other related algorithms [29]. More challengingly, the inherent parallelism of these algorithms calls for parallel computing capacity much higher than what one-dimensional array can possibly furnish. The first major architecture breakthrough is the two-dimensional systolic arrays for matrix multiplication, inversion, etc., and the reader is referred to [26] for an overall review.

3.3 Systolic Array

Systolic processors are a new class of digital architectures that offers a new dimension of parallelism. The principle of systolic structure is an extension of "pipelining" into more than one dimension. According to Kung and Leiserson [27], "A systolic system is a network of processors which rhythmically compute and pass data through the system". For example, it is shown in [27] that some basic "inner product" PE's ($Y \leftarrow Y + A * B$) can be locally connected to perform FIR filtering in a manner similar to the transversal filter. Furthermore, two-dimensional systolic arrays (of the inner product PE's) can be constructed to efficiently execute matrix multiplication, L-U decomposition, and other matrix operations.

The basic principle of systolic design is that all the data, while being "pumped" regularly and rhythmically across the array, can be effectively used in all the PE's. The systolic array features the important properties of modularity, regularity, local interconnection, highly pipelined, and highly synchronized multiprocessing. They require no control and overlap I/O and computation, and hence, speed up computer-bound computation without increasing I/O requirements [26].

For example, a two dimensional hexagonal array will be a natural topology for the matrix multiplication problem. As shown in Fig. 5, the input data (from matrices A and B) are pre-arranged in an orderly sequence. The output data (of the matrix C) will be pumped from the other side of the array, meeting the right data and collecting all the desired "products". The detailed description of data movements and computations is often furnished in terms of the "snapshots" of the the activities [27]. Interestingly, the LU decomposition (a critical procedure for matrix inversion) can be computed in a very similar fashion. The "snapshots" for the LU decomposition are just the reversal of those for the matrix multiplication. For more details, the reader is referred to [27, 26].

However, there are several unresolved controversial issues regarding systolic arrays. First, (pure) systolic arrays tend to equalize the time units for different operations. As an example, for the convolution systolic arrays in [27], a local data transfer consumes the same time delay as a multiply-and-add, i.e. one full time unit. This often results in unnecessary wastage of processing time (cf. Sec. 3), since the data transfer time as needed is almost negligible. This motivates what we called multi-rate systolic array [32]. More critically, systolic arrays require global synchronization, i.e. global clock distribution. This may cause clock-skew problems in high order VLSI system implementations. Another issue of concern is parallel language description and ease of programmability for complex data flows. These aforementioned problems motivate a revisit to a well known asynchronous data-driven scheme, commonly adopted in data-flow machines.

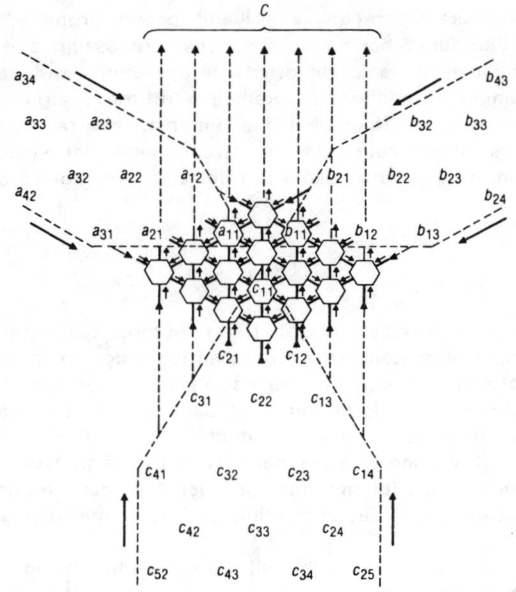

Figure 5: Systolic Array for Matrix Multiplication

3.4 Multi-rate Systolic Array [32]

A pure systolic array has to equalize the time units for different basic operations. For example, a data transfer operation usually requires much less time than a multiply-and-add operation. However, in a systolic system a data transfer alone is required to consume the same full time unit as a multiply-and-add, which is wasteful and unnecessary. There is a solution to this problem by using multi-rate systolic array. A multi-rate systolic array is a generalized systolic array, where different data streams may be pumped through the array at different rates, thus allowing different basic operations to consume different time units.

3.5 Wavefront Array [30]

A wavefront array is a programmable array processor [30], which combines the features of the asynchronous, data-driven properties in data flow machines and the regularity, modularity, and local communication properties in systolic arrays. It has two distinct advantages: (1) it avoids the need for global synchronization; (2) it offers an effective space-time programming language.

While the original systolic concept focuses on the data-movements between processors in the array; unfortunately, it does not naturally lend itself to a simple programming language structure. The need for a powerful description tool is further aggravated when more complex algorithms such as eigenvalue or singular value decompositions are encountered. Fortunately, the wavefront model utilizes the data-driven property of data flow machines and the locality and regularity of systolic arrays. This allows a simple mechanism tracing complex sequences of interactions and data movements. Several examples will be given in the following sections.

The concept of wavefront computing originates from algorithmic analysis, which will lead to a coordinated language and architecture design. In fact, the algorithmic analysis of, say, the matrix multiplication operations will lead first to a notion of two-dimensional computational wavefronts; see Fig. 6. The analysis begins with decomposing the operation into a sequence of recursions. A wavefront corresponds to a state of recursion. Successively pipelining a sequence of computational wavefronts leads to a continuously advancing wave of data and computational activity.

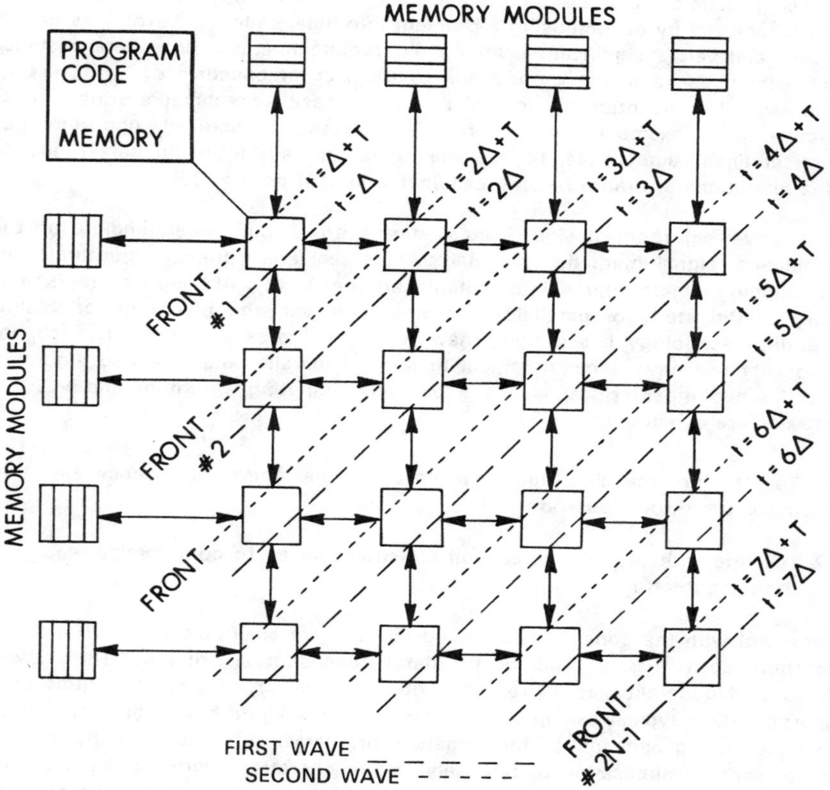

Figure 6: Two-Dimensional Wavefront Array

A general configuration of computational wavefronts traveling down the processor array is exemplified in Fig. 6. This computational wavefront is similar to electromagnetic wavefronts, (they both obey Huygen's principle) since each processor acts as a secondary source and is responsible for the propagation of the wavefront. It may be noted that wave-propagation implies localized data flow as well as data driven computing. The pipelining is feasible because the wavefronts of two successive recursions will never intersect (Huygen's wavefront principle), as the processors executing the recursions at any given instant will be different, thus avoiding any contention problems. The correctness of the sequencing of tasks is ensured by the data-driven nature of wavefront processing. The detailed illustration of the sequencing of mathematical recursions will be discussed in Section 4.

4 Algorithm Analysis to Architecture Design

Since the concept of massive concurrency will be integrated into the modern signal processing techniques, a new algorithmic design methodology will be needed. Concurrency is often achieved by decomposing a problem into independent subproblems or into pipelined subtasks and varies significantly among different techniques. In several instances, several "fast" techniques have lost ground due to the lack of concurrency. Conversely, VLSI has made several conventionally inefficient or impossible techniques rather desirable and attractive. For example, since there is an extensive (parallel) algorithmic analysis on numerical linear algebra [44, 18, 25], the signal processing techniques reducible to basic matrix operations are more favored candidates in VLSI design [29].

An effective algorithm design should start with a full understanding of the problem specification, signal mathematical analysis, (parallel and optimal) algorithmic analysis, and then mapping of algorithms into suitable architectures. Although there is a rather long history of the study of parallel processing; however, the revolution of VLSI device and computing technology has aroused new research interests. The preference on regularity and locality will have a major impact in deriving parallel and pipelined algorithms. In our work, the two most critical issues - parallel computing algorithm and VLSI architectural constraint-are considered:

1. To structure the algorithm to achieve the maximum concurrency and, therefore, the maximum throughput-rate.

2. To cope with the communication constraint so as to compromise least in processing throughput-rate.

To conform with the constraints imposed by VLSI, we shall now look into a special class of algorithms, i.e. recursive[5] and locally (data) dependent algorithms. This, however, incurs little loss of generality, as a great majority of signal processing algorithms possess these properties. One typical example is a class of matrix algorithms, which are most useful for signal processing and applied mathematical problems. Very significantly, these algorithms involve repeated application of relatively simple operations with regular and localized data flow, pointing to VLSI architectures consisting of homogeneous computing networks.

---

[5] In a recursive algorithm, all processors do nearly identical tasks, and each processor repeats a fixed set of tasks on sequentially available data.

4.1 Recursive Algorithm

A recursive algorithm is said to be of local-type if the space index separations incurred in two successive recursions are within a given limit. Otherwise, if the recursion involves globally separated indices, the algorithm will be said to be of global-type; and it will always call for globally interconnected computing structures.

A typical example of global-type recursion is the FFT computations. The principle of the (decimation-in-time) FFT is based on successively decomposing the data, say $\{x(i)\}$, into even and odd parts. This partitioning scheme will result in a necessarily global communication between data. More precisely, the FFT recursions can be written as (using the "in-place" computing scheme [41])

$$x^{(m+1)}(p) = x^{(m)}(p) + w_N^r\, x^{(m)}(q)$$

$$x^{(m+1)}(q) = x^{(m)}(p) - w_N^r\, x^{(m)}(q),$$

with p, q, r varying from stage to stage. The "distance" of the global communication involved will be proportional to $|p - q|$. At the last stage, the maximum distance is $|p - q| = N/2$, cf. Fig. 3. (For example, the maximum distance will be 512 units for a 1024-point FFT.) Obviously, an FFT can not be mapped into local-type computing structures, such as systolic or wavefront arrays.

As we have mentioned earlier, FIR filtering can be converted to one with only local communication. This will be shown below. In fact, we are going to treat a more general (and more interesting) design example of ARMA (IIR) filters. Generally, an IIR (infinite impulse response) filter is defined by a transfer function

$$H(z) = \frac{\sum_{i=1}^{N} b_i z^{-i}}{1 - \sum_{i=1}^{N} a_i z^{-i}}.$$

(Note that the FIR filtering, linear convolution, or transversal filtering, are simply a special case when $a_i = 0$, $i = 1, \ldots, N$. The corresponding recursive algorithm is often given in terms of the difference equation:

$$y(k) = \sum_{m=1}^{N} x(k-m)b(m) + \sum_{m=1}^{N} y(k-m)a(m). \tag{1}$$

This direct form involves global indexing, and thus leads to the global-type realization in the canonic form design (shown in Fig. 7(a)) [43, 22].

However, this design can be easily converted into one of the local-type and, therefore, so can the algorithm represented by Eq. (1).

A modified design is shown in Fig. 7(b). To verify that Fig. 7(b) yields the same transfer function as Fig. 7(a), we simply check the transfer functions $V^{(N)}(z)/U^{(N)}(z)$, $W^{(N)}(z)/U^{(N)}(z)$. Obviously they remain unchanged by the modification. Thus, by induction, so do the

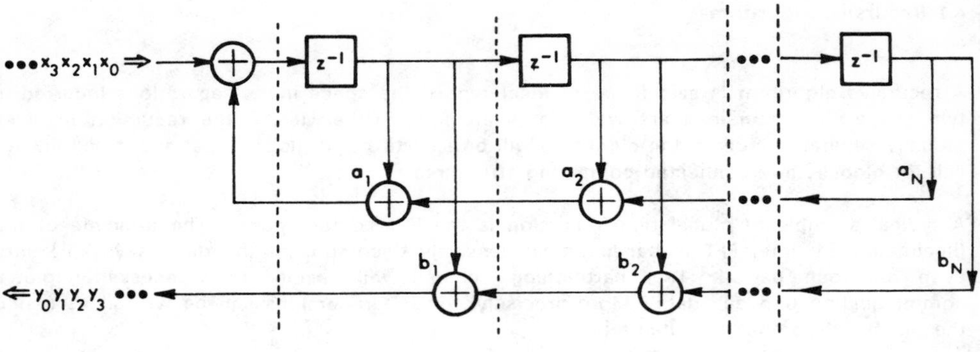

Figure 7(a): Direct Form Design of ARMA (IIR) Filter (Global Inter-connection Required)

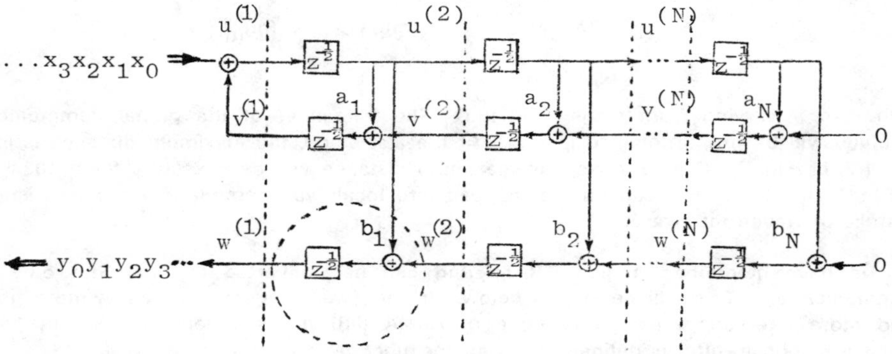

Figure 7 (b): Modified Direct Form with Localized Interconnection

transfer functions $V^{(i)}(z)/U^{(i)}(z)$, $W^{(i)}(z)/U^{(i)}(z)$, for $i = N-1, N-2, \ldots, 1$, and the proof is completed. Note that, in the modified direct form, there are always delays inserted between all the summing nodes, therefore, only localized communications are needed.

4.2 Systolic Array for ARMA (IIR) Filter

We shall show below that the modified form is in fact equivalent to the systolic array design (cf. Fig. 8), except for a rescaling of time unit. Since the half-time delay unit ($z^{-1/2}$) is not common, our next step in Fig. 8(a) is to rescale the time unit by setting $z^{-1/2} = z'^{-1}$, i.e. to renormalize the time unit. Therefore, the delay z^{-1}, the multiplier, and the adder in each single section (defined by means of dashed circles) in Fig. 7(b) are all merged into an "inner product" processor [4], as shown in Fig. 8(b). An overall systolic array configuration for IIR filter is shown in Fig. 8(c). Note that due to the rescaling of time units, the input data $\{x_i\}$ have to be interleaved with "blank" data (cf. Fig. 8(c)), and the throughput rate becomes $0.5T^{-1}$.

Now, having a localized array, we can in turn derive the localized version of the algorithm in Eq. (1). Assigning the superscript $\{\kappa\}$ as the time variable and the subscript $\{n\}$ as the

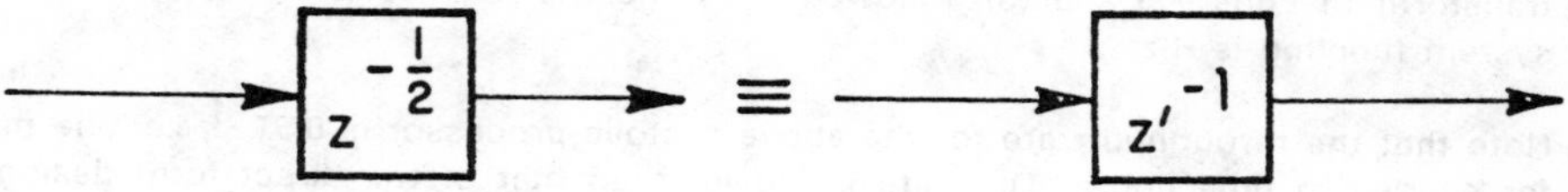

Figure 8**(a):** $z^{-1/2} \quad = \quad z'^{-1}$, Rescaling of Time Unit

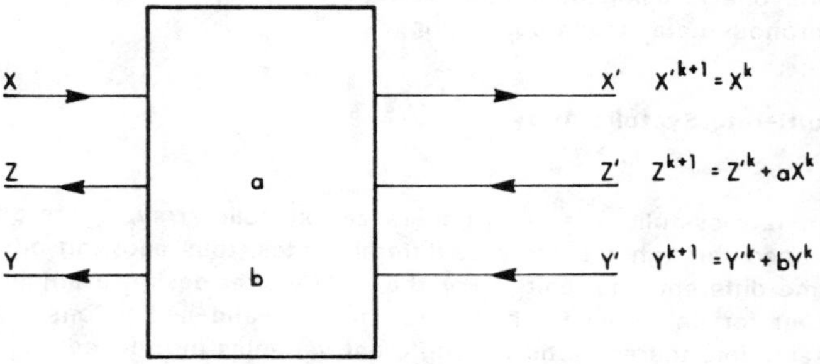

Figure 8**(b):** Systolic Processing Element (Note the one-unit-time delay from X to X')

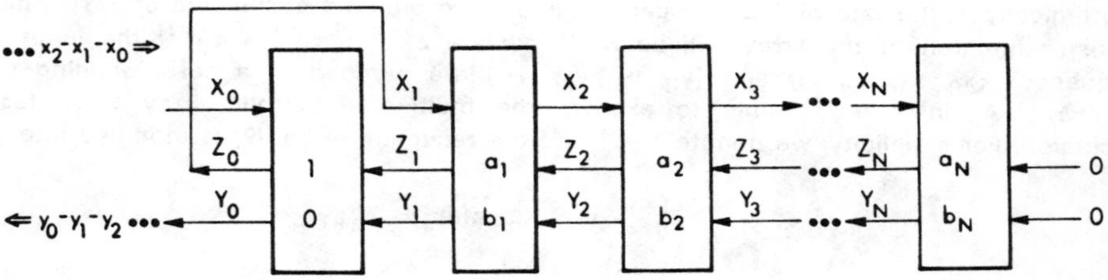

Figure 8 **(c):** A Systolic Array for ARMA (IIR) Filter

space variable for the data movements corresponding to the above systolic array, we have the following local-type recursions (setting a(0) = b(0) = 0):

$$Z_n^{k+1} = X_n^{k+1} a(n) + Z_{n+1}^k \tag{2a}$$

$$Y_n^{k+1} = X_n^{k+1} b(n) + Y_{n+1}^k \tag{2b}$$

$$X_{n+1}^{k+1} = X_n^k \tag{2c}$$

Except for n = 0, the leading PE, where

$$X_1^{k+1} = X_0^k + Z_0^k \tag{3}$$

Interestingly, the correctness of the systolic array can be easily verified by taking the z-transform of Eqns. (2) and (3), followed by some trivial algebra. (This will show that the system function is $H(z^2)$.

Note that the throughput rate for the above systolic processor is $0.5T^{-1}$, i.e. one output data for every two time units. This rate is slower than that of the direct form design ($1.0T^{-1}$), because the data transfer (X → X') alone is required to consume the same full time unit as multiply-and-add, which is wasteful and unnecessary. There are two solutions to this problem: one is using multi-rate systolic array and the other is wavefront array based on asynchronous data-driven computing.

4.3 Multi-rate Systolic Array

A multi-rate systolic array is a generalized systolic array, where different data streams may be pumped through the array at different rates, thus allowing different basic operations to consume different time units. For the ARMA filter design example, we can assign Δ as the time unit for data transfer and T for multiply-and-add. Consequently, in the Z-transform representation, there will be two different variables introduced: $z_1^{-1} = \Delta$ and $z_2^{-1} = T$, and in the circuit representation in Fig. 7(b), we replace $z^{-1/2}$ on the feedforward path (for X) by z_1^{-1}, and $z^{-1/2}$ on the feedback paths (for Y and Z) by z_2^{-1}. These modifications lead to a multi-rate systolic array as shown in Fig. 9(a), 9(b). Here each datum X is pumped rhythmically at the rate of 1/Δ and data Y and Z are pumped at the rate of 1/T. Since the transfer function of the array will be $H(z'')$, where $z''^{-1} = z_1^{-1} + z_2^{-1}$; the input/output sequences $\{x_0, x_1, x_2, ...\}$ and $\{y_0, y_1, y_2, ... \}$ are pumped at a corresponding rate of 1/(T+Δ). Again it is possible to express the multi-rate systolic array by a localized recursion: For simplicity, we denote T = 1 and the recursion in Eq. (2) is modified into

$$Z_n^{k+1} = X_n^{k+1} a(n) + Z_{n+1}^k \tag{2a}$$

$$Y_n^{k+1} = X_n^{k+1} b(n) + Y_{n+1}^k \tag{2b}$$

$$X_{n+1}^{k+\Delta} = X_n^k \tag{2c}$$

Except for n = 0, the leading PE, where

$$X_1^{k+1} = X_0^k + Z_0^k . \tag{3}$$

Again applying the two-dimensional z-transform [28] on the above equation will show that the system transfer function is indeed $H(z'')$. (The reader might want to try it out as an exercise.)

By a multi-rate system realization theory, (see e.g. Kung et al. [28]), corresponding to the same transfer function $H(z'')$, there are many non-unique multi-rate array structures. This is worth noting because certain structures will be numerically better in their digital implementation [28].

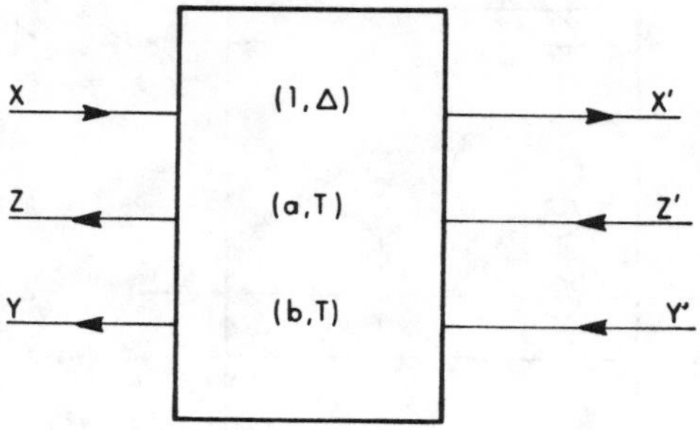

Figure 9**(a):** A Multi-rate Systolic Processing Element

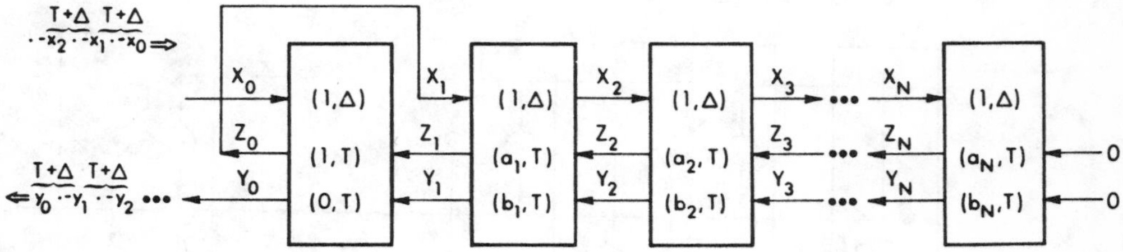

Figure 9**(b):** A Multi-rate Systolic Array for ARMA (IIR) Filter

In fact, a multi-rate systolic array is nothing more than a synchronized version of the wavefront array (see the next section). A potential application of such an array is in mixed bit-wise and word-wise systolic/wavefront processing. A good example is the (synchronized) cordic wavefront array for QR decomposition [30]. Here, in a square array, the vertical wavefront propagation will be word by word, while the horizontal propagation will be bit by bit. (In this case, the bits to be pumped horizontally will be the cordic control bits for angle rotation [11].)

4.4 Wavefront Array for ARMA (IIR) Filter

When the issue of system synchronization becomes critical, then the wavefront approach, using asynchronous data-driven computing, becomes especially powerful. The strategy now is to replace the globally synchronized computing in the systolic array by an asynchronous data driven model, as shown in Fig. 10. Therefore, at each node in Fig. 10, the operation is executed when and only when the required operands are available. Since it uses asynchronous processing, the timing references become unnecessary. An immediate advantage of this model is that the data transfer (X --> X') will use only negligible time compared to the time needed for their arithmetic processing. More precisely, the throughput rate achieved by the WAP is approximately $1.0T^{-1}$, (i.e. twice that of the pure systolic array in Fig. 8).

Now we have another way to represent the recursion. Here, we reassign the superscripts k (in parentheses) as the wavefront number (or the recursion number) since the timing index is no longer applicable. The new representation of Eq. (1) is (setting a(0) = b(0) = 0):

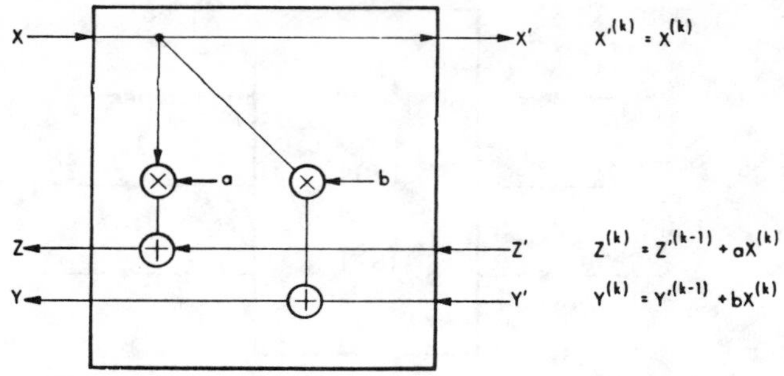

Figure 10**(a)**: Wavefront Processing Element (Asynchronous, data-driven model, i.e. operations take place only on availability of appropriate data.)

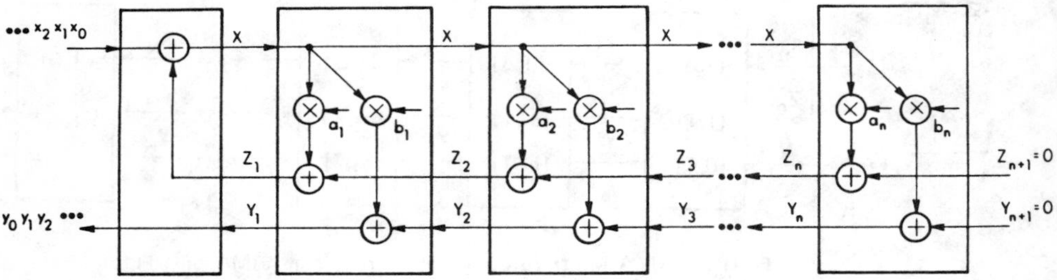

Figure 10 (**b**): Wavefront Array for ARMA (IIR) Filter

$$Y_n^{(k)} = X_n^{(k)} b(n) + Y_{n+1}^{(k-1)} \quad (4a)$$

$$Z_n^{(k)} = X_n^{(k)} a(n) + Z_{n+1}^{(k-1)} \quad (4b)$$

$$X_{n+1}^{(k)} = X_n^{(k)} \quad (4c)$$

Except for n = 0, the leading PE, where

$$X_1^{(k)} = X_0^{(k)} + Z_1^{(k-1)} \quad (4d)$$

Under the wavefront notion, $X^{(k)}$ is initiated at the leading PE(n = 0) according to Eq. (4c), and then propagated rightward across the processor array, activating the operations (4a), (4b) in all the data-driven PE's. As is shown in Eq. (4), the updated data $\{Y_{n+1}^{(k)}, Z_{n+1}^{(k)}\}$ are fedback leftward ready for the next recursion (or wavefront).

Another important feature is that the data driven model allows a simple data flow language

for the description of data movement. A program for the ARMA wavefront array, written in MDFL- Matrix Data Flow Language, can be found in another paper [35].

5 Wavefront Array Processor

In this section, we shall concentrate on the algorithmic analysis, which will lead to a coordinated language and architecture design, needed for a two-dimensional wavefront array. In fact, the algorithmic analysis of the matrix operations will lead first to a notion of two-dimensional computational wavefronts: All these above algorithms are decomposible into a sequence of recursions, and can be mapped into corresponding wavefronts on a homogeneous computing network. (A wavefront corresponds to a state of recursion.) Successively pipelining a sequence of computational wavefronts leads to a continuously advancing wave of data and computational activity.

5.1 Sequencing of Mathematical Recursions

We shall illustrate the sequencing of mathematical recursions, by means of a matrix multiplication example. Let

$$A = [a_{ij}], \quad B = [b_{ij}]$$

and

$$C = A \times B$$

all be N x N matrices. The matrix A can be decomposed into columns A_i and the matrix B into rows B_j and therefore,

$$C = [A_1 B_1 + A_2 B_2 + \ldots + A_N B_N].$$

The matrix multiplication can then be carried out in N recursions,

$$C_{i,j}^{(k)} = C_{i,j}^{(k-1)} + a_i^{(k)} b_j^{(k)} \tag{5a}$$

$$a_i^{(k)} = a_{ik} \tag{5b}$$

$$b_j^{(k)} = b_{kj} \tag{5c}$$

for k = 1, 2, ... , N and there will be N sets of wavefronts involved.

5.2 Pipelining of Computational Wavefronts [30]

The computational wavefront for the first recursion in matrix multiplication will now be examined. Suppose that the registers of all the processing elements (PEs) are initially set to zero:

$$c_{ij}^{(0)} = 0 \quad \text{for all } (i, j);$$

the entries of A are stored in the memory modules to the left (in columns), and those of B in the memory modules on the top (in rows). The process starts with PE (1, 1):

$$c_{11}^{(1)} = c_{11}^{(0)} + a_{11} * b_{11}$$

is computed. The computational activity then propagates to the neighboring PE's (1, 2) and (2, 1), which will execute in parallel

$$c_{12}^{(1)} = c_{12}^{(0)} + a_{11} * b_{12},$$

and

$$c_{21}^{(1)} = c_{21}^{(0)} + a_{21} * b_{11}.$$

The next front of activity will be at PE's (3, 1), (2, 2) and (1, 3), thus creating a computational wavefront traveling down the processor array. This computational wavefront is similar to electromagnetic wavefronts, (they both obey Huygen's principle) since each processor acts as a secondary source and is responsible for the propagation of the wavefront. It may be noted that wave-propagation implies localized data flow. Once the wavefront sweeps through all the cells, the first recursion is over (cf. Fig. 6).

As the first wave propagates, we can execute an identical second recursion in parallel by pipelining a second wavefront immediately after the first one. For example, the (i,j) processor will execute

$$c_{ij}^{(2)} = c_{ij}^{(1)} + a_{i2} * b_{2j}$$

and so on. The pipelining is feasible because the wavefronts of two successive recursions will never intersect (Huygen's wavefront principle), as the processors executing the recursions at any given instant will be different, thus avoiding any contention problems. Here let us be reminded that it is possible to have wavefront propagating in several different fashions. The only critical factor is that the order of task sequencing must be correctly followed. This rule is ensured by the data-driven nature of wavefront processing. Therefore, the actual propagation pattern is practically of no consequence. As a matter of fact, when local clocks are used then the entire wavefront pattern may be crooked and yet yield accurate sequencing and computation.

5.3 LU Decomposition

In the LU decomposition, a given matrix C is decomposed into

$$C = A \times B, \tag{6}$$

where A is a lower- and B an upper- triangular matrix The recursions involved are

$$C_{ij}^{(k)} = C_{ij}^{(k-1)} - a_i^{(k)} \quad b_j^{(k)}, \tag{7a}$$

$$a_i^{(k)} = \frac{1}{C_{kk}^{(k)}} C_{i\,k}^{(k-1)} \tag{7b}$$

$$b_j^{(k)} = C_{k\,j}^{(k-1)} \quad , \tag{7c}$$

$$\text{for} \quad k = 1, 2, \ldots, N; \quad k \leq i \leq N;$$

$$k \leq j \leq N.$$

Verifying the procedure by tracing back Eq. (7a), we note that

$$C = C_{ij}^{(0)} = \sum_{k=1}^{N} a_i^{(k)} \, b_j^{(k)} \tag{8}$$

where $A = \{a_{mn}\} = \{a_m^{(n)}\}$, and $B = \{b_{mn}\} = \{b_n^{(m)}\}$ are the outputs of the array processing. (Compare with Eq. (6).)

Comparing with (5), Eq. (7) is basically a reversal of the matrix multiplication recursions. In fact, in wavefront processing for the LU decomposition, Eq. (7) will exhibit a similar wavefront propagation pattern as Fig. 6. Except for the following:

(i) Just like the matrix multiplication, the data $a_i^{(k)}$ and $b_j^{(k)}$ are to be propagated rightward and downward respectively. However, $a_i^{(k)}$ has to be derived from an arithmetic operation, causing extra delay, cf. Eq. (7b), (while $b_j^{(k)}$ is directly available from the previous recursion, cf. Eq. (7c)).

(ii) In a very straight forward scheme, the second recursion may start at the PE (2, 2), (the third at PE (3, 3), and so on). However, there is a hardware advantage to have all recursions initiated in the PE (1, 1). In any event, the active area of array processing shrinks from one recursion to another. (This inevitably causes some wastage in the processor utilization.) [30]

5.4 Wavefront Language and Architecture

The wavefront concept provides a firm theoretical foundation for the design of highly parallel array processors and concurrent languages, and it appears to have two distinct features:

(i) The wavefront processing notion leads to a wavefront architecture as sketched in Fig. 11. The architecture preserves Huygen's principle and ensures that wavefronts never intersect. More precisely, the information transfer is by mutual convenience between each PE and its immediate neighbors. Whenever the data is available, the transmitting PE informs the receiver of the fact, and the receiver accepts the data when it needs it. It then conveys to the sender the information that the data has been used. This scheme can be implemented by means of a simple handshaking protocol [30, 31]. The wavefront architecture can provide asynchronous waiting capability, and consequently, can cope with timing uncertainties, such as local clocking, random delay in communications and fluctuations of computing-times, [31, 13, 47]. In short, the notion lends itself to a (asynchronous) data-driven computing structure that conforms well with the local interconnection and local clocking constraints of VLSI.

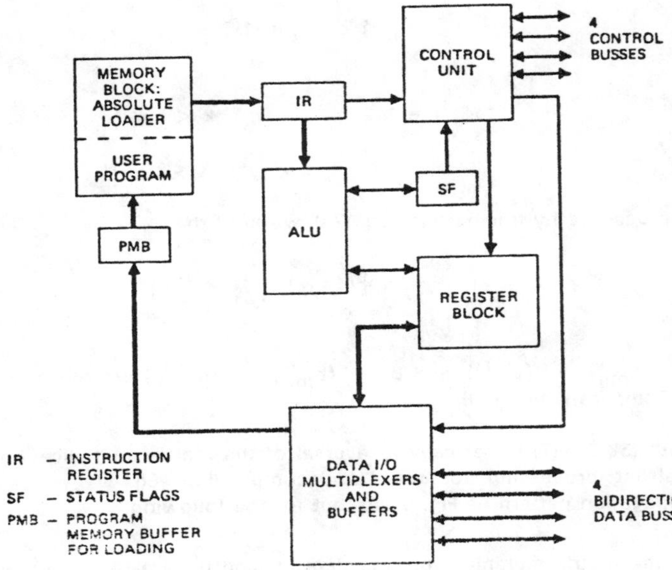

Figure 11: Wavefront processing element

(ii) As to the language aspect, the wavefront notion helps greatly reduce the complexity in the description of parallel algorithms. The mechanism provided for this description is a special purpose, wavefront-oriented language, termed Matrix Data Flow Language (MDFL) [30]. The wavefront language is tailored towards the description of computational wavefronts and the corresponding data flow in a large class of algorithms (which exhibit the recursivity and locality mentioned earlier). Rather than requiring a program for each processor in the array, MDFL allows the programmer to address an entire front of processors. In contrast to the heavy burden of scheduling, resource sharing, as well as the control of processor interactions as often encountered in programming a general purpose multiprocessor- the wavefront notion can facilitate the description of parallel and pipelined algorithms and drastically reduce the complexity of parallel programming.

A complete list of the MDFL instruction repertoire and the detailed syntax can be found in an earlier publication [30]. For an example, a complete (global) MDFL program for matrix multiplication follows:

Array Size: N x N

Computation: C = A x B

$$k^{th} \text{ wavefront: } c_{ij}^{(k)} = c_{ij}^{(k-1)} + a_{ik}b_{kj}$$

$$k = 1, 2, \ldots, N$$

Initial: Matrix A is stored in the Memory Module (MM) on the left (stored row by row). Matrix B is in MM on the top and is stored column by column.

Final: The result will be in the C registers of the PEs.

```
BEGIN
SET COUNT N;
REPEAT;
  WHILE WAVEFRONT IN ARRAY DO
    BEGIN
        FETCH  A, LEFT;
        FETCH  B, UP;
        FLOW   A,  RIGHT;
        FLOW   B,  DOWN;
             (* Now form C: = C + A x B)
        MULT  A, B, D;
        ADD   C, D, C;
     END;
  DECREMENT COUNT;
UNTIL TERMINATED;
ENDPROGRAM.
```

Global MDFL Program for Matrix Multiplication

5.5 Applications of Wavefront Processing

The power and flexibility of the wavefront array and the MDFL programming are best demonstrated by the broad range of the applicational algorithms [30]. Such algorithms can be roughly classified into three groups:

1. Basic Matrix Operations: such as (a) Matrix multiplication, (b) Banded-Matrix Multiplication, (c) Matrix-Vector multiplicaton, (d) LU Decomposition, (e) LU Decomposition with localized pivoting, (f) Givens algorithm, (g) Back substitution, (h) Null space solution, (i) Matrix inversion, (j) Eigenvalue decomposition, and (k) Singular value decomposition.

2. Special Signal Processing Algorithms: (a) Toeplitz system solver, (b) 1-D and 2-D Linear Convolution, (c) Circular Convolution, (d) ARMA and AR Recursive filtering, (e) Linear phase Filtering, (f) Lattice filtering, (g) DFT, and (h) 2-D correlation (image matching).

3. Other Algorithms: PDE (partial difference equation) solution.

Our design approach is to (1) define the application/algorithm domain (2) develop a language tailored to the application, and (3) design a (language-based) wavefront architecture. To maximize the application algorithm domain (with minimum hardware overhead), we are launching a major software development project: Our objective is to develop a complete set of software library for all the systolic/wavefront-type parallel processing. More importantly, combining force with the development in silicon compilers and VLSI cell libraries, it will pave the way to an ultimate top-down-integrated design automation. Suggestions and new (parallel) algorithms from all the interested readers are highly welcome. Only with joint and cohesive efforts from all the related disciplines can this goal be realized.

6 Wavefront Arrays Based on Givens Rotation for Least-Square and Eigenvalue Systems [14]

A special class of matrix computations including eigenvalue and singular value decomposition (SVD), least-square error solvers, Toeplitz system solver, etc. has merged as a very powerful computational tool in many signal and image processing applications. This class of computations are all solvable by means of matrix algorithms based on (Givens) rotation operators. By successfully mapping the mathematical recursions of these algorithms into their corresponding computational wavefronts (of the rotating operations), it is demonstrated all these algorithms can be efficiently carried out by the programmable Wavefront Array Processor, recently developed for the purpose of high-speed parallel signal processing.

For many signal and image processing applications, such as high resolution spectral estimation, image data compression, etc. eigenvalue and singular value decompositions have emerged as extremely powerful and efficient computational tools. In filtering, parameter estimation and identification, determining a least square solution to the system is widely used. In this section, we shall deal with wavefront arrays that address the eigenvalue and singular value decompositions of matrices, the least square (LS) solution, and a related Toeplitz system solution. The factor that is common to all of these is the ability to achieve a solution to the problem by means of rotation operators, among which the Givens Rotator is one of the most useful and offers most numerical stability of the computation. In the present paper, the Givens rotation will be utilized towards decomposing a matrix into its QR components to obtain the LS solution of a linear system, triangularizing a matrix in the application of the repetitive triangularization algorithm for determining its SVD, tridiagonalizing a symmetric matrix for calculation of its eigenvalues and, finally, bidiagonalizing a matrix for its SVD computation. Tridiagonalizing and bidiagonalizing a matrix will be shown to be especially applicable to linear arrays and will be dealt with in their respective sections. The same array can also be reprogrammed to solve Toeplitz systems, arising frequently in many high resolution spectral estimation applications. For all these, we need to first introduce a notion of computational wavefronts of Givens Rotators (GR). It is then used as a building block to develop the wavefront array approach to solving the problems mentioned above.

6.1 A Least-Square Error Solver

A common problem in estimation methods in signal processing is solving an overdetermined system in accordance with some "best solution" criteria. As an example, in the tomographical signal processing in projection radiography, the cross-sectional imaging may be reconstructed by a stochastic estimation technique based on linear measurement. This method can be expressed by

$$Y = H X + V ;$$

where Y = vector of measurements, X = object vector, H is the projection matrix and V is the measurement noise vector. The minimum variance estimator is basically a least-square error solution of the above-mentioned equation. Such operations will require trillions of operations, i.e. multiplications and additions. Therefore, fast and parallel methods have to be exploited for achieving any reasonable processing rates.

Given an $M \times N$ matrix A, where $M \geq N$, and an $M \times 1$ vector b, we seek a vector x such that Ax is the best approximation to b [7]. The Linear LS solution is that which minimizes the Euclidean norm of the error vector $z = b - Ax$. When the columns of A are linearly independent, the LS solution is unique, and can be computed via the $N \times M$ pseudo inverse of A, A^+:

$$x = A^+ b, \tag{9}$$

where

$$A^+ = (A^T A)^{-1} A^T \tag{10}$$

Solution of the normal equations (10) involves the risk of encountering an ill-conditioned matrix A^TA, which will lead to numerical difficulties. It is, therefore, advisable to refrain from forming the product A^TA where possible.

An alternate means of solving the LS problem is through the QR decomposition of a matrix A. When the columns of A are linearly independent, then A can uniquely be decomposed into the product of two matrices, $A = Q^TR$, where the columns of Q are orthonormal and R is an upper triangular matrix [46]. The least square solution can now be obtained by applying:

$$QA = \begin{pmatrix} R \\ 0 \end{pmatrix}, \quad Qb = \begin{pmatrix} c \\ d \end{pmatrix} \tag{11}$$

then solving:

$$Rx = c \tag{12}$$

This is the approach we will implement using the wavefront array. In doing so, we will make efficient use of a full square array of processing elements.

The Givens Rotation

Of the methods used to transform a real matrix into a diagonal, triangular, tridiagonal or bidiagonal form, the Householder transformation method is probably one of the most popular. However, this transformation does not utilize the computational capabilities of the wavefront array efficiently. The reason for this is that the Householder transformations require limited, but nevertheless definite, global communications.

An alternate scheme of transforming the matrix A is by means of Givens rotations. The main advantage of these operators is that they require only local communications for the generation of the rotation parameters. The Givens algorithm is based on applying a orthogonal operator, $Q^{(q,\ p)}$, which performs a plane rotation of the matrix, A, in the (q, p) plane and annihilates the element a_{qp}. In all cases of application to the wavefront array, the rotation is constructed based on adjacent elements of the processing array. When transforming a matrix into upper triangular form, the rotations are applied so as to annihilate the elements $a_{M,j}$, $a_{M-1,j}$, ... $a_{j+1,j}$, j = 1 ... N and in that order. Thus, the elements of the first column are dealt with first, then the elements of the second column, and so forth. For an upper triangularization procedure, we have:

$$QA = R,$$

where R is an upper triangular matrix, and

$$Q = Q(N)*Q(N-1)*...*Q(2)*Q(1)$$

and

$$Q(p) = Q^{(p+1,p)}*Q^{(p+2,p)}*...*Q^{(N,p)}$$

$Q^{(q,p)}$ has the form:

$$Q^{(q,p)} = \begin{bmatrix} 1 & & & & & & \\ & 1 & & & & & \\ & & \cdot & & & & \\ & & & C(q,p) & S(q,p) & & \\ & & & -S(q,p) & C(q,p) & & \\ & & & & & 1 & \\ & & & & & & \cdot \\ & & & & & & & 1 \end{bmatrix} \begin{matrix} \\ \\ \\ q-1 \\ q \\ \\ \\ \\ \end{matrix} \tag{13}$$

(Columns: q−1, q; Rows: q−1, q)

such that:

$$C(q,p) = \frac{a_{q-1,p}}{[a^2_{q-1,p} + a^2_{q,p}]^{1/2}}$$

$$S(q,p) = \frac{a_{q,p}}{[a^2_{q-1,p} + a^2_{q,p}]^{1/2}}$$

The matrix product $A' = [Q^{(q,p)}]*A$ is then:

$$a'_{q-1,k} = C(q,p)*a_{q-1,k} + S(q,p)*a_{q,k} \tag{14a}$$

$$a'_{q,k} \;=\; -S(q,p)*a_{q-1,k} + C(q,p)*a_{q,k} \tag{14b}$$

$$a'_{j,k} \;=\; a_{j,k}\,; \quad j \neq q-1,q$$

$$\text{for all}\;\; k = 1 \ldots N$$

Several important notes should be made here: (1) We shall first cascade A and b, and apply Q to both of them, i.e., perform the operation Q[A : b]. (2) Matrix Q is M x M, R is a triangular N x N matrix, vectors c and x are N x 1 and d is (M−N) x 1. (3) In forming R from A, the GR generate C(q, p) and S(q, p), rather than $Q^{(q,\,p)}$ itself. Applying the rotations on the row elements of A (resp. b) is equivalent to executing $A' = [Q^{(q,\,p)}]*A$ (resp. $b' = [Q^{(q,\,p)}]*b$). Thus, the final QR procedure outcome in the processor array will be the triangular matrix, R and $\binom{c}{a}$. (In doing so, Q need not be retained.) (4) The vector d represents the residual, so that $\|d\|^2 = \|z\|^2$. In the following, we shall discuss how to map the recursions into corresponding computational wavefronts in the wavefront array: Once C(q, p) and S(q, p) have been generated, they are propagated through rows q−1 and q of the array and influence those two rows only. Thus, the Givens rotation involves two distinct tasks: generation of the rotation parameters C and S, and modification of the elements in the two impacted rows through the rotations of Eqs. (14a,b). In general, the first task is carried out by the PE of the array that contains $a_{q,p}$ (the matrix element that is to be annihilated). The second task is executed by the remaining PEs in the same rows.

In order to describe the implementation of the QR decomposition via the WAP, we assume that A is a square matrix. (Expanding this to the overdetermined case is simple and straightforward.)

The wavefront nature can be seen in Fig. 12, which traces the fronts of activity relating to the row operations involved in annihilation of the elements of the first column. These fronts will be called row wavefronts. For descriptive purposes, let us assume a one-to-one mapping of matrix elements onto processing elements, and let each PE include a rotation processor. The wavefront starts at PE(N,1), fetching $a_{N-1,1}$ from above and performing the computation for generating the rotation parameters C(N,1) and S(N,1) which annihilate a_{N1}. Upon completing this task, it will further trigger the processor to the right PE(N,2) and the processor above, PE(N−1,1): (1) The rotation parameters will propagate to PE(N,2), and then PE(N,3), etc., each of which will then perform the rotation operations as in Eqs. (14a,b). (Note that one of the operands is fetched from above, and the updated result will be returned to the PE above.); (2) Almost simultaneously, PE(N−1,1) is triggered to generate its own rotation parameters, and continues to trigger its successor PEs in a similar fashion. In short, the computation activities are propagated upwards and sideways by the first column PEs, and down the rows by all other PEs. Taking a simplified perspective, we can say that the first wavefront activity is started at processing element PE(N,1). PE(N,1) propagates the rotational parameters to PE(N,2) and also triggers the activity of PE(N−1,1), thus forming the second front. They, in turn, activate PE(N,3), PE(N−1,2) and PE(N−2,1) which represent the third front, and so on.

So far we have only described the wavefront propagation for Q(*, 1). The Q(*, 2) wavefront will be triggered as soon as the elements a(N,2) and a(N−1,2) are updated (by Q(N,1) and Q(N−1,1)), and the Q(*, 2) wavefront will follow the Q(*, 1) wavefront in a steady distance.

Similarly, the third wavefront (for Q(*, 3)) will follow the second wavefront in the same distance, and so on. As long as all the N wavefronts are initiated and sweep across the array, the triangularization task is completed.

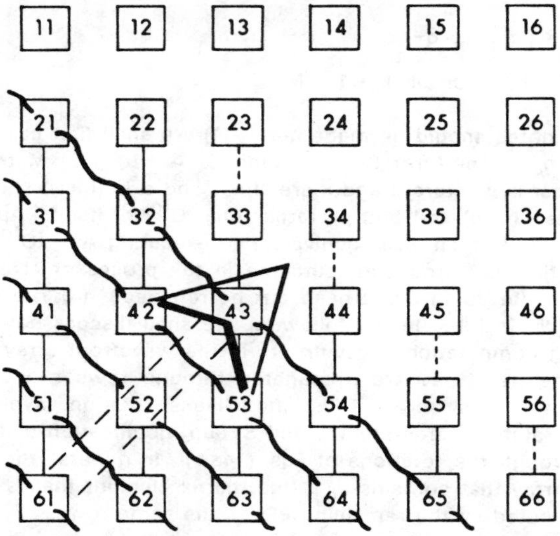

Figure 12: Givens rotation: row wavefronts for least-square and eigenvalue problems

It is easy now to estimate the time required for processing the QR factorization. Note that, for the LS solver, the second computational wavefront need not wait until the first one has terminated its activity. In fact, the second wavefront can be initiated as soon as the matrix elements $a_{N,2}$ and $a_{N-1,2}$ are updated by the first wavefront. This occurs three task time intervals after the generation of that first wavefront. (It is assumed throughout the chapter that generation of the rotation parameters as well as the rotation operation itself require one task time interval.) Thereafter, the annihilating wavefronts can be pipelined with only three time intervals separating them one from the other. The total processing time for the QR factorization would, therefore, be O(3N) on a square N x N array. An MDFL program for the triangularization of a matrix and a detailed discussion of implementation of QR in wavefront array can be found [30].

6.2 Eigenvalue Decomposition of a Symmetric matrix [33]

According to Parlett [42], "the QL and QR algorithms ... have emerged as the most effective way of finding all the eigenvalues of a small symmetric matrix. A full matrix is first reduced to tridiagonal form by a sequence of reflections and then the QL [QR] algorithm swiftly reduces the off diagonal elements until they are negligible. The algorithm repeatedly applies a complicated similarity transformation to the result of the previous transformation, thereby producing a sequence of matrices that converges to a diagonal form. What is more, the tridiagonal form is preserved". Therefore, the QR algorithm can be regarded as the best <u>sequential</u> algorithm available to date. The question is whether or not the QR algorithm may retain that same effectiveness when mapped into a parallel algorithm on a square or linear multiprocessor array.

In the following section, we shall offer an answer to this question using the computational wavefront notion. First, we shall demonstrate that it is advantageous to perform the tridiagonalization of the original matrix by means of a linear array. As the tridiagonalization process requires $O(N^3)$ time on a sequential computing machine, a processing time of $O(N^2)$, using N processing elements in the array, is called for. Secondly, the iteration of the tridiagonal matrix is especially attractive for implementation by means of a linear configuration of the wavefront array, as the volume of data is linear, i.e. O(2N). Both above operations can conveniently be performed involving local communications only.

6.3 Linear Array Tridiagonalization of a Symmetric Matrix

The basic tridiagonalization of a symmetric matrix is implemented by means of the similarity transform:

$$W = Q * A * Q^T$$

where W is tridiagonal and Q is orthogonal. Usually, Q consists of the product of N-2 orthogonal matrices $Q^{(p)}$ such that: $Q = Q^{(N-2)} * Q^{(N-3)} * ... * Q^{(2)} * Q^{(1)}$ and $Q^{(p)}$ causes the (N-p-1) lower elements in the pth column of A to be set to zeros. Similarly, $[Q^{(p)}]^T$ causes the N-p-1 rightmost elements in the pth row of A to be set to zeros. The same constraint of localized communications discussed above motivates use of Givens rotations on the matrix for tridiagonalization (rather than, e.g. a Householder transformation). In essence, the operator $Q^{(p)}$, described above, is again broken down into a sequence of finer operators $Q^{(q,p)}$, where each operator annihilates the element $a_{q,p}$. Thus, $Q^{(p)} = Q^{(p+2,p)} * Q^{(p+3,p)} * ... * Q^{(N,p)}$. Each operator $Q^{(q,p)}$ is of the form shown in Eq. 13.

Of major importance are the following facts: (1) The premultiplication of A by $Q^{(q,p)}$ modifies only rows q-1 and q of A. The elements of those two rows assume the values given by eqs. (14a,b) after applying the rotation, and $a'_{q,p} = 0$. (2) The effect of postmultiplying $A' = [Q^{(q,p)} * A]$ by $[Q^{(q,p)}]^T$ is to modify the elements of columns q-1 and q of A' to assume the following values:

$$a''_{k,q-1} = C(q,p) * a'_{k,q-1} + S(q,p) * a'_{k,q} \tag{15a}$$

$$a''_{k,q} = -S(q,p) * a'_{k,q-1} + C(q,p) * a'_{k,q} \tag{15b}$$

As A was symmetric, this operation is largely a repetition of many of the row operations affected in the Q*A process. The exceptions are the four elements located at the junction of rows and columns q and q-1.

Choosing Between Linear and Square Array Eigenvalue Solvers

When taking the wavefront viewpoint of the operations, two types of waves are discernible. The first is an advancing wave, related to the row operations up to the diagonal elements and referred to as the "row wavefronts". The second involves computation in the junction regions and the column operations and can be seen as a reflected wavefront along the diagonal (cf. Fig. 13). These are dubbed the "column wavefronts".

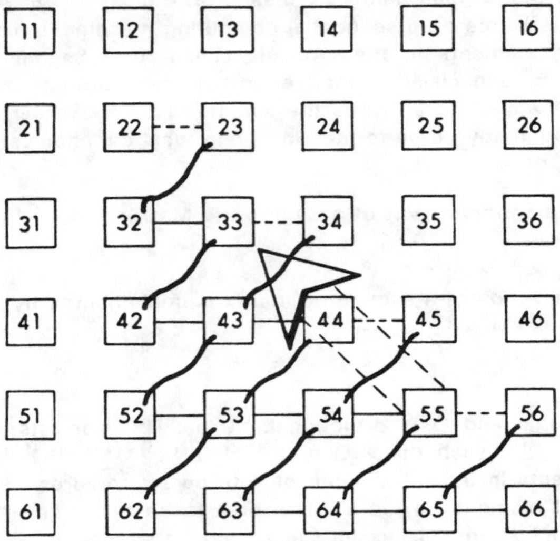

Figure 13: Givens rotation: column wavefronts for eigenvalue computation

In order to take advantage of the symmetry of the symmetric eigenvalue problem, let us delete those PEs above the main diagonal, retaining a triangular array. Since the subdiagonal elements are still producing the same results as before, and the superdiagonal elements are simply their transposition, no information will be lost.

We now pose a most important question: can the square (or triangular) array be utilized with reasonable efficiency in solving the symmetric eigenvalue problem, through Givens type rotations? Our answer to that question is: NO. This critical decision leads us to conclude that, in general the linear array is the better choice.

The reasons supporting this claim can be made clear by a closer examination of the column wavefronts. Fig. 13 shows the sequencing of these column wavefronts and their propagation. (There are several variants of the propagation pattern. This one, however, appears to be the simplest and most representative.) The first column wavefront can be initiated when and only when the first row wave reaches the end of its travel, i.e. the last two elements of the last two rows. Its first task corresponds to iterating columns N and N−1 through operator $[Q^{(N,1)}]^T$. The column wave can advance by one stage when the row wave has operated on the last elements of rows N−1 and N−2. In the evolution of the computations, row operations applied to rows q and q−1 must terminate before the corresponding column operations are initiated. This is due to the fact that column operations require data that is the outcome of the row operations. By the same token, the column operations corresponding to annihilation of the (N−p−1) elements of row p (column wave p) must terminate before the row operations relating to the annihilation of column p+1 (row wavefront (p+1)) may commence. On the basis of these observations we claim two facts:

(1) Unlike the LSS problem, row wavefront (p+1) cannot be initiated until the column wavefront p has reached and updated the values of elements $a_{N,p+1}$ and $a_{N-1,p+1}$. As each wavefront requires O(N−p+1) time to propagate, and there are (N−2) waves of each kind necessary to annihilate the N−2 columns and rows, the total processing time is $O(N^2)$. Utilization of N^2 PEs in a square array (or even half that number in the triangular array) is extremely inefficient and not cost effective, when compared to the single PE execution time of $O(N^3)$.

(2) From Fig. 13 one can also see that essentially at most two PEs in each column are actively executing rotation oriented operations at any time instance. We, therefore, propose to apply the same procedure described above, utilizing a linear or bi-linear array of processing elements. By the above argument, we note that the (bi)-linear array will yield the same $O(N^2)$ execution time as the square (or triangular) array, thus proving that they are unnecessary.

It should be noted that, although the physical configuration of the processor array has changed from square to linear, the nature of the computational wavefront has not, and the theoretical propagation of computational activity is retained. Thus, we have a square array virtual configuration mapped into a bi-linear array actual machine.

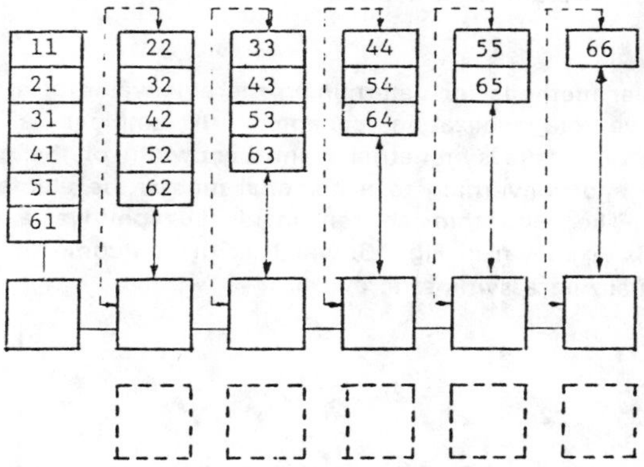

Figure 14: Bi-Linear Wavefront Array Configuration

Is shown in Fig. 14, a bi-linear array is proposed for tridiagonalizing the symmetric matrix. The two rows of processors are identical. The jth PE in either row will have a FIFO stack of depth N−j+1, which will originally contain the j diagonal and below-diagonal elements of the matrix. The stacking scheme replaces the need for indexing the column element values, as was done in the linear array. Each PE in one row "pushes" data onto the stack of the other row, and, of course, each PE "pops" data from its own stacks. The push-pop pair of instructions replace the normal flow-fetch pair in the vertical direction, and recreate the wait conditions typical of the wavefront array. The computational tasks are assigned to the PEs so that first row PEs perform parameter generation and row rotations, while second row PEs execute column rotations.

To facilitate the analysis of the tridiagonalization time, we again make the simplifying assumption that each rotation takes one task time unit for execution, and that data transfer time is negligible (i.e. zero time units). The critical factor in execution time is the inherent delay between fronts p and p+1 (which eliminate columns p and p+1, respectively). To this end, note that: (1) The first column wavefront can start one time unit after the rotation parameters have been generated (as the parameter transfer time through the array is neglected); (2) The second row front can begin when the first column front has updated the values of $a_{N,2}$ and $a_{N-1,2}$. This occurs N−1 time units after the generation of the first column front, and N+1 time units after the beginning of the first row wavefront. In general, the pth wave starts N+3−p time units after the $(p-1)^{th}$ wave (for p = 2,..,N−2); thus totalling up to an overall processing time of approximately $N^2/2$.

The two schemes presented above have several advantages. First, the final values of the tridiagonal matrix are, upon terminating the procedure, already in their proper placement within the processor array. This allows for pipelining the second phase of eigenvalue determination immediately after the first phase. Thus, once the first and second column annihilation has been completed, PE(1) and PE(2) can commence the activities required by the QR iterations. There is no activity gap between the two execution phases. Secondly, the schemes require only local communications and are, therefore, well suited for wavefront array implementation.

6.4 Determining the Eigenvalues of a Symmetric Tridiagonal Matrix

Among the most popular methods for determining the eigenvalue of a symmetric tridiagonal matrix is the iterative diagonalization scheme. It employs a series of similarity transformations which retain the symmetricity and bandwidth of the matrix, while reducing the off-diagonal norm and converging to a diagonal matrix, the elements of which are the sought eigenvalues. The algorithm chosen involves repetitive application of the QR algorithm to the matrix A shown in Fig. 15, which is the outcome of the first computation phase, that of tridiagonalizing a symmetric matrix.

$$
\begin{bmatrix}
a_1 & b_1 & & & & & \\
b_1 & a_2 & b_2 & & & \phi & \\
 & b_2 & a_3 & b_3 & & & \\
 & & b_3 & a_4 & b_4 & & \\
 & & & \ddots & \ddots & \ddots & \\
 & \phi & & & b_{n-2} & a_{n-1} & b_{n-1} \\
 & & & & & b_{n-1} & a_n
\end{bmatrix}
$$

Figure 15: Tridiagonal structure of the matrix to which QR is applied

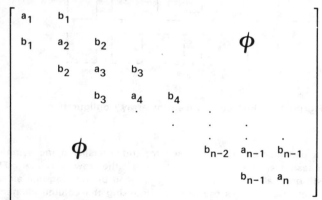

In the basic QR algorithm, the matrix A is decomposed into the product of an orthogonal

matrix, Q, and an upper triangular matrix, R, such that $A = Q^T * R$. Thus, $R = Q * A$. Postmultiplying R by Q^T creates $A' = R * Q^T = Q * A * Q^T$, so that A′ is similar to the original A. Rather than generating the decomposing orthogonal matrix Q in a single operation, we choose, as before, to create Q as a product of orthogonal matrices, $Q = Q^{N-1} * \ldots * Q^2 * Q^1$, where each $Q^{(p)}$ represents a rotation operator of the type shown in Eq. 13 designed to annihilate a single subdiagonal element. The order of application of premultiplications and postmultiplications is, to some extent, flexible. Assume, for the moment, that all premultiplications (row operations) are executed first. The resulting A′ is of the form given in Fig. 16. The values of $\{d_i\}$ do not have to be computed, as they are redundant. It can clearly be seen that implementation of these iterations involves local dependence only, as the updated diagonal, sub- and super-diagonal values are generated from the original element values in the same and adjacent locations.

$$\begin{bmatrix} x & x & d_1 & & & & & \\ 0 & x & x & d_2 & & & \phi & \\ & 0 & x & x & d_3 & & & \\ & & 0 & x & x & d_4 & & \\ & & & \cdot & \cdot & \cdot & & \\ & & & & \cdot & \cdot & x & d_{n-2} \\ & \phi & & & & 0 & x & x \\ & & & & & 0 & 0 & x \end{bmatrix}$$

Figure 16: Matrix of Fig. 15 after row modification

The second phase of the algorithm requires column oriented multiplication which will convert the matrix back to a symmetric, tridiagonal form. The operation involved is similar to that of the row operations described above. Thus the problem is defined by means of an algorithm which adheres to the locality constraint of the wavefront array.

6.5 Computation of the SVD by Means of a Linear Array

The Singular Value Decomposition (SVD) has become an extremely useful tool in numerical analysis, signal and image processing and other fields. We will deal, here, with the SVD relating to real matrices, although there is nothing to prevent generalization to the complex case. The SVD involves decomposing an M x N matrix A into the product of three matrices:

$$A = U\,S\,V$$

where U is a (real) M X N matrix with orthonormal columns, V is a (real) N x N matrix with orthogonal rows, and S is a diagonal matrix, consisting of the singular values of A.

There are a number of SVD methods well known and popularly used, due to their efficient manipulation of the CPU resources - memory and time. Several of these algorithms will be described and their implementations on linear or square arrays briefly discussed here.

One of the better known algorithms for SVD is the Golub-Reinsch method. This method involves bidiagonalizing the matrix A, then performing the SVD on the resulting bidiagonal matrix [15, 16]. The process of bidiagonalizing A involves nullifying the elements of A below the main diagonal through repetitive premultiplying A by a sequence of matrices, {P}, and the elements above the secondary diagonal (the diagonal which consists of elements $a_{i, i+1}$) through repetitive postmultiplying A by a set of orthogonal matrices, {Q}. The order of pre- and postmultiplying A is critical. It is this ordering that presents a major weakness in the implementation of the Golub-Reinsch by means of a full square array of PEs. Another drawback for a square array implementation is the fact that, after the diagonalization has been completed, we are left with only (2N-1) nonzero data elements, which would naturally only call for O(N) PEs, rather than the N^2 PEs in the square array. This line of thought is further developed below.

We now use a linear array to bidiagonalize the given matrix A, in emulation of the Golub-Reinsch algorithm [16]. The bidiagonalization procedure is identical, in most respects, to the tridiagonalization routine used above in the symmetric eigenvalue problem. A basic difference lies in the fact that the original matrix is, in general, not symmetric and frequently not even square. This does not pose a major problem. Assuming a rectangular matrix of dimensions (M x N), with M > N, the matrix can first be reduced to a square configuration by means of repetitive row operations that annihilate all of the elements in the last M-N rows [5]. By efficient pipelining of the operations in the linear array, this reduction requires (M-N) task time units. Then, the alternating row and column operations are applied. The sequence of operations is as shown in Fig. 17. The kth sweep, in which the k-1 subdiagonal elements of the kth column and the k-2 rightmost elements of the kth row are annihilated, requires 3k-3 task time units, leading to an overall, square matrix, bidiagonalization time of $(3N^2-3N-2)/2$.

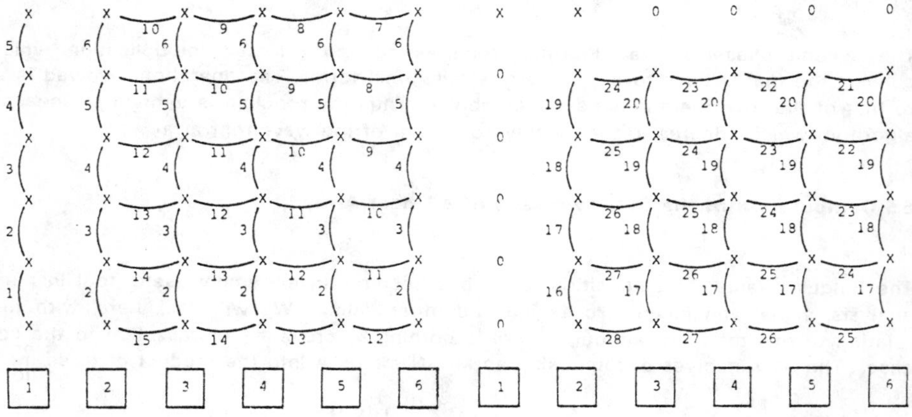

Figure 17: Sequence of Bidiagonalization Computation Activity

Once the original matrix A has been transformed into a bidiagonal matrix, the number of data elements active in the processing has been reduced to 2N-1. The Golub-Reinsch "skipping" sequence [16] is now implemented on these elements by means of Givens rotations. Due to the reduced number of operands, a linear processor array can effectively be used at this stage to converge to the singular values. Thus, the original square matrix is largely impotent and of no use As convergence time is generally considered to be O(N), it again seems to be more beneficial to implement the entire Golub-Reinsch procedure by means of a linear array, in $O(N^2)$ time. The application of the linear array to execution of the diagonalization is, also, very similar to that of the symmetric tridiagonal matrix described above, and will not be further dwelt upon here.

A Note on SVD with Square Array

A second popular approach to SVD is that presented by Nash [40], Hestenes [19] and others. This scheme was modified by Finn et al. [12] and implemented using a triangular systolic array of processors. The drawback of their approach was that the scheme required "wrap-around" communication ability, i.e. the Nth column transferred data to the first. Another method is proposed by Gal-Ezer [14], called Repetitive Triangularization (RT) algorithm. Numerous simulations have shown that the RT method converges in the same order of time as Finn's algorithm. Moreover, the theoretical convergence analysis for the RT algorithm can be found in [14]. More crucially, this algorithm requires only localized communications and is, therefore, very suitable for wavefront processing.

In summary, in the eigenvalue and singular value decomposing schemes, there is strong evidence that a linear array structure can, very often and quite effectively, rival the processing speed achieved by the square array. Some facts supporting these claims are: (1) Tridiagonalizing a symmetrical matrix can be performed efficiently in $O(N^2)$ time with a linear array. Thereafter, reducing the tridiagonal matrix can be accomplished in O(N) iterations. (2) Other, non-tridiagonalizing methods, such as RT or modified Hestenes schemes involve full matrix manipulations. These, in general, involve a linear number of iterations of $O(N^3)$ operation each [42]. Thus, a square array of processors can, in general, only achieve $O(N^2)$[6] processing time, which is the same as the linear array.

7 Wavefront Array for Solving Toeplitz Linear Systems

The least-square error estimations often arise in the context of stationary signal processing, where the final solutions hinge upon solving a covariance matrix equation

$$Rx = y. \tag{16}$$

Here R has a special structure that $R(i,j) = R(|i-j|)$ and is in general termed symmetric Toeplitz matrix. The most popular sequential algorithms solving a symmetric Toeplitz system is the well-known Levinson algorithm [36, 21]. However, for parallel execution of the Levinson algorithm on a linear processor array with N processing elements, the parallelism will be severely hampered by the bottleneck of the inner product operations [34]. Consequently, the total parallel computing time amounts to $O(N \text{ Log}_2 N)$ on a linear processor array (of O(N) processors). This is of course not satisfactory. In this section, we shall develop an algorithm which avoids the inner product operations and therefore achieves much higher parallelism. More precisely, this algorithm will attain a processing time of O(N) time units (as opposed to $O(N \text{ Log}_2 N)$) on a linear array of N processing elements. More surprisingly, the procedure again involves applying (a special type of) rotations on two-rows of data in the linear array, thus the array processor proposed for eigenvalue decomposition can be easily reprogrammed to solve Toeplitz systems.

---

[6] There is recently developed some methods achieving O(NlogN) processing time, but some global communications will be needed.

7.1 Concurrent Algorithm [35]

In this section, we shall present a highly concurrent algorithm. Mathematically, this algorithm can find its roots back to the now classical Schur's algorithm [45, 10] (For a mathematical derivation, the reader is referred to an earlier chapter by Kailath in this book). Our derivation has two distinct features: (a) the parallelism and the (localized) data dependency inherent in the algorithm are explicitly exposed; and (b) there surfaces a natural topological mapping from the mathematical algorithm to the computing structure to be discussed in the next section.

Once more, the major function in solving the Toeplitz system is to perform a triangular decomposition on matrix R, that is[7],

$$R = \bar{U}^T D \bar{U} = \bar{U}^T U \qquad (U = D \bar{U})$$

where D is a diagonal matrix. Then the solution $\underline{x}$ of (16) can be solved explicitly with back substitution:

$$\underline{x} = R^{-1}\underline{y} = U^{-1} (\bar{U}^T)^{-1}\underline{y} \tag{17}$$

For simplicity, we use a 4 x 4 matrix example. The problem is to find the elements $\{\ell_{ij}\}$ and $\{u_{ij}\}$ such that

$$\begin{bmatrix} 1 & 0 & 0 & 0 \\ \ell_{21} & 1 & 0 & 0 \\ \ell_{31} & \ell_{32} & 1 & 0 \\ \ell_{41} & \ell_{42} & \ell_{43} & 1 \end{bmatrix} R = \begin{bmatrix} u_{11} & u_{12} & u_{13} & u_{14} \\ 0 & u_{22} & u_{23} & u_{24} \\ 0 & 0 & u_{33} & u_{34} \\ 0 & 0 & 0 & u_{44} \end{bmatrix} \tag{18}$$

denoted as

$$\tilde{L} R = U \qquad \text{where } \tilde{L} \triangleq (\bar{U}^T)^{-1}$$

Let us start with the following equation

$$\begin{bmatrix} 1 & 0 & 0 & 0 \\ 0 & 1 & 0 & 0 \end{bmatrix} R = \begin{bmatrix} t_0 & t_1 & t_2 & t_3 \\ t_1 & t_0 & t_1 & t_2 \end{bmatrix}$$

Now perform row operations on both sides of the above equation:

[7] The overbar "-" of the triangular matrix U indicates that U has unities along its diagonal.

$$\begin{bmatrix} 1 & K^{(2)} & 1 & 0 & 0 & 0 \\ K^{(2)} & 1 & 0 & 1 & 0 & 0 \end{bmatrix} R = \begin{bmatrix} \underline{v}^{(2)} \\ \underline{u}^{(2)} \end{bmatrix}$$

where the so called "reflection coefficients" $K^{(2)}$ is computed as

$$K^{(2)} = -\frac{-t_1}{t_0} .$$

This is rewritten as

$$\begin{bmatrix} 1 & K^{(2)} & 0 & 0 \\ K^{(2)} & 1 & 0 & 0 \end{bmatrix} R = \begin{bmatrix} v_0^{(2)} & 0 & v_2^{(2)} & v_3^{(2)} \\ 0 & u_1^{(2)} & u_2^{(2)} & u_3^{(2)} \end{bmatrix} \tag{19}$$

Compare the second row of the RHS (right hand side) of Eq(18) with that of Eq. (19), i.e. $\underline{u}^{(2)}$, it is clear that a zero is created by the row operation and the desired second rows of $\tilde{L}$ and U are obtained :

$$\tilde{L}_2 = [\ell_{21}\ 1\ 0\ 0] = [K^{(2)}\ 1\ 0\ 0]$$

$$U_2 = 0\ u_{22}\ u_{23}\ u_{24} = 0\ u_1^{(2)}\ u_2^{(2)}\ u_3^{(2)}$$

To compute the third rows of the $\tilde{L}$ and U matrices, the same strategy can be repeated. To ready for the next recursion, we first right-shift the second rows in the both sides of Eq. (19), i.e.

$$[K^{(2)}\ 1\ 0\ 0] \rightarrow [0\ K^{(2)}\ 1\ 0],$$

$$\begin{bmatrix} 1 & K^{(2)} & 0 & 0 \\ 0 & K^{(2)} & 1 & 0 \end{bmatrix} \mathbf{R} = \begin{bmatrix} v_0^{(2)} & 0 & v_2^{(2)} & v_3^{(2)} \\ u_{-1}^{(2)} & 0 & u_2^{(1)} & u_2^{(2)} \end{bmatrix} \tag{20}$$

Note that by using the Toeplitz structure of R matrix, we have $\underline{u}^{(2)}$ in Eq. (19) right-shifted accordingly and the only new term is $u_{-1}^{(2)}$, which is equal to $v_2^{(2)}$, since $u_{-1}^{(2)} = K^{(2)}\ t_1 + t_2 =$ v(2)

2

Note that through this shift operation, the two "0" created in the previous recursion on the RHS are re-aligned into the same column. They will remain uneffected by the linear combination of the two rows in the next recursion. With this arrangement, a similar procedure as in the previous recursion can now be repeated:

$$\begin{bmatrix} 1 & K^{(3)} \\ K^{(3)} & 1 \end{bmatrix} \begin{bmatrix} 1 & K^{(2)} & 0 & 0 \\ 0 & K^{(2)} & 1 & 0 \end{bmatrix} R = \begin{bmatrix} \underline{v}^{(3)} \\ \underline{u}^{(3)} \end{bmatrix} = \begin{bmatrix} v_0^{(3)} & 0 & 0 & v_3^{(3)} \\ 0 & 0 & u_2^{(3)} & u_3^{(3)} \end{bmatrix}$$

where

$$K^{(3)} = -\frac{v_2^{(2)}}{u_1^{(2)}}$$

By comparing $\underline{u}^{(3)}$ with the third row on the RHS of Eq. (18), clearly, the third rows of the $\tilde{L}$ and U matrices are obtained:

$$\tilde{L}_3 = \ell_{31} \quad \ell_{32} \quad 1 \quad 0$$

$$= K^{(3)}, \ (K^{(3)} K^{(2)} + K^{(2)}), \ 1 \quad 0$$

$$U_3 = 0 \quad 0 \quad u_{33} \quad u_{34} = 0 \quad 0 \quad u_2^{(3)} \quad u_3^{(3)}$$

This completes the second recursion. By induction, the future recursions can be carried out in the same manner until all the rows of the $\tilde{L}$ and U matrices are computed. Summarizing the above procedure, several observations can be made:

1. The shift operation in each recursion is natural due to the Toeplitz structure of R matrix, since it retains the zeros produced in the previous recursion. The purpose of the shift is to realign these zeros with those of the auxiliary vector $\underline{v}$ (cf. Eq. (20)), such that these zeros will remain unaffected by the upcoming row operations. This also explains the purpose and the necessity of computing for the auxiliary vectors $\underline{v}$ in each recursion.

2. This new formula completely avoids the need of an inner product operation; therefore, the bottleneck incurred in parallel execution of the Levinson algorithm no longer exists. Furthermore, the reflection coefficient computation and row operations in the formulation are very suitable for parallel execution; hence, this new algorithm is inherently highly concurrent. As a consequence, it has the following parallel formulation:

<u>Main Algorithm</u>

INITIAL CONDITIONS :

$v_k^{(1)} = u_k^{(1)} = t_k \quad (0 \leq k \leq N)$

FOR i = 1 UNTIL N DO BEGIN

IN PARALLEL DO BEGIN

$$K^{(i+1)} = -v_i^{(i)} [u_{i-1}^{(i)}]^{-1}$$

END IN PARALLEL DO;

IN PARALLEL FOR $0 \leq k \leq N$ DO BEGIN

$$v_k^{(i+1)} = v_k^{(i)} + K^{(i+1)} \quad u_{k-1}^{(i)}$$

$$u_k^{(i+1)} = u_{k-1}^{(i)} + K^{(i+1)} \quad v_k^{(i)}$$

END IN PARALLEL DO;
OUTPUT;
END FOR LOOP;

3. Based on above formulation, it is clear that with O(N) processing elements connected in a linear array, the parallel computing time for each recursion can take as little as 2 time units, (one for reflection coefficient computation, the other for row operation). For N recursions, this amounts to a total parallel processing time of O(N) time units as opposed to O(N Log_2N) in the Levinson algorithm. In the next section, a pipelined computing structure for concurrent processing of symmetric Toeplitz systems solutions will be discussed.

7.2 Pipelined Lattice Processor

In this section, we consider the implementation of the parallel algorithm on a VLSI chip. The major computation in the algorithm lies in the rotations. Therefore, a wavefront computing structure (parallel lattice processor [34]) with a linear processor array as depicted in Fig. 18 is proposed, which uses only nearest neighbor interconnections.

During each recursion, the reflection coefficient is first computed in the divider cell and then broadcast to all the lattice cells through the global (horizontal) interconnections. Then the row operation are performed simultaneously in all the lattice cells. Upon completion, the result in each upper PE is left-shifted to its immediate left neighbor preparing for the next recursion[8]. Meanwhile, the contents of the lower PEs which corresponds to a row of the U matrix, can be output and the recursion is thus completed.

The operation stops after N such recursions. Let τ_1 denote the time interval needed for division, and τ_2 denote the time interval for each lattice operation (multiplication and addition), then the total computing time will be $N(\tau_1+\tau_2)$.

Computational Wavefront

To achieve maximal parallelism in this locally connected computing network, we must resort to a pipelined operation which renders efficient and smooth data flow, which naturally leads to wavefront array.

As mentioned earlier, a computational wavefront in a computing structure corresponds to

---

[8] This corresponds to the shift operation discussed earlier. Note that since only the relative position between the data in the upper and lower PEs are of importance, the left-shift for the upper PEs ($\underline{v}$ vector) is equivalent to the right-shift for the lower PEs ($\underline{u}$ vector) as described earlier.

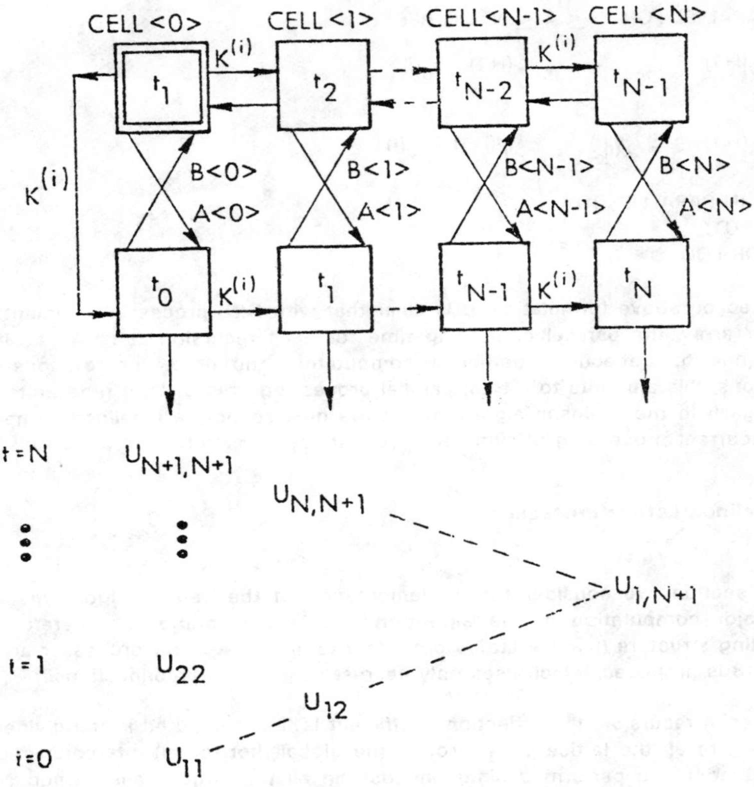

Figure 18: Parallel lattice processor

the computational activity incurred in one recursion step in a recursive (parallel) algorithm. As an example, the computational wavefront of the first recursion is examined below. Suppose that the data are initially placed in the registers of the PEs such that: $B_{<0>} = t_0$ and $A_{<m-1>} = B_{<m>} = t_m$ for m = 1, 2, . . , N, where $A_{<m>}$ register is in the m-th upper PE and $B_{<m>}$ in the m-th lower PE (c.f. Fig. 18). The process starts with the <0> cell (divider cell) where the reflection coefficient is computed and stored in register $C_{<0>}$, that is

$$C_{<0>} \Longleftarrow A_{<0>} / B_{<0>}$$

The computation activity then propagates to the <1> cell (after propagating $C_{<0>}$ to $C_{<1>}$) where the following computations are executed simultaneously in upper-half and lower half of the PEs[9]:

---

[9]The lower PE of the <0> cell should also perform the operation $B_{<0>} \Longleftarrow B_{<0>} - C_{<0>} \times A_{<0>}$ at this moment. This accounts for the computation of the D matrix

$$A_{<1>} \Leftarrow A_{<1>} - C_{<1>} \times B_{<1>} \quad \text{(upper PE)}$$

$$B_{<1>} \Leftarrow B_{<1>} - C_{<1>} \times A_{<1>} \quad \text{(lower PE)}$$

Upon completion of the execution, the <1> cell propagates its new content in the upper PE,$A_{<1>}$, to its left neighbor, $A_{<0>}$, in order to prepare for the next recursion. Meanwhile, it also sends the content of $C_{<1>}$ to $C_{<2>}$ so that the computation activity continues propagating to the <2> cell. The next fronts of activity will be at <3> cell, then <4> cell, and so on.

As the above mentioned first computational wavefront propagates, the second recursion can be executed concurrently by pipelining a second wavefront as soon as the content in $A_{<1>}$ is sent to $A_{<0>}$ register in <0> cell. Therefore, the time-interval between the first and the second wavefront is estimated to be $\tau_1 + \tau_2$. (Note that it takes $\tau_1 + \tau_2$ time interval before $A_{<1>}$ is made available if data transfer time is considered negligible).

The same pipelining scheme can be repeated for the third and eventually all the recursions. When all the N wavefronts are generated and operations executed, the parallel algorithm is completed. Since the wavefronts are generated consecutively at a rate of one wavefront per $\tau_1 + \tau_2$ time interval, the total computing time will be $N(\tau_1 + \tau_2)$. Therefore, the pipelined operation takes the processing time, $N(\tau_1 + \tau_2)$, which is the same as that of the parallel operation even if global communication is used. In other words, it has satisfied the design goal of using only local interconnections and yet not sacrificing any degree of parallelism attainable.

7.3 Complete Toeplitz System Solvers

So far, we have used the so-called Pipelined Lattice Processor(PLP) [11] to decompose a Toeplitz system into lower and upper triangular matrices, i.e., $T = u^t D^{-1} U$ in 0(N) time units (assume that T is symmetric). To completely solve the Toeplitz system, however, an explicit solution x of Eq. (16) has to be derived. Therefore, subsequent operations are needed and should also be performed in 0(N) time units.

<u>Back Substitution Method.</u> This method involves computing x via (17) (for the symmetric case):

$$x = T^{-1} y = U^{-1} D(u^T)^{-1} y$$

which can be separated in two back-substitution steps:

$$g = D(u^T)^{-1} y$$

and

$$x = U^{-1} g$$

Back substitution is a standard matrix operation for solving linear systems, which also enjoys a pipelined operation. Therefore, it can be implemented with a locally connected linear processor array. A complete Toeplitz system solver is composed of a PLP and a (pipelined) back-substitution processor. For a more detailed description and VLSI design block-diagram, the readers are referred to [34].

7.4 Extension to Band Toeplitz Matrices

The extension of the Toeplitz system solver to include banded Toeplitz matrix T_B is straightforward. As in the previous discussion, we perform a triangular decomposition of that:

$$L_B\, T_B \quad = \quad [U_B]$$

The strategy used to construct the matrices L_B and U_B is essentially the same as what discussed earlier. Now we note that the indexes of the nonzero elements of the L and U matrices are shifted towards the right. Therefore we can, along the same lines, derive the same structure for Toeplitz band matrices, except for that system is composed of a PLP of M (instead of N) lattice sections where M is the band width. Again we will need a a back substitution processor and LIFO stacks [34]. The only caution is that each of the LIFO stacks needs to hold N rows of data.

Remarks. The design, fabrication, and testing of a VLSI chip, implementing the wavefront architecture for Toeplitz System Solver, has recently been completed under a joint project between the University of Southern California and Hughes Research Laboratory, Malibu. For more detailed discussion about the Toeplitz System Chip, please see [39, 34]. This wavefront architecture can be immediately applied to some modern spectrum estimation methods such as maximum entropy or maximum likelihood methods. More interestingly, the same algorithm and architecture can be extended to solve the eigenvalue of a Toeplitz system [20]. The latter has found an important application to deriving the (super-resolution) Pisarenko spectral estimation [17], closely related to the eigenvalue-based direction finding methods. In the context of statistical time series analysis, we note that the wavefront architecture is mathematically related to the structure of adaptive lattice filtering(see e.g. [17]), commonly used in the speech processing. Thus a minor modification on the wavefront lattice processor will make it amenable to adaptive signal processing environment.

8 Conclusion

In order to test the feasibility of the systolic concept, a synchronous systolic array test bed was constructed [3] at NOSC, San Diego. (The test bed is constructed in conjunction with an ONR SRO Project at USC, L.A.) To cope with the difference between the boundary and interior PE's, a programmable systolic PE is constructed using commercially available LSI circuits. A systolic PE is composed of a control microprocessor, a floating point Arithmetic Processing Unit(APU), a data storage RAM, an EPROM (eletrically programmable ROM), and an I/O module. While the machine has already been successfully programmed to do some simple matrix problems, such as matrix multiplications. A better programming technique is yet to be developed, for more complicated tasks such as QR decomposition, SVD, etc.

There are numerous exciting and important topics remaining in the future development of VLSI Array Processor. For examples, when several hundred thousand gates are condensed into a chip, the gate speed increases tremendously, almost inversely proportional to the scaling of the feature size. If the gate switching time is in the order of .5 nsec, the corresponding clock rate will be around the order of 300 MHz. Then synchronization over a large array (say of 100 PEs or more) may become a serious problem. Simply trying to distribute a master clock across the board to all PEs is itself a nontrivial task at the aforementioned clock rate. Among many attributes causing timing uncertainties, the wire lengths and loads may be slightly different along all the clock distribution paths.

A data-flow multiprocessor [8] is an asynchronous, data-driven multiprocessor which runs programs expressed in data-flow notations. Since the execution of its instructions is "data driven", i.e. the triggering of instructions depends only on the availability of operands and resources required, unrelated instructions can be executed concurrently without interference. However, for a general purpose data flow multiprocessor the interconnection and memory conflict problems remain very critical. Such problems can be eliminated if the notion of regularity and locality is elegantly inserted into data-flow processing [30]. In fact, this perception has motivated the notion of wavefront processing.

The Illiac IV system (a SIMD computer implemented as an array of arithmetic processors) has been used as a typical example for the study of array computers [2]. The instructions of these systems are stored in a global main memory, together with data. The central control unit directs the operation of all PEs, communicating with them via a global broadcasting network. As in the systolic array, the processors are all synchronized. Due to large random clock skew and expensive communication in VLSI, the adoption of the SIMD scheme in future multiprocessors has to be carefully reviewed. In contrast to the Illiac IV, the WAP employs local instruction storage, data-flow based control, local communication and needs no global synchronization, all of which make the WAP very appealing for VLSI. Despite of the very simple structure of the wavefront array, the wavefront array is applicable to the algorithms with inherent locality and recursivity, which represent a rather broad class of applications.

Language and software design is crucial to the array processor design as it provides a description tool for programming the parallel data movements and operations [6]. It is also very important to develop software packages for simulation and verification prior to the hardware implementation. Ultimately, it is desirable, though ambitious, to develop a methodology which maps a parallel language structure, such as MDFL, into silicon compilers, see e.g. [9]. This of course depends upon further advance in modern CAD (computer aided design) technology for VLSI systems.

Acknowledgments

The author wishes to thank his colleagues in the VLSI signal processing group at the University of Southern California, for their very valuable contributions to the Wavefront Array Software/Hardware (WASH) Project.

References

1. Anceau, F. and Reis, R. Design Strategy for VLSI. In VLSI Architecture, Randell, B. and Treleaven, P.C., Eds., Prentice-Hall, Inc., 1983, pp. 128-137.

2. Barnes, G. H., Brown, R. M., Kato, M., Kuck, D. J., Stonick, D. L., and Stokes, R. A. "The Illiac IV Computer." IEEE Trans. Computers C-17 (Aug 1968), 746-757.

3. Bromley, K., Symanski, J.J., Speiser J.M., and Whitehouse, H.J. Systolic Array Processor Developments. VLSI Systems and Computations, Ed. H.T. Kung et al.,, Rockville, MD, 1981.

4. Capon, J. "High-Resolution Frequency-Wavenumber Spectrum Analysis." Proc. IEEE 57 (August 1969), 1408-1418.

5. Chan, T. C. On Computing the Singular Value Decomposition. Computer Science Department, Stanford University, Feb, 1977.

6. Cremers, A. B. and Kung, S. Y. On Programming VLSI Concurrent Array Processors. Proc. IEEE Workshop on Languages for Automation, Chicago, 1983, pp. 205-210. also in INTEGRATIONS, the VLSI Journal, Vol. 2, No. 1, March, 1984.

7. Dahlquist, G., Bjorck, A., and Anderson, N.. Numerical Methods. Prentice Hall, Inc., Englewood Cliffs, NJ, 1974.

8. Dennis, J. B. Data Flow Supercomputers. IEEE Computer, Nov, 1980, pp. 48-56.

9. Denyer, P. B. and Renshaw, D. Case Studies in VLSI Signal Processing Using a Silicon Compiler. Proceedings, IEEE ICASSP, Boston, 1983, pp. 939-942.

10. Dewilde, P., Vieira, A., and Kailath, T. "On a Generalized Szego-Levinson Realization Algorithm for Optimal Linear Predictors Based on a Network Synthesis Approach." IEEE Trans. CAS 25, 9 (1978), 663-675.

11. Dewilde, P. et al. Parallel and Pipelined VLSI Implementation of Signal Processing Algorithms. to appear in VLSI and Modern Signal Processing, S. Y. Kung et al. (eds.), Prentice-Hall, Inc. 1984

12. Finn, A., Luk, F., and Pottle, C. Systolic Array Computation of the Singular Value Decomposition. Proc. SPIE International Symposium East, Real Time Signal Processing V, Arlington, Virginia, May, 1982.

13. Franklin, M. and Wann, D. Asynchronous and Clocked Control Structures for VLSI Based Interconnection Networks. The 9-th Annual Symposium on Comput. Architecture, April, 1982. Austin, TX

14. Gal-Ezer, R. J. The Wavefront Array Processor. Ph.D. Th., Dept. of Electrical Engineering, University of Southern California, Dec 1982. Ph.D. Dissertation

15. Golub, G. and Kahan, W. "Calculating the Singular Values and Pseudo-Inverse of a Matrix." J. SIAM Numerical Analysis 2, 2 (1965). Series B

16. Golub, G. H. and Reinsch, C. "Singular Value Decomposition and Least Square Solutions." Numerical Math. 14 (1970), 403-420.

17. Haykin, S. S.. Nonlinear Methods of Spectral Analysis. Springer-Verlag, New York:, 1979.

18. Heller, D. "A Survey of Parallel Algorithms in Numerical Linear Algebra." SIAM Review 20, 4 (Oct 1978), 740-777.

19. Hestenes, M. R. "Inversion of Matrices by Biorthogonalization and Related Results." Soc. Ind. Appl. Math. 6 (1958), 51-90.

20. Hu, Y. H. and Kung, S. Y. A Fast and Highly Concurrent Algorithm for Toeplitz Eigenvalue System Solver. Proceedings, ICCASP, IEEE, Boston, Apr, 1983.

21. Kailath, T. "A View of Three Decades of Linear Filtering Theory." IEEE Trans. Inform. Theory IT_20, 2 (Mar 1974), 145-181.

22. Kailath, T.. Linear System. Prentice-Hall, Inc., Englewood Cliffs, NJ, 1980.

23. Kinniment, D. J. VLSI and Machine Architecture. In VLSI Architecture, Randell, B. and Treleaven, P.C., Eds., Prentice-Hall, Inc., 1983, pp. 24-33.

24. Kooij, T. Fast Processing Technologies for Undersea Surveillance Optical and Systolic Processors. DARPA, 1982.

25. Kung, H. T. The Structure of Parallel Algorithms. In Advances in Computers, Academic Press, 1980, pp. 70-111.

26. Kung, H. T. "Why Systolic Architectures." IEEE, Computer 15, 1 (Jan 1982).

27. Kung, H. T. and Leiserson, C. E. Systolic Arrays (for VLSI). Sparse Matrix Symposium, SIAM, 1978, pp. 256-282.

28. Kung, S. Y., Levy, B. C., Morf, M., and Kailath, T. "New Results in 2-D Systems Theory, Part II." Proceedings of IEEE 65, 6 (June 1977), 945-961.

29. Kung, S. Y. VLSI Array Processor for Signal Processing. Conf. on Advanced Research in Integrated Circuits, MIT, Cambridge, MA, Jan, 1980.

30. Kung, S. Y., Arun, K. S., Gal-Ezer, R. J. and Bhaskar Rao, D. V. "Wavefront Array Processor: Language, Architecture, and Applications." IEEE Transactions on Computers, Special Issue on Parallel and Distributed Computers C-31, 11 (Nov 1982), 1054-1066.

31. Kung, S. Y. and Gal-Ezer, R. J. Synchronous vs. Asynchronous Computation in VLSI Array Processors. Proceedings, SPIE Conference, 1982. Arlington, VA

32. Kung, S. Y. From Transversal Filter to VLSI Wavefront Array. Proc. Int. Conf. on VLSI 1983, IFIP, Trondheim, Norway, 1983.

33. Kung, S. Y. and Gal-Ezer, R. J. Eigenvalue, Singular Value and least Square Solvers via the Wavefront Array Processor. In Algorithmically Specialized Computer Organizations, L. Snyder et al., Eds., Academic Press, 1983.

34. Kung, S. Y. and Hu, Y. H. "A Highly Concurrent Algorithm and Pipelined Architecture for Solving Toeplitz Systems." IEEE Transactions on ASSP ASSP-31, No.1 , pp.66-76 (Feb. 1983).

35. Kung, S. Y. VLSI Signal Processing: From Transversal Filtering to Concurrent Array Processing. to appear in VLSI and Modern Signal Processing, S. Y. Kung, H. J. Whitehouse and T. Kailath (eds.), Prentice-Hall Inc. Englewood Cliffs, NJ 1985

36. Levinson, N. "The Wiener RMS (Root-Mean-Square) Error Criterion in Filter Design and Prediction." J. Math. Phys. 25 (Jan 1947), 261-278.

37. Love, H. H. Reconfigurable Parallel Array Systems. In Designing and Programming Modern Computers and Systems, S. P. Kartashev and S. I. Kartashev,, Eds., Prentice-Hall, Inc., 1982.

38. Mead, C. and Conway, L.. Introduction to VLSI Systems. Addison-Wesley, 1980.

39. Nash, J. G., Hansen, S., and Nudd, G. R. VLSI Processor Array for Matrix Operations and Linear Systems Solution. Hughes Research Laboratories, Malibu, March, 1982.

40. Nash, J. G.. Compact Numerical Methods for Computers: Linear Algebra and Function Minimization. John Wiley and Sons, 1981.

41. Oppenheim, A. and Schafer, R.. Digital Signal Processing. Prentice-Hall, Inc., Englewood Cliffs, NJ, 1975.

42. Parlett, B. N.. The Symmetric Eigenvalue Problem. Prentice-Hall, Inc., Englewood Cliffs, NJ, 1980.

43. Rabiner, L. R. and Gold, B.. Theory and Application of Digital Signal Processing. Prentice-Hall, Inc., Englewood Cliffs, N.J., 1975.

44. Sameh, A. Numerical Parallel Algorithm - A Survey. In High Speed Computer and Organization, Academic Press, 1977, pp. 207-228.

45. Schur, I. "Uber Potenzreihen, die in Innern des Einheitskreises Beschrankt Sind." J. fur die Reine und Angewandte Mathematik 147 (1917), 205-232.

46. Stewart, G. W.. Introduction to Matrix Computations, Academic Press, 1973.

47. Wann, D. F. and Franklin, M. A. "Asynchronous and Clocked Control Structures for VLSI Based Interconnection Networks." IEEE Transactions on Computers C-32, 3 (March 1983).

48. Whalen, H. H. VLSI Applications and Testing. In Very Large Scale Integration, D. F. Barbe, Ed.,Springer-Verlag, 1982.

Index